An Introduction to
THE THEORY OF
AEROELASTICITY

Y. C. FUNG

University of California, San Diego

DOVER PUBLICATIONS, INC.

NEW YORK

ACKNOWLEDGMENT

Acknowledgment is made to the following bodies for permission to reproduce figures from their publications: The National Advisory Committee for Aeronautics (Figs. 1.11, 5.5a, 5.5b, 9.2, 9.4, 9.7–9.10, 10.4–10.6, 15.7, 15.8) the Aeronautical Research Council of Great Britain (Figs. 9.1, 9.6), the Institute of Aeronautical Sciences (Figs. 5.1, 5.2, 13.4), the Max-Planck Institut für Strömungsforschung (Figs. 15.2–15.6), the Royal Society (Fig. 2.1), and the McGraw-Hill Book Company (Fig. 8.4). The photographs following page 62 are courtesy of F. B. Farquharson; those following page 312 are reproduced here courtesy of (respectively) National Physical Laboratory, England, and Von Karman Gas Dynamics Facility, ARO, Inc.

COPYRIGHT

Published in Canada by General Publishing Company, Ltd., 30 Lesmill Road, Don Mills, Toronto, Ontario.
Published in the United Kingdom by Constable and Company, Ltd., 3 The Lanchesters, 162–164 Fulham Palace Road, London W6 9ER.

BIBLIOGRAPHICAL NOTE

This Dover edition, first published in 1993, is an unabridged republication of the slightly revised 1969 Dover reprint of the work first published in 1955 by John Wiley & Sons, Inc., New York. The Preface to the Dover Edition has been updated for this edition.

LIBRARY OF CONGRESS CATALOGING-IN-PUBLICATION DATA

Fung, Y. C. (Yung-cheng), 1919–
    An introduction to the theory of aeroelasticity / Y. C. Fung.
        p.·   cm.
    Previously published: New York : Dover, 1969. With new updated preface.
    Includes bibliographical reference and index.
    ISBN 0-486-67871-7 (pbk.)
    1. Aeroelasticity. I. Title.
TL574.A37F8 1993
629.132′3—dc20   93-6371
                                                                          CIP

Manufactured in the United States of America
Dover Publications, Inc., 31 East 2nd Street, Mineola, N.Y. 11501

PREFACE TO THE DOVER EDITION, 1993

Aeroelasticity deals with the combined features of fluid mechanics and solid mechanics. Aircraft designers are concerned with aerodynamic performance of an elastic aircraft. Designers of bridges and skyscrapers need to know what the wind will be doing to their structures. Designers of artificial heart valves and students of medicine want to know how blood flows in very flexible vessels. Naturalists and environmentalists are interested in the locomotion of birds, fish, and mammals, or the swaying of trees and fluttering of leaves. Scientific study of these problems has to focus on flow in regions with deformable boundaries, and on the deformation of solids subjected to fluid loading, which varies with the deformation itself. The dynamics of these systems are often intricate, surprising, and important for the survival of man, machine, animals, and plants. I was involved in each of these areas at different periods of my life.

I entered college at the time of World War II and chose to study aeronautics because it seemed to be a topic relevant to national survival. After graduation I worked in the Chinese Institute of Aeronautical Research in Chengdu. One of my assignments was the preparation of a design manual of the control surfaces, and in that job I read about flutter. Then I received a scholarship and entered the California Institute of Technology in Pasadena. When my mentor, Professor Ernest Sechler, asked me what I would like to work on for my Ph.D. thesis, I said flutter. Dr. Sechler said that our Dr. Theodor von Kármán used to have a research project on the flutter of suspension bridges sponsored by the State of Washington Bureau of Highways and Bridges after the wind-driven failure of the Tacoma Narrows Bridge. The cause of the failure was believed to be the von Kármán vortex street. When Dr. von Kármán retired, Dr. Maurice A. Biot took over the project. When Dr. Biot went back to Belgium, Dr. Louis Dunn continued in this research. But Dr. Dunn went to the Jet Propulsion Laboratory. "We have a wind tunnel built specially for the testing of suspension bridges. See what you can do with it." So I played with models in that wind tunnel, focused on airplanes, and wrote my thesis, which was expanded later and became this book.

Then came the jets and spacecraft. I worked on flutter, control, wind and gust load, clear-air turbulence, stability and fatigue of aircraft structures, fuel sloshing in space vehicles, landing impact, and ground shocks. I earned consulting fees on the analysis of wind load on the cantilevered roof of the University of Washington stadium in Seattle, and on methods to damp out wind-generated vibrations of a restaurant suspended under two arches at the Los Angeles Airport. By the 1960s, my interest was shifted to biomechanics. In 1966, I resigned my post at Caltech and went to the University of California at San Diego to start a bioengineering program in the School of Medicine. Many research problems in biomechanics are aeroelastic in nature. Indeed, at a meeting of the American Society of Zoologists, I found many zoologists are familiar with this book on aeroelasticity because it is a convenient reference for the non-stationary wing theory. For readers who are interested in biological applications of aeroelasticity, may I offer the following three books written by Y. C. Fung, and published by Springer-Verlag, New York: *Biomechanics: Mechanical Properties of Living Tissues,* 1st ed., 1981, 2nd ed., 1993. *Biodynamics: Circulation,* 1984. *Biomechanics: Motion, Flow, Stress, and Growth,* 1990. Also, the following book may be helpful: Y. C. Fung, *A First Course in Continuum Mechanics, for Physical and Biological Engineers and Scientists,* Prentice Hall, Englewood Cliffs, New Jersey, 3rd ed., 1993.

Y. C. FUNG, La Jolla

EDITORS' PREFACE

The previous volumes of the GALCIT * series have included two on aerodynamics and two on structural analysis and elasticity. The present work combines elements from these two fields. The subject of aeroelasticity has received a rapidly increasing amount of attention during the past few years, and the recent literature has become very voluminous. However, practically all of the contributions to this literature have been in the form of scientific or technical papers dealing with specific problems. Very few books dealing with the subject as a whole have appeared, and it is believed that a work of this type would now be useful, even though the subject is still expanding rapidly. The present volume has been prepared with this in mind, and it is hoped that it will prove valuable both in connection with academic instruction and also to scientists and engineers working in the field.

CLARK B. MILLIKAN
For the Editors

May 1955

* Guggenheim Aeronautical Laboratory, California Institute of Technology.

PREFACE TO THE FIRST EDITION

Trees sway in the wind; so may smokestacks. Flags and sails flutter; so may airplane wings and suspension bridges. The wind played ancient aeolian harps; so it plays the electric transmission lines, making them "sing" or even "gallop." These, and other phenomena that reveal the effect of aerodynamic forces on elastic bodies form the subject matter of aeroelasticity. It is a subject of growing importance in many fields of engineering, particularly in aeronautics, where, in the last decade or so, with the ever-increasing aircraft size and speed, aeroelasticity has become one of the most important considerations in aircraft design.

Although aeroelastic phenomena occur in everyday life, the attempts to develop a theory toward their understanding were made essentially by aeronautical engineers. A serious study of aeroelasticity started in the early '20's and, at present, it is still a young science making rapid progress. Nevertheless, part of the theory has reached a classical stage, and a general treatment is now possible.

This book owes its existence to a course on aeroelasticity which has been given to students in aeronautics at the California Institute of Technology since 1948. It is intended primarily as a textbook, but it should be useful also to designers and flutter engineers, for it gives a composite picture of various aspects of aeroelastic problems.

As the subject relies so heavily on aerodynamics, elasticity, and mechanical vibrations, most of the fundamental concepts in these related branches of mechanics are briefly explained in Chapter 1. Those having a rudimentary knowledge of aerodynamics and strength of material should have no difficulty in reading it. In this brief review, an effort is made to stress a number of points of importance in aeroelasticity, such as the arbitrariness contained in the usual definitions of shear center, the meaning of elastic axis, the conditions under which influence coefficients are unsymmetrical, the spanwise phase shift in the torsional vibration of a cantilever beam with viscous damping. However, no attempt is made to present, in a detailed manner, all the methods of analyzing the elastic deformation of complicated structures, or their natural vibration modes. The basic reason is a desire to move as quickly as possible to the main topics of aeroelasticity, to bring out aeroelasticity's main features in contrast to those usually found in treatises on aerodynamics, elasticity, and mechanical vibrations.

The main body of the text is divided into two parts. Chapters 2 to 11 contain a survey of aeroelastic problems, their historical background, basic physical concepts, and the principles of analysis. Chapters 12 to 15 contain the fundamentals of oscillating airfoil theory, and a brief summary of experimental results.

The selection of material perhaps requires some explanation. In Chapter 2, some problems of common occurrence in civil and mechanical engineering are outlined.. The main character of these problems is that the structures concerned are not streamlined and are unamenable to theoretical treatment. One must rely chiefly on experimental results to understand the nature of aerodynamic forces. A wide variety of phenomena is described. These phenomena are likely to appear also in aeronautical engineering when unfavorable conditions are encountered, such as when a wing is stalled, or when it is situated in a turbulent flow. An understanding of these phenomena not only is important in itself but also provides a natural background for the linearized theory to be discussed in the following chapters.

Principal problems of aeronautical interest are developed in Chapters 3 to 11. The steady-state problems and flutter theory are treated along conventional lines. But in discussing the dynamic stresses in aircraft due to gusts or other dynamic loading, as well as buffeting and stall flutter, the statistical aspects are emphasized. I believe that some statistical concepts and techniques are of vital importance to a proper understanding of many dynamic-stress problems.

The theory of aeroelasticity is concluded in Chapter 11 with a general formulation and a brief discussion of the basic mathematical characteristics. It is hoped that this discussion will attract the attention of mathematicians toward the important field of non-self-adjoint equations, which, owing to certain inherent difficulties, are very much neglected in the mathematical literature. Although in the past most of the practically important problems in physics were governed by self-adjoint equations, it is now shown that the entire field of aeroelasticity would have to be based on non-self-adjoint equations. In aeroelasticity those problems that are reducible to self-adjoint systems are really exceptional. Thus, a new impetus exists for the study of non-self-adjoint equations.

The aerodynamics of oscillating airfoils presented in Chapters 12 to 15 is of an introductory nature, and includes only those topics that are required in reading this book. Many aspects of the unsteady airfoil theory are still undergoing rapid development, and it seems yet too early to give a complete account.

As this book is not intended to be a compendium or a handbook, attention is directed only to the fundamental principles. The physical assumptions involved in the mathematical formulation of a problem are always emphasized, so that the degree of approximation relative to the real physical system can be seen. In this way the reader will realize the directions in which future improvement may lie.

In order to limit the size of the book, I have refrained from going into detailed discussions on such topics as flight-flutter testing, control and stability of aircraft, swept-wing analysis, stochastic theory of buffeting, and applications of digital and analog computing machines. An attempt is made, however, to provide sufficient guidance to the existing literature, so that the reader may find proper references to the particular subjects of his own interest.

The list of references in this book is by no means complete. Only those that are believed to be readily accessible to the general reader are quoted. Papers published expressly for limited circulations are generally omitted. A few important papers appeared too late to be included in the bibliography. But the reader can easily obtain the most recent references from the journal *Applied Mechanics Reviews*.

In preparation of this book, I am indebted to many of my colleagues at the California Institute of Technology. The constant advice of Professor E. E. Sechler is gratefully acknowledged. Doctors H. Dixon, A. Kaplan, A. Roshko, W. H. Wittrick, C. M. Cheng, Y. J. Wu, and Professors M. L. Williams and G. Housner read part of the manuscript and offered many valuable comments. Mrs. Dorothy Eaton, Mrs. Betty Wood, Mrs. Virginia Boughton, Mrs. Virginia Sloan, and Mrs. Gerry Van Gieson helped the preparation of the manuscript. To them and to many other friends I wish to express my sincere thanks. To Dr.-Ing. H. Drescher of the Max-Planck Institute I am especially grateful for permitting me to use some illustrations from his research work. My wife, Luna Yu, helped in reading the manuscript and proofs, checking the equations, and working out examples. I have derived infinite encouragement from her enthusiasm and patience in working on this book.

This book is dedicated to the memory of my father.

Y. C. FUNG

Pasadena
May 1955

CONTENTS

INTRODUCTION

Aeroelasticity is the study of the effect of aerodynamic forces on elastic bodies.

The classical theory of elasticity deals with the stress and deformation of an elastic body under prescribed external forces or displacements. The external loading acting on the body is, in general, independent of the deformation of the body. It is usually assumed that the deformation is small and does not substantially affect the action of external forces. In such a case we often neglect the changes in dimensions of the body and base our calculations on the initial shape. Even in problems of bending and buckling of columns, plates, or shells, either the external loading or the boundary constraints are considered as prescribed. The situation is different, however, in most significant problems of aeroelasticity. The aerodynamic forces depend critically on the attitude of the body relative to the flow. The elastic deformation plays an important role in determining the external loading itself. The magnitude of the aerodynamic force is not known until the elastic deformation is determined. In general, therefore, the external load is not known until the problem is solved.

One of the interesting problems in aeroelasticity is the stability (or rather instability) of a structure in wind. Since, for a given configuration of the elastic body, the aerodynamic force increases rapidly with the wind speed, while the elastic stiffness is independent of the wind, there may exist a critical wind speed at which the structure becomes unstable. Such instability may cause excessive deformations, and may lead to the destruction of the structure.

A major problem is the *flutter* of structures such as airplanes or suspension bridges, when small disturbances of an incidental nature induce more or less violent oscillations. It is characterized by the interplay of aerodynamic, elastic, and inertia forces, and is called a problem of *dynamic aeroelastic instability*. The particular case of an oscillation with zero frequency, in which in general the inertia force may be neglected, is called the *steady-state*, or *static*, *aeroelastic instability*.

Quite different from the above are the *response* problems in which the response of an aeroelastic system to an externally applied load is to be found. The external load may be caused by a deformation of the elastic body, such as a displacement of the control surfaces of an airplane, or by disturbances such as gusts, landing impacts, or turbulences in the flow.

1

The response to be found may be the displacement, the motion, or the stress state induced in the elastic body. Again the response problems may be classified into the *steady-state* or *static* problems, in which the inertia forces may be neglected, and the *dynamic* problems, in which the aerodynamic, elastic, and inertia forces all enter into the picture.

There is a close relationship between the stability problems and the response problems. Mathematically, most stability problems can be described by a system of homogeneous equations, which are satisfied by a trivial solution of zero displacement (or zero motion), meaning that nothing happens at all. On the other hand, a response problem is represented by a nonhomogeneous system; i.e., the initial conditions and the external forces are such as to cause the governing equations to be nonhomogeneous, and to admit a solution not vanishing identically. A response problem generally associates with a stability problem. As an example, consider the response of an airplane wing to atmospheric turbulences. We can formulate the problem of flutter by asking the following questions: Is there a critical speed of flight at which the airplane structure becomes exceedingly sensitive to the atmospheric turbulence; i.e., does there exist a speed at which the structure may have a motion of finite amplitude, even in the limiting case of an atmospheric turbulence of zero intensity? This is equivalent to the following formulation which is usually made in flutter analysis: Is there a critical speed at which the aeroelastic system becomes neutrally stable, at which motion of the structure is possible without any external excitation? Thus the response of an airplane structure to atmospheric turbulence and the flutter problem are linked together. When the response of the structure to a finite disturbance is finite, the structure is stable, and flutter will not occur. When the structure flutters at a critical speed of flow, its response to a finite disturbance becomes indefinite.

This alternative theorem is true in practically all corresponding response and stability problems. Either the homogeneous system has a nontrivial solution while the corresponding nonhomogeneous system has no solution, or the nonhomogeneous system has a solution while the corresponding homogeneous system has no solution other than the trivial one.* It is thus proper to discuss the response and stability problems together as two phases of the same phenomenon.

There exists, however, a very important distinction between the response and stability problems, in regard to the justification of the linearization process often used in the mathematical formulation of a physical problem.

---

* In exceptional cases both the nonhomogeneous and the corresponding homogeneous system may have a solution. But such exceptions ordinarily have little engineering significance.

In the stability problems, the amplitude of the elastic deformation is indeterminate, and only the modes of deformation (not their absolute magnitude) are of interest; hence, it is logical to consider the elastic deformation as infinitesimal in the neighborhood of an equilibrium state. Therefore the small deflection theory in elasticity and aerodynamics is applicable, and linearization of the governing equations can be justified. On the other hand, the absolute magnitudes of the deformation and stress in a structure are of primary interest in the response problems. Hence, it is necessary to consider finite deformations. As the fundamental equations of fluid and solid mechanics are often nonlinear, it is necessary to consider the effects of nonlinearity, whenever the response reaches a finite amplitude. Thus the justification of linearization of the fundamental equations is always open to question.

Of course it is desirable to treat the nonlinear equations per se, but the mathematical difficulties are generally insurmountable. Generally we are forced to linearize, in order to reach a practical solution. Then it must always be remembered that the justification of the linearization remains to be shown.

In this book, attention will be directed mainly to the stability problems, not because the response problems are less important, but because they are well-known in engineering philosophy. On the other hand, the stability aspect of aeroelasticity is novel.

Generally speaking, aeroelasticity includes the study of all structures in a flow. But those problems in which the elastic deformation plays no significant role in the determination of the external loading will not be discussed in this book. For example, the problem of the distribution of wind load on a building will be excluded.

A survey of the field of aeroelasticity is given in Chapters 1 through 11. Important problems are discussed from the physical point of view. The chief aim is to provide an elementary treatment of the basic problems and to point out the essential parameters involved in their solution. The aerodynamic problems are discussed in greater details in Chapters 12 through 15.

Chapter 1

PRELIMINARIES

The discussion of aeroelasticity requires certain preliminary information on the theory of elasticity, aerodynamics, and mechanical vibrations. There exist a number of excellent textbooks on these subjects. Therefore we shall review only briefly some of the fundamental facts in this chapter and explain the notations and sign conventions that will be used in the text.

The reader is urged, however, to read carefully §§ 1.2 and 1.3, concerning the definitions of shear center, elastic axis, flexural line, etc., because these terms have been used somewhat ambiguously in the engineering literature. In § 1.4 the influence functions are explained, and in § 1.6 the generalized coordinates and Lagrange's equations are reviewed and illustrated by several examples. These subjects must be understood thoroughly.

Throughout this book a vector will be printed in boldface type, as, for instance, a velocity vector $\mathbf{v}$, a force vector $\mathbf{F}$. A vector in a three-dimensional space has three components, which are indicated by subscripts. Thus a force $\mathbf{F}$ referred to a system of rectangular Cartesian coordinates x, y, z has three components F_x, F_y, F_z. Sometimes it is more convenient to label the xyz coordinates as $x_1 x_2 x_3$ coordinates and to indicate F_x as F_1, F_y as F_2, F_z as F_3. The vector $\mathbf{F}$, being specified by the three components F_1, F_2, F_3, may also be identified simply by writing F_i $(i = 1, 2, 3)$.

A relation among several vectors may be expressed either by a single vector equation or by a system of equations expressing the relations among the components of the vector. For example, let $\mathbf{a}$ (with components a_1, a_2, a_3) be the acceleration of a particle, m its mass, and $\mathbf{F}$ (with components F_1, F_2, F_3) the force acting on the particle. Then Newton's law of motion for this particle can be written either as

$$\mathbf{F} = m\mathbf{a} \tag{1}$$

or as

$$F_1 = ma_1$$
$$F_2 = ma_2 \tag{2}$$
$$F_3 = ma_3$$

Equations 2 may be shortened into the following form

$$F_i = ma_i \qquad (i = 1, 2, 3) \tag{3}$$

We shall consider Eqs. 1 and 3 as entirely equivalent expressions.

This notation will be extended to tensor equations and matrix equations by means of multiple subscripts.

One of the most important simplifying conventions in all mathematics is the *summation convention*: *to use the repetition of an index to indicate a summation over the total range of that index.* For example, if the range of the index i is 1 to 5, then

$$a_ib_i \equiv \sum_{i=1}^{5} a_ib_i = a_1b_1 + a_2b_2 + \cdots + a_5b_5 \tag{4}$$

If $a_i = \mathbf{a}$ and $b_i = \mathbf{b}$ are two vectors, the product a_ib_i is the *scalar product* of $\mathbf{a}$ and $\mathbf{b}$:

$$\mathbf{a} \cdot \mathbf{b} = a_ib_i \tag{5}$$

As another example, if $i, j = 1, 2, 3$, then

$$C_{1j}F_j \equiv \sum_{j=1}^{3} C_{1j}F_j = C_{11}F_1 + C_{12}F_2 + C_{13}F_3$$

$$C_{2j}F_j \equiv \sum_{j=1}^{3} C_{2j}F_j = C_{21}F_1 + C_{22}F_2 + C_{23}F_3 \tag{6}$$

$$C_{3j}F_j = \sum_{j=1}^{3} C_{3j}F_j = C_{31}F_1 + C_{32}F_2 + C_{33}F_3$$

The system of Eqs. 6 may be simply written as

$$C_{ij}F_j \equiv \sum_{j=1}^{3} C_{ij}F_j \qquad (i = 1, 2, 3) \tag{7}$$

This summation convention will be used in this book.

1.1 ELEMENTARY BEAM THEORY

Consider a cylindrical beam of uniform isotropic material. In each cross section of the beam, two mutually perpendicular *principal axes*, passing through the centroid of the cross section, can be determined, about which the *second moments* (the moments of inertia) of the beam cross-sectional area assume stationary values with respect to rotation of the centroidal axes, and the *product of inertia* of the area vanishes. The plane containing one of the principal axes of all cross sections is called a *principal plane.* If the beam is acted on by a *bending moment M* in a principal plane, the beam will deflect in that plane. Let $1/R$ represent the change of curvature of the beam in that plane; then, within the elastic limit of the beam,

$$\frac{1}{R} = \frac{M}{EI} \tag{1}$$

where E is Young's modulus of the material, and I is the moment of inertia* of the beam cross section about a principal axis perpendicular to the principal plane in which M acts. Let y denote the distance from the neutral plane; then the bending stress is given by

$$\sigma = \frac{My}{I} \tag{2}$$

Equations 1 and 2 are applicable, approximately, also to a straight beam with nonuniform cross sections subjected to distributed external loads, provided that E, I, $1/R$, and M are the local values and that the variation of the beam cross section is gradual. They are, however, not directly applicable to curved beams.

Equations 1 and 2 are derived under the assumptions that the displacement of the beam is small, that the Hooke's law between stress and strain holds, and that the plane cross sections of the beam remain plane during deformation. They are referred to as *engineering beam formulas.*

When a system of external forces acts on a beam, it produces *shear* and *bending moment* in the beam. The loading (lateral force per unit length) p, the shear S, and the moment M are connected by the equations

$$\frac{dM}{dx} = -S$$
$$\frac{dS}{dx} = -p \tag{3}$$

where x denotes distance measured along the axis of the beam. For a given beam, after a given direction has been chosen for the coordinate x, the signs (i.e., the positive senses) of p, S, and M may be *consistently* chosen by verifying Eqs. 3.

If a *twisting* moment whose vector is parallel to the beam axis is applied on the beam, the cross sections will *rotate* about the beam elastic axis.

---

* For semimonocoque thin-walled box beams part of the skin may be buckled under a compressive stress. The contribution of such buckled panels to the bending stiffness can be accounted for by reducing the actual width of the skin panels to their "effective width." In this case the factor I in Eq. 1 is the "effective" moment of inertia, computed on the basis of the effective width of skin.[1,3]

The rate of change of the *angle of twist* θ (radians) along the length of the beam is given by the formula

$$\frac{d\theta}{dx} = \frac{T}{GJ} \tag{4}$$

where T is the twisting moment about the shear center of a section at x, and *GJ denotes the torsional rigidity. G is the shear modulus of rigidity*, but J stands for the quantities as shown in Table 1.1, according to various cross sections.[1.27]

In Eq. 4, the positive sense of the vector θ is chosen as that of the coordinate axis x. However, the positive sense of the torque T, like that of the shearing stresses, cannot be determined until the positive side of the surface on which the torque acts has been chosen. The sign convention is as follows: Consider a beam element of length dx, which is bounded on both ends by normal cross sections of the beam. Let normal vectors of the cross sections be drawn from inside of the element, (the so-called outer normals). If a torque T acting on the end of the beam element agrees in its vector sense with that of the outer normal, then T is positive; otherwise it is negative.

If the beam is subjected to a system of distributed twisting moment of intensity m per unit length, then the twisting moment T is variable across the span. Let us define m as positive if its vector sense agrees with that of x, which we shall assume to be pointing to the right. Then on an element of length dx, there acts a torque $-T$ on the left-hand side, a torque $T + dT$ on the right-hand side, and a torque $m\,dx$ on the element. (See Fig 1.21 on p. 48.) Thus the condition of equilibrium of the element demands that

$$-T + m\,dx + T + dT = 0$$

or

$$\frac{dT}{dx} = -m \tag{5}$$

Combining Eqs. 4 and 5, we obtain the following relation for a beam subjected to a system of distributed twisting moments:

$$\frac{d}{dx}\left(GJ\frac{d\theta}{dx}\right) = -m \tag{5a}$$

In engineering beam theory, the beam deflection is assumed to be *infinitesimal*. Let w be the deflection of the beam; then, approximately,

$$\frac{1}{R} = \frac{d^2w}{dx^2} = \frac{M}{EI} \tag{6}$$

Table 1.1

Beam Cross Section	J
Circular cylinder	Polar moment of inertia $= \dfrac{\pi}{32}(d^4 - d_i^4)$
Elliptic cylinder	$\dfrac{Aa^2b^2}{4(a^2 + b^2)}$, $A = \text{area} = \pi\,\dfrac{ab}{4}$
Rectangular section	$\dfrac{A^4}{40 I_p}$ (approx.) $I_p = $ polar moment of inertia about centroid
	$\dfrac{At^2}{3}$, $A = $ area of cross section
Single-bay thin-walled tube	$4A^2 \Big/ \displaystyle\oint \frac{ds}{t}$, $A = $ area enclosed by tube walls
Double-bay thin-walled tube	$4\left[\dfrac{a_{20}A_1^2 + a_{12}(A_1 + A_2)^2 + a_{01}A_2^2}{a_{01}a_{12} + a_{12}a_{20} + a_{20}a_{01}}\right]$, where $a_{ij} = \displaystyle\int \frac{ds}{t}$, the integral being taken along boundary between A_i and A_j.

When a positive sense is chosen for w, the *positive sense* of M must be checked against Eq. 6. Such a check of signs should always be made in order to avoid confusion in the calculations.

From Eqs. 3 and 6, we obtain

$$\frac{d^2}{dx^2}\left(EI\,\frac{d^2w}{dx^2}\right) = p \tag{6a}$$

where w is the deflection and p is the loading per-unit length acting on the beam. The positive senses of w and p agree with each other.

When the external forces acting on the beam do not lie in a principal plane, the forces should be resolved into components lying in each of the two principal planes. The deflection of the beam can then be computed in these two planes separately and then added vectorially. Similarly, if an external couple acting on the beam is inclined to the beam axis, the couple should be resolved into a bending moment and a twisting moment, and the induced displacements computed separately.

In solid beams, the deflection is essentially induced by the bending moments. The deflection caused by the shear S can be neglected. But in thin-walled box beams the *shear deflection* can become quite important, particularly in calculating the higher-order vibration frequencies and modes.[1.64]

Accompanying the application of external load on the beam, elastic strain energy is stored in the beam. For a solid beam the strain energy due to transverse shear stresses is usually negligible in comparison with that due to bending and torsion. If the deflection $w(x)$ and the torque $T(x)$ are measured at the shear center of a section at x, then the strain energy can be written as

$$V = \frac{1}{2}\int_0^l EI\left(\frac{d^2w}{dx^2}\right)^2 dx + \frac{1}{2}\int_0^l GJ\left(\frac{d\theta}{dx}\right)^2 dx \tag{7}$$

or

$$V = \frac{1}{2}\int_0^l \frac{1}{EI}\,M^2(x)\,dx + \frac{1}{2}\int_0^l \frac{1}{GJ}\,T^2(x)\,dx \tag{8}$$

where l is the length of the beam. For thin-walled structures, the strain energy due to transverse shear is not negligible. A term of the following form,

$$V = \frac{1}{2}\int_0^l K(x)\,S^2(x)\,dx \tag{9}$$

should be added, where $K(x)$ is a function of the cross-sectional shape and the material of the structure.

The elementary theories of torsion and bending are based on assumptions that are usually violated in actual aircraft wing structures. The elementary torsion theory is valid for a shell of constant cross section, subjected to a torque at each end in the form of a shear flow that is distributed in the section in accordance with the theory, and that leaves the end sections free to warp out of their original planes. An actual wing has a variable section and is subjected to distributed torque loads; as a result, the tendency to warp, in general, differs from section to section, and secondary stresses are set up by the resulting interference effects. The normal stresses so induced are called the "bending stresses due to torsion." Similarly, the elementary bending theory is strictly valid if the applied load is a pure bending moment. In actual wing structures, the bending moments are produced by transverse loads, and, in general, the shear strains in the beam produced by these loads violate the assumption that plane cross sections remain plane. As in the torsion case, interference effects between adjacent sections may produce secondary stresses. In the particular case of thin-walled box beams, the effect of the shear strain on the distribution of normal stresses is called the "shear lag."

A general theory of bending and torsion of beams of variable cross sections, subject to a variable loading, is very complicated. A practical solution exists for thin-walled cylinders under the assumption that cross-sectional shape of the cylinders is maintained by diaphragms, which are infinitely rigid against deformation in their own planes but are perfectly flexible for deflection normal to their planes. For box beams of closed sections the effect of shear lag is important with respect to the stress distribution, but is insignificant with respect to the deflection of the beam except for swept wings. This is because the deviations of the stresses from those predicted by the elementary theories are local, and local disturbances are smoothed out by the integration process necessary to calculate deflections. The deviations of the deflections from those predicted by the elementary theories are therefore much smaller than the stress deviations. For this reason adequate accuracy can often be achieved for the deflection (and the influence-coefficient) calculations, even when highly simplified theories are used. On the other hand, the effect of shear lag and restrained warping is very large for thin-walled structures with open cross sections (the portion of a wing with a large cutout may be regarded as open sections). See articles listed in the bibliography at the end of this chapter for the treatment of special problems. See, in particular, Ref. 1.70.

In applying the results of the elementary beam theory to airplane structures, considerable engineering judgment is often necessary because the effects of cutouts, shear lag, differential bending of the spars, discontinuous changes of section properties, etc., must be properly accounted for. Occasionally a wing or a tail cannot at all be considered as a "beam" or a "torque tube." In such cases a more comprehensive analysis of the elastic deformation is necessary. In aeroelasticity the most convenient scheme of describing the elastic properties of a structure is to specify its influence functions. The calculation of the influence functions for a structure other than a simple beam may be very difficult, but it is a prerequisite for aeroelastic analyses.

1.2 SHEAR CENTER

Consider a straight thin-walled beam of uniform cross section, which is subjected to a shearing force S as shown in Fig. 1.1. Let a set of ortho-gonal Cartesian coordinates xyz be chosen as shown in the figure, where

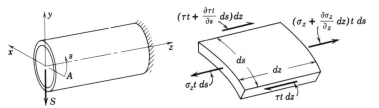

Fig. 1.1. Equilibrium of forces acting on an element of a
thin-walled section.

the z axis is parallel to the axis of the beam, and the x and y axes are the principal axes of the cross section. Let us assume that at any point in the cross section, the stress distribution over the wall thickness is uniform. It is convenient to introduce a curvilinear coordinate s along the circum-ferential direction of the cross section, and to speak of the "shear flow," meaning the shearing force per unit of circumference

$$q = \tau t \qquad (1)$$

where τ is the shearing stress, and t is the thickness of the wall. The equation of equilibrium of the forces in the axial direction acting on a small element of area $ds\,dz$ is, according to the diagram on the right-hand side of Fig. 1.1, where σ_z denotes the axial stress:

$$\left(\sigma_z + \frac{\partial \sigma_z}{\partial z}\,dz\right)t\,ds - \sigma_z t\,ds + \left(\tau + \frac{\partial \tau}{\partial s}\,ds\right)t\,dz - \tau t\,dz = 0 \qquad (2)$$

Passing to the limit $ds,\ dz \to 0$, one obtains

$$\frac{\partial \sigma_z}{\partial z} + \frac{\partial \tau}{\partial s} = 0 \qquad (3)$$

or, multiplying by the thickness t, if it is constant,

$$t\frac{\partial \sigma_z}{\partial z} + \frac{\partial q}{\partial s} = 0 \tag{4}$$

Thus the shear flow q depends on the rate of change of the axial stress σ_z. A few simple cases will be considered below.

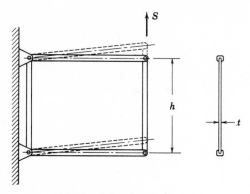

Fig. 1.2.   A flat shear web.

(a) *Flat Shear Webs.* A flat "shear web" is one that is subjected to pure shear; i.e., it is defined by the condition $\sigma_z = 0$. Such a web is approximated by the one shown in Fig. 1.2. In this case, Eq. 4 implies that

$$q = \text{const} \tag{5}$$

The constant is easily evaluated from statics:

$$q = \frac{S}{h} \tag{6}$$

where S is the total shear force over a section, and h is the height of the web.

(b) *Curved Shear Webs.* According to Eq. 4, the condition $q = \text{const}$ prevails also for a "curved shear web," over which $\sigma_z = 0$. As shown in Fig. 1.3, the shear flow follows the curvature of the cross section. The resultant force in the y direction is

$$S = \int q \cos \theta \, ds \tag{7}$$

where θ is the angle between the vector of the shear flow q and the y axis, and the integration is taken over the entire section. Obviously

$$S = \int q \, dy = q h \tag{8}$$

or

$$q = \frac{S}{h} \qquad (9)$$

which is the same as Eq. 5. The shear flow is equal to the total shear force divided by the height of the beam.

It is important to find the location of the resultant shear. For this purpose the moment about a point O in the plane of the cross section

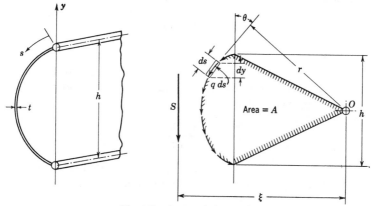

Fig. 1.3. A curved shear web.

produced by the shear flow may be computed. Let r be the perpendicular distance from O to the tangent at s where q acts. From Fig. 1.3 the moment about O is

$$M_O = \int qr\, ds \qquad (10)$$

Since q is a constant, this may be written as

$$M_O = q \int r\, ds = 2qA \qquad (11)$$

where A is the area of a sector enclosed between the curved shear web and two radius vectors with origin at O as shown in the figure. But this moment is also equal to the resultant shear S times its moment arm ξ about the origin O,

$$M_O = S\xi$$

Hence, from Eqs. 9 and 11,

$$\xi = \frac{M_O}{S} = \frac{2A}{h} \qquad (12)$$

It can be easily verified that the spatial location of the line of action of the resultant shear is independent of the location of the origin O.

The web of a two-flange type of beam approximates in practice the shear web described above. Such a beam can transmit a shear force, but only if the force is applied at the location specified by Eq. 12 and parallel to the plane of the two flanges, as shown in Fig. 1.3. If it is applied along any other line of action, it will cause an unbalanced twisting moment that the beam alone cannot resist.

(c) *Open Thin-Walled Sections.* In general, the axial stress σ_z does not vanish when the beam is subjected to a transverse loading. In many cases, however, it is sufficiently accurate to use the elementary beam formula for a bending moment M due to the transverse loading

$$\sigma_z = -\frac{My}{I} \tag{13}$$

where y is the distance from the neutral plane, provided that the y axis is a principal axis of the cross section, and that the shear load vector is parallel to the y axis. Hence,

$$\frac{\partial \sigma_z}{\partial z} = -\frac{y}{I}\frac{dM}{dz} = \frac{y}{I}S \tag{14}$$

in accordance with Eq. 3 of § 1.1. If we assume again that the shearing stress is uniformly distributed over the wall thickness, the shear flow is obtained from Eq. 4 by an integration:

$$q = q_0 - \int_{s_0}^{s} t\frac{\partial \sigma_z}{\partial z}\,ds$$

where q_0 is the value of q at s_0. Using Eq. 14, one obtains

$$q = q_0 - \frac{S}{I}\int_{s_0}^{s} yt\,ds \tag{15}$$

The shear flow is determined when q_0 is known. For the open thin-walled sections, the average value of the shearing stress across the wall

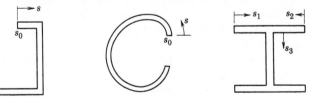

Fig. 1.4.   Open thin-walled sections.

thickness vanishes at the ends of the section, in accordance with St. Venant's theory of torsion. Hence, if s_0 is taken as one of the ends, then $q_0 = 0$. This is illustrated in Fig. 1.4.

In the last case of Fig. 1.4, there is a junction where three members meet. Let q_1, q_2, q_3 be the shear flow *toward* the junction in the three members, respectively; then a condition of equilibrium is

$$\sum_{i=1, 2, 3} q_i = 0 \tag{16}$$

This can be seen from the free-body diagram of a small element at the junction as shown in Fig. 1.5. Equation 16 represents the condition of equilibrium of the forces in the axial direction.

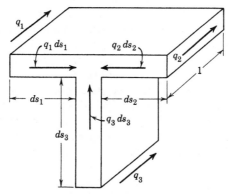

Fig. 1.5. Shear flow at a joint.

For a multi-flanged beam, the material concentrated at the flanges resists the bending moment (Fig. 1.6). Equation 15 can be modified to give the change of shear flow across each of the flanges. Let ΔA be the

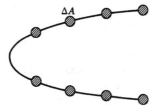

Fig. 1.6. A multi-flanged beam.

area of a flange, then the difference of q on the two sides of the flange is

$$\Delta q = -\frac{S}{I} y \, \Delta A \tag{17}$$

The equations given above are sufficient to determine the shear flow in any open thin-walled section. When q is known, the resultant force is given by Eq. 7, and the resultant moment about the origin O is given by Eq. 10. The location of the line of action of the resultant shear is therefore at a distance ξ from the origin:

$$\xi = \frac{M_O}{S} = \frac{1}{S} \int qr \, ds \tag{18}$$

where the symbols r, s, and y have the same meaning as shown in Fig. 1.3, and the integral is taken over the entire cross section. The shear flow q is, in general, a function of s and cannot be taken outside of the integral.

Similar calculation can be made if a shear acts in the direction of the other principal axis of the beam cross section. The distance from the origin, η, of the line of action of the resultant shear which is parallel to x axis, can be determined by an expression similar to Eq. 18. The intersection of these two lines of action, the point with coordinates ξ, η referred to a set of centroidal principal axis, is called the *shear center* of the cross section. Since an arbitrary shear load can be resolved into components parallel to the two principal directions, it becomes evident that, if a shear vector acts through the shear center of the cross section, the bending stress distribution will be given by the elementary beam formula, and no twisting of the beam will occur. This is the significance of the shear center.

If the resultant shear does not pass through the shear center, it can be resolved into a shear of equal magnitude that does pass through the shear center and cause torsionless bending, and a couple that will cause twisting.

(d) *Closed Thin-Walled Sections and Solid Sections.* The equations derived in the previous paragraph apply as well to closed thin-walled sections, except that it is now no longer evident that there exists a point s_0 at which q_0 vanishes. The exact values of q depend on the line of action of the resultant shear force.

It is convenient to define the *shear center* as a point through which acts a shear force that will produce a "pure" or "torsion-free" bending. The beam subjected to a shear force that acts through the shear center will be "torsion-free." A shear force that acts in a line that does not pass through the shear center will cause both bending and torsion.

The analytical formulation of the above definition depends, however, on the precise meaning of the term "pure bending" or "torsion-free bending." Based on the classical theories of bending and torsion, several different but equally convincing definitions have been proposed for "pure bending." A detailed discussion is presented in Appendix I, which covers also solid sections and thick-walled sections. For a given definition of the term "pure bending," a corresponding location of the shear center can be computed. Fortunately, shear-center positions of airplane structures based on various definitions differ very little from each other, and in practical applications any one of these definitions may be used.

It may be noted that the shear-center locations calculated in previous paragraphs under cases *b* and *c* are in agreement with the definition just given.   No ambiguity exists for the concept of torsion-free bending in an open thin-walled section.   This clear-cut feature is a result of the assumption that the shearing stress is distributed uniformly over the thickness of the section.

1.3 THE ELASTIC AXIS

In the preceding Section, the shear center of a straight, cylindrical beam is defined.   Since the analysis is based on St. Venant's theory of bending and torsion, the exact boundary conditions at the ends of the beam cannot always be satisfied.   The shear center so defined is a section property. For each cross-sectional shape, there associates a shear-center location.

The concept of shear center can be extended to a curved beam.   One imagines that, at any point on the beam axis, a cylinder tangent to the given beam, and having the same normal cross section, is prescribed.   The shear center of this cylinder is then taken as the shear center of the curved beam at that point.

The locus of shear centers of the cross sections of a beam is called *the elastic axis* of that beam.*   The elastic axis is a natural reference line in describing the elastic deformation of the beam; for the resulting differential equations are the simplest.

By using the elastic axis as defined above, it is possible to derive the classical Bernoulli-Euler equations of bending for a curved beam in analogy with Eq. 1 of § 1.1.   The theory of a curved beam whose elastic axis coincides with the centroidal axis of the beam has been developed completely in Chapter 18, § 289 of Love's book.[1,2]   If the shear centers do not coincide with the centroids of the cross sections, it is necessary to define a set of local coordinate axes parallel to the *generalized principal torsion-flexure axes* defined by Love, but with the origin located at the shear center.   The "torsion" and curvature of the elastic axis can then be determined in the same manner as those of the centroidal axis, and the relation between the bending moments and torque and the change of curvatures and "torsion" at a section can be derived.   The resulting equations are useful for "thin" beams, the cross-sectional dimensions of which are, by definition, much smaller than the radius of curvature of the beam.

---

* Some authors define elastic axis as the locus of flexural centers or as a flexural line. This is not the sense to be used in this book.

The practical application of the elastic axis to aircraft structures is, however, subject to severe limitations. Owing to the effects of sweep angle, shear lag, restrained torsion, cutouts, buckling of skin panels, etc., the shear center often cannot be defined without ambiguity. In other words, the concepts of shear center and elastic axis lose their simplicity or usefulness when a structure other than a simple cylindrical beam is considered. In any case of doubt, it is better to use influence functions which describe the deformation pattern of the structure due to a unit load acting on the structure. An influence function is a well-defined quantity. It can be measured on an existing structure, and it may be calculated by refined structural theories in special cases.

In the present book, the elastic axis will be used only when it is a straight line and is associated with a structure that behaves like a simple beam.

In engineering literature, the terms *flexural center, center of twist, and flexural line* are often used. They are defined as follows: Consider a

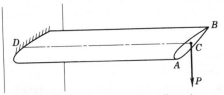

Fig. 1.7. Flexural center of a cantilever beam.

cantilever beam of uniform cross section. If a load P is applied at a point A on the free end (see Fig. 1.7), the section ACB will rotate counterclockwise in its own plane. If P is applied at B, the section ACB will rotate clockwise. Somewhere between A and B is a point (say C) at which the load P can be applied without causing rotation of the section ACB in its own plane. This point is called the *flexural center of the section ACB*.

More generally, for a slender, curved, cantilever beam, the *flexural center of a cross section* is defined as a point in that section, at which a shear force can be applied without producing a *rotation of that section* in its own plane. Closely associated is the *center of twist* which is a point in a cross section that remains stationary when a torque is applied in that section. If the supporting constraint of the beam is perfectly rigid, the flexural center coincides with the center of twist. (See Ref. 1.15, or p. 29 of Ref. 1.9.) It is important to notice that these definitions are referred to particular sections. For example, when a force is applied at the flexural center of a section, that section will not rotate; but, unless the beam is homogeneous, cylindrical, and clamped in a plane normal to its axis, other sections of the beam may rotate. In other words, whether other sections of the beam rotate or not is irrelevant to the definition of a flexural center. This fact makes the locus of flexural centers useless except when it is a straight line; then it coincides with the elastic axis.

For a given loading, a *flexural line* is defined as a curve on which that loading may be applied, so that there results no twist at any section of the beam. In general, different load distributions correspond to different flexural lines, and there exist load distributions that do not have a flexural line. A simple example is a concentrated load acting on a curved beam. This property renders the flexural line useless in aeroelasticity except in the simplest cases.

1.4 THE INFLUENCE COEFFICIENTS AND INFLUENCE FUNCTIONS

In aeroelasticity, the most concise description of the elastic property of a structure is obtained by means of influence functions. If the structure is perfectly elastic, meaning that its load-deflection relationship is linear and that it returns to the initial configuration after all the loads are removed, then the influence functions can be uniquely determined. In most of the problems discussed below, not only the material elements of the structure are assumed to be elastic (subject to stresses below the proportional limit), but the entire structure also must be supposed to remain elastic. In judging the elasticity of the structure, we must specify the range of loads to be considered in a given problem. For example, consider an aircraft wing of sheet-metal construction. When the air load gradually increases, certain structural members may become buckled and go out of action. But the wing may remain elastic for a range of additional loads, even though the rigidity of the wing changes before and after the buckling.

Consider a perfectly elastic structure, *rigidly supported*,* and subjected to a set of forces Q_1, Q_2, $\cdots Q_n$ at points 1, 2, $\cdots$, n respectively.

---

* We shall assume tacitly that the structure is rigidly supported in a specific manner when the influence functions are measured. However, this does not prevent their usefulness in applications to an airplane in free flight, because all that is necessary is to measure the elastic displacements (or influence functions) of the airplane structures with respect to a set of rectangular coordinates attached to the airplane, with the origin located at a convenient point, e.g., a point on the center line of the main spar, or the airplane center of gravity. The structure may be regarded as clamped with respect to the coordinate system at its origin. The change in direction of the coordinates and the motion of the origin can then be determined by the free-body motion of the airplane.

According to Hooke's law, the deflection q_1 at the point 1, due to the set of forces $\{Q_i\}$ may be written as

$$q_1 = a_{11}Q_1 + a_{12}Q_2 + \cdots + a_{1n}Q_n \qquad (1a)$$

The proportional constants a_{11}, $a_{12} \cdots$ are independent of the forces $Q_1, Q_2, \cdots, Q_n$. Similarly, the deflections at points 2, 3, $\cdots$, etc., are

$$q_2 = a_{21}Q_1 + a_{22}Q_2 + \cdots + a_{2n}Q_n \qquad (1b)$$
$$\cdots \cdots$$
$$q_n = a_{n1}Q_1 + a_{n2}Q_2 + \cdots + a_{nn}Q_n$$

If the summation convention is used, the above equations may be expressed simply either as

$$q_i = \sum_{j=1}^{n} a_{ij}Q_j \qquad (i = 1, 2, \cdots, n) \qquad (2)$$

or as

$$q_i = a_{ij}Q_j \qquad (i, j = 1, 2, \cdots, n) \qquad (3)$$

For a rigidly supported perfectly elastic body, the elastic deformation is a unique function of the forces acting on the body. For, if there exists a different set of deflections $q'_1, q'_2, \cdots, q'_n$, corresponding to the same $Q_1, Q_2, \cdots, Q_n$, we may first apply the loads $Q_1, \cdots, Q_n$ to deform the body into $q_1, q_2, \cdots, q_n$ at points 1, 2, $\cdots$, n; then apply the loads $-Q_1, -Q_2, \cdots, -Q_n$ to obtain deflections $-q'_1, -q'_2, \cdots, -q'_n$ at points 1, 2, $\cdots$, n. The final configuration has deflections $q_1 - q'_1$, $q_2 - q'_2, \cdots, q_n - q'_n$ at the specified points. But the external loads are now completely removed; it follows from the definition of perfect elasticity that $q_1 - q'_1$, $q_2 - q'_2$, etc., must all vanish; i.e., $q_1 = q'_1$, $q_2 = q'_2$, etc. Hence a contradiction is obtained, and we must conclude that q_i and Q_i are in one-to-one correspondence.

The uniqueness of the force-deflection relationship implies not only that the constants a_{ij} are unique functions of the elastic body (because a_{ij} may be interpreted as the deflection at point i due to a unit load acting at point j), but also that the set of Eqs. 3 can be solved for Q_j and that the solution is unique:

$$Q_i = \sum_{j=1}^{n} K_{ij}q_j \qquad (i = 1, 2, \cdots, n) \qquad (4)$$

In other words, the determinants $|a_{ij}|$ and $|K_{ij}|$ do not vanish.

The constants a_{ij} and K_{ij} are called, respectively, *flexibility-influence coefficients* and *stiffness-influence coefficients*. The flexibility-influence coefficients are generally called simply *influence coefficients*.

The physical interpretation of a stiffness-influence coefficient K_{ij} is the force that is required to act at the point i due to a unit deflection at the point j, while all points other than j are held fixed.* In the case of a single degree of freedom, the stiffness-influence coefficient is the familiar spring constant.

It is often convenient to regard Eqs. 3 and 4 as matrix equations, $\{q_i\}$, $\{Q_i\}$ being column matrices and $\{a_{ij}\}$, $\{K_{ij}\}$ being square matrices (§ 3.6). We have shown that $\{a_{ij}\}$, $\{K_{ij}\}$ are nonsingular and that they are the inverse of each other.

When a set of forces is applied to an elastic body, work is done by the forces. Let us define a displacement *corresponding* to a force as the component of the displacement under the point of application of the force and in the direction of the force. If Δq_1 *corresponds* to Q_1, the *work done* by Q_1 through a displacement Δq_1 is equal to $Q_1 \Delta q_1$.

Let $q_1, q_2, \cdots, q_n$ denote the displacements corresponding to the forces $Q_1, Q_2, \cdots, Q_n$ which act at the points $1, 2, \cdots, n$, respectively. If all the forces are applied very slowly so that equilibrium is maintained at all times, the total work done by the forces will be

$$W = \int_0^{q_1} Q_1 \, dq_1 + \int_0^{q_2} Q_2 \, dq_2 + \cdots + \int_0^{q_n} Q_n \, dq_n \tag{5}$$

The evaluation of the integrals in Eq. 5 is extremely simple if the ratios $Q_1 : Q_2 : \cdots : Q_n$ are maintained while the absolute values of the forces gradually increase. In this case Q_i is proportional to q_i. Writing $Q_i = K_i q_i$, we have

$$W = \sum_{i=1}^{n} \int_0^{q_i} K_i q_i \, dq_i = \sum_{i=1}^{n} \frac{1}{2} K_i q_i^2 \tag{6}$$

i.e.,

$$W = \frac{1}{2} \sum_{i=1}^{n} Q_i q_i$$

The last expression, however, must depend solely on the final stress status of the body, i.e., on the final loading and displacements; for, otherwise, by performing a loading cycle in which the body is loaded by applying $Q_1, Q_2, \cdots, Q_n$ in certain order and unloading by $-Q_1, -Q_2, \cdots,$

---

* To readers familiar with Southwell's relaxation method, the above interpretation of the stiffness-influence coefficients K_{ij} at once suggests an identification of K_{ij} with the "relaxation patterns." Since for plate-like thin wings K_{ij} can be determined easily from the governing differential equations, D. Williams[1.71] proposed that the influence coefficients $\{a_{ij}\}$ should be obtained from an inversion of the $\{K_{ij}\}$ matrix.

$- Q_n$ in a different order, one will be able to extract energy from the elastic body, in violation of the principle of conservation of energy.

Thus the total work done is independent of the manner in which the final configuration is reached. This work is stored in the body in the form of elastic strain, and is called the *strain energy*. By using Eqs. 3 and 4, the strain energy V (numerically equal to W) may be written as

$$V = \frac{1}{2} \sum_{i=1}^{n} \cdot \sum_{j=1}^{n} K_{ij} q_i q_j \qquad (7)$$

or

$$V = \frac{1}{2} \sum_{i=1}^{n} \sum_{j=1}^{n} a_{ij} Q_i Q_j \qquad (8)$$

where K_{ij} and a_{ij} now refer to "corresponding" forces and displacements.

If we now load the body first by a force Q_1 at point 1 and then a force Q_2 at point 2, the strain energy may be written as

$$V = \tfrac{1}{2} a_{11} Q_1{}^2 + \tfrac{1}{2} a_{22} Q_2{}^2 + Q_1 (a_{12} Q_2) \qquad (9)$$

If the order of application of Q_1 and Q_2 is reversed, we have

$$V = \tfrac{1}{2} a_{22} Q_2{}^2 + \tfrac{1}{2} a_{11} Q_1{}^2 + Q_2 (a_{21} Q_1) \qquad (10)$$

In order that Eqs. 9 and 10 represent the same quantity for arbitrary values of Q_1, Q_2, we must have

$$a_{12} = a_{21} \qquad (11)$$

The same argument can be applied to any pair of forces and the corresponding displacements to show that the flexibility-influence coefficients are symmetric; i.e.,

$$a_{ij} = a_{ji} \qquad (i, j = 1, 2, \cdots, n) \qquad (12)$$

Similarly,

$$K_{ij} = K_{ji} \qquad (i, j = 1, 2, \cdots, n) \qquad (13)$$

Castigliano's theorem can be derived from Eqs. 7 and 8. Differentiating Eq. 7 with respect to q_i, we obtain

$$\frac{\partial V}{\partial q_i} = K_{ii} q_i + \frac{1}{2} \sum_{j \neq i} K_{ij} q_j + \frac{1}{2} \sum_{j \neq i} K_{ji} q_j = \sum_{j=1}^{n} K_{ij} q_j$$

Hence,

$$\frac{\partial V}{\partial q_i} = Q_i \qquad (14)$$

Similarly,

$$\frac{\partial V}{\partial Q_i} = q_i \qquad (15)$$

Equation 15 states that, if the strain energy is expressed as a quadratic function of the loading, then its partial derivative with respect to the load at a point gives the corresponding deflection at that point. Similar statement can be made for Eq. 14.

The above results can be generalized in several directions. First, let us consider a three-dimensional body subjected to a system of forces. Let a system of rectangular Cartesian coordinates be chosen so that a displacement **u**, resolved in the direction of coordinate axes, has components (u_1, u_2, u_3), and a force **F** the components (F_1, F_2, F_3). Let $f_i(\xi, \eta, \zeta) d\xi\, d\eta\, d\zeta$ be the ith component of the force acting over an element of volume $d\xi\, d\eta\, d\zeta$ at the point (ξ, η, ζ), $f_i(\xi, \eta, \zeta)$ being a force density. Then in analogy with Eq. 2, we may write the ith component of the displacement **u** at the point (x, y, z):

$$u_i(x, y, z) = \int \int \int \sum_{j=1}^{3} G_{ij}(x, y, z;\ \xi, \eta, \zeta) f_j(\xi, \eta, \zeta)\, d\xi\, d\eta\, d\zeta \qquad (16)$$

where the integration is taken over the entire volume of the elastic body. The function $G_{ij}(x, y, z;\ \xi, \eta, \zeta)$ is the *influence function* of the displacement at (x, y, z) due to a force at (ξ, η, ζ). **u** and **f** being vectors, $G_{ij}(x, y, z;\ \xi, \eta, \zeta)$ is a tensor of rank 2 in a 3-dimensional Euclidean space.

The expressions for strain energy can be obtained in analogy with Eqs. 7 and 8. For example,

$$V = \frac{1}{2} \int \cdots \int \sum_{i=1}^{3} \sum_{j=1}^{3} G_{ij}(x, y, z;\ \xi, \eta, \zeta) f_i(x, y, z)$$
$$f_j(\xi, \eta, \zeta)\, dx\, dy\, dz\, d\xi\, d\eta\, d\zeta \qquad (17)$$

Furthermore, the symmetry argument can be generalized to show that

$$G_{ij}(x, y, z;\ \xi, \eta, \zeta) = G_{ji}(\xi, \eta, \zeta;\ x, y, z) \qquad (i, j = 1, 2, 3) \quad (18)$$

In other words, the *influence functions are symmetrical*: The ith component of displacement at **x** due to the jth component of a unit force at **y** is equal to the jth component of displacement at **y** due to the ith component of a unit force at **x**. This is the *reciprocal theorem* of Maxwell and of Betti and Rayleigh.

In practical applications, the external forces acting on an elastic body often take the form of concentrated forces and moments. In such cases the linear and angular displacements *under the points of application and in the direction of* the forces and moments are of particular interest. It is convenient to define a *generalized force* as either a concentrated force or a couple, and, correspondingly, a *generalized displacement* as either a linear or an angular displacement. A concentrated couple can be regarded as the limiting case of two equal and opposite forces approaching each other but maintaining a constant moment. The extension of the above results to cover the generalized forces and displacements is obvious. The notation can be simplified as follows: Let Q_i and q_i denote, respectively, the generalized force and the generalized displacement at the point

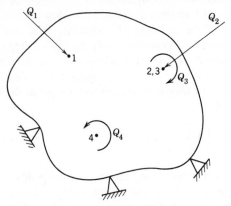

Fig. 1.8. Generalized forces.

i.* Q refers to either a force or a couple; q refers to either a linear or an angular displacement. Then

$$q_i = \sum_{j=1}^{n} c_{ij}Q_j \qquad (i = 1, 2, \cdots, n) \tag{19}$$

The constants c_{ij} are called the *influence coefficients of the (generalized) displacement at i due to a unit (generalized) force at j.* The inverse of the $\{c_{ij}\}$ matrix defines the stiffness-influence coefficients $\{K_{ij}\}$ in the generalized sense.

The symmetry property (Eq. 12, 13, or 18) can be generalized to show that, if c_{ij} relates the *corresponding* generalized displacements and forces, then

$$c_{ij} = c_{ji}, \qquad K_{ij} = K_{ji} \qquad (i, j = 1, 2, \cdots, n) \tag{20}$$

It is very important to notice the word "corresponding" in the statement

---

* As shown in Fig. 1.8, a force and a couple acting at the same point may be counted on as two generalized forces, indicated by two different subscripts.

of the reciprocal relations.   A generalized force *corresponds* to a generalized displacement if their product gives *exactly* the work done.   Thus, for a beam, the change of slope corresponds to a moment, the deflection corresponds to a force, and the twisting angle corresponds to a torque. Although Maxwell's reciprocal relation asserts that the *deflection* at P_1 due to a unit couple at P_2 is equal to the *rotation* at P_2 due to a unit force at P_1, it is completely wrong to assert that the rotation at P_1 due to a unit force at P_2 must be equal to the rotation at P_2 due to a unit force at P_1. In the latter case the force and rotation do not correspond to each other.

Examples of Influence Functions and Influence Coefficients.   The influence coefficients, being displacements under a unit load acting at some point on the structure, may be determined experimentally or computed according to the principles of elasticity.   There are many efficient methods of calculating the elastic displacements.   The reader is referred to books on the theory of structures.   A few examples will be given below:

Example 1.   *A Cantilever Beam Clamped at $x = 0$* (Fig. 1.9).   Let the stiffness EI be a constant, and let the load and displacement be both

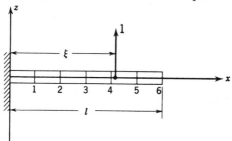

Fig. 1.9.   Example 1.

parallel to the z axis; so the indices i, j in Eq. 16 are both 3 and can be omitted.   The influence function is given by the deflection curve under a unit load at $x = \xi$.

(i) For $0 \leq x \leq \xi \leq l,$ $\dfrac{d^2w}{dx^2} = \dfrac{M}{EI} = \dfrac{1}{EI}(\xi - x)$

Therefore

$$w = G(x, \xi) = \frac{x^2}{6EI}(3\xi - x) \tag{21}$$

(ii) For $0 \leq \xi \leq x \leq l,$ $w = w(\xi) + \left(\dfrac{dw}{dx}\right)_{\xi}(x - \xi)$

$$w = G(x, \xi) = \frac{\xi^2}{6EI}(3x - \xi) \tag{22}$$

Note that $G(x, \xi)$, $\partial G/\partial x$, and $\partial^2 G/\partial x^2$ are continuous throughout the range $(0, l)$ but $\partial^3 G/\partial x^3$ is discontinuous at the point $x = \xi$.

Example 2. *A Beam of Torsional Rigidity GJ Subject to a Torsional Moment* (Fig. 1.10). The differential equation is

$$\frac{d\theta}{dx} = \frac{T}{GJ} \tag{23}$$

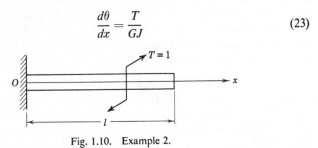

Fig. 1.10. Example 2.

For a unit torque applied at ξ, the rotation $\theta(x, \xi)$ is the influence function

$$G(x, \xi) = \theta(x, \xi) = \int_0^x \frac{1}{GJ(y)}\, dy \qquad (0 \le x \le \xi \le l)$$

$$= \int_0^\xi \frac{1}{GJ(y)}\, dy \qquad (0 \le \xi \le x \le l) \tag{24}$$

Note that $G(x, \xi)$ is continuous in the range $(0, l)$ but $\partial G/\partial x$ is discontinuous at $x = \xi$.

Example 3. Consider the uniform cantilever beam of Ex. 1. Let the beam be divided into six equidistant sections as marked in Fig. 1.9. The deflection at the points $1, 2, \cdots, 6$ due to a unit load acting at point 6 is given by Eqs. 21 and 22 by taking $\xi = l$ and $x = l/6$, $l/3$, etc. For example,

$$w_1 = c_{16} = \left(\frac{1}{6}\right)^2 \frac{17}{6} \frac{l^3}{6EI}, \qquad w_2 = c_{26} = \left(\frac{2}{6}\right)^2 \frac{16}{6} \frac{l^3}{6EI}$$

In this way the entire matrix of flexibility-influence coefficients may be obtained:

$$\{c_{ij}\} = \frac{l^3}{6^4 EI} \begin{Bmatrix} 2 & 5 & 8 & 11 & 14 & 17 \\ 5 & 16 & 28 & 40 & 52 & 64 \\ 8 & 28 & 54 & 81 & 108 & 135 \\ 11 & 40 & 81 & 128 & 176 & 224 \\ 14 & 52 & 108 & 176 & 250 & 325 \\ 17 & 64 & 135 & 224 & 325 & 432 \end{Bmatrix} \tag{25}$$

An inversion of the matrix $\{c_{ij}\}$ gives the stiffness-influence coefficients $\{K_{ij}\}$.

$$\{K_{ij}\} = \frac{6^4 EI}{l^3} \left(\begin{array}{ccc} 3.1384\ 1565 & -\ 1.9844\ 5597 & 0.7994\ 0766 \\ -\ 1.9844\ 5597 & 2.4455\ 9557 & -\ 1.7979\ 2676 \\ 0.7994\ 0766 & -\ 1.7979\ 2676 & 2.3923\ 0068 \\ -\ 0.2131\ 7531 & 0.7461\ 1348 & -\ 1.7712\ 8020 \\ 0.0532\ 9384 & -\ 0.1865\ 2846 & 0.6928\ 2076 \\ -\ 0.0088\ 8231 & 0.0310\ 8807 & -\ 0.1154\ 7008 \end{array} \right.$$

$$\left. \begin{array}{ccc} -\ 0.2131\ 7531 & 0.0532\ 9384 & -\ 0.0088\ 8231 \\ 0.7461\ 1348 & -\ 0.1865\ 2846 & 0.0310\ 8807 \\ -\ 1.7712\ 8020 & 0.6928\ 2076 & -\ 0.1154\ 7008 \\ 2.3390\ 0970 & -\ 1.5847\ 5411 & 0.4307\ 9176 \\ -\ 1.5847\ 5411 & 1.6461\ 9015 & -\ 0.6076\ 9728 \\ 0.4307\ 9176 & -\ 0.6076\ 9728 & 0.2679\ 4926 \end{array} \right\} \quad (26)$$

If a desk calculator is used, the inversion can be best done by Crout's method (see footnote in § 3.6, p. 102). The matrix $\{K_{ij}\}$, as given in Eq. 26, carries more significant figures than its numerical accuracy warrants from the physical point of view, but it is given here for arithmetic reasons. If one wishes to verify that the product $\{K_{ij}\}\{c_{ij}\}$ gives the unit matrix up to six significant figures, it is imperative to carry at least eight significant figures in the process of computation.

The determinant of the matrix $\{c_{ij}\}$ is

$$|c_{ij}| = 1.330\ 244 \times 10^{-9} \left(\frac{l^3}{6EI} \right)^6 \tag{27}$$

Its numerical smallness is caused by the near proportionality of the deflection modes when the points of application of two unit loads are close to each other. (Compare, for example, the last two columns of $\{c_{ij}\}$ in Eq. 25.) Thus, although $|c_{ij}|$ never vanishes, the inversion of $\{c_{ij}\}$ may become difficult as more stations are taken across the span.

1.5 ELEMENTARY AERODYNAMICS

A real fluid is viscous and compressible. But, if the speed of flow is much less than the speed of propagation of sound, the variation of density caused by the motion of a body in the flow is so small that the fluid may be regarded as *incompressible*. Furthermore, for fluids like water and air, the effects of viscosity are felt only in a thin layer (the boundary layer) next to the solid wall of the body. Outside the boundary layer the fluid may be regarded as *nonviscous*. A nonviscous and incompressible fluid is called a *perfect fluid*. In many problems of aeroelasticity, it is sufficient to consider the fluid as a perfect fluid.   However, there are cases in which the viscosity, however small, has profound effects, for it controls the boundary layer which may become detached from part of the solid body, and thus affects the macroscopic picture of the flow.

The force, exerted by the fluid on a body situated in a flow, does not depend on the absolute velocity of either the fluid or the body, but only on the *relative velocity* between them.   The aerodynamic force consists of two components: the *pressure force* normal to the surface of the body, and the *skin friction*, or *shearing force*, tangential to the surface of the body.   The latter is often negligible in aeroelastic problems.

In order to determine the parameters, on which depends the force that acts on a body situated in a flow, a dimensional analysis can be made. Obviously the force depends on the geometry of the body and its attitude relative to the flow;   these, for geometrically similar bodies, can be characterized by a typical length l and a typical angle α.   The force will also depend on the density of the fluid ρ, the viscosity of the fluid μ, the speed of flow U, the compressibility of the fluid, and the nonstationary characteristics of the flow.   The last item can be expressed in a clear-cut way if the conditions of flow happen to be periodic.   In a periodic oscillation, the frequency ω characterizes the nonstationary feature.   The compressibility of the fluid may be expressed in various ways.   A simple index of the compressibility is the speed of propagation of sound in the fluid, because sound is propagated as longitudinal elastic waves.

Let the speed of sound propagation be denoted by c.   Then, in an oscillating flow of a compressible fluid, the force experienced by a solid body will depend on the following variables:

$$l, \ \alpha, \ \rho, \ U, \ \mu, \ \omega, \ c$$

A dimensional analysis shows that, for geometrically similar bodies, the force F acting on the body can be expressed as

$$F = f \left(\alpha, \frac{Ul\rho}{\mu}, \frac{\omega l}{U}, \frac{U}{c} \right) \frac{1}{2} \rho U^2 l^2 \tag{1}$$

where f is a function of the variables contained inside the parentheses. It is easy to verify that the parameters $Ul\rho/\mu$, $\omega l/U$, and U/c are all dimensionless numbers.

The following notations will be used throughout this book:

$$R = \frac{Ul\rho}{\mu} = \frac{Ul}{\nu} = \text{Reynolds number}$$

$$k = \frac{\omega l}{U} = \text{reduced frequency or Strouhal number} \tag{2}$$

$$M = \frac{U}{c} = \text{Mach number}$$

$$q = \tfrac{1}{2}\rho U^2 = \text{dynamic pressure}$$

The parameters R, k, and M are named in honor of O. Reynolds, V. Strouhal, and E. Mach, respectively. The factor $\nu = \mu/\rho$ is called the *kinematic viscosity*.

The speed of sound propagation in a gas is given by the equation

$$c = \sqrt{\frac{\gamma p}{\rho}}$$

where p is the static pressure, ρ is the density, and γ is the ratio of the specific heat at constant pressure to the specific heat at constant volume. For dry air, $\gamma = 1.4$. Using the equation of state, we obtain

$$c = \sqrt{\gamma R_g T}$$

where R_g denotes the gas constant. For air, the above equation becomes

$$c = 49.1\sqrt{T}$$

where T is in degrees Rankine. At the standard conditions at sea level, $c = 1130$ ft per sec.

For air under standard conditions of temperature and pressure:

$\rho = 0.002378$ slug per ft^3 at 15° C and 760 mm Hg

$\nu = 0.0001566$ ft^2 per sec at 15° C and 760 mm Hg

For water under the same conditions, $\nu = 1.228 \times 10^{-5}$ ft^2 per sec.

More complete data of the physical properties of air and water can be found in Refs. 1.51, and 2.25. For approximate mental calculation, the following formulas may be useful:

Dynamic pressure at sea level:

$$q \doteq 25\left(\frac{U \text{ in mph}}{100}\right)^2 \text{lb per ft}^2$$

Reynolds number in air at sea level:

$$R \doteq 10{,}000 \,(U \text{ in mph})\,(l \text{ in ft})$$

In aeroelasticity we are concerned primarily with two components of force and one component of moment that act on a body. These are:

> Lift $= L =$ force perpendicular to the direction of motion

Drag $= D =$ force in the direction of motion, positive when the force acts in the downstream direction

Pitching moment $= M =$ moment about an axis perpendicular to both the direction of motion and the lift vector, positive when it tends to raise the leading edge of the body

When airplane wings or complete airplanes are considered, the mean chord c of the wing is usually taken as the characteristic length, and the wing area S the characteristic area. The three primary airplane coefficients are:

$$
\begin{aligned}
C_L &= \text{lift}/(qS) & &= \text{lift coefficient} \\
C_D &= \text{drag}/(qS) & &= \text{drag coefficient} \qquad\qquad (3) \\
C_M &= \text{pitching moment}/qSc &&= \text{pitching-moment coefficient}
\end{aligned}
$$

C_L, C_D, and C_M are functions of the Reynolds number, Mach number, Strouhal number, and the body's shape and attitude with respect to the flow. For an airfoil, the attitude is described by the angle between the direction of motion U and a reference axis called the chord line. This angle, denoted by α, is called the *angle of attack*.

In a steady flow of an incompressible fluid, the Strouhal number and the Mach number both vanish, and C_L, C_D, C_M depend on R and α alone. In the remainder of this section, unless mentioned otherwise, we shall consider only the steady, incompressible case.

The variation of the coefficients C_L, C_D, and C_M with α is illustrated for a typical airfoil in Fig. 1.11. When α is small, C_L increases linearly with α. The proportional constant is called the *lift-curve slope* and is denoted by a. When α becomes larger, the lift curve begins to level off and finally drops downward. The wing is then said to have *stalled*. The lift-curve slope a is nearly independent of the Reynolds number.

Let v be the fluid velocity at any point along a closed path of integration, dl an element of length along the path, and θ the angle between v and dl. Then the circulation Γ around the given path is defined as the line integral of $v \cos \theta$:

$$
\Gamma = \oint v \cos \theta \, dl \qquad\qquad (4)
$$

The fundamental importance of the concept of circulation may be roughly stated by the fact that, in a perfect fluid, with constant total pressure and in steady motion, the circulation is identical around every simple closed path enclosing a given set of solid bodies. Thus the circulation around a body has a unique and well-defined meaning, independent of the choice

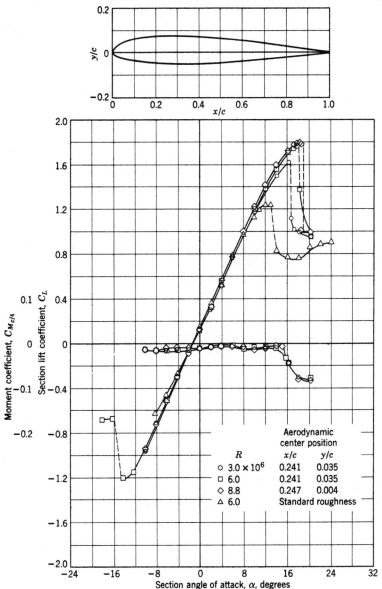

Fig. 1.11. Lift and moment characteristics for the *NACA* 23012 airfoil. The moment is taken about a point located at $^1/_4$-chord length behind the leading edge. The aerodynamic center location is computed from C_L and $C_{Mc/4}$ data. The Reynolds number is seen to affect mainly the maximum lift coefficient. (From Abbott, von Doenhoff, and Stivers, *NACA Rept.* **824.** Courtesy of the *NACA.*)

that may be made of the particular path used for calculating the value of the circulation.

A type of singularity in a fluid called a *vortex line* is a curve around which the circulation is a constant. A vortex line cannot end itself in a fluid; it must either form a closed curve or extend to infinity, or else end on a free surface or a solid boundary. A *free vortex*, one on which no external force acts, is transported with the fluid; i.e., it moves with the flow. On the other hand, a *bound vortex*, one that moves relative to the flow, requires the action of external forces to maintain such relative motion. As mentioned before, in a steady flow, the circulation about a solid body is a constant. For a steady, two-dimensional flow of a perfect fluid about an infinitely long cylindrical body of any cross sectional shape whatever, the following can be proved theoretically:

(*a*) The drag is zero.

(*b*) In the absence of circulation around the body, the lift is zero.

(*c*) If there is a circulation of magnitude Γ around the body, and if the body moves with a rectilinear velocity U relative to the fluid at infinity, then a lift exists, whose magnitude per unit length perpendicular to the flow is given by Joukowsky's theorem:

$$L = \rho U \Gamma \tag{5}$$

where ρ is the density of the fluid. The direction of the lift is normal both to the velocity U and the axis of the cylinder.

It is clear that, as far as the lift is concerned, the solid body may be replaced by a bound vortex line. For an unstalled airfoil, the circulation Γ around it varies linearly with the angle of attack α. If α is measured from the zero-lift attitude which corresponds to $\Gamma = 0$, then $\Gamma \sim L \sim \alpha$, and the lift coefficient can be expressed as

$$C_L = a_0 \alpha \tag{6}$$

The angle of attack thus measured is called the *absolute angle of attack*. The word "absolute" is sometimes omitted. It is understood throughout this book that, when we speak of an angle of attack, we mean the absolute angle of attack. If α is measured in radians (1 radian = 57.3°), hydrodynamic theory gives the lift-curve slope

$$a_0 = 2\pi \quad \text{(theory, incompressible fluid)} \tag{7}$$

for thin airfoils in a two-dimensional flow. Experimental values of the lift-curve slope may be expressed in the form

$$a_0 = 2\pi\eta \tag{8}$$

The correction factor η is called the *airfoil efficiency factor*, which is less than 1 (≈ 0.9) for conventional airfoil sections, but is greater than 1 for NACA low-drag sections.

According to the theory of thin airfoils in a two-dimensional incompressible fluid, the center of pressure of the additional lift due to change of α is located at $^1/_4$-chord aft of the leading edge. This point is called the *aerodynamic center*. If the moment coefficient is computed about the aerodynamic center, it does not vary with C_L. The symbol C_{M0} or $C_{Mc/4}$ is used to denote the moment coefficient referred to an axis located at the $^1/_4$-chord point. The aerodynamic center remains close to the $^1/_4$-chord point in a compressible fluid as long as the flow is *subsonic*; but it moves close to the mid-chord point if the flow is entirely *supersonic*.

The subscript 0 of the lift-curve slope a_0 signifies that a_0 is the value pertaining to an airfoil of infinitely long span. For wings of *finite span*

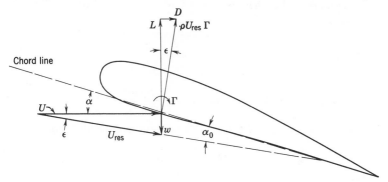

Fig. 1.12. Downwash and induced angle.

the lift-curve slope is smaller. In Prandtl's finite-wing theory, a wing is replaced by a *vortex line*. Since a vortex line cannot end at the wing tip, it must continue laterally out of the wing and become a *free vortex* (trailing vortex) in the fluid. The vertical velocity w induced by the trailing vortices is called the *induced velocity*. Because of the induced velocity, the direction of flow at the airfoil is changed by an amount ε, which is given by the relation

$$\tan \varepsilon = w/U \qquad (9)$$

From Fig. 1.12 it is seen that the effective angle of attack α_0 is smaller than the geometric angle of attack α. Assuming that w is infinitesimal compared with U, we may write

$$\alpha_0 = \alpha - \varepsilon \doteq \alpha - \frac{w}{U} \qquad (10)$$

The resultant force $\rho U_{\text{res}}\Gamma$ acts in a direction normal to the resultant velocity vector U_{res}. It can be resolved into a lift component L perpendicular to the velocity of flow U and a drag component (induced drag) D in the direction of U. Prandtl assumes that the circulation Γ is proportional to the effective angle of attack α_0. By using Eq. 6 relating C_L (and hence Γ) with α_0, and expressing the downwash w in terms of the spanwise-lift distribution, a relation between the lift coefficient C_L and the geometrical angle of attack α can be derived.

The final result takes a particularly simple form if the wing planform is an elongated ellipse and is untwisted (having a constant angle of attack across the span). In this case the lift distribution across the span is elliptical, with a principal axis at the mid-span. The special properties associated with the elliptic lift distribution are that (1) the downwash is constant across the span, and (2) for a given total lift, span, and speed of flow the induced drag has its lowest possible value. For an elliptic lift distribution,

$$\frac{w}{U} = \frac{C_L}{\pi R} \tag{11}$$

where

$$R = \text{aspect ratio} = \frac{(\text{span})^2}{\text{wing area}} \tag{12}$$

Hence,

$$\alpha_0 = \alpha - \frac{C_L}{\pi R} \tag{13}$$

From the relations

$$C_L = a_0\alpha_0 = a\alpha \tag{14}$$

and Eq. 10, one obtains the *lift-curve slope* of a wing of *finite span*:

$$a = \frac{a_0}{1 + (a_0/\pi R)} \quad \text{(elliptic lift distribution)} \tag{15}$$

For nonelliptic lift distribution, the downwash is no longer constant across the span. Equation 15 is modified as follows:

$$a = \frac{a_0}{1 + (a_0/\pi R_e)(1 + \tau)} \quad \text{(nonelliptic)} \tag{16}$$

where τ is a small correction factor depending on the deviation of the lift distribution from the ideal elliptical form. Figure 1.13 shows the value of τ as calculated by Glauert for straight-tapered, untwisted wings of aspect ratio $R = 2\pi$. R_e is the effective aspect ratio. If the lift is symmetrically distributed over the span, R_e is given by Eq. 12. If the lift is antisymmetrically distributed over the span (as induced by an anti-symmetrical deflection of the ailerons), with zero lift at the mid-span, then R_e should be taken as half of the actual R. Hence,

$$R_e = R \quad \text{(for symmetrical lift distribution)}$$

$$R_e = R/2 \quad \text{(for antisymmetrical lift distribution)} \qquad (17)$$

Such a correction of aspect ratio is necessary because Eq. 16 is derived for an untwisted wing. The correction (Eq. 17) is plausible, because each of the two halves of the antisymmetrical lift distribution appears similar to an elliptic distribution with a span equal to half of the actual wing span.

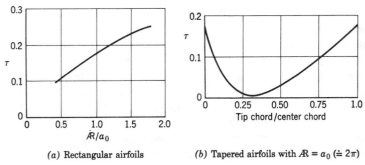

(a) Rectangular airfoils   (b) Tapered airfoils with $R = a_0 \; (\doteq 2\pi)$

Fig. 1.13.   Glauert's correction factor τ.

Some important characteristic quantities of airfoils, such as the profile drag coefficient C_{D0}, the maximum lift coefficient $C_{L\,\mathrm{max}}$, the moment coefficient at zero lift C_{M0}, the angle of zero lift α_{L0}, are only of minor importance in aeroelasticity. They do not appear in most of the problems. On the other hand, the question of *spanwise lift distribution* corresponding to a twisted airfoil is of great importance in aeroelasticity. For mathematical simplicity, we shall often use the so-called *strip theory* as a first approximation. In this, one assumes that the local lift coefficient $C_l(y)$ is proportional to the local geometric angle of attack $\alpha(y)$:

$$C_l(y) = a\alpha(y) \qquad (18)$$

The effect of finite span is then to be corrected by using a value of a corrected for aspect ratio.*

* The "local" lift coefficient at any point (x,y) on a lifting surface is defined as the limit of dL/qdS as $dS \to 0$, where dS is a surface element enclosing the point (x,y) and dL is the lift force acting on dS. The local lift coefficient $C_l(y)$ used in Eq. 18 is defined over a chordwise section, so that $dS = cdy$. We shall use lower case subscript to indicate a local aerodynamic coefficient such as C_l, whereas a capital letter subscript will be used for coefficients referred to the entire wing.

Finally, the principle of superposition must be mentioned. In aero-dynamics, the fundamental equations are essentially nonlinear, and so the superposition principle is valid only in special cases. However, in the airfoil theory, it is often assumed that the disturbance in the flow due to the presence of an airfoil is infinitesimal; i.e., the velocity induced by the solid body is infinitesimal in comparison with the speed of the undisturbed flow, and the hydrodynamic equations can be linearized. In this case the superposition is permissible.

1.6 GENERALIZED COORDINATES AND LAGRANGE'S EQUATIONS

In deriving the equations of motion for many problems in aeroelasticity, generalized coordinates and Lagrange's equations are often used. The ideas of generalized coordinates are developed in the classical mechanics, and are associated with the great names of Bernoulli, Euler, d'Alembert, Lagrange, Hamilton, Jacobi, and others. There are many excellent text-books on this subject,[1.54, 1.68, 1.69] and so we shall not explain the method in great detail. Instead, we shall survey the fundamental principles briefly, and illustrate their meaning by several examples.

The foundation of the mechanics of a single particle is *Newton's second law of motion*

$$\mathbf{F} = \frac{d\mathbf{p}}{dt} \tag{1}$$

where $\mathbf{F}$ is the total force acting on the particle, and $\mathbf{p}$ is the *linear momentum* of the particle defined by the product of mass m and velocity $\mathbf{v}$:

$$\mathbf{p} = m\mathbf{v} \tag{2}$$

In applying this law to a system of N particles, we must distinguish between the *external forces* acting on the particles due to sources outside the system and *internal forces* on a particle i due to all other particles in the system. Thus the equation of motion for the ith particle is*

$$\sum_{j=1}^{N}{}' \mathbf{F}_{ij} + \mathbf{F}_i^{(e)} = \dot{\mathbf{p}}_i \qquad (i = 1, 2, \cdots, N) \tag{3}$$

where $\mathbf{F}_i^{(e)}$ is the external force, $\mathbf{F}_{ij}$ is the internal force on the ith particle due to the jth particle, and $\dot{\mathbf{p}}_i$ is $d\mathbf{p}_i/dt$. If we assume that $\mathbf{F}_{ij}$ (like $\mathbf{F}_i^{(e)}$)

---

* A prime over the summation sign is introduced to indicate that the term $j = i$ must be deleted from the sum.

obeys Newton's third law that the forces the two particles exert on each other are equal and opposite and lie along the line joining the particles, then by a summation over all the particles we can show that (a) the center of mass moves as if the total external forces were acting on the entire mass of the system concentrated at the center of mass, (b) the time derivative of the total angular momentum about an origin is equal to the moment of the external force about the same point, (c) the total angular momentum about a fixed point O is the angular momentum about O of the system concentrated at the center of mass, plus the angular momentum about the center of mass, (d) the kinetic energy, like the angular momentum, also consists of two parts: the kinetic energy obtained as if all the mass were concentrated at the centre of mass, plus the kinetic energy of motion about the center of mass.

The problems in mechanics would have been reduced to solving the set of differential equations 3 if no *constraints* which limit the motion of the system were present. But constraints do occur. For example, the distance between any two particles in a rigid body remains unchanged; the root of a cantilever wing is restrained against relative motion with respect to the clamping wall; the aileron must remain attached at the hinge; the fuel must move only *inside* of the container. Two difficulties are introduced by the constraints. First, the coordinates of the particles, $\mathbf{r}_i$ ($i = 1, 2, \cdots, N$), are no longer all independent, since they are connected by the equations of constraint. Second, the forces of constraint, e.g., the forces exerted by the clamping wall of a cantilever beam, are not known *a priori*; they are among the unknowns of the problem and in many cases are of no direct interest. To overcome these difficulties, the problems of mechanics may be formulated in terms of independent coordinates and in a form in which the forces of constraint do not appear. This is the purpose of the Lagrangian formulation in terms of generalized coordinates.

Constraints may be classified as follows: If the conditions of constraint can be expressed as equations connecting the coordinates of the particles and the time, having the form

$$f(\mathbf{r}_1, \mathbf{r}_2, \mathbf{r}_3, \cdots, t) = 0 \qquad (4)$$

then the constraints are said to be *holonomic*. Otherwise they are *nonholonomic*. The constraints of the particles in a rigid body are holonomic, because they can be expressed as

$$(\mathbf{r}_i - \mathbf{r}_j)^2 - c_{ij}^2 = 0$$

The walls of fuel tank limiting the motion of fuel molecules are nonholonomic because the constraint cannot be written in the form of Eq. 4.

When the constraints are *holonomic*, a number of coordinates can be eliminated. In terms of Cartesian coordinates, a system of N particles, free from constraints, has $3N$ independent coordinates or *degrees of freedom*. If there exist holonomic constraints expressed in k equations, we may use these equations to eliminate k of the $3N$ coordinates, and we are left with $3N - k$ independent coordinates, and the system is said to have $3N - k$ degrees of freedom. This elimination of the dependent coordinates can be expressed in another way, by introducing $3N - k$ new independent variables $q_1, q_2, \cdots, q_{3N-k}$, in terms of which the old coordinates $\mathbf{r}_1, \mathbf{r}_2, \cdots, \mathbf{r}_N$ are expressed by equations of the form

$$\mathbf{r}_1 = \mathbf{r}_1(q_1, q_2, \cdots, q_{3N-k}, t)$$

.

.   (5)

.

$$\mathbf{r}_N = \mathbf{r}_N(q_1, q_2, \cdots, q_{3N-k}, t)$$

containing the constraints in them implicitly. These are equations of *transformation* connecting the set of variables $(\mathbf{r}_i)$ with the new variables (q_i) which are all independent.

By transformation of variables from $\mathbf{r}_i$ to q_i, it can be shown (see Ref. 1.54, 1.56, 1.68, or 1.69) that the equations of motion 3 can be transformed into the form

$$\frac{d}{dt}\left(\frac{\partial T}{\partial \dot{q}_j}\right) - \frac{\partial T}{\partial q_j} = Q_j \qquad (j = 1, 2, \cdots, n) \qquad (6)$$

where n is the number of degrees of freedom of the system. T is the kinetic energy, which is one half of the sum of the mass times velocity squared of all the particles of the system. When the velocity of each particle is expressed in terms of the rate of change of the independent coordinates q_i, T can be expressed in the form

$$T = \frac{1}{2}\sum_{i,j} M_{ij}\dot{q}_i\dot{q}_j$$

The constant coefficients M_{ij} are called the generalized masses. In Eq. 6, $Q_j (j = 1, 2, \cdots, n)$ are the components of the *generalized force* defined as

$$Q_j = \sum_i \mathbf{F}_i \cdot \frac{\partial \mathbf{r}_i}{\partial q_j}$$

where $\mathbf{F}_i$ is the force applied on the ith particle. Note that the virtual work done by the forces through virtual displacements $\delta\mathbf{r}_i$ (or the corresponding δq_j) is

$$\sum_i \mathbf{F}_i \cdot \delta\mathbf{r}_i = \sum_{i,j} \mathbf{F}_i \cdot \frac{\partial \mathbf{r}_i}{\partial q_j} \delta q_j = \sum_j Q_j \delta q_j \qquad (7)$$

Often it is more convenient to calculate the generalized force Q_i by Eq. 7. Note also that just as the q's need not have the dimensions of length, so the Q's do not necessarily have the dimensions of force; but $Q_j \delta q_j$ must always have the dimensions of work.

It may not be superfluous to remark that a (infinitesimal) virtual displacement of a system refers to a change in the configuration of the system as the result of any arbitrary infinitesimal change of the coordinates $\delta\mathbf{r}_i$ or δq_i, *which obeys the constraints imposed on the system at the given instant t.* It is called "virtual" in order to be distinguished from an actual displacement of the system occurring in a time interval dt. By means of a virtual displacement, we can compare the kinetic energy of a system with that of a "neighboring" system which differs only infinitesimally from the actual one, yet not obeying the laws of motion. The Lagrange equations of motion can be deduced from such a comparison.

When the generalized force can be derived from a scalar potential so that

$$Q_j = -\frac{\partial V}{\partial q_j} \quad \text{(conservative system)} \qquad (8)$$

Eq. 6 can be written as

$$\frac{d}{dt}\left(\frac{\partial T}{\partial \dot{q}_j}\right) - \frac{\partial(T - V)}{\partial q_j} = 0$$

The potential V is a function of position only, and must be independent of the generalized velocities $\dot{q}_j$. Hence, one can include a term V in the partial derivative with respect to $\dot{q}_j$:

$$\frac{d}{dt}\left(\frac{\partial(T - V)}{\partial \dot{q}_j}\right) - \frac{\partial(T - V)}{\partial q_j} = 0 \qquad (9)$$

Or, if we define a new function, the *Lagrangian L*, as

$$L = T - V \qquad (10)$$

then

$$\frac{d}{dt}\left(\frac{\partial L}{\partial \dot{q}_j}\right) - \frac{\partial L}{\partial q_j} = 0 \quad \text{(conservative systems)} \qquad (11)$$

Equations 11 are usually called *Lagrange's equations.* In aeroelasticity, the forces derived from the elastic deformation are conservative, and their potential can be identified with the strain-energy function. The aero-dynamic forces, on the contrary, cannot be derived from a potential. Hence, we must retain the general equation 6. In general, we may write the equations of motion as

$$\frac{d}{dt}\left(\frac{\partial L}{\partial \dot{q}_j}\right) - \frac{\partial L}{\partial q_j} = Q_j \qquad (j = 1, 2, \cdots, n) \qquad (12)$$

where Q_j represents that part of the generalized forces that are *not* derived from a potential, and L contains the potential of the conservative forces as before.

Example 1. A Weightless Beam with a Concentrated Mass at the Mid-span. Consider a weightless beam with a concentrated mass at the mid-

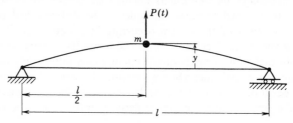

Fig. 1.14.   Example 1.

span (Fig. 1.14). A force $P(t)$ acts on the mass. The geometrical con-figuration is determined completely by the displacement y of the mass m. Here

$$T = \tfrac{1}{2}m\dot{y}^2, \qquad V = \tfrac{1}{2}Ky^2, \qquad Q = P(t)$$

where V is the strain energy. Substituting into Lagrange's equation,

$$\frac{d}{dt}\left[\frac{\partial}{\partial \dot{y}}\left(\frac{1}{2}m\dot{y}^2\right)\right] - \frac{\partial}{\partial y}\left(\frac{1}{2}m\dot{y}^2 - \frac{1}{2}Ky^2\right) = P(t)$$

we obtain the equation of motion

$$m\ddot{y} + Ky = P(t)$$

Example 2. A Weightless Cantilever Beam with a Mass, that Has a Finite Moment of Inertia, Attached at the Free End (Fig. 1.15). Let us take the displacement of m as q_1, and the rotation of the mass moment of inertia I as q_2; then

$$T = \tfrac{1}{2}m\dot{q}_1{}^2 + \tfrac{1}{2}I\dot{q}_2{}^2$$
$$V = \tfrac{1}{2}K_{11}q_1{}^2 + \tfrac{1}{2}K_{22}q_2{}^2 + K_{12}q_1q_2$$

where K_{ij} are the stiffness-influence coefficients. Hence, from Lagrange's equations, we obtain

$$m\ddot{q}_1 + K_{11}q_1 + K_{12}q_2 = P$$
$$I\ddot{q}_2 + K_{12}q_1 + K_{22}q_2 = Q$$

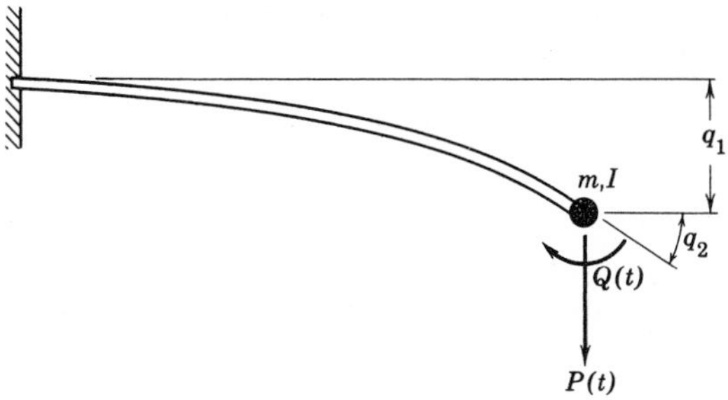

Fig. 1.15. Example 2.

Example 3. An Airplane Idealized into a Weightless Elastic Beam Connecting a Number of Concentrated Masses and Moments of Inertia. Let us take the displacements of the masses and the rotations of the inertias as the generalized coordinates (as shown in Fig. 1.16). Then

$$T = \frac{1}{2}\sum_{i=1}^{5} m_i \dot{q}_i^2 + \frac{1}{2}\sum_{i=6}^{10} I_i \dot{q}_i^2$$

$$V = \frac{1}{2}\sum_{i=1}^{10}\sum_{j=1}^{10} K_{ij}q_i q_j \quad \text{(strain energy)}$$

$$L = T - V$$

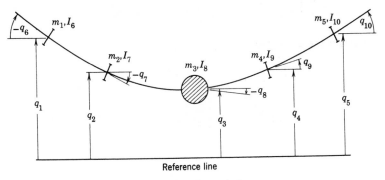

Fig. 1.16. Example 3.

The equations of motion can be obtained from Lagrange's equation. If the gravitational force need be considered, the gravitational potential should be added to V.

Example 4.   *A Simply Supported Beam of Uniform Cross Section* (Fig. 1.17).   Since the deflection curve of the beam must always remain con-

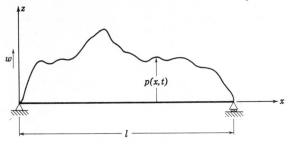

Fig. 1.17.   Example 4.

tinuous, it can be developed in a Fourier series.   Thus the deflection w can be written as

$$w = \sum_{n=1}^{\infty} a_n \sin \frac{n\pi x}{l}$$

The configuration of the beam is completely determined by the Fourier coefficients a_n.   Hence, a_n may be taken as the generalized coordinates:

$$q_i = a_i \qquad (i = 1, 2, 3, \cdots)$$

When the beam vibrates, w and hence $q_i = a_i$ are functions of time.

$$T = \frac{1}{2} \int_0^l m\dot{w}^2 \, dx \qquad (m = \text{mass per unit length} = \text{const})$$

$$= \frac{m}{2} \int_0^l \left(\sum_1^{\infty} \dot{a}_n(t) \sin \frac{n\pi x}{l} \right)^2 dx = \frac{ml}{4} \sum_1^{\infty} \dot{a}_n^2(t)$$

The strain energy is given by

$$V = \frac{1}{2} \int_0^l EI \left(\frac{\partial^2 w}{\partial x^2} \right)^2 dx \qquad (EI = \text{bending rigidity} = \text{const})$$

$$= \frac{1}{2} EI \int_0^l \left(\sum_1^{\infty} a_n(t) \frac{n^2 \pi^2}{l^2} \sin \frac{n\pi x}{l} \right)^2 dx$$

$$= \frac{\pi^4}{4} \frac{EI}{l^3} \sum_1^{\infty} n^4 a_n^2(t)$$

If there is a system of lateral loads acting on the beam, the generalized forces can be calculated as follows: Let the generalized coordinate a_n be given a virtual displacement δa_n. The beam configuration undergoes a virtual displacement $\delta w(x) = \delta a_n \sin (n\pi x/l)$. The work done by the external loads $p(x, t)$, the positive sense of which is defined as the same as that of w, is

$$Q_n \, \delta a_n = \int_0^l p(x, t) \, \delta w(x) \, dx = \int_0^l p(x, t) \, \delta a_n \sin \frac{n\pi x}{l} \, dx$$

But δa_n is arbitrary; hence,

$$Q_n = \int_0^l p(x, t) \sin \frac{n\pi x}{l} \, dx$$

The equations of motion can then be written down according to Lagrange's equation:

$$\frac{d}{dt}\left(\frac{\partial T}{\partial \dot{a}_n}\right) + \frac{\partial V}{\partial a_n} = Q_n$$

i.e.,

$$\frac{ml}{2}\ddot{a}_n + \frac{\pi^4 EI}{2l^3} n^4 a_n = \int_0^l p(x, t) \sin \frac{n\pi x}{l} \, dx \qquad (n = 1, 2, 3, \cdots)$$

Example 5. A Cantilever Beam of Uniform Cross Section. The deflection of the beam can be expanded into a series

$$w(x, t) = \sum_{n=1}^{\infty} q_n(t) f_n(x)$$

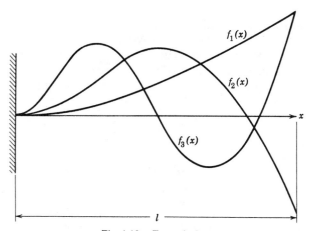

Fig. 1.18. Example 5.

where $f_n(x)$ is the nth mode of the undamped free vibration of the beam (Fig. 1.18). The functions $f_n(x)$ are orthogonal and can be normalized so that

$$\int_0^l f_\nu(x)\, f_n(x)\ dx = \begin{cases} 1 & \text{when } \nu = n, \\ 0 & \text{when } \nu \neq n. \end{cases}$$

Hence, following the same procedure as in Ex. 4, we have

$$T = \frac{ml}{2} \sum_{n=1}^{\infty} \dot{q}_n{}^2$$

$$V = \frac{EIl}{2} \sum_{n=1}^{\infty} \kappa_n{}^4 q_n{}^2$$

$$Q_n = \int_0^l p(x, t) f_n(x)\ dx$$

where κ_n are the solutions of the equation $\cos \kappa l \cdot \cosh \kappa l + 1 = 0$. Therefore, under a lateral load $p(x, t)$ the equations of motion are

$$ml\ddot{q}_n + EIl\kappa_n{}^4 q_n = \int_0^l p(x, t) f_n(x)\ dx$$

1.7 NORMAL COORDINATES

The last two examples of the preceding section point out a very important fact: By introducing the undamped free-vibration modes of a structure as the basis of generalized coordinates, the equations of motion can be simplified.

The theory of small free oscillations of an elastic body about an equilibrium configuration has been well developed. In particular, it suggests that, if $\boldsymbol{\phi}_n(\mathbf{x})$ represents the oscillation mode* associated with a frequency ω_n, and if the frequency spectrum is so arranged that $\omega_n \geqslant \omega_{n-1}, (n > 1)$, then an arbitrary disturbed configuration $\mathbf{u}$ can be represented by a series

$$\mathbf{u}(\mathbf{x}) = \sum_{n=0}^{\infty} q_n \boldsymbol{\phi}_n(\mathbf{x}) \tag{1}$$

with

$$q_n = \int \rho(\mathbf{x})\, \mathbf{u}(\mathbf{x}) \cdot \boldsymbol{\phi}_n(\mathbf{x})\, d\tau(\mathbf{x}) \Big/ \int \rho(\mathbf{x})\, \boldsymbol{\phi}_n(\mathbf{x}) \cdot \boldsymbol{\phi}_n(\mathbf{x})\, d\tau(\mathbf{x}) \tag{2}$$

* That is, the amplitude of the displacement from the equilibrium configuration at the point $\mathbf{x}$. $\mathbf{u}$ and $\boldsymbol{\phi}_n$ may be regarded as vectors having three components, each a function of the position vector $\mathbf{x}$, if a three-dimensional elastic body is considered. In Eq. 2 et seq., the product $\mathbf{u}(\mathbf{x}) \cdot \boldsymbol{\phi}_n(\mathbf{x})$ denotes the scalar product of $\mathbf{u}(\mathbf{x})$ and $\boldsymbol{\phi}_n(\mathbf{x})$. Similarly $\boldsymbol{\phi}_n(\mathbf{x}) \cdot \boldsymbol{\phi}_m(\mathbf{x})$ is a scalar product.

where $\rho(x)$ represents the density of the body and $d\tau(\mathbf{x})$ represents an element of volume at $\mathbf{x}$, the integration being taken over the entire body. Moreover, $\boldsymbol{\phi}_n(\mathbf{x})$ are orthogonal and can be normalized so that

$$\int \rho(\mathbf{x}) \, \boldsymbol{\phi}_\nu(\mathbf{x}) \cdot \boldsymbol{\phi}_n(\mathbf{x}) \, d\tau(\mathbf{x}) = \begin{cases} 1 & \text{if } \nu = n \\ 0 & \text{if } \nu \neq n \end{cases} \tag{3}$$

We shall call $\boldsymbol{\phi}_n(\mathbf{x})$ the *normal modes* and q_n the *normal coordinates*. Then the kinetic energy and the elastic strain energy can be expressed, respectively, as

$$T = \frac{1}{2} \sum_{n=0}^{\infty} m_n \dot{q}_n{}^2$$

$$V = \frac{1}{2} \sum_{n=0}^{\infty} m_n \omega_n{}^2 q_n{}^2 \tag{4}$$

The constants m_n are called generalized masses:

$$m_n = \int \rho(\mathbf{x}) \, \boldsymbol{\phi}_n(\mathbf{x}) \cdot \boldsymbol{\phi}_n(\mathbf{x}) \, d\tau(\mathbf{x}) \tag{5}$$

The generalized force is

$$Q_n = \int \mathbf{F}(\mathbf{x}) \cdot \boldsymbol{\phi}_n(\mathbf{x}) \, d\tau(\mathbf{x}) \tag{6}$$

where the integral is again taken over the entire body, $\mathbf{F}(\mathbf{x})$ being the force acting at $\mathbf{x}$.

It is clear that, when T and V are expressible in the form of Eq. 4, the inertia and elastic "couplings" between the various generalized coordinates are absent. The equations of motion can be reduced to a set of independent equations, each containing one q_n, provided that all Q_n are independent of the coordinates q_n.

We shall not discuss the methods of calculating vibration modes and frequencies in this book.*

1.8 COMPLEX REPRESENTATION OF HARMONIC MOTIONS

In the discussion of periodic phenomena, it is convenient to use the complex representation. Consider a quantity x which varies periodically with frequency ω (radians per second) according to the rule:

$$x = A \cos(\omega t + \psi) \tag{1}$$

This is a simple-harmonic motion. A is called the amplitude and ψ the

---

* For aircraft structures, see Scanlan and Rosenbaum.[1.57] For general principles, see Den Hartog,[1.52] Timoshenko.[1.59] Myklestad's book[1.55] treats the numerical aspects of the analysis.

phase angle. It is customary to regard x as the projection of a rotating vector on the real axis (see Fig. 1.19). The length of the vector is A and its angle with the x axis is $\omega t + \psi$. The vector rotates counterclockwise with angular speed ω. Since such a vector is specified by two components, it can be represented by a complex number. For example, the vector in Fig. 1.19 can be specified by the components $x = A \cos (\omega t + \psi)$, and $y = A \sin (\omega t + \psi)$, and hence by the complex number $x + iy$. But

$$e^{i(\omega t + \psi)} = \cos (\omega t + \psi) + i \sin (\omega t + \psi) \qquad (2)$$

so the rotating vector can be represented by the complex number

$$x + iy = Ae^{i(\omega t + \psi)} \qquad (3)$$

The quantity x given by Eq. 1 is the real part of the quantity given by

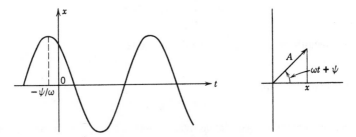

Fig. 1.19. Complex representation of a harmonic motion.

Eq. 3, which is the *complex representation* of Eq. 1. The imaginary part of Eq. 3 represents another simple harmonic motion.

The vector representation is very convenient for "composing" several simple-harmonic oscillations of the same frequency. For example, if

$$x = A_1 \cos (\omega t + \psi_1) + A_2 \cos (\omega t + \psi_2) = A \cos (\omega t + \psi) \qquad (4)$$

then x is the resultant of two vectors as shown in Fig. 1.20. The component $A_2 \cos(\omega t + \psi_2)$ is said to lead the component $A_1 \cos(\omega t + \psi_1)$, if, as shown in the figure, $\psi_2 > \psi_1$.

Hereafter we shall write x given by Eq. 1 in the complex form

$$x = (a + ib)e^{i\omega t} = Ae^{i(\omega t + \psi)} \qquad (5)$$

where

$$A = \sqrt{a^2 + b^2}, \qquad \tan \psi = \frac{b}{a}$$

It is understood that the physical quantity is given by the real part of the complex representation.

Since the vector i leads the vector 1 by an angle $\pi/2$, a multiplication of a given vector by the imaginary number i simply means a change of phase angle to lead by $\pi/2$.

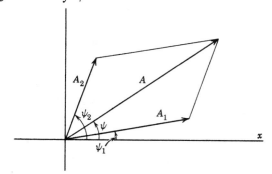

Fig. 1.20.   Vector sum of two simple harmonic motions of the same frequency.

As an illustration, let us consider the forced oscillation of a single particle, with elastic restraint and with damping, excited by a simple-harmonic force. The equation of motion, written in real form, is

$$m\frac{d^2x}{dt^2} + \beta\frac{dx}{dt} + Kx = F_0 \sin \omega t \qquad (6)$$

The solution may be written as

$$x = x_1 + x_2$$

where, if $\beta^2 < 4Km$,*

$$x_1 = C_1 e^{-(\beta/2m)t} \cos \sqrt{\frac{K}{m} - \frac{\beta^2}{4m^2}}\, t + C_2 e^{-(\beta/2m)t} \sin \sqrt{\frac{K}{m} - \frac{\beta^2}{4m^2}}\, t \qquad (7)$$

$$x_2 = F_0 \frac{(K - m\omega^2) \sin \omega t - \beta\omega \cos \omega t}{(K - m\omega^2)^2 + \beta^2\omega^2} \qquad (8)$$

where C_1, C_2 are integration constants to be determined from the initial conditions. If $\beta > 0$, $x_1 \to 0$ as $t \to \infty$. So, as time increases, x tends to the *steady* oscillation, x_2, which can be written also as

$$x_2 = \frac{F_0 \sin (\omega t - \psi)}{\sqrt{(K - m\omega^2)^2 + \beta^2\omega^2}} \qquad (9)$$

ψ being a phase angle defined by $\tan \psi = \beta\omega/(K - m\omega^2)$.

* If $\beta^2 > 4Km$, the argument of the trigonometric functions in Eq. 7 becomes imaginary. The solution can be written in the real form if one replaces the sine and cosine in Eq. 7 by hyperbolic sine and hyperbolic cosine, and write $\sqrt{\beta^2 - 4Km}/2m$ as the argument. The motion is then an exponential decay when $\beta > 0$.
If $\beta^2 = 4Km$, the solution is of the form $(C_1 + C_2 t)e^{-(\beta/2m)t}$.

These results can be derived by using the complex representation as follows. Consider the equation

$$m\frac{d^2x}{dt^2} + \beta\frac{dx}{dt} + Kx = F_0 e^{i\omega t} \tag{10}$$

Since $F_0 e^{i\omega t} = F_0 \cos \omega t + iF_0 \sin \omega t$, the real part of x will give the solution for a periodic force $F_0 \cos \omega t$, whereas the imaginary part that of $F_0 \sin \omega t$. Putting x proportional to $e^{i\omega t}$, we obtain

$$m\frac{d^2x}{dt^2} + \beta\frac{dx}{dt} + Kx = [m(i\omega)^2 + \beta i\omega + K]x \tag{11}$$

Hence we obtain the steady oscillation

$$x = \frac{F_0 e^{i\omega t}}{Z(i\omega)} \tag{12}$$

where

$$Z(i\omega) = m(i\omega)^2 + \beta i\omega + K = K - m\omega^2 + \beta i\omega \tag{13}$$

The modulus of $Z(i\omega)$ is $\sqrt{(K - m\omega^2)^2 + \beta^2\omega^2}$, and its phase angle is $\psi = \tan^{-1}[\beta\omega/(K - m\omega^2)]$. Since x is the ratio of two complex numbers, and since we divide two complex numbers by dividing their moduli and subtracting their arguments, we obtain

$$x = \frac{F_0 e^{i(\omega t - \psi)}}{\sqrt{(K - m\omega^2)^2 + \beta^2\omega^2}} \tag{14}$$

1.9 TORSIONAL OSCILLATIONS OF A CANTILEVER BEAM

The equation of static equilibrium of a straight, cantilever beam in torsion is (Fig. 1.21), according to Eq. 5a of § 1.1,

$$\frac{\partial}{\partial y}\left(GJ\frac{\partial\theta}{\partial y}\right) = \frac{\partial T}{\partial y}$$

where θ is the angle of twist, T the torque, GJ the torsional rigidity, and y the axial coordinate. If the beam is in free oscillation, $\partial T/\partial y$ arises

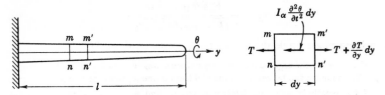

Fig. 1.21. Moments acting on an element of a beam in free torsional oscillation.

from the inertia, and is $I_\alpha(\partial^2\theta/\partial t^2)$, I_α being the mass moment of inertia about the shear center of the cross section. Hence, the equation of motion is

$$\frac{\partial}{\partial y}\left(GJ\frac{\partial\theta}{\partial y}\right) = I_\alpha\frac{\partial^2\theta}{\partial t^2} \tag{1}$$

For a cantilever beam, the boundary conditions are (1) no rotation at the root, and (2) no twisting moment at the tip:

$$\theta = 0 \quad \text{at} \quad y = 0$$
$$GJ\frac{\partial\theta}{\partial y} = 0 \quad \text{at} \quad y = l \tag{2}$$

A Beam of Constant Cross Section. Consider the simplest case when GJ and I_α are constants. Assuming a solution of the form

$$\theta(y, t) = \phi(y)e^{i\omega t} \tag{3}$$

we obtain, from Eq. 1

$$\frac{d^2\phi}{dy^2} + \kappa^2\phi = 0 \tag{4}$$

where

$$\kappa^2 = \frac{I_\alpha}{GJ}\omega^2 \tag{5}$$

The general solution of Eq. 4 is

$$\phi(y) = A \sin \kappa y + B \cos \kappa y \tag{6}$$

where A and B are arbitrary constants. From the boundary conditions (Eqs. 2), we see that a nontrivial solution ($\phi(y)$ not identically vanishing) exists only if

$$B = 0, \qquad \cos \kappa l = 0, \qquad A = \text{arbitrary constant}$$

This requires that

$$\kappa l = n\frac{\pi}{2} \qquad (n = 1, 3, 5, \cdots) \tag{7}$$

i.e.,

$$\omega = \pm n\frac{\pi}{2l}\sqrt{\frac{GJ}{I_\alpha}} \qquad (n = 1, 3, 5, \cdots) \tag{8}$$

The corresponding function $\phi(y)$ is then

$$\phi_n(y) = A \sin \frac{n\pi}{2l}y \qquad (n = 1, 3, 5, \cdots) \tag{9}$$

A general solution of Eq. 1 can be obtained by combining Eqs. 9 and 3. Writing the positive root of Eq. 8 as ω_n and sum over n, we obtain

$$\theta(y, t) = \sum_n \sin \frac{n\pi y}{2l} (A_n e^{i\omega_n t} + B_n e^{-i\omega_n t}) \tag{10}$$

where A_n and B_n are arbitrary constants. The real part of Eq. 10 gives the general solution as a real-valued function in the form

$$\theta(y, t) = \sum_n \sin \frac{n\pi y}{2l} (C_n \cos \omega_n t + D_n \sin \omega_n t) \tag{11}$$

The particular values ω_n are called the *eigenvalues* (or the *characteristic values*) of the problem, and the corresponding $\phi_n(y)$ are the *eigenfunctions* (or the *characteristic functions*). As long as the coefficient A in Eq. 9 is arbitrary, the eigenfunctions have unspecified magnitudes. When some special rule is given, so that the amplitudes can be definitely defined, the eigenfunctions are said to be "normalized." For example, we may take A to be 1. Or we may define A so that $\phi_n{}^2(y)$ integrated over the span l is equal to 1; or in such a way that $\phi_n(y)$ becomes 1 at $y = l$; etc.

It should be observed that all points along the span reach their maximum or minimum amplitude at the same time; i.e., they are "*in phase*." The *period* of the motion is $T_n = 2\pi/\omega_n$, and the *number of oscillations per second is* $n = \omega_n/2\pi$.

Torsional Oscillations of an Arbitrary Beam with Damping. The general case of torsional oscillation of an arbitrary beam with damping force proportional to the angular velocity can be solved approximately by a number of methods. Here we shall not discuss the methods of solution, but will investigate a general feature of the oscillation of such a beam. The equation of motion is

$$\frac{\partial}{\partial y} \left(GJ \frac{\partial \theta}{\partial y} \right) - \beta \frac{\partial \theta}{\partial t} - I_\alpha \frac{\partial^2 \theta}{\partial t^2} = 0 \tag{12}$$

where GJ, β, and I_α are functions of y. Equation 12 can be solved by assuming, as in Eq. 3,

$$\theta(y, t) = \phi(y) e^{i\omega t} \tag{13}$$

and determining the eigenvalue ω from the solution $\phi(y)$ and the boundary conditions. Depending on the sign of β, the motion is either damped or amplified as time increases. The eigenvalue ω is, in general, a complex number. If we substitute Eq. 13 into Eq. 12, and cancel the factor $e^{i\omega t}$, the result may be written as

$$\frac{d}{dy} \left(GJ \frac{d\phi}{dy} \right) + [\psi_1(y) + i \, \psi_2(y)] \phi = 0 \tag{14}$$

where $\psi_1(y)$ and $\psi_2(y)$ are real functions. Since this is an equation with complex coefficients, the solution $\phi(y)$ will be complex and can be written as

$$\phi(y) = \phi_1(y) + i\,\phi_2(y) \qquad (15)$$

where $\phi_1(y)$ and $\phi_2(y)$ are real functions. Therefore a solution of (12) that satisfies the boundary conditions is

$$\theta_1 = [\phi_1(y) + i\,\phi_2(y)]e^{i\omega t} \qquad (16)$$

But Eq. 12 is an equation with real coefficients; hence, there must exist another solution, conjugate to θ_1, which also satisfies the boundary conditions because the real and imaginary parts of θ_1 satisfy these conditions separately. The general solution is the sum of these two particular solutions multiplied by two arbitrary constants. The solution can be written in the real form

$$\theta(y, t) = \{[c_1\,\phi_1(y) - c_2\,\phi_2(y)]\cos qt - [c_1\,\phi_2(y) + c_2\,\phi_1(y)]\sin qt\}e^{pt} \quad (17)$$

where

$$i\omega = p + iq$$

Let

$$\Phi(y) = \sqrt{\phi_1{}^2(y) + \phi_2{}^2(y)}, \qquad \Psi(y) = \arctan\frac{\phi_2(y)}{\phi_1(y)}$$

then Eq. 17 can be written as

$$\theta(y, t) = 2\Phi(y)e^{pt}\{c_1\cos[qt + \Psi(y)] - c_2\sin[qt + \Psi(y)]\} \qquad (18)$$

The meaning of the function $\Psi(y)$ can be clarified by considering the time at which $\theta(y, t)$ becomes zero; i.e., when the cross section at y passes through the equilibrium position. Setting $\theta(y, t)$ equal to zero and solving for t, we obtain

$$t = \frac{1}{q}\left[\arctan\frac{c_1}{c_2} - \Psi(y)\right] + \frac{n\pi}{q} \qquad (19)$$

where n is any integer. The right-hand side of Eq. 19 depends on y, which means that different cross sections of the beam pass through the equilibrium position at different instants of time; the motion of the beam is *out of phase*.

If $\Psi(y)$ is a constant, Eq. 19 will be independent of y; the beam then oscillates in phase. $\Psi(y)$ is a constant if the ratio $\psi_2(y)/\psi_1(y)$ is a constant. It can be shown that the necessary and sufficient condition for an oscillation of the beam to be in phase is that the ratio of the damping coefficient $\beta(y)$ to the sectional moment of inertia $I_\alpha(y)$ be constant along the entire span.[6,13]

1.10 COUPLED TORSION-FLEXURE OSCILLATIONS OF A CANTILEVER BEAM

We shall show that the oscillation of a beam whose elastic axis and line of centers of gravity do not coincide is always "coupled"; i.e., it is a combination of flexure and torsion. Let x_α be the distance between the elastic axis and the line of center of gravity (Fig. 1.22). Assume that the elastic axis is a straight line and that the deflection of the elastic axis is

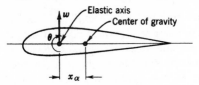

Fig. 1.22.   Flexure-torsional oscillation, notations.

restrained to the vertical direction only. Let the vertical deflection be w and the rotation of the beam cross sections about the elastic axis be θ. Then the equations of motion are, according to Eqs. 5a and 6a of § 1.1,

$$\frac{\partial^2}{\partial y^2}\left(EI\frac{\partial^2 w}{\partial y^2}\right) + m\frac{\partial^2 w}{\partial t^2} - mx_\alpha\frac{\partial^2\theta}{\partial t^2} = 0$$

$$\frac{\partial}{\partial y}\left(GJ\frac{\partial\theta}{\partial y}\right) + mx_\alpha\frac{\partial^2 w}{\partial t^2} - I_\alpha\frac{\partial^2\theta}{\partial t^2} = 0 \tag{1}$$

with the usual boundary conditions of a cantilever beam:

$$\text{At}\quad y = 0, \qquad w = \frac{\partial w}{\partial y} = \theta = 0$$

$$\text{At}\quad y = l, \qquad \frac{\partial^2 w}{\partial y^2} = \frac{\partial^3 w}{\partial y^3} = \frac{\partial\theta}{\partial y} = 0 \tag{2}$$

These equations reduce to two independent ones for the purely torsional and purely flexural oscillations if $x_\alpha = 0$.

Again we look for particular solutions in the form

$$w(y, t) = A f(y)e^{i\omega t}, \qquad \theta(y, t) = B \phi(y)e^{i\omega t} \tag{3}$$

Substituting Eq. 3 into Eq. 1, we obtain

$$A\left[\frac{d^2}{dy^2}\left(EI\frac{d^2 f}{dy^2}\right) - m\omega^2 f\right] + B\,mx_\alpha\omega^2\phi = 0$$

$$-A\omega^2 mx_\alpha f + B\left[\frac{d}{dy}\left(GJ\frac{d\phi}{dy}\right) + I_\alpha\omega^2\phi\right] = 0 \tag{4}$$

with the corresponding boundary conditions obtained by replacing w by f, and θ by ϕ in Eqs. 2. Let us first remark that, by a generalization of the Sturm-Liouville's theorem, it can be shown that there exist real valued eigenvalues ω, and hence real-valued eigenfunctions $f(y)$ and $\phi(y)$. Without proving this, we may verify it a posteriori by assuming that real ω, f, ϕ exist, and then carry through the computations to show that we can obtain them.

Let $f(y)$ and $\phi(y)$ be solutions of Eqs. 4. Multiply the first and the second of Eqs. 4 by f and ϕ respectively, and integrate the results with respect to y from 0 to l. The following equations are obtained:

$$(a_{11} - c_{11}\omega^2)A + c_{12}\omega^2 B = 0$$
$$c_{21}\omega^2 A + (a_{22} - c_{22}\omega^2)B = 0 \tag{5}$$

where

$$a_{11} = \int_0^l \frac{d^2}{dy^2}\left(EI\frac{d^2f}{dy^2}\right)f\,dy, \qquad c_{22} = \int_0^l I_\alpha\phi^2\,dy$$

$$a_{22} = -\int_0^l \frac{d}{dy}\left(GJ\frac{d\phi}{dy}\right)\phi\,dy, \qquad c_{12} = \int_0^l mx_\alpha\phi f\,dy \tag{6}$$

$$c_{11} = \int_0^l mf^2\,dy, \qquad c_{21} = c_{12}$$

An integration by parts and an application of the boundary conditions (Eqs. 2) transform a_{11} and a_{22} into the following form:

$$a_{11} = \int_0^l EI\left(\frac{d^2f}{dy^2}\right)^2 dy, \qquad a_{22} = \int_0^l GJ\left(\frac{d\phi}{dy}\right)^2 dy \tag{7}$$

For a nontrivial solution, A and B must not both vanish. In order that a significant solution may exist, the determinant of the coefficients of Eqs. 5 must be zero:

$$\begin{vmatrix} a_{11} - c_{11}\omega^2 & c_{12}\omega^2 \\ c_{21}\omega^2 & a_{22} - c_{22}\omega^2 \end{vmatrix} = 0 \tag{8}$$

i.e.,

$$P\omega^4 - Q\omega^2 + R = 0 \tag{9}$$

where

$$P = c_{11}c_{22} - c_{12}{}^2, \qquad Q = a_{11}c_{22} + a_{22}c_{11}, \qquad R = a_{11}a_{22} \tag{10}$$

The solution of Eq. 9 is

$$\omega^2 = \frac{Q \pm \sqrt{Q^2 - 4PR}}{2P} \tag{11}$$

It can be verified that the right-hand side of Eq. 11 is always positive. We obtain, therefore, four real roots for ω. Hence, four particular solutions are obtained, each representing a simple-harmonic motion. The ratio A/B, corresponding to each root ω, is obtained from Eqs. 5:

$$\frac{A}{B} = \frac{c_{12}\omega^2}{c_{11}\omega^2 - a_{11}} = \frac{c_{22}\omega^2 - a_{22}}{c_{12}\omega^2} \tag{12}$$

Thus, if $f(y)$ and $\phi(y)$ were known, the constants a_{11}, a_{12}, etc., and P, Q, R can be computed according to Eqs. 6, 7, 10; and the frequency ω and the ratio A/B can be obtained from Eqs. 11 and 12. To determine $f(y)$ and $\phi(y)$, the *method of successive approximation* can be used. Let $f_0(y)$ and $\phi_0(y)$ be two arbitrary functions which satisfy the "rigid" boundary conditions* at the clamped end $y = 0$, which are given by the first of Eqs. 2. Considering f_0 and ϕ_0 as approximate solutions, we determine the corresponding approximate values of a_{11}, a_{12}, $\cdot\cdot\cdot$ and ω^2 and A/B. Now, by a formal successive integration, Eqs. 4 may be written as

$$\frac{1}{\omega^2}f(y) = \int_0^y dy_4 \int_0^{y_4} \frac{dy_3}{EI(y_3)} \int_{y_3}^l dy_2 \int_{y_2}^l dy_1 \left[m f(y_1) - \frac{B}{A} m x_\alpha \phi(y_1) \right] dy_1$$

$$\frac{1}{\omega^2}\phi(y) = \int_0^y \frac{dy_2}{GJ(y_2)} \int_{y_2}^l dy_1 \left[I_\alpha \phi(y_1) - \frac{A}{B} m x_\alpha f(y_1) \right] \tag{13}$$

So the process of approximation can be carried out as follows: Substituting f_0 and ϕ_0 into the integrands on the right-hand sides of Eqs. 13, we obtain two new functions $\frac{1}{\omega^2}f(y)$ and $\frac{1}{\omega^2}\phi(y)$, which we shall call $f_1(y)$ and $\phi_1(y)$. Using $f_1(y)$, $\phi_1(y)$ as the first approximation, we determine a_{11}, a_{12}, etc., and ω^2 and A/B. If we substitute f_1, ϕ_1, and the new value of A/B into the integrands of Eqs. 13 and integrate, the result may be labeled f_2 and ϕ_2. The process can be repeated with f_2 and ϕ_2 as the starting approximations. When f_n and ϕ_n converge satisfactorily, ω^2 and A/B must also converge to their true values. The rate of convergence depends on the choice of f_0 and ϕ_0. If f_0 and ϕ_0 were chosen as the fundamental modes of uncoupled flexural and torsional oscillations of a uniform beam, a sufficiently accurate result can usually be achieved in three or four cycles.

Let us assume that, in the process indicated above, we always take the smaller of the two ω^2 values of Eq. 11. It can be shown[1.58] that, for

---

* It is necessary only to satisfy the boundary conditions imposed by constraints against displacement. Such boundary conditions are called the "rigid" boundary conditions. The boundary conditions at an end where no constraint is imposed, such as those at the free end of the beam, are called the "natural" boundary conditions. The natural boundary conditions will be satisfied in the limit by the method of successive approximations.

arbitrary choice of f_0, ϕ_0 (which satisfies the boundary conditions), the process of successive approximation converges to the fundamental mode of the coupled flexure-torsional oscillation, i.e., the mode with the lowest frequency.

If the larger of the two ω^2 values of Eq. 11 is taken in the successive approximation, a different set of values A/B, $f(y)$ and $\phi(y)$ will be obtained, which would converge to the second mode of the flexure-torsional oscillation. The physical meaning of the two modes can be clarified by the following illustration. Consider a beam whose center of gravity at each section lies exactly on the elastic axis ($x_\alpha = 0$). Such a beam can oscillate in purely flexural and purely torsional modes. If the center of gravity is moved away slightly from the elastic axis, the modes will change slightly, and the oscillation will be a combined flexure-torsional motion. With small values of x_α, it is expected to have one of the combined flexural-torsional mode vary but little from the purely torsional mode, and the other vary but little from the purely flexural mode. Accordingly, one of the coupled oscillations is said to be predominantly flexural, and the other predominantly torsional.

It might appear that, if the center of gravity is moved further away, the frequencies may approach each other and finally coincide. For a cantilever beam, however, this will never happen, because the frequencies can coincide only if $Q^2 - 4PR = 0$, and it can be verified that $Q^2 - 4PR$ is always positive if $x_\alpha \neq 0$. Hence, the roots must be discrete.

In order to obtain higher modes by the method of successive approximation, the orthogonality relations among the normal modes may be used.[1.52–1.65] But this subject will not be pursued further here.

The natural frequencies can be found experimentally by a resonance test. The oscillations may be excited, for example, by a rotating eccentric weight driven by a motor through a flexible shaft. Intense oscillations occur when the frequency of the motor coincides with one of the natural frequencies of the wing. As discussed above, the excited resonance oscillation is always coupled unless the elastic axis coincides with the inertia axis. But one of the modes will be predominantly flexural, and the other predominantly torsional. It is customary to call the first one flexural oscillation and the second one torsional oscillation.

BIBLIOGRAPHY

General principles of structural analysis are presented in the following books. Extensive lists of references can be found in Refs. 1.3 and 1.13.

1.1 Barton, M. V.: *Fundamentals of Aircraft Structures*. Prentice-Hall, New York (1948).

1.2 Love, A. E. H.: *A Treatise on the Mathematical Theory of Elasticity*, 1st Ed. Cambridge Univ. Press (1927). Reprinted by Dover Publications, New York (1945).

1.3 Niles, A. S., and J. S. Newell: *Airplane Structures*, 4th Ed. John Wiley & Sons, New York (1954).

1.4 Peery, D. J.: *Aircraft Structures*. McGraw-Hill, New York (1950).

1.5 Pippard, A. J. S., and J. F. Baker: *The Analysis of Engineering Structures*, 2d Ed. E. Arnold & Co., London (1943).

1.6 Sechler, E. E., and L. G. Dunn: *Airplane Structural Analysis and Design*. John Wiley & Sons, New York (1942).

1.7 Sechler, E. E.: *Elasticity in Engineering*. John Wiley & Sons, New York (1952).

1.8 Shanley, F. R.: *Basic Structures*. John Wiley & Sons, New York (1944).

1.9 Southwell, R. V.: *Theory of Elasticity*. Oxford Univ. Press, London (1936), 2d Ed. (1941).

1.10 Steinbacher, R., and G. Gerard: *Aircraft Structural Mechanics*. Pitman Publishing Corp., New York (1952).

1.11 Timoshenko, S., and J. N. Goodier: *Theory of Elasticity*, 2d Ed. McGraw-Hill, New York (1951).

1.12 Timoshenko, S., and D. H. Young: *Theory of Structures*, 2d Ed. McGraw-Hill, New York (1965).

1.13 Tye, W., J. H. Argyris, and P. C. Dunne: *Structural Principles and Data*. No. 1 of *Handbook of Aeronautics*. Published by the Royal Aeronautical Society. Pitman Publishing Corp., New York (1952).

For an interesting discussion on the design of structures to meet the requirements of aeroelasticity, see

1.14 Williams, D.: Some Novel Structural Properties of Stressed Skin Wings. *Aeronaut. Conf. London*, 407–444 (1947). Royal Aeronautical Society, London.

For shear center, elastic axis, flexural line, etc., see

1.15 Duncan, W. J., and D. L. Ellis: The Flexural Center and the Center of Twist of an Elastic Cylinder. *Phil. Mag.* **16**, 201–235 (1933).

1.16 Goodier, J. N.: A Theorem on the Shearing Stress in Beams with Applications to Multicellular Sections. *J. Aeronaut. Sci.* **11**, 272–280 (1944).

1.17 Hatcher, R. S.: Rational Shear Analysis of Box Girders. *J. Aeronaut. Sci.* **4**, 233–238 (1936).

1.18 Kuhn, P.: *Stresses in Aircraft and Shell Structures*. McGraw-Hill, New York (1956).

Papers of particular interest for aircraft wing analysis:

1.19 Argyris, J. H., and P. C. Dunne: The General Theory of Cylindrical and Conical Tubes under Torsion and Bending Loads. *J. Roy. Aeronaut. Soc.* **51**, 199–269, 757–784, 884–930 (1947); **53**, 461–483, 558–620 (1949).

1.20 Benscoter, S. U., and M. L. Gossard: Matrix Methods for Calculating Cantilever Beam Deflections. *NACA Tech. Note* **1827** (1949).

1.21 Benscoter, S. U., and R. H. MacNeal: Introduction to Electrical-Circuit Analogues for Beam Analysis. *NACA Tech. Note* **2785** (1952).

1.22 Benscoter, S. U., and R. H. MacNeal: Equivalent Plate Theory for a Straight Multicell Wing. *NACA Tech. Note* **2786** (1952).

1.23 Benscoter, S. U.: Secondary Stresses in Thin-Walled Beams with Closed Cross Sections. *NACA Tech. Note* **2529** (1951).

1.24 Duberg, J. E.: A Numerical Procedure for the Stress Analysis of Stiffened Shells. *J. Aeronaut. Sci.* **16**, 451–462 (Aug. 1949).

1.25 Fung, Y. C.: Bending of Thin Elastic Plates of Variable Thickness. *J. Aeronaut. Sci.* **20**, 455–468 (July 1953).

1.26 Reissner, E., and M. Stein: Torsion and Transverse Bending of Cantilever Plates. *NACA Tech. Note* **2369** (1951).

1.27 Trayer, G. W., and H. W. March: The Torsion of Members Having Sections Common in Aircraft Construction. *NACA Rept.* **334** (1930).

For structural measurements, see

1.28 Broadbent, E. G., and D. L. Woodcock: The Measurement of Structural Stiffness of Aircraft. *Aeronaut. Research Council R. & M.* **2208** (1945).

1.29 Hetenyi, M. I.: *Handbook of Experimental Stress Analysis.* John Wiley & Sons, New York (1950).

For swept or delta wing analysis, see the following papers. Extensive bibliography can be found in Refs. 1.36 and 1.38.

1.30 Dixon, H. H.: Stresses and Deflections of Unswept and Swept Thin-Walled Beams. *U.S. Air Force Tech. Rept.* **5761**, part 10 (Jan. 1952). Wright Air Development Center, Dayton, Ohio.

1.31 Fung, Y. C.: On the Stresses and Deflections of Swept Thick-Walled Box Beams. *U.S. Air Force Tech. Rept.* **5761**, part 12 (Sept. 1950). Wright Air Development Center, Dayton, Ohio.

1.32 Hall, A. H.: An Experimental Study of the Static Stiffness and Deformation of Swept Wings with Uniform Chord. *3d Anglo-Am. Aeronaut. Conf. Brighton,* 525–544 (1951). Royal Aeronautical Society, London.

1.33 Lang, A. L., and R. L. Bisplinghoff: Some Results of Sweptback Wing Structural Studies. *J. Aeronaut. Sci.* **18**, 705–717 (Nov. 1951).

1.34 Levy, S.: Computation of Influence Coefficients for Aircraft Structure with Discontinuities and Sweepback. *J. Aeronaut. Sci.* **14**, 547–560 (Oct. 1947).

1.35 Schuerch, H. U.: Structural Analysis of Swept, Low Aspect Ratio, Multispar Aircraft Wings. *Aeronaut. Engg. Rev.* **11**, 34–41 (Nov. 1952).

1.36 Sechler, E. E., M. L. Williams, and Y. C. Fung: Summary, Design Principles for the Static Analysis of Swept Wing Structures. *U.S. Air Force Tech. Rept.* **5761**, part 18 (July 1952). Wright Air Development Center, Dayton, Ohio.

1.37 Stein, M., J. E. Anderson, and J. M. Hedgepath: Deflection and Stress Analysis of Thin Solid Wings of Arbitrary Planform with Particular Reference to Delta Wings. *NACA Tech. Note* **2621** (Feb. 1952).

1.38 Williams, M. L.: A Review of Certain Analysis Methods for Swept Wing Structures. *J. Aeronaut. Sci.* **19**, 615–629 (Sept. 1952).

1.39 Wittrick, W. H.: Torsion and Bending of Swept and Tapered Wings with Rigid Chordwise Ribs. *Australia Aeronaut. Research Com., Dept. Supply, Rept.* **ACA-51** (Sept. 1950).

1.40 Zender, G. W., and W. A. Brooks: An Approximate Method of Calculating the Deformations of Wings Having Swept, *M*, or *W*, Λ, and Swept-Tip Plan Forms. *NACA Tech. Note* **2878** (Oct. 1953).

The following are some of the books on aerodynamics. Extensive lists of references can be found in Refs. 1.41 and 1.51.

1.41 Durand, W. F.: *Aerodynamic Theory.* J. Springer, Berlin (1934).

1.42 Ferri, A.: *Elements of Aerodynamics of Supersonic Flows*. Macmillan Co., New York (1949).

1.43 Glauert, H.: *The Elements of Aerofoil and Airscrew Theory*, 2d Ed. Cambridge Univ. Press, and Macmillan Co., London (1947).

1.44 Hilton, W. F.: *High-Speed Aerodynamics*. Longmans, Green & Co., New York (1951).

1.45 Kuethe, A. M., and J. D. Schetzer: *Foundations of Aerodynamics*, 2d Ed. John Wiley & Sons, New York (1959).

1.46 Liepmann, H. W., and A. Roshko: *Elements of Gasdynamics*. John Wiley & Sons, New York (1957).

1.47 Miles, E. R. C.: *Supersonic Aerodynamics*. McGraw-Hill, New York (1950).

1.48 Millikan, C. B.: *Aerodynamics of the Airplane*. John Wiley & Sons, New York (1941).

1.49 Mises, R. von: *Theory of Flight*. McGraw-Hill, New York (1945).

1.50 Prandtl, L., and O. G. Tietjens (translated by L. Rosenhead): *Fundamentals of Hydro- and Aeromechanics*. McGraw-Hill, New York (1934).

1.51 Prandtl, L.: *Essentials of Fluid Dynamics*. Blackie & Son, London (1952).

For principles of vibration analysis, see one of the following books.

1.52 Den Hartog, J. P.: *Mechanical Vibrations*, 3d Ed. McGraw-Hill, New York (1947).

1.53 Duncan, W. J.: Mechanical Admittances and Their Applications to Oscillation Problems. *Aeronaut. Research Council Monograph R & M* **2000** (1946).

1.54 Kármán, Th. von, and M. A. Biot: *Mathematical Methods in Engineering*, McGraw-Hill, New York (1940).

1.55 Myklestad, N. O.: *Vibration Analysis*. McGraw-Hill, New York (1944).

1.56 Lord Rayleigh: *Theory of Sound*, 1st Ed. (1877). Reprinted, Dover Publications, New York (1945).

1.57 Scanlan, R. H., and R. Rosenbaum: *Introduction to the Study of Aircraft Vibration and Flutter*. Macmillan Co., New York (1951).

1.58 Temple, G. F., and W. G. Bickley: *Rayleigh's Principle and Its Applications to Engineering*. Oxford Univ. Press, London (1933).

1.59 Timoshenko, S.: *Vibration Problems in Engineering*. D. Van Nostrand, New York (1937).

For vibration of aircraft structures, see

1.60 Anderson, R. A., and J. C. Houbolt: Determination of Coupled and Uncoupled Modes and Frequencies of Natural Vibration of Swept and Unswept Wings from Uniform Cantilever Modes. *NACA Tech. Note* **1747** (1948).

1.61 Beskin, L., and R. M. Rosenberg: Higher Modes of Vibration by a Method of Sweeping. *J. Aeronaut. Sci.* **13**, 597–604 (Nov. 1946).

1.62 Trail-Nash, R. W.: The Anti-Symmetric Vibrations of Aircraft. *Aeronaut. Quart.* **3**, 145–160 (Sept. 1951).

1.63 Walker, P. B.: Mechanical Vibration and Aeroelasticity. *J. Roy. Aeronaut. Soc.* **51**, 417–423 (1947).

The following paper is of special interest for thin-walled structures.

1.64 Trail-Nash, R. W., and A. R. Collar: Effects of Shear Flexibility and Rotary Inertia on the Bending Vibrations of Beams. *Quart. J. Mech. & Applied Math.* **6**, 2, 186–222 (June 1953).

A very powerful method of analysis, particularly suitable when it is desired to obtain the higher frequencies and modes of oscillaton, is

1.65 Lanczos, C.: An Iteration Method for the solution of the Eigenvalue Problem of Linear Differential and Integral Operators. *U.S. Bur. Standards, J. Research* **45**, RP 2133, 255–282 (1950).

Matrix iteration methods are given in Ref. 1.54 and the following papers. Rasof's paper explains the reasoning behind every step of analysis in great detail and is helpful to beginners.

1.66 Biot, M. A.: Vibration Analysis of a Wing Carrying Large Concentrated Weights. *GALCIT Flutter Rept.* **1** (Aug. 1941). California Institute of Technology.

1.67 Rasof, B.: Vibration Analysis of a Wing Mounting Flexibly Suspended Engines. *GALCIT Flutter Rept.* **10** (Nov. 1943). California Institute of Technology.

A good reference to generalized coordinates and Lagrange's equations is Ref. 1.69, which indulges in the philosophical implications as well as the fundamental techniques. For shorter references, see 1.54 and 1.68.

1.68 Goldstein, H.: *Classical Mechanics.* Addison-Wesley Press, Cambridge, Mass. (1950).

1.69 Lanczos, C.: *Variational Principles of Mechanics.* Univ. of Toronto Press, Toronto, Canada (1949).

The effects of restrained warping on the torsion of thin-walled open sections can be found in

1.70 Argyris, J. H.: The Open Tube—a study of Thin-Walled Structures Such as Interspar Wing Cut-Outs and Open-Section Stringers. *Aircraft Engg.* **26**, 102–112 (April 1954).

The use of stiffness-influence coefficients is discussed in

1.71 Williams, D. : Recent Developments in the Structural Approach to Aeroelastic Problems. *J. Roy. Aeronaut. Soc.* **58**, 403–421 (June 1954). *Aircraft Engg.* **26**, 303–307 (Sept. 1954).

An introductory text in continuum mechanics is

1.72 Fung, Y. C.: *A First Course in Continuum Mechanics.* Prentice-Hall, New Jersey (1969).

A basic text in solid mechanics with an extensive bibliography is

1.73 Fung, Y. C.: *Foundations of Solid Mechanics.* Prentice-Hall, New Jersey (1965).

Chapter 2

SOME AEROELASTIC PROBLEMS IN CIVIL AND MECHANICAL ENGINEERING

Aeroelastic oscillations are generally sustained by aerodynamic forces induced by the structure itself. Many types of structures may develop excessively large elastic deformation, or suffer sustained or divergent oscillations in certain ranges of wind speeds. For a structure designed to maintain static equilibrium such aeroelastic trouble can be serious.

The failure of the original Tacoma Narrows Bridge is an example of aeroelastic oscillation. If the wind speed and the mode and frequency of the structural oscillation are such that energy can be absorbed from the wind by the structure, and if the energy absorbed is larger than that dissipated by the structural damping, the amplitude of oscillation will continue to increase and will finally lead to destruction. The Tacoma Narrows Bridge failed at a wind speed of 42 mph, whereas the structure as built should have been able to resist a steady wind of at least 100 mph if no oscillation had occurred. Since all structures exposed to wind oscillate under some disturbances, it becomes apparent that it is essential for the designer to predict the critical wind speed at which the structure may become aeroelastically unstable.

In this chapter we shall outline the general principles of aeroelasticity for some civil and mechanical engineering applications. The main character of the problems discussed in this chapter is that the structures concerned are mainly *unstreamlined* and hence unamenable to theoretical treatment. We must rely heavily on experimental results to understand the nature of the aerodynamic forces.

Several typical examples of aeroelastic oscillations are given in § 2.1. Some aerodynamic considerations are given in § 2.2. The flow around a cylinder is then reviewed in § 2.3. The aeroelastic oscillation of a cylindrical body is treated in § 2.4, and that of an H-shaped section is discussed in § 2.5. Finally, means of preventing aeroelastic instabilities are discussed in § 2.6.

2.1 SOME TYPICAL EXAMPLES OF AEROELASTIC OSCILLATIONS

Whoever has rowed a boat must have observed the trail of vortices leaving the oar.   One can easily perform an experiment by moving a stick in water with sufficient speed and observe the wake.   In the wake the flow is turbulent, but a vortex pattern can be seen (Fig. 2.1).   These vortices are "shed" alternately from each side of the stick.   This shedding of vortices induces a periodic force in the direction perpendicular to the line of motion, and the stick wobbles back and forth.   Similar action occurs on any cylindrical body with a blunt nose or tail.   The frequency

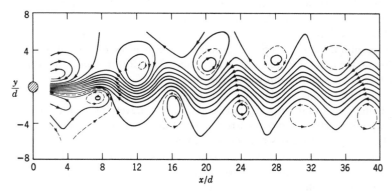

Fig. 2.1. The wake behind a circular cylinder. Reynolds number 56. Measurements by Kovasznay, Ref. 2.26. Figure shows the streamline pattern viewed relative to the undisturbed flow at infinity. The development and decay of the vortices can be seen. The lines correspond to differences in the stream function $\Delta \psi = 0.1\ Ud$; the dotted lines are half-values between two full lines. (Figure reproduced by courtesy of the Royal Society of London.)

of the wake vortices is determined by the geometry of the body and the speed of flow.   If the frequency is close to the natural vibration frequency of the body, resonance will set in.

Among examples of engineering significance, the most familiar is probably the oscillation of telephone wires,[2.7] with high frequency and small amplitude, producing musical tones.   Smokestacks,[2.3] submarine periscopes,[2.1] oil pipe lines,[2.2] television antennas,[2.1] and other cylindrical structures[2.7] often encounter vibrational troubles of aeroelastic origin. These may be cured either by stiffening the structures so that the natural frequency is much higher than the frequency of the vortex shedding in wind or by introducing vibration dampers into the system to absorb the energy.*

A different type of vibration is the *"galloping" of transmission lines*.[2.4, 2.5] During a sleet storm a transmission line may vibrate in a strong wind. The cable span oscillates sometimes as a whole, but more frequently with one or more nodes in a span. When the oscillations become severe, the cables move irregularly, but freely, through vertical distances of as much as 20 to 35 ft in a span of 500 ft. The phenomenon cannot be observed every day, nor can it be seen at any specific place; it appears and disappears suddenly. Once started, it is very persistent. Sometimes it may continue for 24 hours. The cause of galloping has been shown to be the sleet on the conductors. The ice forms a cross section of a more or less elliptical shape, with the major axis perpendicular to the wind direction. Such a section is unstable in an airstream: the aerodynamic force exerts a "negative damping" component so that, once the oscillation is started, it will continue to build up. The observed frequencies of oscillation are close to the natural frequencies of the span. The vibration will stop when the ice is broken and thrown off the line.

Since vibrations of this type are originated from unfavorable aerodynamic configurations, they can be avoided by preventing such unfavorable configurations from occurring. For transmission lines, however, a satisfactory solution has not yet been found.

The two types of aeroelastic oscillations considered above are characterized by a *separated flow* in the rear of the body, i.e., a flow that does not follow the contour of the solid body. Let us now consider another type of self-excited oscillation which does not necessarily involve flow separation. This is the *flutter*. The best example occurs in the field of aeronautics where streamlining is a rule and flow separation is avoided. In civil engineering, suspension bridges sometimes have sufficiently clean contours so that the flow may be considered as essentially unseparated; but, as a rule, separation does occur over part of the body or during part of the oscillation cycle; hence the name *stall flutter*. The failure of the original Tacoma Narrows Bridge[2.10] is generally believed to be due to stall flutter. This bridge was a suspension structure, with a center span of 2800 ft, two side spans of 1100 ft each, and a width of 39 ft center to center of cables. The cables had a sag ratio of 1/12. The stiffening structure was of the plate-girder type. Vertical oscillations of considerable amplitudes were first observed during the erection of the suspended

---

* This type of oscillation, due to periodic shedding of vortices, may be called the oscillations of the Aeolian harp type. The principle of playing strings by the wind was recognized in ancient times. King David, according to Rabbinic records, used to hang his kinnor (Ki'' ra) over his bed at night, where it sounded in the midnight breeze.

Stall flutter of the original Tacoma Narrows Bridge, Puget Sound, Washington. (Courtesy of F. B. Farquharson)

The failure of the original Tacoma Narrows Bridge, Nov. 7, 1940.
(Courtesy of F. B. Farquharson)

floor and continued, at intervals, until the day of failure, but no damage was done. On November 7, 1940, four months after the bridge was opened to traffic, suddenly at a wind speed of 42 mph, the center span developed a torsional movement with a node at mid-span. (There was, however, considerable motion at the mid-point.) The frequency of oscillation suddenly changed from 37 to 14 cycles per minute. The motion grew violently, and failure occurred half an hour later.

A lesson learned from the Tacoma Bridge is the recognition of the importance of aeroelastic investigations in structural design. The conventional design procedure focuses mostly on the *strength* of a structure, whereas the aeroelastic design focuses on the *rigidity*, *damping characteristics*, and the *aerodynamic shape*. Hence there exists a different point of view.

The three types of oscillations illustrated above can occur in a uniform flow without external disturbance. For that reason they are often said to be *self-excited*.

2.2 GENERAL CONSIDERATIONS IN AERODYNAMICS AND THE DIMENSIONLESS COEFFICIENTS

In order to discuss aeroelastic oscillations associated with the shedding of vortices, it is necessary to consider the effect of the viscosity of the fluid. In the flow of a nonviscous fluid the possibility of slip between the fluid and a solid body in contact with the fluid may be assumed. But such a relative velocity is impossible in a viscous fluid: The velocity of the fluid in contact with the solid must be exactly the same as that of the solid itself. This nonslip condition persists, no matter how small the viscosity of the fluid, as long as the characteristic length of the body is much larger than the mean free path of the fluid molecules. But, when the viscosity is small, its effect is felt only in a thin layer next to the solid body: the *boundary layer*, in which the velocity of flow changes rapidly from the velocity of solid surface to that of the free stream. Outside the boundary layer, the fluid may be regarded as nonviscous, (see Fig. 2.2 in which the scale normal to the solid wall is greatly exaggerated).*

* The boundary layer thickness δ in Fig. 2.2 is defined by the condition that, when the distance from the solid wall is δ, the retardation of the velocity of flow due to the skin friction is 1 per cent of the potential flow velocity; i.e., at δ, the velocity attains 0.99 of the potential value. For a given body, the boundary-layer thickness decreases as the Reynolds number increases. At higher Reynolds number, the flow in the boundary layer is laminar near the leading edge, but becomes turbulent behind a transition point. As shown in Fig. 2.2c, the transition point moves forward as the Reynolds number increases. A turbulent boundary layer can resist an adverse pressure gradient better than a laminar one, and separation over an airfoil can be delayed.

The flow in the boundary layer is pulled forward by the free stream, but is retarded by friction at the solid wall. It is also retarded by an adverse pressure gradient, if such a gradient exists. If the adverse pressure gradient is sufficiently large, the flow may be interrupted entirely, and the

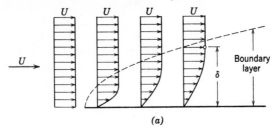

(a)

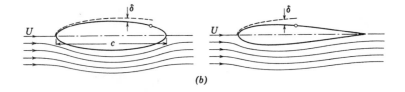

(b)

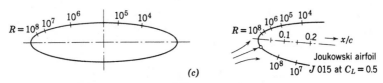

(c)

Fig. 2.2. Boundary layer. (a) Mean velocity distribution in a boundary layer. (b) Laminar boundary layer over an ellipse and an airfoil. The boundary layer thickness δ is magnified by a factor $\sqrt{R}/100$, where $R = Uc/\nu$ is the Reynolds number based on the chord length. The small circle over the upper surface marks the separation point. (c) Points of transition from laminar to turbulent flow in the boundary layer, as a function of the Reynolds number. (From H. Schlichting, *NACA Tech. Memo.* **1218.**)

boundary layer may detach from the solid wall. The flow is then said to be *separated*.

An adverse pressure gradient exists in a flow around a bluff body, such as a cylinder or a sphere. For a streamlined body, such as an airfoil, the rear part of the body is very thin and has very gentle curvature; the adverse pressure gradient is small, and separation can be prevented. A streamlined body maintains a smooth flow when it is properly situated in the flow; but, when the angle between the chord and the undisturbed stream is sufficiently large, separation may still occur.

The above physical picture explains why the Reynolds number is so important in determining the condition of flow around a bluff body and the force acting on it, because the Reynolds number expresses a ratio between the inertia force and the friction force. A flow pattern is determined by the interplay among the pressure gradient, friction, and inertia. As these three forces are subject to the condition of equilibrium, only two of them are independent, and we may select the friction and inertia as independent influences. The inertia force has components of the form

$$-\rho \frac{du}{dt} = -\rho \left(u \frac{\partial u}{\partial x} + \cdots \right)$$

and, hence, for a given flow pattern, must be proportional to the products $\rho U^2/l$, where U is a characteristic velocity, and l a characteristic length. On the other hand, the friction force has components of the form $\mu(\partial^2 u/\partial y^2)$, and, hence, for a given flow pattern, must be proportional to $\mu U/l^2$. ($\partial^2 u$ means a small difference of velocity of the second order, and is therefore proportional to the velocity U, whereas ∂y^2 is the square of a small difference in length and is proportional to l^2.) The ratio between these two forces is therefore

$$\frac{\rho U^2}{l} \div \frac{\mu U}{l^2} = \frac{\rho U l}{\mu}$$

which is the Reynolds number.* A small value of Reynolds number means that friction forces predominate; a large value that inertia forces predominate.

2.3 THE FLOW AROUND A CIRCULAR CYLINDER

The theoretical investigation of the vortex pattern which is observed in the wake of a cylinder was originated by von Kármán[2.24] who considers a double row of vortices in a two-dimensional flow (Fig. 2.3). The equilibrium configurations (the patterns that can maintain themselves in the

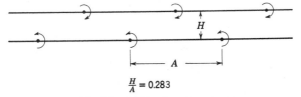

$$\frac{H}{A} = 0.283$$

Fig. 2.3. A Kármán vortex street.

* The explanation is given in this form by Prandtl.[1.51]

flow) and the stability of the equilibrium configurations against infinitesimal disturbances, (displacements of individual vortices), are studied. It is found that the double row of vortices is stable only if the vortices in one row are opposite to points half way between the vortices in the other row, and if the distance between the rows is 0.281 times the distance between two consecutive vortices in each row. Such an arrangement of vortices is known as a *Kármán vortex street.*

The theoretical treatment of the vortex street based on the assumptions of a perfect fluid cannot reveal the Reynolds-number effect, whereas measurements do indicate definite effect of the Reynolds number on the flow pattern.

In a flow past a long circular cylinder, a great variety of changes occur with an increasing Reynolds number* R. At low values of R the flow is smooth and unseparated, but the fluid at the back of the cylinder is appreciably retarded. At higher values of R two symmetrical standing vortices are formed at the back. With increasing Reynolds number these vortices stretch farther and farther downstream from the cylinder. Eventually the standing vortices become considerably elongated and distorted. When R reaches a number of order 40, the vortices become asymmetrical, detach from the obstacle, and move downstream as if they were discharged alternately from the two sides of the cylinder. In this way an eddying motion in the wake is set up. As the flow moves downstream, the eddying motion is gradually diffused and "decays" into a general turbulence. For R in the range of 40 to 150, the "shedding" of vortices is regular. The eddying motion in the wake is periodic both in space and time. The flow can be approximated by a Kármán vortex street. The range of R between 150 and 300 is a transition range, in which the vortex shedding is no longer so regular as before. Its frequency appears to be somewhat erratic. For $R > 300$, the vortex shedding is "irregular." A predominant frequency can be easily determined, but the amplitude appears to be more or less random. In addition, "background" random fluctuations, similar to turbulences behind a grid in a wind tunnel, becomes more appreciable. The kinetic energy of fluctuations in the wake is partly contained in the periodic motion and partly carried by the turbulences. As R increases, more and more of the energy is transferred to the turbulences. Finally, at R of order 3×10^5, the separation point of the boundary layer moves rearward on the cylinder. The drag coefficient of the cylinder decreases appreciably owing to this important change (see Fig. 2.4). The flow in the wake at these large Reynolds numbers becomes so turbulent that the vortex street pattern is no longer recognizable.

---

* The diameter of the cylinder is taken as the characteristic length.

The geometry of the wake, when the Reynolds number is in a range in which vortices may be regarded as "shedding," is as follows: The frequency at which the vortices are shed, expressed nondimensionally as the *Strouhal number*[2.23] k, is a function of the Reynolds number. Here the Strouhal number k shall be defined* as $\omega d/U$, where ω is the frequency in radians per second and d is the diameter of the cylinder. The results of Relf and Simmons[2.25], Kovasznay[2.26], and Roshko[2.27] are given in Fig. 2.4, from which the value of k for each R can be found.† The number

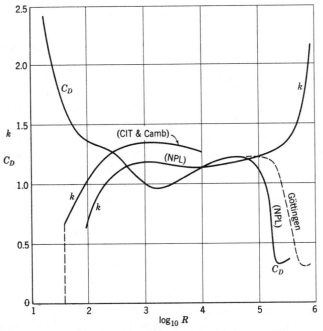

Fig. 2.4. Variation of the Strouhal number and drag coefficient against Reynolds number for a circular cylinder. C_D and R are based on the diameter of the cylinder. Sources of data are: NPL; Relf and Simmons, *Aeronaut. Research Com. R. & M.* **917** (1924). Cambridge; Kovasznay, *Proc. Roy. Soc. A.* **198** (1949). CIT; Roshko, *NACA Tech. Note* **2913** (1953). Göttingen; *Ergebnisse AVA Göttingen*, **2** (1923).

* In some books, the Strouhal number is defined by replacing ω by the number of cycles per second. For uniformity in notation, we shall not distinguish the reduced frequency and the Strouhal number in this book.

† There are some quantitative differences between Kovasznay's and Relf and Simmons' results, probably reflecting the difference between the wind tunnels they used.

of vortices $n = \omega/2\pi$, shed from each side of the cylinder, is

$$n = \frac{kU}{2\pi d} \text{ per second} \qquad (1)$$

The distance A between two consecutive vortices in a row is

$$A \doteq \frac{U-v}{n} \qquad \left(\frac{v}{U} \sim 0.25 \text{ to } 0\right) \qquad (2)$$

where v is the relative velocity of the vortices with respect to the free stream. The ratio of the distance H between the rows of vortices to the distance A is approximated by von Kármán's theoretical formula $H = 0.281A$ at small distance from the cylinder, but H/A increases as the distance from the cylinder increases. At large distance H/A is of order 0.9. The intensity of the velocity fluctuations is small close behind the cylinder. The maximum intensity of the fluctuations occurs in the vicinity of 7 diameters downstream. Thus it appears that the vortices are not really "shed" from the cylinder, but are developed gradually.

The range of Reynolds numbers of interest in aeroelasticity is the range in which the flow fluctuates, i.e., for $R > 40$. At a large Reynolds number, say $R > 3 \times 10^5$, or at a distance far downstream of the cylinder, although no distinct vortex pattern can be observed, the flow is turbulent and can still excite oscillations in an elastic structure.

In the subcritical Reynolds number range ($R < 3 \times 10^5$), a periodic lift force (perpendicular to the flow) acts on the cylinder; the root-mean-square value of the lift coefficient (lift force/unit span)/($\frac{1}{2}\rho U^2 \cdot$ diameter), $\sqrt{\overline{C_L^2}}$, is of the order of 0.45. In the Reynolds number range 0.3×10^6 to 3×10^6, the lift is no longer a periodic function of time, but becomes random, the root-mean-square value of the lift coefficient ranges from 0.03 if the cylinder surface is highly polished and the section is far away from a tip, to 0.13 if the surface is not polished. Near a tip of a structure a large variation of lift may occur. Thus a rocket with a spherical nose may have 4 or 5 times as much lift as one with a conical nose. The unsteady component of the drag coefficient at such large Reynolds number has a root-mean-square value of the order of 0.03. At Reynolds number greater than 3×10^6, periodicity reappears in the wake and the mean drag coefficient rises again to 0.7 or 0.8.

2.4 THE OSCILLATION OF CYLINDRICAL STRUCTURES IN A FLOW

In the preceding section, it is seen that asymmetry develops itself in a flow around a circular cylinder if the Reynolds number is sufficiently

large. A lift force, perpendicular to the direction of flow, is created by
the unsymmetric flow. Should the circular cylinder be stationary, there
will be no exchange of energy between the cylinder and the flow. But, if
the cylinder oscillates in phase with the lift force, the fluid will do work
on the cylinder through the lift force, and energy will be extracted from
the flow and imparted to the cylinder. The oscillation of the cylinder can
then be built up.

Let us consider a two-dimensional cylinder of unit length supported by
a spring and a dashpot (Fig. 2.5). Let the spring constant be K, the mass

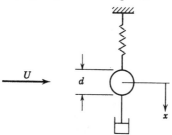

Fig. 2.5. Oscillation of a cylinder.

per unit length m, and the ratio of the damping constant to the critical
damping γ.* If the cylinder oscillates in still air, the natural frequency
ω_0 (radians per second) will be given approximately by the equation

$$\omega_0{}^2 = \frac{K}{m}$$

The equation of motion of the cylinder in a flow is then

$$\frac{d^2x}{dt^2} + 2\gamma\omega_0 \frac{dx}{dt} + \omega_0{}^2x = \frac{1}{m} F(t) \qquad (1)$$

where $F(t)$ is the aerodynamic force per unit length acting on the cylinder,
and x is the transverse displacement (perpendicular to the direction of
flow) of the cylinder. Now $F(t)$ consists of the lift force acting on the

* For a free-vibration system governed by the equation

$$m\ddot{x} + \beta\dot{x} + Kx = 0 \qquad (\beta \geq 0)$$

the motion will be periodic if $\beta < \beta_{cr}$, and aperiodic if $\beta > \beta_{cr}$, where the *critical
damping* β_{cr} is given by

$$\beta_{cr} = 2\sqrt{mK} = 2m\omega_0$$

$\omega_0 = \sqrt{K/m}$ being the frequency (radians per second) of the system without damping.
It is often convenient to write β as $\gamma\beta_{cr}$. Then, if we divide through by m, the
equation of motion can be written as

$$\ddot{x} + 2\gamma\omega_0\dot{x} + \omega_0{}^2x = 0$$

cylinder due to shedding of vortices and the *apparent mass* force of the air surrounding the cylinder. The latter is equal to $\rho \dfrac{\pi d^2}{4} \dfrac{d^2x}{dt^2}$ where ρ is the density of the air. Let us assume that the apparent mass of air is negligible in comparison with the mass of the cylinder m. (Otherwise we may simply modify m to be $m' = m + \rho\pi d^2/4$.) $F(t)$ is then regarded as the lift force alone. (The drag has no effect on the transverse oscillation.) By dimensional analysis, $F(t)$ can be written as

$$F(t) = \tfrac{1}{2}\rho U^2 C_L d \quad \text{(per unit length of cylinder)} \tag{2}$$

Now the lift coefficient C_L excited by the shedding of vortices can be written in the complex form

$$C_L = C_{L0}e^{i\omega t} \tag{3}$$

where ω is the frequency (radians per second) of the vortices on each side of the cylinder.* ω is related to the Strouhal number by

$$\omega = Uk/d \tag{4}$$

Hence, we may write

$$\frac{d^2x}{dt^2} + 2\gamma\omega_0 \frac{dx}{dt} + \omega_0{}^2 x = \frac{1}{2m}\rho U^2 \, dC_{L0} \, e^{i\omega t} \tag{5}$$

The steady-state solution of this equation is, according to § 1.8,

$$x(t) = Ae^{i(\omega t - \psi)} \tag{6}$$

where

$$A = \frac{\rho \, dC_{L0}}{2m\omega_0{}^2} \frac{U^2}{[(1 - \Omega^2)^2 + (2\gamma\Omega)^2]^{1/2}}$$

$$\Omega = \frac{\omega}{\omega_0} = \frac{\text{forcing frequency}}{\text{undamped natural frequency}} \tag{7}$$

$$\psi = \text{arc tan}\left(\frac{2\gamma\Omega}{1 - \Omega^2}\right)$$

The amplitude of the response $x(t)$ is given by A, to which the stress in the spring is proportional. Now, by Eqs. 4 and 7,

$$\Omega = \frac{U}{\omega_0 d}k \tag{8}$$

---

* We shall assume that the Reynolds number is in the range of vortex shedding with a clearly defined predominant frequency, $40 < R < 3 \times 10^5$.

But k is a function of the Reynolds number R:

$$R = \frac{Ud}{\nu}$$

Hence the dependence of A on U is rather complicated.

Let us consider a numerical example of a pipe line with $d = 30$ in. and $\omega_0 = 2.5$ cycles per second. Then $R = 16 \times 10^3 U$. Using the data given by Relf and Simmons (Fig. 2.4), we obtain the amplitude response

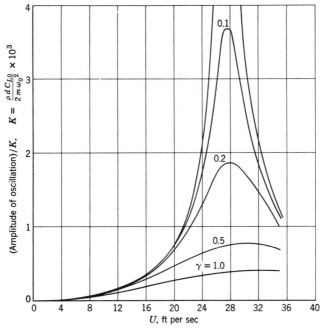

Fig. 2.6. The amplitude of oscillation as a function of wind speed. ($d = 30$ in., $\omega_0 = 2.5$ cycles per second).

A as a function of U and γ by substituting Ω into Eq. 7. The result is shown in Fig. 2.6. It is seen that maximum amplitude is reached when U is approximately 28 ft per sec or 19 mph while the frequency ratio Ω is approximately 1.

The calculated amplitude response near and beyond the wind speed at which the maximum response occurs is somewhat doubtful, because in that Reynolds number range the flow is turbulent, and the power spectrum of the wake is no longer a sharp line. In other words, the wake frequency is no longer sharply defined. The flow and the lift force are stochastic processes and must be analyzed accordingly. The real response curves are probably flatter than those given by Fig. 2.6.

The value of C_{L0} is of the order of 0.63 if the Reynolds number of flow lies between 40 and 3×10^5.

The above calculation is based on the experimental values of k obtained on a *stationary* cylinder. Hence, the calculation is valid only when the amplitude of oscillation is infinitesimal. If the amplitude is finite, additional aerodynamic force associated with the shifting of the points of separation on the cylinder during the oscillation will become important. Such additional lift is, in general, a nonlinear function of the amplitude. An example of how the motion of the structure may affect the shedding of vortices and the characteristics of aeroelastic oscillations will be given in the next section in connection with the H-shaped sections.

Resonance Condition. The example given above shows that the maximum amplitude of oscillation is reached in the neighborhood of $\Omega = 1$; i.e., when the frequency of shedding vortices agrees with the natural frequency of the structure. This gives an easy rule for estimating the character of the structure as follows: Let us define a Strouhal number of the structure:

$$k_{\text{stru}} = \frac{\omega_0 d}{U_{\max}} \tag{9}$$

where ω_0 is the fundamental natural frequency (radians per second) of the structure in still air, d is the characteristic length (here the diameter of the cylinder), and $U_{\max}$ the expected maximum speed of flow (i.e., the highest wind speed). Determine the maximum Reynolds number $R = U_{\max} d/\nu$. From Fig. 2.4 find the corresponding k of the shedding vortices. Let this k be denoted by k_{cr}; then large amplitude resonant oscillation will not occur if k_{stru} is greater than k_{cr} by a sufficiently large margin.

Applications to Three-Dimensional Structures. The two-dimensional model treated above may be regarded as a typical section of a three-dimensional structure. For an example of an overground pipe line, m may be taken as the mass per unit length at the center span, ω_0 the fundamental natural frequency of the pipe line, and γ the ratio of the actual damping to the critical damping when the pipe line oscillates in the fundamental mode. For a smokestack, the typical section may be taken at three fourths of its length above ground.

Generalization of the analysis to three-dimensional structures can be made without difficulty. The simplest approach is to use generalized coordinates and Lagrange's equations. However, the effect of a free tip on the fluctuating aerodynamic force is both profound and difficult to predict. Rash calculations without due account of the tip-effect can be dangerous.

2.5 THE H-SHAPED SECTIONS

A series of interesting experiments on oscillating H-shaped sections was performed by von Kármán and Dunn.[2.21, 2.22] (The original Tacoma Bridge section is H-shaped.)   When an H-shaped section is suspended in a flow, vortices are shed from the leading and trailing edges.   Three types of self-excited oscillations are of interest: (1) vertical, (2) torsional, (3) coupled vertical and torsional.   For the first type quantitative results are available.   Some salient features will be described below, although, because of the limitation of space, the details of the experiments cannot be presented.   Attention should be directed to the qualitative features that reveal the effects of intrinsic nonlinearity of a separated flow.   A satisfactory theory has not yet been advanced.   Nor have the experiments been comprehensive enough to permit sweeping generalizations.   The problem is, at present, one of the most challenging and worthy fields of research.

Vertical Oscillations.   It appears that the vertical oscillation is determined by the following conditions:

1. While the structure is at rest, the vortex frequency is controlled by the wind.

2. At certain discrete wind speeds—which we designate as the "critical speeds"—the vortex frequency will either coincide with or be a multiple of one of the frequencies of motion of the structure.   Such coincidence results in self-excited oscillations.

3. Beyond the critical wind speeds, the oscillating structure and not the wind speed controls the vortex frequency.   In such instances the oscillation extends over certain finite ranges of wind speed.   The lower limit of each range is a critical speed.   The upper limit is not so well defined. However, between the upper limit and the next critical speed the structure is practically at rest.

These statements are based on wind-tunnel experiments on an H-shaped section model suspended by coil springs[2.21] (see Fig. 2.7).   In these experiments vortex-frequency measurements were made with a hot-wire anemometer, both when the model was held stationary and when it was allowed to oscillate.*

A typical result is shown in Fig. 2.8, which indicates that, as the vertical motion starts, the vortex frequency has a well-defined constant value; in other words, here the model controls the frequency.   It appears that this constant vortex frequency exists only over a limited range of wind speed;

---

* The hot-wire was placed immediately behind the upstream girder (about one inch aft and slightly above the upper edge of the girder.   The dimensions of the model are shown in Fig. 2.7.   The Reynolds number based on the chord length is of order 10^5.

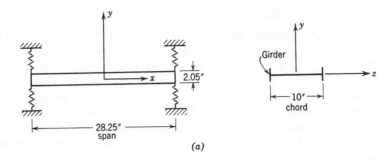

(a)

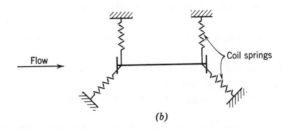

(b)

Fig. 2.7. Experimental arrangements in Dunn's tests.

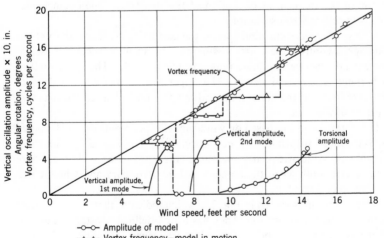

Fig. 2.8. Vortex frequency vs. wind speed. (Courtesy of Dr. L. Dunn of the Ramo-Wooldridge Corp. Formerly of the California Institute of Technology.)

i.e., as the wind speed is increased, a point is reached at which the model is no longer capable of controlling the vortex frequency, and the wind again becomes the controlling factor. Similarly, in torsional oscillations the vortex frequency also becomes constant as motion starts; but in this instance it shifts suddenly to a higher value as the wind speed exceeds certain limits.

Figure 2.9 shows amplitude-response curves for a different set of suspending springs, supporting a similar model in an arrangement as shown in Fig. 2.7. Here n_0 denotes the frequency in which the model moves vertically as a single unit, and n_1, n_2 denote those of rotations about the z and x axes, respectively (see Fig. 2.7a). The figure shows that, when the wind speed is less than 6 ft per sec, no motion of the model is noticeable. As the wind speed U is increased gradually beyond 6 ft per sec, the

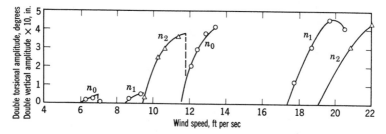

Fig. 2.9. Amplitude response of a model with three modes of oscillation
(coil springs supported). (Courtesy of Dr. L. Dunn.)

amplitude of the vertical motion gradually increases, while the model oscillates at the frequency n_0. At about 7 ft per sec, however, the vertical motion stops. As U is increased to 8.5, rotational oscillation about the z axis with frequency n_1 starts and gradually increases its amplitude as U increases. Then it is shifted to rotational oscillation about the x axis at frequency n_2 at $U = 9.4$. This torsional oscillation seems to increase indefinitely with U but may be stopped by external damping. At $U = 12$, vertical oscillation with frequency n_0 appears again, which is followed in succession by rotational oscillations about the z and x axes, as shown in the figure. Thus it indicates that, for a given frequency, there are two critical speeds. With increasing wind speed various modes of oscillation appear and disappear in succession. Then the cycle repeats as the wind speed is increased further.

The wind speeds at which the amplitude-response curves first intersect the abscissa in Fig. 2.8 or 2.9 may be called "critical wind speeds." As functions of frequency the critical speeds are shown in Fig. 2.10. It is seen that a linear relationship exists for each of the two values of the critical speed. Furthermore, the first critical speed is just half of the second. The critical wind speed divided by the product of frequency and chord length (proportional to the slope of the straight lines in Fig. 2.10) is the critical reduced speed. Its inverse is the critical reduced frequency. The higher of the two critical reduced frequencies computed from Fig. 2.10 agrees with that of the vortex shedding, which is a function of the geometry of the body, and is independent of the elasticity of the supporting structure. The fact that there exist two distinct values of critical reduced

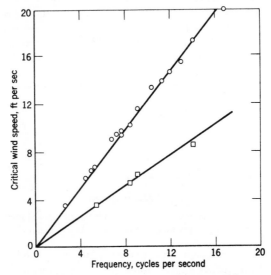

Fig. 2.10. The critical wind speeds as a function of frequency for vertical oscillations. (Courtesy of Dr. L. Dunn.)

speeds and that they differ by a multiple of 2 indicates that the oscillations are of the nature of a subharmonic resonance.

Torsional Oscillations. The most important feature of the self-excited torsional oscillations (about the x axis in Fig. 2.7) of an elastically supported H-shaped section is the possibility of being "catastropic." That is, at wind speeds in excess of certain critical value, the amplitude of oscillation can become very large.

Available data, however, are not sufficiently comprehensive to explain the basic phenomena.

Application to Suspension Bridges. Of all the oscillations that a girder-stiffened suspension bridge (H-shaped section) may suffer, the torsional oscillation is the most dangerous, because, once the critical reduced speed is exceeded, the destabilizing aerodynamic force may become very large.

Since H-shaped sections are susceptible to torsional instability, their use should be avoided whenever possible. However, the critical reduced speed can be raised considerably if the girder depth of the H-shaped section is reduced to a small value.

On the other hand, a truss-stiffened suspension-bridge section behaves more or less like a flat plate, and torsional instability develops when the angle of attack of the wind to the bridge roadbed is close to the stalling angle of the bridge section (or order 7 to 10°), whereas at smaller angles of attack, flutter of the classical type, involving coupled vertical and torsional motions, may occur.

Considerable amount of theoretical and experimental data have been obtained by Farquharson, Vincent, Bleich, and others.[2.12–2.16]

2.6   PREVENTION OF AEROELASTIC INSTABILITIES

Aeroelastic oscillations cannot always be prevented. But it is desirable to limit the amplitude of oscillation within a safe bound for the entire range of flow speeds. For a cylindrical body, the resonance oscillation between the structure and the shedding vortices can be avoided by providing the structure with sufficient *stiffness*, so that the natural frequency of the structure is much higher than the frequency of the vortices. For an unstable aerodynamic section, the only way of preventing large amplitude oscillations is to provide sufficient *damping* in the system. For a suspension bridge, instability occurs when the reduced frequency of the bridge falls below a critical value, and can be stabilized by raising its natural vibration frequency. Stiffening and damping are the two most useful methods in controlling aeroelastic oscillations. Dampers, however, must be used with care, for they can be destabilizing. Cf. § 6.11, p. 242.

One of the most important factors in aeroelasticity is the *geometrical shape* of the structure. In a civil engineering structure, aerodynamic forces being undesirable (unlike in airplane structures), an ideal section is one that produces no lift and small drag.

The application of this idea can be illustrated by the design of the Second Tacoma Narrows Bridge.[2.11, 2.12] The new design uses deep open trusses as the stiffening members (instead of plate girders), open trussed floor beams (instead of solid), and streamlined rail sections. Trusses, with small frontal area, are clearly aerodynamically ineffective. Tests in the laboratory for the new design showed complete stability even at higher angles of attack (up to 15°). These tests also revealed that the concrete deck, fitted with open steel grid slots of varying widths between each of the four traffic lanes and at the curb, has remarkable benefit.

On the other hand, one may try to design the structure streamlined so that no separation occurs.   Then, flutter, if any, will be of the "classical type" whose critical speed is higher than that of the stall flutter and can be predicted with good accuracy.

It is generally beneficial for aeroelastic stability to design the structure so as to have the least projected frontal area against wind.   Decreasing the projected area decreases the magnitude of the aerodynamic forces.   This follows from the fact that the aerodynamic forces are proportional to the vorticity strength, which in turn is proportional to the profile drag.   A reduction in projected frontal area reduces the profile drag, and hence reduces the effective aerodynamic force.

BIBLIOGRAPHY

For a description of oscillations of pipe lines, smokestacks, and related problems, see

2.1   Den Hartog, J. P.: Recent Technical Manifestations of von Kármán's Vortex Wake. *Proc. Natl. Acad. Sci.* **40**, 3, 155–157 (1954).
2.2   Housner, G. W.: Bending Vibrations of a Pipe Line Containing Flowing Fluid. *J. Applied Mech.* **19**, 205–208 (June 1952).
2.3   Pagon, W. W.: What Aerodynamics Can Teach the Civil Engineer. *Engg. News-Record*, 348, 411, 456, 814 (1934); 582, 665, 742, 601 (1935).

For the singing and galloping of transmission lines and vanes, see

2.4   Davison, A. E.: Dancing Conductors. *Trans. AIEE* **49**, 1444–1449 (1930).
2.5   Den Hartog, J. P.: Transmission Line Vibration Due to Sleet. *Trans. AIEE* **51**, 1074–1076 (1932). See also his book, *Mechanical Vibrations*, McGraw-Hill, New York (1947).
2.6   Gongwer, C. A.: A Study of Vanes Singing in Water. *J. Applied Mech.* **19**, 432–438 (Dec. 1952).
2.7   Harris, R. G.: Vibrations of Rafwires. *Aeronaut. Research Com. R. & M.* **759** (1921).
2.8   Kerr, W., J. F. Shannon, and R. N. Arnold: The Problem of the Singing Propeller. *Proc. Inst. Mech. Engrs. London* **144**, 54–76 (1940).
2.9   Krall, G.: Forced or Self-Excited Vibration of Wires. *Proc. 7th Intern. Congr. Applied Mech.* **4**, 221–225 (1948).

For descriptions of aeroelastic failures of suspension bridges, see

2.10   Ammann, O. H., Th. von Kármán, and G. B. Woodruff: The Failure of the Tacoma Narrows Bridge. A report to the Administrator, Federal Works Agency (Mar. 28, 1941).   Reprinted as *Texas Engg. Exptl. Sta. Bull.* **78** (1944). Texas A & M, College Station, Texas.
2.11   Andrew, C. E.: Unusual Design Problems—Second Tacoma Narrows Bridge. *Proc. ASCE* **73**, 10, 1483–1497 (1947).
2.12   Farquharson, F. B. et al.: Aerodynamic Stability of Suspension Bridges. *Univ. Wash. Engg. Expt. Sta. Bull.* **116**. A series of reports issued since 1949. *Part I.* Investigations Prior to October, 1944 (by Farquharson). *Part II.* Mathematical Analysis of the Natural Modes of Vibration of the Bridge and the Models (by F. C. Smith and G. S. Vincent). *Part III.* Investigation of Models of the Original Tacoma Narrows Bridge (by Farquharson). *Part IV.*

Model Investigations Which Influenced the Design of the New Tacoma Narrows Bridge (by Farquharson). *Part V.* Extended Studies: Logarithmic Decrement, Field Damping, Prototype Predictions, Four Other Bridges (by G. S. Vincent).

2.13 Finch, J. K.: Wind Failures of Suspension Bridge or Evolution and Decay of the Stiffening Truss. *Engg. News-Record*, 74 (Mar. 13, 1941).

2.14 Steinman, D. B.: Design of Bridges against Wind. (*a*) Historical Background, *Civil Engg.* 15, 501–504 (1945); (*b*) Elementary Explanation of Aerodynamic Instability, 558 (1945); (*c*) Wind-Tunnel Tests Reveal Serious Inadequacy of Present Bridge Specifications (Oct. 1947); (*d*) Simple Model Tests Predict Aerodynamic Characteristics of Bridges (Jan., Feb., 1947).

For vibration and flutter analysis of suspension bridges, see

2.15 Bleich, F., C. B. McCullough, R. Rosecrans, and G. S. Vincent: *The Mathematical Theory of Vibration in Suspension Bridges.* U.S. Govt. Printing Office, Washington, D.C. (1950).

2.16 Bleich, F.: Dynamic Instability of Truss-Stiffened Suspension Bridges. *Trans. ASCE* 114, 1177–1232 (1949).

2.17 Pinney, E.: Aerodynamic Oscillations in Suspension Bridges. *Trans. ASME* 70 (1948); *J. Applied Mech.* 151.

2.18 Reissner, H.: Oscillations of Suspension Bridges. *Trans. ASME* 65 (1943); *J. Applied Mech.*, A-23.

2.19 Steinman, D. B.: Aerodynamic Theory of Bridge Oscillations. *Proc. ASCE* 75, 1147–1183 (1949). *Trans. ASCE* 115, 1180–1260 (1950). Rigidity and Aerodynamic Stability of Suspension Bridges. *Trans. ASCE* 110, 439–580 (1945). Suspension Bridges. *Am. Scientist*, 42, 397–438 (July 1954).

2.20 Westergaard, H. M.: On the Method of Complementary Energy. *Proc. ASCE* 67, 199–227 (1941).

The following two unpublished reports are of special interest:

2.21 Dunn, L. G.: A Report on the Aerodynamic Investigation of the Bending Oscillations of the Original Tacoma Narrows Bridge (Apr. 1943).

2.22 Kármán, Th. von, and L. G. Dunn: Preliminary Report on the Aerodynamic Investigations for the Design of the Tacoma Narrows Bridge. Submitted to the Board of Consulting Engineers (May 30, 1942).

Wind-tunnel tests on self-excited oscillations of airfoils are also discussed in Chapter 9. Strouhal number was first introduced in the following paper:

2.23 Strouhal, V.: Über eine besondere Art der Tonerregung. *Wied. Ann. Physik u. Chem.*, Neue Folge, Band V. 216–251 (1878).

The investigation of the stability of vortex patterns is given by von Kármán:

2.24 Kármán, Th. von: Flüssigkeits-u. Luftwiderstand, *Physik. Z.* 13, 49 (1911). Also, *Nachr. Ges. Wiss. Göttingen*, 547 (1913). The investigation is only given in outline in these papers. Detailed presentation is given by Lamb, H.: *Hydrodynamics*, 6th Ed., p. 225. Cambridge, London (1932).

Results of researches on the flow behind a circular cylinder, before 1938, have been reviewed in Ref. 2.25. Some additional references are given below:

2.25 Goldstein, S.: *Modern Development in Fluid Dynamics*, 2 Vols. Oxford Univ. Press, London (1938).

2.26   Kovasznay, L. S. G.: Hot-Wire Investigation of the Wake behind Cylinders at Low Reynolds Numbers. *Proc. Roy. Soc. London, A*, **198**, 174 (1949). See also *NACA Tech. Memo.* **1130** (1947).

2.27   Roshko, A.: On the Development of Turbulent Wakes from Vortex Streets. *NACA Tech. Note* **2913** (1953).

2.28   Townsend, A. A.: Diffusion in the Turbulent Wake of a Cylinder. *Proc. 7th Intern. Congr. Applied Mech.* **2**, 227–248 (1948). Also *Proc. Cambridge Phil. Soc.* **43**, 560 (1947).

2.29   Fung, Y. C.: Fluctuating Lift and Drag Acting on a Cylinder in a Flow at Super-critical Reynolds Numbers. *J. Aerospace Sciences* **27**, 801–814 (1960).

2.30   Schmidt, L. V.: Measurements of Fluctuating Air Loads on a Circular Cylinder. *J. Aircraft* **2**, 49–55 (1965).

Chapter 3

DIVERGENCE OF A LIFTING SURFACE

The central problem in steady-state aeroelasticity is the effect of elastic deformation on the lift distribution over lifting surfaces such as airplane wings and tails.   At lower speeds of flight, the effect of elastic deformation is small.   But at higher speeds of flight, the effect of elastic deformation may become so serious as to cause a wing to be unstable, or to render a control surface ineffective, or even to reverse the sense of control.

In this chapter, we shall treat in detail the problem of wing divergence, which is probably the simplest of all aeroelastic problems.   Many fundamental concepts and methods of solution can be illustrated in this connection.

In § 3.1, the phenomenon of divergence is explained by a two-dimensional example.   In § 3.2, the problem of the divergence of an idealized three-dimensional wing is formulated in a general form.   A "strip" assumption regarding the aerodynamic force is introduced to simplify the governing equations.   The mathematical nature of the problem is then pointed out and illustrated by an example.   Several methods of solution are discussed.   First, in § 3.3, a solution based on a semirigid assumption is given.   Second, in § 3.4, the semirigid assumption is reconsidered in the light of generalized coordinates.   Third, in § 3.5, a process of successive approximation is discussed.   The last method is mathematically more satisfactory; its convergence can be proved, and the relation between the successive approximations can readily be seen.   Finally, in order to provide a practical numerical method of solution, the method of matrices is suggested.   The basic concepts and definitions of matrix calculus is outlined in § 3.6, and the reduction of differential and integral equations into matrix equations according to the method of finite differences is explained in § 3.7.   Some remarks regarding further refinements are made in the last section, § 3.8.

The unfortunate wing failure of Professor S. P. Langley's famous monoplane in 1903, shortly before the Wright brothers' successful flight, could probably be attributed to wing divergence.[3.5]   Monoplane designs in early aeronautical history often had divergence trouble.[3.6]   For modern aircraft, the critical speeds of flight at which divergence sets in are usually higher than those of flutter or other aeroelastic instabilities.

Hence the divergence speed itself is often of minor importance. However, it is a convenient reference quantity for other aeroelastic phenomena; it enters into expressions for the effect of elastic deformation on lift distribution, static and dynamic stability of the airplane, etc. Moreover, since the calculation of the divergence speed is relatively simple, it is generally made in the course of airplane design.

3.1 A TWO-DIMENSIONAL EXAMPLE

Wing divergence is a simple example of the steady-state aeroelastic instability. If a wing in steady flight is accidentally deformed, an aerodynamic moment will generally be induced which tends to twist the wing. This twisting is resisted by elastic moment. However, since the elastic stiffness is independent of the speed of flight, whereas the aerodynamic

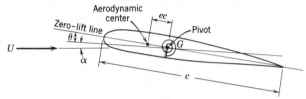

Fig. 3.1. A two-dimensional airfoil.

moment is proportional to the square of the flight speed, there may exist a *critical speed*, at which the elastic stiffness is barely sufficient to hold the wing in a disturbed position. Above such a critical speed, an infinitesimal accidental deformation of the wing will lead to a large angle of twist. This critical speed is called the *divergence speed*, and the wing is then said to be *torsionally divergent*.

As a two-dimensional example, let us consider a strip of unit span of an infinitely long wing of uniform cross section. As shown in Fig. 3.1, let the elastic restraint imposed on this strip be regarded as a torsional spring with an axis at a point G which is fixed in space. The airfoil can rotate only about G. Assume that the spring is linear so that the torque is directly proportional to the angle of twist. Initially the "zero-lift" line of the airfoil coincides with the direction of the undisturbed flow. Let the entire system be first rotated through an angle α as a rigid body; then let the constraint be released to allow the wing to deflect elastically through an additional angle θ. It is desired to find the equilibrium position of the wing (i.e. the angle θ) in a flow of speed U.

The action of the aerodynamic force on the airfoil can be represented by a lift force, acting through the aerodynamic center, and a moment about the same point.   Let us write the distance from the aerodynamic center to the axis of the torsional spring as ec, c being the chord length and the factor e being a ratio expressing the *eccentricity* of the aerodynamic center (positive if the spring lies behind the aerodynamic center).   Now the lift coefficient C_l is proportional to the angle of attack, whereas the coefficient of moment about the aerodynamic center C_{m0} is practically independent of the angle of attack.*   Hence, the lift and moment per unit span acting on the airfoil are, respectively,

$$L' = qC_l c = qca(\theta + \alpha)$$
$$M_0' = qC_{m0}c^2 \quad \text{(about aerodynamic center)} \tag{1}$$

where a is the lift-curve (C_l vs. α) slope, q is the dynamic pressure ($\frac{1}{2}\rho U^2$), α is the initial angle of attack, and θ is the angle of twist.

The aerodynamic moment per unit span about the axis of the spring is therefore

$$M_a' = M_0' + L'ec = qc^2 C_{m0} + qec^2 a(\theta + \alpha)$$
$$= qec^2 a\theta + qec^2 a(\alpha + C_{m0}/ea) \tag{2}$$

In the above equation α is measured from the zero-lift line of the airfoil. If we define a direction corresponding to an angle $-\alpha_0 = C_{m0}/ea$ as the "zero-moment" line, and redefine the angle α as measured from the zero-moment line (i.e., take α as zero when the free stream direction coincides with the zero-moment line), then Eq. 2 becomes

$$M_a' = qec^2 a(\theta + \alpha) \tag{3}$$

When equilibrium prevails, the aerodynamic moment is balanced by the elastic restoring moment.   Let K_α be the spring constant, so that the elastic restoring moment per unit span, corresponding to the twisting angle θ, is $K_\alpha \theta$.   On equating this with the aerodynamic moment given by Eq. 3, and solving for θ, we obtain

$$\theta = \frac{qec^2 a\alpha}{K_\alpha - qec^2 a} \tag{4}$$

For a given nonvanishing α, the angle θ will increase when dynamic pressure q increases.   When q is so large that the denominator tends to

---

* We assume that the total angle of attack remain so small that C_l is linearly proportional to the angle of attack.   This requires that the initial angle of attack do not approach the stalling angle.   If the initial angle of attack is large, the additional elastic twist may cause the wing to stall.   Divergence near stalling angle is an interesting nonlinear problem, which is closely related to stall flutter and buffeting.

zero, the angle θ becomes indefinitely large, and the airfoil is "divergent." Hence, the condition of divergence is

$$K_\alpha - qec^2a = 0 \tag{5}$$

The dynamic pressure at divergence q_{div} and the divergence speed U_{div} are given by the equations

$$q_{\text{div}} = \frac{K_\alpha}{ec^2a} \quad \text{and} \quad U_{\text{div}} = \sqrt{\frac{2K_\alpha}{\rho ec^2a}} \tag{6}$$

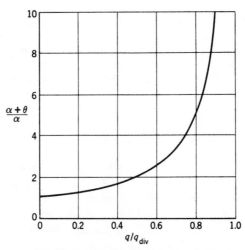

Fig. 3.2.   Ratio of angles of twist of an elastic wing to that of a rigid wing.

If the wing were infinitely rigid, $K_\alpha \to \infty$; then $\theta = 0$. The ratio of the total equilibrium angle of attack $\alpha + \theta$ of an elastic wing to that of a rigid wing is, according to Eqs. 4 and 6,*

$$\frac{\alpha + \theta}{\alpha} = \frac{1}{1 - q/q_{\text{div}}} \tag{7}$$

This ratio is illustrated in Fig. 3.2. It is seen that the deflection of the elastic wing tends to be very large when $q \to q_{\text{div}}$. When the speed of

* Equation 7 may be written as $\dfrac{1}{\theta} = \dfrac{q_{\text{div}}}{\alpha}\left(\dfrac{1}{q} - \dfrac{1}{q_{\text{div}}}\right)$. Hence, a curve showing $1/\theta$ plotted against $1/q$ will be a straight line. The divergence dynamic pressure can be found either from the slope of this line, q_{div}/α, or from its intercept on the $1/q$ axis, $1/q_{\text{div}}$. This relation can be used for an experimental determination of the divergence speed.

flight is 80 per cent of the divergence speed, $q/q_{div} = 0.64$, the wing twist is already 1.8 times the angle of attack of a rigid wing.

We may also take a different point of view as follows: Suppose that the airfoil of Fig. 3.1 be in equilibrium, so that θ is given by Eq. 4. Let us ask whether this equilibrium configuration is stable. An equilibrium configuration is stable if there is a tendency to return to the equilibrium position, should the equilibrium be slightly disturbed. To test the stability of the equilibrium configuration described by Eq. 4, let θ be given an additional infinitesimal displacement $\Delta\theta$. Then the change in the aerodynamic moment is, from Eq. 2,

$$\Delta M_a' = qec^2a\,\Delta\theta \tag{8}$$

and that of the elastic restoring moment is

$$\Delta M_e' = K_\alpha\,\Delta\theta \tag{9}$$

Now, if the change in the elastic restoring moment is larger than that of the aerodynamic moment, the wing will tend to return to its original position. So the equilibrium is stable if

$$\Delta M_e' > \Delta M_a'$$

If we follow the same argument, the wing will be unstable if the inequality sign reverses. The critical condition is then

$$\Delta M_e' = \Delta M_a' \tag{10}$$

i.e.,

$$qec^2a\,\Delta\theta = K_\alpha\,\Delta\theta$$

which gives the same critical speed as Eq. 6.*

Note that $\Delta M_a'$ and $\Delta M_e'$ given by Eqs. 8 and 9 are independent of α and θ. Therefore the critical condition (Eq. 10) can be derived by assuming α and θ to be zero. *Hereafter we shall interpret θ, M_a', M_e' as the changes from the initial equilibrium values*; the symbol Δ will be omitted.

Equation 6 shows that the critical divergence speed increases with increasing rigidity of the spring and decreasing chord length and eccentricity.

3.2   DIVERGENCE OF AN IDEALIZED CANTILEVER WING

Consider an idealized unswept cantilever monoplane wing which has a straight elastic axis normal to a fuselage that is fixed in space, as in a

---

* This shows that, although Eq. 7 gives a finite value of the torsional deflection when the speed of flow exceeds the divergence speed, that equilibrium position is unstable. Note that the equilibrium value of θ when $q > q_{div}$ is greater than α in magnitude and opposite in sign; i.e., the wing twists to opposite direction of α.

wind-tunnel testing.   For such a wing the action of a general distributed load can be considered as produced by a distributed load acting through the elastic axis and a distributed twisting moment about the elastic axis.   The former produces pure bending which does not affect the angle of incidence, whereas the latter produces rotation of the sections about the elastic axis.

The aerodynamic lift acts through the line of aerodynamic centers. The moment about the aerodynamic centers does not change appreciably with the angle of attack and can be neglected.   Let $L'(y)$ be the lift per unit span and ce the eccentricity of the elastic axis from the aerodynamic center (positive when the elastic axis lies behind the latter).   Then the aerodynamic moment about the elastic axis is $L'(y)c(y)e(y)$ per unit span.

The equation of divergence for a wing as shown in Fig. 3.3 can now be

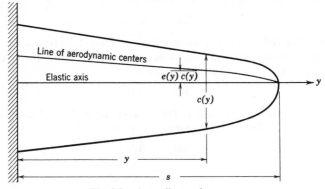

Fig. 3.3.   A cantilever wing.

derived.   The fuselage of the airplane is assumed fixed in space.   At the critical divergence speed, the aerodynamic moment about the elastic axis just balances the elastic moment due to twisting.   For the first approximation, let us make the *strip assumption* that the aerodynamic lift on a chordwise strip, of width dy, due to a change of angle of attack $\theta(y)$, is given by

$$L'(y)\,dy = q\,c(y)\,dy \cdot a\,\theta(y)$$

where a is the lift-curve slope corrected for aspect ratio.*   The aerodynamic moment about the elastic axis is therefore

$$M'(y)\,dy = qa\,e(y)\,c^2(y)\,\theta(y)\,dy \qquad (1)$$

If $G(x, y)$ represent the influence function for wing rotation at x about the

---

* To account for the finite span effect, a should be corrected for the aspect ratio, as given by Eq. 15 of § 1.5 et seq.; see also § 4.3.   With this correction, the resulting q_{div} is quite accurate.   Cf. Miles.[3.12]

elastic axis due to a unit couple at y, then, the wing being in static equilibrium, the total angle of rotation $\theta(x)$ at x is

$$\theta(x) = \int_0^s G(x, y)\, M'(y)\, dy \tag{2}$$

where s is the semispan of the wing.   According to the strip assumption, Eq. 2 becomes

$$\theta(x) = q_{\text{div}} a \int_0^s G(x, y)\, e(y)\, c^2(y)\, \theta(y)\, dy \tag{3}$$

Note that Eq. 3 is satisfied only at the divergence speed, because at other speeds the wing will not be in static equilibrium if it is displaced from an equilibrium configuration, and inertia forces must be considered.   This fact is indicated by writing q_{div} instead of q in the equation above.

Equation 3 may be replaced by a differential equation.   Since it is shown in § 1.4 that, for a simple cantilever beam,

$$G(x, y) = \begin{cases} \displaystyle\int_0^y \frac{d\xi}{GJ(\xi)} & \text{for} \quad (0 \leqslant y \leqslant x \leqslant s) \\[12pt] \displaystyle\int_0^x \frac{d\xi}{GJ(\xi)} & \text{for} \quad (s \geqslant y \geqslant x \geqslant 0) \end{cases} \tag{4}$$

so

$$\frac{\partial G(x, y)}{\partial x} = \begin{cases} 0 & \text{for} \quad (0 \leqslant y < x \leqslant s) \\[12pt] \dfrac{1}{GJ(x)} & \text{for} \quad (s \geqslant y \geqslant x \geqslant 0) \end{cases} \tag{5}$$

Differentiating Eq. 3 with respect to x and substituting Eq. 5, we obtain

$$\frac{d\theta(x)}{dx} = \frac{q_{\text{div}} a}{GJ(x)} \int_x^s e(y)\, c^2(y)\, \theta(y)\, dy \tag{6}$$

Multiplying both sides of Eq. 6 by $GJ(x)$ and differentiating once, we obtain

$$\frac{d}{dx}\left[GJ(x)\, \frac{d\theta(x)}{dx} \right] = -\, q_{\text{div}}\, a\, e(x)\, c^2(x)\, \theta(x) \tag{7}$$

The boundary conditions for Eq. 7 must correspond with Eq. 4; these are

$$\theta(x) = 0 \quad \text{when} \quad x = 0$$

$$\frac{d\theta(x)}{dx} = 0 \quad \text{when} \quad x = s \tag{8}$$

Equation 3 or Eqs. 7 and 8 are the fundamental equations for divergence of the idealized wing specified above.   The simplification introduced by choosing the elastic axis as a reference line for measuring the moment and twisting of the wing should be noted. If an arbitrary line is chosen as a reference line, and the aerodynamic action is resolved into a lift force acting through this reference line and a moment about it, the change of angle of incidence of the wing would have to be expressed by two influence functions: one giving the rotation at x due to a unit force at y acting on the reference line, and the other giving the rotation at x due to a unit moment at y about the reference line. The final result is equivalent to the above equations. Therefore, whenever an elastic axis can be defined and is a straight line, its use as a reference line simplifies the derivation. In general, however, for aircraft structures an elastic axis may not be easily defined because of the effect of sweep angle, cutout, or restrained warping; then the use of an arbitrary reference line together with the corresponding influence functions may become imperative. These influence functions may be computed by more advanced structural theory or by experiments.

Note further that, if the wing aspect ratio is small, or if the deformation in the root region of a swept or a delta wing is aerodynamically important, the chordwise sections may not be considered as rigid. Then it is no longer pertinent to speak of the "rotation" of the wing sections. The deformation would have to be specified by influence functions describing the deflection *surface* of the wing due to a unit force or moment, and the aerodynamic forces would have to be computed by the "lifting-surface" theory. In such complicated cases the formulation of the problem in the form of integral equations in terms of influence functions is still straightforward, whereas a formulation in the form of differential equations may become very cumbersome. These more general cases will be discussed later.

The assumption of a fixed fuselage may be removed by allowing proper rigid-body rotation of the entire airplane. Such a refinement generally causes only a slight correction on the divergence speed.

Return now to the idealized case, and consider Eq. 3. An obvious solution is the trivial one: $\theta = 0$. Then the wing is not twisted at all. In general, this is the unique continuous solution.* However, for particular discrete values of the parameter q, there may exist some nontrivial solutions. Such particular values of q are called the *eigenvalues* (or *characteristic values*) of the integral equation. The corresponding solutions are called the *eigenfunctions* (or *characteristic functions*). Since Eq. 3 is homogeneous in $\theta(y)$, the absolute magnitude of the eigenfunction is undefined. If $\theta(y)$ is a solution of Eq. 3, $k\theta(y)$ is another solution, where

---

* The requirement that the solution $\theta(y)$ should be continuous in the range $0 \leqslant y \leqslant s$ is imposed by physical reasons.

k is any constant. $\theta(y)$ is said to be *normalized* when some definite rule is specified so that its absolute value can be determined.

The physical meaning of the existence of eigenfunctions is that wing torsional divergence becomes possible when the dynamic pressure reaches certain critical values. The smallest positive eigenvalue is the critical divergence pressure which has engineering significance. For q less than the smallest eigenvalue, a disturbed wing is stable, whereas, for q greater than that eigenvalue, a disturbed wing is unstable (cf. §§ 6.3–5). Since $q = \frac{1}{2}\rho U^2$ is a nonnegative quantity, a negative eigenvalue, if it exists, has no physical significance. If all eigenvalues are negative, the wing is stable, and neutral equilibrium in a disturbed position is impossible.

The solution for a rectangular wing can be obtained easily, provided that the torsional stiffness GJ, the chord length c, and the eccentricity e are constant across the span. In this particular case Eq. 7 becomes

$$\frac{d^2\theta}{dx^2} = -\mu^2\theta \tag{9}$$

where

$$\mu^2 = \frac{1}{GJ}aec^2 q_{\text{div}} \tag{10}$$

The general solution is

$$\theta = A \sin \mu x + B \cos \mu x \tag{11}$$

where A and B are arbitrary constants which must be determined from the boundary conditions (Eqs. 8). The first condition requires that

$$B = 0 \tag{12}$$

and the second condition requires that

$$A\mu \cos \mu s = 0 \tag{13}$$

Two obvious solutions of Eq. 13 are $A = 0$ and $\mu = 0$. But both lead to the trivial solution $\theta(x) = 0$. The nontrivial solutions are, therefore, possible only when μ assumes the special values for which

$$\cos \mu s = 0 \tag{14}$$

i.e., when

$$\mu s = \pm (2n + 1)\frac{\pi}{2} \qquad (n = 0, 1, 2, \cdots) \tag{15}$$

The μ given by Eq. 15 are the eigenvalues of the differential equation 9 with the boundary conditions 8. The corresponding eigenfunction $A \sin \mu x$ can be normalized by taking $A = 1$.

Hence, by Eqs. 10 and 15,

$$q_{\text{div}} = (2n + 1)^2\pi^2 \frac{GJ}{4aec^2s^2} \qquad (n = 0, 1, 2, \cdots) \qquad (16)$$

the lowest critical value q_{div} is given by $n = 0$:

$$q_{\text{div}} = \frac{\pi^2 GJ}{4aec^2s^2} \qquad (17)$$

The lowest critical speed U_{div} is

$$U_{\text{div}} = \frac{\pi}{cs} \sqrt{\frac{GJ}{2\rho ae}} \qquad (18)$$

The corresponding "fundamental" *divergence mode* is

$$\theta(x) = \theta_0 \sin \frac{\pi x}{2s} \qquad (19)$$

3.3   A SOLUTION BASED ON THE SEMIRIGID ASSUMPTION

The solution of the example in the last section is simple because of the extremely simple geometry of the wing. In the general case, the calculations will become more complicated. Therefore, let us ask whether a simple approximation can be made to determine the critical-divergence speed.

If, at the critical speed, the distribution of angle of twist along the span is known approximately, an estimation of q_{div} can be made easily. A theory in which one assumes a definite deformation pattern that is "invariable" with respect to the load distribution is called a *semirigid* theory.

For a semirigid wing, the deformation $\theta(y)$ can be written as

$$\theta(y) = \theta_0 f(y) \qquad (1)$$

where $f(y)$ is a specified function. The only quantity that varies with the loading is the amplitude θ_0.

With the semirigidity assumption it is possible to treat a three-dimensional wing as a two-dimensional one at a single reference section. This possibility can be based on the *influence coefficients* as follows. Let a_{12} be the influence coefficient of rotation at section 1 due to a unit torque at section 2. Then the angle of rotation at section 1 due to a moment M_2 acting at 2 is

$$\theta_{12} = a_{12}M_2$$

Now the same angle θ_{12} can be produced by a moment M_1 acting at the section 1, if M_1 is determined from the equation

$$\theta_{12} = a_{11}M_1$$

where a_{11} is the influence coefficient of angle of rotation at 1 due to a unit moment at the same section.   Comparing the above equations, we see that

$$M_1 = \frac{a_{12}}{a_{11}} M_2 \tag{2}$$

When a distributed twisting moment acts on a wing, the total rotation at section 1 can be obtained by replacing the moments acting at other sections by fictitious moments acting at section 1, according to the rule given by Eq. 2.

By applying this result to the idealized cantilever wing of § 3.2 (Fig. 3.3), it is seen that the angle of rotation produced at a *reference section* $y = r$ by the moment $L'ce\,dy$, acting on an element dy located at y, is equal to the angle caused by a moment

$$\frac{a_{ry}}{a_{rr}} L'ce\,dy$$

acting at the reference section itself.   The total angle of rotation of the reference section can then be obtained by computing the angle of rotation caused by a fictitious concentrated couple of magnitude

$$\int_0^s \frac{a_{ry}}{a_{rr}} L'ce\,dy$$

acting at the reference section.

According to Maxwell's reciprocal theorem, $a_{ry} = a_{yr}$.   But a_{yr}/a_{rr} is precisely the ratio of the angle of rotation at y to that at r due to a unit couple acting at r.   This is known for a semirigid wing, since $a_{yr}/a_{rr} = f(y)/f(r)$, and the function $f(y)$ (cf. Eq. 1) is assumed to be known and invariant.   Hence, as far as the rotation of the reference section is concerned, the distributed load over the entire wing can be replaced by a single couple acting at the reference section, of magnitude

$$\int_0^s \frac{f(y)}{f(r)} L'(y)\, c(y)\, e(y)\, dy \tag{3}$$

By strip assumption, writing $L'(y) = qca\theta$, we obtain the representative concentrated couple at the reference section:

$$M_a = q\theta_0 a \int_0^s \frac{f^2(y)}{f(r)}\, c^2(y)\, e(y)\, dy \tag{4}$$

At the critical-divergence condition, this couple is balanced by the elastic restoring moment.   Let K be the *stiffness* constant of the wing defined by the ratio

$$K = \frac{\text{moment acting at the reference section}}{\text{angle of rotation at that section}} \tag{5}$$

According to Ex. 4, § 1.5, we have, for a cantilever wing,

$$\frac{1}{K} = \int_0^r \frac{1}{G J(y)} \, dy$$

The elastic moment corresponding to the angle $\theta(r) = \theta_0 f(r)$ at the reference section is

$$M_e = K\theta_0 f(r) \tag{6}$$

Hence the critical condition $M_a = M_e$ yields the following equation for the determination of q_{div}:

$$K = q_{\text{div}} a \int_0^s c^2(y) \, e(y) \frac{f^2(y)}{f^2(r)} \, dy \tag{7}$$

Consider again the rectangular wing for which c, a, and e are constant along the span.   Let the reference section be taken at the wing tip where $y = s$, and let the semirigid mode of the wing be defined by the function

$$f(y) = \sin \frac{\pi y}{2s} \tag{8}$$

Then, from Eq. 7,

$$K = q_{\text{div}} c^2 ae \int_0^s \sin^2 \frac{\pi y}{2s} \, dy = \frac{1}{2} q_{\text{div}} c^2 aes$$

The stiffness K in this particular case is

$$K = \frac{GJ}{s} \tag{9}$$

Hence,

$$q_{\text{div}} = \frac{2GJ}{aec^2 s^2} \tag{10}$$

Note that the q_{div} so obtained does not agree with the exact solution, even though the assumed mode $f(y)$ is exact.   This is caused by the incompatibility between the assumed mode (Eq. 8) and the deflection mode of the wing under a concentrated torque at the reference section, from which K is derived.

On the other hand, if the semirigid mode of the wing is taken as*

$$f = \frac{y}{s} \tag{11}$$

and the reference section is again taken at the wing tip, then Eq. 7 gives

$$K = q_{\text{div}} a \int_0^s c^2 e \left(\frac{y}{s}\right)^2 dy$$

Hence we obtain, for a rectangular wing,

$$q_{\text{div}} = \frac{3GJ}{aec^2s^2} \tag{12}$$

The exact formula (Eq. 17) of § 3.2 gives

$$q_{\text{div}} = \frac{\pi^2 GJ}{4aec^2s^2} \tag{13}$$

The dynamical pressures found by these three methods are in the ratios $2 : \frac{\pi^2}{4} : 3$. The fundamental divergence speeds are approximately in the ratios $0.9 : 1 : 1.1$. The divergence speed found by the semirigid assumption Eq. 8 is about 10 per cent too low, whereas that found by the assumption Eq. 11 is about 10 per cent too high.

The stiffness constant K may be measured in practice by applying a torque on the wing at the reference section and measuring the angle of rotation. It is perhaps needless to remark that the deflection pattern obtained in this experiment (a concentrated torque) may be different from the assumed divergence mode.

3.4   SOLUTION BY GENERALIZED COORDINATES

An alternative procedure based on the semirigid assumption is to apply the concept of generalized coordinates. This point of view provides a new ground for refinements in the calculation.

Let us again consider the idealized problem of § 3.2. Assume that the angle of twist along the span is a known function of y with unspecified amplitude. Let us write

$$\theta(y) = \theta_0 f(y) \tag{1}$$

---

* For a uniform beam, the angle of twist due to a concentrated torque applied at the tip (reference) section varies linearly along the span. So the assumption (Eq. 11) is consistent with the influence coefficients, and with the stiffness constant K given by Eq. 9.

where θ_0 is an undetermined constant and $f(y)$ is a given function. We shall consider θ_0 as the generalized displacement of the wing. The strain energy due to torsion is

$$V = \frac{1}{2} \int_0^s GJ \left(\frac{d\theta}{dy}\right)^2 dy = \frac{\theta_0^2}{2} \int_0^s GJ \left(\frac{df}{dy}\right)^2 dy \qquad (2)$$

The equation of equilibrium may be obtained from Lagrange's equation

$$\frac{d}{dt} \left(\frac{\partial T}{\partial \dot{\theta_0}}\right) + \frac{\partial V}{\partial \theta_0} = Q \qquad (3)$$

where the generalized force Q is to be found as follows: The aerodynamic twisting moment about the elastic axis, per unit span, is, according to the *strip assumption*,

$$M_a' = qc^2 ae\theta = qc^2 ae\theta_0 f(y)$$

By permitting a variation of the amplitude $\delta\theta_0$, the corresponding virtual work is

$$Q\,\delta\theta_0 = \int_0^s M_a'(y)\,\delta\,\theta(y)\,dy = \int_0^s qc^2 ae\theta_0 f(y)\,\delta\theta_0 f(y)\,dy$$

Hence,

$$Q = q\theta_0 \int_0^s c^2 ae\, f^2(y)\,dy \qquad (4)$$

At the critical-divergence speed, the wing is neutrally stable. A disturbed configuration can be maintained without causing any motion. Hence the kinetic energy T vanishes, and Eq. 3 implies

$$\int_0^s GJ \left(\frac{df}{dy}\right)^2 dy = q_{\text{div}} \int_0^s c^2 ae\, f^2(y)\,dy \qquad (5)$$

Therefore,

$$q_{\text{div}} = \frac{\displaystyle\int_0^s GJ \left(\frac{df}{dy}\right)^2 dy}{\displaystyle\int_0^s c^2 ae\, f^2(y)\,dy} \qquad (6)$$

Example 1.   Let GJ, c, a, e be constants, and assume $f(y) = y/s$. Then

$$\int_0^s \left(\frac{df}{dy}\right)^2 dy = \frac{1}{s}$$

$$\int_0^s f^2(y)\,dy = \frac{1}{s^2}\frac{y^3}{3}\bigg|_0^s = \frac{s}{3}$$

Hence,

$$q_{\text{div}} = \frac{3GJ}{c^2 a e s^2}$$

which agrees with the result in § 3.3.

Example 2.   Let GJ, c, a, e be constants, and assume

$$f(y) = \sin\frac{\pi y}{2s}$$

Then

$$\int_0^s \left(\frac{df}{dy}\right)^2 dy = \int_0^s \frac{\pi^2}{4s^2}\cos^2\frac{\pi y}{2s}\,dy = \frac{\pi^2}{4s^2}\frac{s}{2} = \frac{\pi^2}{8s}$$

$$\int_0^s f^2(y)\,dy = \int_0^s \sin^2\frac{\pi y}{2s}\,dy = \frac{s}{2}$$

Hence,

$$q_{\text{div}} = \frac{\pi^2 GJ}{4c^2 a e s^2}$$

which agrees with the exact result.   In the present approach, a correct assumption of the semirigid mode yields an exact solution.

The procedure can be generalized as follows: Let

$$\theta(y) = \sum_{i=0}^n \theta_i f_i(y) \tag{7}$$

where $f_i(y)$ $(i = 0, 1, 2, \cdots, n)$ are known functions, and θ_i are generalized coordinates.   Then

$$V = \frac{1}{2}\int_0^s GJ\left(\sum_{i=0}^n \theta_i \frac{df_i}{dy}\right)^2 dy \tag{8}$$

$$Q_i = \int_0^s M_a'(y) f_i(y)\,dy \qquad (i = 0, 1, 2, \cdots, n)$$

At the critical-divergence speed,

$$\frac{\partial V}{\partial \theta_i} = Q_i \qquad (i = 0, 1, 2, \cdots, n) \tag{9}$$

Equation 9 gives a set of homogeneous linear simultaneous equations in θ_i.   Since the θ_i are, by assumption, not all zero, the determinant of the coefficients of these equations must vanish.   This determinantal equation involves the dynamic pressure q_{div}. Its real positive roots are the critical values.

In calculating the aerodynamic moment $M_a{}'$ in Eq. 8, a theory may be selected, such as Prandtl's lifting-line theory or Weissinger's theory. If we use the strip assumption, then

$$M_a{}' = q_{\text{div}}c^2ae \sum_{i=0}^{n} \theta_i f_i(y) \tag{10}$$

Finally, let us remark that, if the functions $f_i(y)$ $(i = 0, 1, 2, \cdots)$ are infinite in number and form a complete set,* then in the limit $n \to \infty$ an exact solution can be obtained.†

Example. Let GJ, e, c, a be constants across the span, and assume

$$\theta(y) = \theta_1 \frac{y}{s} + \theta_2 \frac{y^2}{s^2}$$

Then

$$V = \frac{GJ}{2}\int_0^s \left(\frac{\theta_1}{s} + \frac{2y\theta_2}{s^2}\right)^2 dy = \frac{GJ}{2s}\int_0^1 (\theta_1 + 2\eta\theta_2)^2 d\eta$$

$$= \frac{GJ}{2s}\left(\theta_1{}^2 + 2\theta_1\theta_2 + \frac{4}{3}\theta_2{}^2\right)$$

By strip theory,

$$M_a{}' = qc^2ae\left(\frac{y}{s}\theta_1 + \frac{y^2}{s^2}\theta_2\right)$$

Hence,

$$Q_1 = qc^2ae\int_0^s \left(\frac{y}{s}\theta_1 + \frac{y^2}{s^2}\theta_2\right)\frac{y}{s}\,dy = qc^2aes\int_0^1 (\eta\theta_1 + \eta^2\theta_2)\eta\,d\eta$$

$$= qc^2aes\left(\frac{\theta_1}{3} + \frac{\theta_2}{4}\right)$$

Similarly,

$$Q_2 = qc^2aes\left(\frac{\theta_1}{4} + \frac{\theta_2}{5}\right)$$

---

* A set of functions $f_i(y)$, $a \leqslant y \leqslant b$, $(i = 0, 1, 2, \cdots)$ is said to be "complete" if an arbitrary function $u(y)$ can be expanded into a generalized "Fourier" series of the form $\sum_{i=0}^{\infty} a_i f_i(y)$.

† If the function $f(y)$ in Eq. 1 is regarded as unknown, the equation governing $f(y)$ can be derived from Eq. 5 by considering the first variation of that equation with respect to $f(y)$. The result is the same as that obtained in § 3.2.

Let

$$\lambda = \frac{1}{GJ} q_{\text{div}} aec^2 s^2$$

Then the equations of equilibrium 9 become

$$(\theta_1 + \theta_2) = \lambda \left(\frac{\theta_1}{3} + \frac{\theta_2}{4} \right)$$

$$\left(\theta_1 + \frac{4}{3} \theta_2 \right) = \lambda \left(\frac{\theta_1}{4} + \frac{\theta_2}{5} \right)$$

A nontrivial solution exists only if the determinantal equation

$$\begin{vmatrix} 1 - \frac{1}{3}\lambda & 1 - \frac{1}{4}\lambda \\ 1 - \frac{1}{4}\lambda & \frac{4}{3} - \frac{1}{5}\lambda \end{vmatrix} = 0$$

is satisfied.   The two roots are $\lambda = 2.48$ and 32.3, corresponding to

$$q_{\text{div}} = \frac{2.48 GJ}{aec^2 s^2} \quad \text{and} \quad \frac{32.3 GJ}{aec^2 s^2}$$

The first one is the "fundamental" divergence dynamic pressure.   Compared with the exact value given by Eq. 17 of § 3.2, the error is 0.6 per cent too large.

3.5   THE METHOD OF SUCCESSIVE APPROXIMATION

From a mathematical point of view, the method of successive approximation possesses a number of advantages.   It is best to illustrate it by an example.

Consider again the rectangular wing, for which a, e, c, GJ are constants. In this example

$$G(x, y) = \begin{cases} \dfrac{1}{GJ} y & \text{for} \quad (0 \leqslant y \leqslant x \leqslant s) \\[3mm] \dfrac{1}{GJ} x & \text{for} \quad (s \geqslant y \geqslant x \geqslant 0) \end{cases} \tag{1}$$

and the fundamental equation 3 of § 3.2 becomes

$$\theta(x) = q_{\text{div}} \frac{aec^2}{GJ} \left[\int_0^x y \, \theta(y) \, dy + \int_x^s x \, \theta(y) \, dy \right] \tag{2}$$

If we introduce the nondimensional parameters

$$\xi = \frac{x}{s}, \qquad \eta = \frac{y}{s} \tag{3}$$

as the spanwise coordinates, Eq. 2 may be written as

$$\theta(\xi) = \lambda \left[\int_0^\xi \eta\theta(\eta)\, d\eta + \xi \int_\xi^1 \theta(\eta)\, d\eta \right] \tag{4}$$

where

$$\lambda = \frac{1}{GJ} q_{\text{div}} aec^2 s^2 \tag{5}$$

Let $\theta^{(1)}(\xi) = \xi$ be the first approximation (a guess) of the solution $\theta(\xi)$. Putting $\theta^{(1)}$ into the integrals on the right-hand side of Eq. 4, we obtain the value of the integrals

$$I^{(1)}(\xi) = \left(\int_0^\xi \eta^2\, d\eta + \xi \int_\xi^1 \eta\, d\eta \right) = \frac{1}{2}\left(\xi - \frac{\xi^3}{3} \right) \tag{6}$$

Since $\theta^{(1)}(\xi)$ is not an exact solution, it does not satisfy Eq. 4. But, if $\theta^{(1)}(\xi)$ approximates closely enough to the correct solution, Eq. 4 must be approximately satisfied. Equating $\lambda I^{(1)}(\xi)$ with $\theta^{(1)}(\xi)$ at one point, say $\xi = 1$, we obtain the approximate critical value $\lambda = 3$.

As a second approximation let us take the result given by Eq. 6, multiplied by a suitable constant, and assume

$$\theta^{(2)}(\xi) = \frac{3}{2}\left(\xi - \frac{\xi^3}{3} \right) \tag{7}$$

Then the integral on the right-hand side of Eq. 4 becomes

$$I^{(2)}(\xi) = \frac{3}{2}\left(\frac{5}{12}\xi - \frac{\xi^3}{6} + \frac{\xi^5}{60} \right) \tag{8}$$

The second approximate solution of λ is again obtained by equating the value of $\lambda I^{(2)}(\xi)$ to $\theta^{(2)}(\xi)$ at $\xi = 1$, which gives $\lambda = 2.5$.

The exact solution is $\lambda = \pi^2/4 = 2.46740$. The error at this stage is 1.30 per cent too large. If we take $\theta^{(3)}(\xi) = \frac{25}{18}(\xi - \frac{2}{5}\xi^3 + \frac{1}{25}\xi^5)$ and substitute into the right-hand side of Eq. 4 to obtain $I^{(3)}(\xi)$, we will have

$$I^{(3)}(\xi) = \frac{425}{272}(\xi - \frac{25}{61}\xi^3 + \frac{3}{61}\xi^5 - \frac{1}{427}\xi^7)$$

and

$$\lambda = 2.4705$$

The error of λ is now only 0.12 per cent too large.

It is quite evident how to proceed further. We substitute $\theta^{(n)}(\xi)$ into the right-hand side of Eq. 4, and evaluate the integral. Let the result be written as $I^{(n)}(\xi)$. If $I^{(n)}(\xi)/\theta^{(n)}(\xi)$ tends to a constant in the whole range $0 \leqslant \xi \leqslant 1$, then $\theta^{(n)}(\xi)$ is an approximate solution and $\lambda \doteq \theta^{(n)}(1)/I^{(n)}(1)$.

In order to compare the degree of approximation, it is convenient to normalize all $\theta^{(n)}(\xi)$ in such a way that they are equal at a reference section. In the above example, we have normalized $\theta^{(n)}(\xi)$ so that $\theta^{(n)}(1) = 1$.

Note that, according to § 3.2, the exact solution is $f(\xi) = \sin\dfrac{\pi}{2}\,\xi$. By series expansion,

$$\sin\frac{\pi}{2}\,\xi = \left[\,\xi - \frac{1}{3!}\left(\frac{\pi}{2}\right)^3\xi^3 + \frac{1}{5!}\left(\frac{\pi}{2}\right)^5\xi^5 - \frac{1}{7!}\left(\frac{\pi}{2}\right)^7\xi^7 + \cdots\right]$$

which may be compared with our successive approximations $\theta^{(1)}(\xi)$, $\theta^{(2)}(\xi)$, $\theta^{(3)}(\xi)$, etc.

The process is actually a repeated integration. It is iterative. So the method of successive approximation is also called the *method of iteration*.

An integral equation of the form Eq. 3 of § 3.2 is known as a *Fredholm integral equation of the second kind*. Since the influence function is symmetric, $G(x, y) = G(y, x)$, the method of iteration will converge to the correct eigenfunction corresponding to the smallest (in absolute value) eigenvalue.

The integration process can be very difficult in practice. Often a simple numerical solution can be obtained by replacing the integrals by a finite sum and transforming the governing equation into a matrix equation. The process of iteration can then be applied to the matrix equation in exactly the same manner as described above.

3.6   MATRICES[3.18–3.20]

We have emphasized above that the simplest formulation of aeroelastic problems usually results in integral equations. The exact solution of such equations is difficult to obtain except in special cases. In general, however, it is permissible to approximate the integral equation by a matrix equation and to obtain the solution by numerical methods. In the present section the meaning of matrices will be explained. Its connection with the finite-differences approximation will be examined in the next section. Practical examples are given in the next chapter.

A table of $m \times n$ numbers arranged in a rectangular array of m rows and n columns is called a *matrix with m rows and n columns*, or a *matrix of order $m \times n$*. If a_{ij} is the element in the ith row and jth column, then the matrix can be written down in the following pictorial form:

$$\mathbf{A} = \begin{pmatrix} a_{11} & a_{12} & \cdots & a_{1n} \\ a_{21} & a_{22} & \cdots & a_{2n} \\ \cdot & \cdot & \cdots & \cdot \\ a_{n1} & a_{n2} & \cdots & a_{nn} \end{pmatrix}$$

We shall denote a matrix either by a single symbol in boldface type such as **A**, or by a symbol (a_{ij}), the first subscript referring to the row and the second to the column. A *square matrix of order* $n \times n$ is a particular case. So is a *single-column matrix* of m rows (order $m \times 1$) or a *single-row matrix* of n columns (order $1 \times n$). A special square matrix denoted by **0**, called a *zero matrix*, is a square matrix, all of whose elements are zero. Another special square matrix denoted by **1**, called a unit matrix, is the following:

$$\mathbf{1} = (\delta_{ij})$$

where

$$\delta_{ij} = 1 \quad \text{if} \quad i = j$$
$$= 0 \quad \text{if} \quad i \neq j$$

These are, respectively,

$$\mathbf{0} = \begin{pmatrix} 0 & 0 & \cdots & 0 \\ 0 & 0 & \cdots & 0 \\ \cdot & \cdot & \cdots & \cdot \\ 0 & 0 & \cdots & 0 \end{pmatrix}, \qquad \mathbf{1} = \begin{pmatrix} 1 & 0 & 0 & \cdots & 0 \\ 0 & 1 & 0 & \cdots & 0 \\ \cdot & \cdot & \cdot & \cdots & \cdot \\ 0 & 0 & 0 & \cdots & 1 \end{pmatrix}$$

Two matrices are *equal* when the corresponding elements of each are equal. The addition, subtraction, multiplication, and division of matrices are defined as follows:

Addition. The sum of two matrices **A** and **B** is written as $\mathbf{A} + \mathbf{B}$ and stands for the matrix with elements $a_{ij} + b_{ij}$.

Subtraction. The matrix $-\mathbf{A}$ is defined as the matrix with elements $-a_{ij}$, and $\mathbf{A} - \mathbf{B}$ is defined as the matrix with elements $a_{ij} - b_{ij}$. For addition and subtraction to be significant, the matrices must have the same number of rows and the same number of columns.

It is clear that the associative law

$$(\mathbf{A} + \mathbf{B}) + \mathbf{C} = \mathbf{A} + (\mathbf{B} + \mathbf{C}) \tag{1}$$

and the commutative law

$$\mathbf{A} + \mathbf{B} = \mathbf{B} + \mathbf{A} \tag{2}$$

both hold for addition of matrices.

Multiplication. The product of the matrices $\mathbf{A} = (a_{ij})$, $\mathbf{B} = (b_{ij})$, written as **AB**, is a matrix **C** with elements c_{ij} given by

$$c_{ij} = a_{ik}b_{kj} = a_{i1}b_{1j} + a_{i2}b_{2j} + \cdots + a_{in}b_{nj} \tag{3}$$

The summation convention (Chapter 1) is used in Eq. 3. In order that the product **AB** of two matrices be well defined, the number of rows in the matrix **B** must be precisely the number of columns in the matrix **A**.

The product is then a matrix, with as many rows as $\mathbf{A}$ and as many columns as $\mathbf{B}$..

The commutative law of multiplication does not necessarily hold even if $\mathbf{A}$ and $\mathbf{B}$ are square. For $\mathbf{BA}$ must be defined as the matrix whose elements are $b_{ik}a_{kj}$, and this will be equal to $a_{ik}b_{kj}$ only in special cases; in general,

$$\mathbf{AB} \neq \mathbf{BA} \tag{4}$$

Pairs of matrices that satisfy $\mathbf{AB} = \mathbf{BA}$ are said to *commute*; those that satisfy $\mathbf{AB} = -\mathbf{BA}$ to *anticommute*.

The associative law

$$(\mathbf{AB})\mathbf{C} = \mathbf{A}(\mathbf{BC}) \tag{5}$$

and the distributive law

$$\mathbf{A}(\mathbf{B} + \mathbf{C}) = \mathbf{AB} + \mathbf{AC} \tag{6}$$

hold, provided the order is maintained and the operations are significant. Consequently, the products in Eq. 5 can be written without parentheses as $\mathbf{ABC}$, since the position of the parentheses is irrelevant. It follows that all positive powers of a given matrix commute; for $\mathbf{A}^2\mathbf{A} = \mathbf{A}\mathbf{A}^2$, and $\mathbf{A}^m\mathbf{A}^n = \mathbf{A}^n\mathbf{A}^m$ (m, n positive integers) follows by induction.

The unit matrix $\mathbf{1}$ has the interesting property that it commutes with all square matrices of the same order. In fact,

$$\mathbf{A1} = \mathbf{1A} = \mathbf{A} \tag{7}$$

The product of two matrices may be a zero matrix without either factor being zero. As an example,

$$\mathbf{A} = (1, 1, 0), \qquad \mathbf{B} = \begin{pmatrix} 0 \\ 0 \\ 1 \end{pmatrix}, \qquad \mathbf{AB} = (1)(0) + (1)(0) + (0)(1) = 0$$

The *transposed matrix* of a matrix $\mathbf{A}$ is the matrix formed from $\mathbf{A}$ by interchanging its rows and columns. We shall denote it by $\mathbf{A}'$ and its elements by a'_{ij}. Then

$$a'_{ij} = a_{ji} \tag{8}$$

Since

$$(\mathbf{AB})_{ij} = a_{ik}b_{kj} = a'_{ki}b'_{jk} = b'_{jk}a'_{ki} = (\mathbf{B}'\mathbf{A}')_{ji} \tag{9}$$

it follows that the transposed matrix of the product $\mathbf{AB}$, denoted by $(\mathbf{AB})'$, is equal to the product $\mathbf{B}'\mathbf{A}'$.

Symmetry Properties. A matrix is said to be *symmetrical* if it is unaltered by interchanging rows and columns; i.e.,

$$a_{ij} = a_{ji} \quad \text{or} \quad \mathbf{A} = \mathbf{A}' \tag{10}$$

It is *antisymmetrical* or *skew-symmetrical* if the sign is changed when rows and columns are interchanged; i.e.,

$$a_{ij} = -a_{ji} \quad \text{or} \quad \mathbf{A} = -\mathbf{A}' \tag{11}$$

A *diagonal matrix* is one all of whose elements are zero except those in the leading diagonal; i.e., a_{11}, a_{22}, $\cdots$, a_{nn}. All pairs of diagonal matrices of the same order commute.

Inverse of a Matrix and the Solution of Linear Equations. The inverse a^{-1}, or reciprocal, of a real number a is well defined if $a \neq 0$. Analogously, if $\mathbf{A}$ is a square matrix of order n and if the determinant $|a_{ij}| \neq 0$, then there exists a unique matrix, written as $\mathbf{A}^{-1}$ in analogy to the inverse of a number, with the properties

$$\mathbf{A}\mathbf{A}^{-1} = 1, \qquad \mathbf{A}^{-1}\mathbf{A} = 1 \tag{12}$$

The matrix $\mathbf{A}^{-1}$, if it exists, is called the *inverse matrix* of $\mathbf{A}$. The necessary and sufficient condition that a matrix $\mathbf{A} = (a_{ij})$ has an inverse is that the associated determinant $|a_{ij}| \neq 0$. The determinant $|a_{ij}|$ is formed by the elements of the square matrix $\mathbf{A}$, and is usually referred to as the *determinant* of the matrix $\mathbf{A}$. If $|a_{ij}|$ vanishes, $\mathbf{A}$ has no reciprocal and is said to be *singular*.

The practical calculation of the inverse of a matrix can be shortened by properly arranging the scheme of computation. The method of Crout* is the best known.

We can now define *division of matrices* as follows: Division by a non-singular matrix is defined as multiplication by its reciprocal, but the quotient depends on the order of the factors as with a product. In general, $\mathbf{A}^{-1}\mathbf{B}$ is not equal to $\mathbf{B}\mathbf{A}^{-1}$.

Since

$$\mathbf{A}\mathbf{B}\mathbf{B}^{-1}\mathbf{A}^{-1} = \mathbf{A}1\mathbf{A}^{-1} = 1 \tag{13}$$

it follows that $\mathbf{B}^{-1}\mathbf{A}^{-1}$ is the reciprocal of $\mathbf{A}\mathbf{B}$, that is $(\mathbf{A}\mathbf{B})^{-1}$. Hence, *in forming the reciprocal of a product, the order of the factors must be inverted.*

The inverse of a matrix has a simple application to the solution of n nonhomogeneous linear algebraic equations in n unknowns $x_1, x_2, \cdots, x_n$.

---

* See P. D. Crout, Marchant Calculating Machine Co. Bulletin MM-182, MM-183 (Sept. 1941). Also *Trans. AIEE*, **60** (1941).

P. S. Dwyer, *Psychometrika* **6**, 101–129 (1941).

W. E. Milne, *Numerical Calculus*, Princeton Univ. Press, Chapter 1 (1949).

Note that, according to Eq. 17, the kth column of $\mathbf{A}^{-1}$ can be obtained by solving Eqs. 16 for a column matrix $\mathbf{B}$ for which every element is zero except the kth element which is 1. Thus the calculation of the inverse of a matrix is equivalent to the numerical solution of a system of linear equations.

A set of linear equations

$$\begin{cases} a_{11}x_1 + a_{12}x_2 + \cdots + a_{1n}x_n = b_1 \\ a_{21}x_1 + a_{22}x_2 + \cdots + a_{2n}x_n = b_2 \\ \cdots\cdots\cdots\cdots\cdots\cdots\cdots\cdots \\ a_{n1}x_1 + a_{n2}x_2 + \cdots + a_{nn}x_n = b_n \end{cases} \tag{14}$$

may be written in the abbreviated form

$$a_{ij}x_j = b_i \tag{15}$$

where i and j run from 1 to n. If we think of (x_i) as a matrix with a single column, Eq. 15 may be written as

$$\mathbf{AX} = \mathbf{B} \tag{16}$$

where $\mathbf{B}$ is also a matrix with a single column. If we assume that the determinant of the matrix $\mathbf{A}$ is not zero, the inverse matrix $\mathbf{A}^{-1}$ will exist, and we shall have by matrix multiplication

$$\mathbf{A}^{-1}(\mathbf{AX}) = \mathbf{A}^{-1}\mathbf{B}$$

Since $\mathbf{A}^{-1}\mathbf{A} = \mathbf{1}$ and $\mathbf{1X} = \mathbf{X}$, we obtain the solution of Eq. 16:

$$\mathbf{X} = \mathbf{A}^{-1}\mathbf{B} \tag{17}$$

Multiplication of Matrices by Numbers. If $\mathbf{A} = (a_{ij})$ is a matrix, not necessarily a square matrix, and c is a number, real or complex, then $c\mathbf{A}$ denotes the matrix (ca_{ij}). This operation of multiplication by numbers enables us to consider matrix polynomials of the type

$$c_0\mathbf{A}^n + c_1\mathbf{A}^{n-1} + \cdots + c_{n-1}\mathbf{A} + c_n\mathbf{1}$$

where $c_0, c_1, \cdots, c_n$ are numbers, $\mathbf{A}$ is a square matrix, and $\mathbf{1}$ is the unit matrix of the same order as $\mathbf{A}$.

Characteristic Equation of a Matrix and the Cayley-Hamilton Theorem. If $\mathbf{A} = (a_{ij})$ is a given square matrix of order n, one can form the matrix $\lambda\mathbf{1} - \mathbf{A}$, which is called the *characteristic matrix* of $\mathbf{A}$. The determinant of this matrix, considered as a function of λ, is a polynomial of degree n in λ, and is called the *characteristic function* of $\mathbf{A}$. More explicitly, let $f(\lambda) = |\lambda\mathbf{1} - \mathbf{A}|$ where $|\lambda\mathbf{1} - \mathbf{A}|$ denotes the determinant of $\lambda\mathbf{1} - \mathbf{A}$, then $f(\lambda)$ has the form $f(\lambda) = \lambda^n + a_1\lambda^{n-1} + \cdots + a_{n-1}\lambda + a_n$. Since $a_n = f(0)$, we see that $a_n = |-\mathbf{A}|$. The algebraic equation of degree n for λ

$$f(\lambda) = 0 \tag{18}$$

is the *characteristic equation of the matrix* **A**, and the roots of the equation are the *characteristic roots of* **A**.*

We shall quote the famous *Cayley-Hamilton theorem* without proof: Let

$$f(\lambda) = \lambda^n + a_1\lambda^{n-1} + \cdots + a_{n-1}\lambda + a_n$$

be the characteristic function of a matrix **A**, and let **1** and **0** be the unit matrix and zero matrix, respectively, with an order equal to that of **A**. Then the matrix polynomial equation

$$\mathbf{X}^n + a_1\mathbf{X}^{n-1} + \cdots + a_{n-1}\mathbf{X} + a_n\mathbf{1} = \mathbf{0} \tag{19}$$

is satisfied by $\mathbf{X} = \mathbf{A}$.

Differentiation and Integration of Matrices Depending on a Numerical Variable. We shall have occasion to differentiate or integrate matrices whose elements depend on a numerical variable. Here we shall give the definitions as follows. Let $\mathbf{A}(t)$ be a matrix depending on a numerical variable t so that the elements $\mathbf{A}(t)$ are numerical functions of t.

$$\mathbf{A}(t) = \begin{pmatrix} a_{11}(t) & a_{12}(t) & \cdots & a_{1n}(t) \\ \cdot\cdot & \cdot\cdot & \cdots & \cdot\cdot \\ a_{m1}(t) & a_{m2}(t) & \cdots & a_{mn}(t) \end{pmatrix}$$

Then the derivative of $\mathbf{A}(t)$, written as $\dfrac{d\mathbf{A}(t)}{dt}$, is

$$\frac{d\mathbf{A}(t)}{dt} = \begin{pmatrix} \dfrac{da_{11}(t)}{dt} & \dfrac{da_{12}(t)}{dt} & \cdots & \dfrac{da_{1n}(t)}{dt} \\ \cdots & \cdots & \cdots & \cdots \\ \dfrac{da_{m1}(t)}{dt} & \dfrac{da_{m2}(t)}{dt} & \cdots & \dfrac{da_{mn}(t)}{dt} \end{pmatrix}$$

Similarly, we *define the integral of* $\mathbf{A}(t)$ *by*

$$\int \mathbf{A}(t)\, dt = \begin{pmatrix} \int a_{11}(t)\, dt & \int a_{12}(t)\, dt & \cdots & \int a_{1n}(t)\, dt \\ \cdots & \cdots & \cdots & \cdots \\ \cdots & \cdots & \cdots & \cdots \\ \int a_{m1}(t)\, dt & \int a_{m2}(t)\, dt & \cdots & \int a_{mn}(t)\, dt \end{pmatrix}$$

3.7 NUMERICAL APPROXIMATIONS — REDUCTION OF INTEGRAL EQUATIONS INTO MATRIX EQUATIONS

In practical numerical calculations, the integral equations of the preceding sections can be replaced by matrix equations. As a physical

---

* Sometimes also called *latent roots*, or *eigenvalues*.

example, consider the divergence of the cantilever wing shown in Fig. 3.4. In analyzing the wing deformation approximately, it is natural to divide the wing into a number of segments. The loading on each segment may be assumed as concentrated at a reference point in the segment. The average angle of rotation in each segment is represented by that measured at the reference point. Let θ_i be the angle of rotation of the ith segment, and M_i the aerodynamic moment in the same. Let the influence coefficient

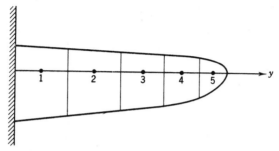

Fig. 3.4. A wing segmented.

for rotation at the ith segment due to a unit moment at the jth segment be denoted by c_{ij}. Then the total rotation at i is obtained by summing over the effects of all the moments:

$$\theta_i = \sum_{j=1}^{n} c_{ij} M_j \tag{1}$$

where n is the total number of segments in which the wing is divided.

It is convenient to write θ_i and M_i as matrices $\boldsymbol{\theta}$ and $\mathbf{M}$:

$$\boldsymbol{\theta} = \{\theta_i\} = \begin{pmatrix} \theta_1 \\ \theta_2 \\ \cdot \\ \cdot \\ \cdot \\ \theta_n \end{pmatrix}, \qquad \mathbf{M} = \{M_i\} = \begin{pmatrix} M_1 \\ M_2 \\ \cdot \\ \cdot \\ \cdot \\ M_n \end{pmatrix} \tag{2}$$

and the influence coefficients c_{ij} as a square matrix $\mathbf{c}$:

$$\mathbf{c} = \{c_{ij}\} = \begin{pmatrix} c_{11} & c_{12} & \cdots & c_{1n} \\ c_{21} & c_{22} & \cdots & c_{2n} \\ \cdot & \cdot & \cdots & \cdot \\ c_{n1} & c_{n2} & \cdots & c_{nn} \end{pmatrix} \tag{3}$$

Then Eq. 1 can be written simply as

$$\boldsymbol{\theta} = \mathbf{c} \cdot \mathbf{M} \quad \text{or} \quad \{\theta_i\} = \{c_{ij}\}\{M_j\} \tag{4}$$

$\mathbf{M}$ is a linear function of $\boldsymbol{\theta}$ and is proportional to the dynamic pressure q. Hence, we may write, in general,

$$\mathbf{M} = q\mathbf{A} \cdot \boldsymbol{\theta} \tag{5}$$

or

$$\{M_i\} = q\{A_{ij}\}\{\theta_j\}$$

where $\{A_{ij}\}$ is a square matrix. The form of the matrix $\mathbf{A}$ depends on the aerodynamic theory used (cf. § 4.5). In particular, if the "strip" assumption is used, $\mathbf{A}$ becomes a diagonal matrix:

$$\mathbf{A} = \begin{pmatrix} ac_1^2 e_1 & 0 & \cdots & 0 \\ 0 & ac_2^2 e_2 & \cdots & 0 \\ \cdot & \cdot & \cdots & \cdot \\ 0 & 0 & \cdots & ac_n^2 e_n \end{pmatrix} \tag{6}$$

where a is the lift-curve slope, c_i the chord length at the ith segment, and e_i the eccentricity, i.e., the ratio of the distance between the aerodynamic center and the elastic axis at the ith segment to the chord length. The condition of divergence can be represented as

$$\boldsymbol{\theta} = q_{\text{div}} \mathbf{c} \cdot \mathbf{A} \cdot \boldsymbol{\theta} \tag{7}$$

from which the eigenvector $\boldsymbol{\theta}$ and the eigenvalue q can be solved. An example will be given in § 4.5. Equation 7 should be compared with Eq. 3 of § 3.2.

That all the methods discussed so far can be transformed into matrix form is evident from recognizing that the matrix formulation amounts to a finite-differences approximation of continuous operations. We represent a continuous function $Y(x)$ of x in the interval (a, b) by a numerical table:

x	$a = x_0$	x_1	x_2	$\cdots$	$x_n = b$
$Y(x)$	Y_0	Y_1	Y_2	$\cdots$	Y_n

If the divisions $(x_i - x_{i-1})$ are sufficiently fine, this table will represent sufficient information for the function $Y(x)$. The value of $Y(x)$ at a point x other than the x_i's can be obtained by interpolation.[3.21–3.23]

By using the method of finite differences, the derivatives and integrals of $Y(x)$ can be replaced by a suitable combination of $Y(x_i)$, and equations governing $Y(x)$ can be written as matrix equations with $Y(x_i)$ as elements. From a mathematical point of view, the introduction of finite differences to replace continuous differential and integral operators is an approximation whose convergence can be rigorously treated. From an engineering point of view, the finite-differences method is the only natural method of specifying any physically measurable quantities. For example, in order to record the deflection curve of a beam under certain specified loading condition, the best an engineer can do is to measure the deflection at as many stations as possible. The result is a numerical table of the deflection function.

An integral is approximated as follows. Let $Y(x)$ be given at the points $x_m = \dfrac{m}{n}(b - a)$ $(m = 0, 1, \cdots, n)$, the points x_m being equally spaced. Let us write

$$Y(x_m) = Y_m$$

Then

$$\int_a^b Y(x)\,dx = (b - a) \sum_0^n s_m Y_m \bigg/ \sum_0^n s_m \tag{8}$$

The "weights" s_m are taken from the well-known rules of integration. For example,

	s_0	s_1	s_2	s_3	$\cdots$	s_{n-2}	s_{n-1}	s_n
1. Trapezoid rule	$\frac{1}{2}$	1	1	1	$\cdots$	1	1	$\frac{1}{2}$
2. Simpson's rule (n, even)	1	4	2	4	$\cdots$	2	4	1

There are other more complicated rules which, however, are often inferior to the simple rules quoted above, save for exceptional cases. Represented graphically, these rules are illustrated in Fig. 3.5. Equation 8 can be written as a matrix product

$$\int_a^b Y(x)\,dx \doteq h\mathbf{s}\mathbf{Y} \tag{9}$$

where $\mathbf{Y}$ is a column matrix, $\mathbf{s}$ is a row matrix of "weights," and h is a constant:

$$h = \frac{b - a}{\displaystyle\sum_{m=0}^{n} s_m}, \quad \mathbf{Y} = \begin{pmatrix} Y_0 \\ Y_1 \\ \cdot \\ \cdot \\ \cdot \\ Y_n \end{pmatrix}, \qquad \mathbf{s} = (s_0, s_1, \cdots, s_n) \tag{10}$$

Similarly, a function of two variables $K(x, y)$ can be represented as a double-entry table. Let the domain of (x, y) be the square $(a \leqslant x, y \leqslant b)$, and let the interval $(a, b)^2$ be divided into n^2 subdivisions by the points

$$x_0 = a, \quad x_1, x_2, \cdots, x_{n-1}, \quad x_n = b$$

$$y_0 = a, \quad y_1, y_2, \cdots, y_{n-1}, \quad y_n = b$$

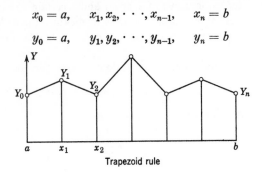

Trapezoid rule

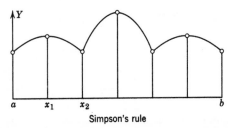

Simpson's rule

Fig. 3.5. The trapezoid rule and the Simpson's rule of integration. A given curve passing through the points $(Y_0, Y_1 \ldots Y_n)$ is replaced by straight-line segments or parabolic arcs. The areas between these curves and the x axis are then given by Eq. 8 with proper values of weighting factors s_m.

Introducing the notation

$$K(x_m, y_n) = K_{mn} \tag{11}$$

we may write

$$\int_a^b K(x, y)\, \phi(y)\, dy \doteq h\mathbf{KS\Phi} \tag{12}$$

where

$$\mathbf{K} = \begin{pmatrix} K_{11} & K_{12} & \cdots & K_{1n} \\ \cdot & \cdot & \cdots & \cdot \\ K_{n1} & K_{n2} & \cdots & K_{nn} \end{pmatrix}, \qquad \mathbf{\Phi} = \begin{pmatrix} \phi_1 \\ \phi_2 \\ \cdot \\ \cdot \\ \cdot \\ \phi_n \end{pmatrix} \tag{13}$$

h is gven by Eq. 10, and S is a diagonal matrix of "weights":

$$S = \begin{pmatrix} s_0 & & & & 0 \\ & s_1 & & & \\ & & \cdot & & \\ & & & \cdot & \\ & & & & \cdot \\ 0 & & & & s_n \end{pmatrix} \tag{14}$$

The integral equation

$$\phi(x) - \lambda \int_a^b K(x, y)\phi(y)\, dy = f(x)$$

is then replaced by the approximate matrix equation

$$\mathbf{\Phi} - \lambda h \mathbf{KS\Phi} = \mathbf{f} \tag{15}$$

where $\mathbf{f}$ is a column matrix with elements $(f_1, f_2, \cdots, f_n)$.

The matrix equation 15 can be made a starting point for a rigorous treatment of integral equations. The mathematical theory has been completed by Volterra, Fredholm, Hilbert, and others.

In Eqs. 8, 9, and 12 it was assumed that the intervals between the subdivisions $(x_i - x_{i-1})$ are equal. In some cases it is advantageous to use

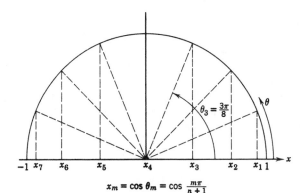

$$x_m = \cos \theta_m = \cos \frac{m\pi}{n+1}$$

Fig. 3.6. Division points in Multhopp's integration formula.
Location of x_m for $n = 7$.

unequal subdivisions, such as in Gauss's integration formula (cf. p. 159 of Ref. 3.21 or p. 115 of Ref. 3.23). However, in aeroelastic problems involving airplane wings, it is usually advantageous to use Multhopp's formula which emphasizes the wing-tip regions.   Multhopp's formula,[4.51] as is known in the theory of lift distribution, is

$$\int_{-1}^{1} Y(x)\, dx = \frac{\pi}{n+1} \sum_{m=1}^{n} Y_m \sin \theta_m \qquad (n \text{ odd}) \qquad (16)$$

where (Fig. 3.6)

$$x_m = \cos \theta_m, \qquad \theta_m = \frac{m\pi}{n+1} \qquad (17)$$

and

$$Y_m = Y(x_m) \qquad (m = 1, 2, \cdots, n)$$

This formula holds exactly so long as the function $Y(x)$ can be represented by a series of the following form:

$$Y(x) = \sqrt{1 - x^2} \sum_{\nu=0}^{2n-1} A_\nu x^\nu = \sum_{\nu=1}^{2n} a_\nu \sin \nu\theta \qquad (18)$$

Clearly, with a suitable modification of the constants s_0, s_1, $\cdots$, s_n, the method of reducing an integral equation into a matrix equation remains the same for unequal subdivisions (cf. sections on "divided differences," pp. 20, 96, 104, Ref. 3.21).

3.8   CONCLUDING REMARKS

Throughout this chapter the aerodynamic moments have been based on the "strip" assumption.   Relaxation of this assumption leads to complicated equations.   For a normal wing without appreciable sweepback angle, the over-all effect of finite-aspect ratio on the divergence speed can be accounted for by taking the lift-curve slope a corrected for aspect ratio.

The effect of compressibility can be included by using proper aerodynamic coefficients which correspond to the Mach number at which divergence occurs.   The calculation can be made by a process of trial and error.

BIBLIOGRAPHY

General review of the field of aeroelasticity:

3.1   Collar, A. R.: Aeroelastic Problems at High Speed. *J. Roy. Aeronaut. Soc.* **51**, 1–23 (Jan. 1947).

3.2   Collar, A. R.: The Expanding Domain of Aeroelasticity. *J. Roy. Aeronaut. Soc.* **50**, 613–636 (Aug. 1946).

3.3   Greidanus, J. H.: A Review of Aeroelasticity. *Applied Mech. Rev.* **4**, 138–140 (1951).

3.4 Russell, A. E.: Some Factors Affecting Large Transport Airplanes with Turbo-prop Engines. *J. Aeronaut. Sci.* **17**, 67–106 (Feb. 1950).

For historical remarks about divergence difficulty, see

3.5 Brewer, G.: The Collapse of Monoplane Wings. *Flight* (Jan. 1913).
3.6 Hill, G. T. R.: Advances in Aircraft Structural Design. *3d Anglo-Am. Aeronaut. Conf. Brighton*, 1–24 (1951). Royal Aeronautical Society, London.

For divergence analysis, see

3.7 Broadbent, E. G.: The Estimation of Wing Divergence Speeds. *Aeronaut. Research Council R. & M.* **2288** (1945).
3.8 Cox, H. Roxbee, and A. G. Pugsley: Stability of Static Equilibrium of Elastic and Aerodynamic Actions on a Wing. *Aeronaut. Research Com. R. & M.* **1509** (1932).
3.9 Diederich, F. W., and B. Budiansky: Divergence of Swept Wings. *NACA Tech. Note* **1680** (1948).
3.10 Flax, A. H.: Aeroelastic Problems at Supersonic Speed. *2d Intern. Aeronaut. Conf. N.Y.*, 322–360 (1949). Institute of Aeronautical Sciences, New York.
3.11 Hall, A. H.: Torsional Divergence with Downwash Correction. *Natl. Research Council, Canada, Aeronaut. Note* **AN-2** (1948).
3.12 Hildebrande, F. B., and E. Reissner: The Influence of the Aerodynamic Span Effect on the Magnitude of the Torsional-Divergence Velocity and on the Shape of the Corresponding Deflection Mode. *NACA Tech. Note* **926** (1944). Comments by J. W. Miles, *J. Aeronaut. Sci.* **16**, 63–64, 126 (1949).
3.13 Pugsley, A. G., and G. A. Naylor: The Divergence Speed of an Elastic Wing. *Aeronaut. Research Com. R. & M.* **1815** (1937).

For chordwise divergence of supersonic airfoils and the analysis of chordwise deformation, see

3.14 Ashwell, D. G.: The Anticlastic Curvature of Rectangular Beams and Plates. *J. Roy. Aeronaut. Soc.* **54**, 708–715 (Nov. 1950).
3.15 Biot, M. A.: Aero-Elastic Stability of Supersonic Wings. (*a*) Chordwise Divergence, the Two-Dimensional Case, *CAL/CM*-427 (1947). (*b*) An Approximate Treatment of Some Three-Dimensional Cases, *CAL/CM*-470 (1948). (*c*) General Method for the Two-Dimensional Case and Its Application to the Chordwise Divergence of a Biconvex Section, *CAL/CM*-506 (1948). (*d*) Some Exact Solutions Based on Plate Theory, *CAL/CM*-580 (1949). (*e*) Stability of Cantilever Solid Wings of Symmetric Cross-Section and without Sweep, *CAL/CM*-730 (1950). (*f*) Solution of the Divergence Problem by the Use of Generalized Coordinates, *CAL/CM*-730, *CAL-1-E-2* (1952). Cornell Aeronautical Laboratory, Buffalo, N.Y.
3.16 Fung, Y. C.: The Aeroelastic Behavior of a Sharp Leading Edge. *J. Aeronaut. Sci.* **20**, 644 (Sept. 1953).
3.17 Fung, Y. C., and W. H. Wittrick: The Anticlastic Curvature of a Strip with Lateral Thickness Variation. *J. Applied Mech.* **21**, 351–358 (Dec. 1954).

A brief account of the theory of matrices can be found in the following books:

3.18 Michal, A. D.: *Matrix and Tensor Calculus.* John Wiley & Sons (1947).
3.19 Jeffreys, H., and B. S. Jeffreys: *Methods of Mathematical Physics.* Cambridge Univ. Press (1946); 2d Ed. (1950).

DIVERGENCE OF A LIFTING SURFACE

A comprehensive treatise emphasizing the practical computational aspects is

.20   Frazer, R. A., W. J. Duncan, and A. R. Collar: *Elementary Matrices.* Cambridge Univ. Press (1938).

For numerical analysis, see

.21   Whittaker, E. T., and G. Robinson: *The Calculus of Observation.* Blackie and Sons, London (1924).
3.22   Scarborough, J. B.: *Numerical Mathematical Analysis.* Johns Hopkins Univ. Press and Oxford Univ. Press (1930).
3.23   Hartree, D. R.: *Numerical Analysis.* Oxford Univ. Press (1952).

Chapter 4

STEADY-STATE AEROELASTIC PROBLEMS IN GENERAL

Generalization of the analysis of the previous chapter to include the more accurate aerodynamic theories will be considered in this chapter together with other steady-state problems. The problem of loss of aileron efficiency and reversal of control will be discussed in §§ 4.1 and 4.2. Although the nature of the reversal problem is entirely different from that of the divergence problem, the methods of solution are analogous. Hence, we shall emphasize only the physical aspects of the problem without going into the details of calculation.

In § 4.3, the aerodynamic-lift distribution over a rigid wing is reviewed. In § 4.4, the effect of elastic deformation on the lift distribution is treated. These are followed by discussions of swept wings in § 4.5, tail efficiency in § 4.6, static longitudinal stability of an airplane in § 4.7, and twisting of propeller blades in § 4.8. It is possible to formulate many other steady-state problems, but the typical methods of analysis are well illustrated by the examples treated here.

4.1 LOSS AND REVERSAL OF AILERON CONTROL— TWO-DIMENSIONAL CASE

Ailerons control the rolling motion of an airplane. When an aileron is displaced downward, the lift over the wing increases, thus producing a rolling moment. But the aileron deflection also creates a nose-down aerodynamic pitching moment which twists the wing in a direction tending to reduce the lift and hence reducing the rolling moment. As the elastic stiffness of the wing is independent of the flight speed, whereas the aerodynamic force varies with U^2, there exists a critical speed at which the aileron becomes completely ineffective. This critical speed is called the *critical aileron-reversal speed*. When the airspeed is higher than the critical reversal speed, the aileron control is reversed; i.e., a downward movement of the aileron on the starboard wing produces a rolling moment which moves the starboard wing tip downward. The closer the speed is to the critical speed, the less effective is the aileron control. The effectiveness of the aileron control may be expressed in terms of the *rolling power* of

113

the wing, which may be defined as the ratio of the steady rolling velocity of an airplane, due to a unit deflection of the aileron, to that of an otherwise identical airplane in the same flight condition but with an infinitely rigid wing.   However, in order to simplify the problem and to obtain a convenient index for the aeroelastic characteristics of the wing itself, we shall consider the aileron control problem in a strictly static sense.   We define the *elastic efficiency of the aileron*, or simply the *aileron efficiency*, as the ratio of the *rolling moment* produced by an aileron deflection to that produced by the same deflection on a hypothetical rigid wing of the same planform, while the wing root is rigidly held against roll.   The situation may be conceived as existing in a wind-tunnel testing.   The airplane

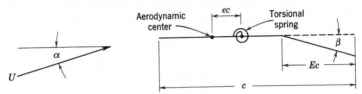

Fig. 4.1.   A two-dimensional wing with aileron.

fuselage is fixed in the tunnel, while the aileron is deflected and the rolling moment measured.

Since the net rolling moment vanishes at the reversal speed, both the "rolling power" and the "elastic efficiency" become zero at the critical condition.

Let α be the geometrical angle of attack (measured from the zero-lift line) at a section of the main airfoil, and β the angle of deflection of the aileron (positive downward) with respect to the main airfoil at that section (see Fig. 4.1).

The lift coefficient and the coefficient of moment about the aerodynamic center can be written in the form

$$C_l = a\alpha + \beta \frac{\partial C_l}{\partial \beta} \tag{1}$$

$$C_m = \beta \frac{\partial C_m}{\partial \beta} + C_{m0} \tag{2}$$

where a and C_{m0} are, respectively, the lift-curve slope and the coefficient of moment about the aerodynamic center of the airfoil with undeflected aileron.   According to Glauert,[1.43] the coefficients $\partial C_l/\partial \beta$ and $\partial C_m/\partial \beta$ for a two-dimensional airfoil in an incompressible fluid are

$$\frac{1}{a} \frac{\partial C_l}{\partial \beta} = \frac{1}{\pi} \left[\text{arc cos}\,(1 - 2E) + 2\sqrt{E(1 - E)} \right] \tag{3}$$

$$\frac{\partial C_m}{\partial \beta} = -\frac{a}{\pi}(1 - E)\sqrt{E(1 - E)} \qquad (\beta \text{ in radians}) \qquad (4)$$

where E is the ratio of the flap chord to the total chord (Fig. 4.1). For a finite wing with partial-span aileron, corrections are needed; but the effect of finite aspect ratio on the combination $\dfrac{1}{a}\dfrac{\partial C_l}{\partial \beta}\left(\dfrac{\partial C_m}{\partial \beta}\right)^{-1}$, which appears in the reversal problem, is small, provided that the lift-curve slope

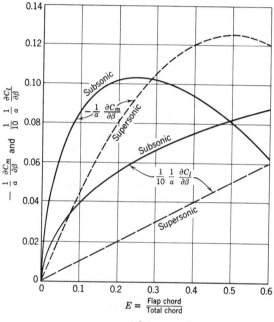

Fig. 4.2. $\dfrac{1}{a}\dfrac{\partial C_m}{\partial \beta}$ and $\dfrac{1}{a}\dfrac{\partial C_l}{\partial \beta}$.

a is properly corrected for finite aspect ratio. Comparison with experimental results[4.18] shows that the mean values of both $(1/a)(\partial C_l/\partial \beta)$ and $\partial C_m/\partial \beta$ at "incompressible" speeds lie at about 80 per cent of the theoretical values. The experimental values are, however, scattered and are influenced by the design details of the flap, such as its nose shape, nose gap, and aerodynamic balance.

The theoretical values of the ratios $(1/a)(\partial C_l/\partial \beta)$ and $(1/a)(\partial C_m/\partial \beta)$ are summarized in Fig. 4.2 for both the subsonic and the supersonic cases.

Let us consider a two-dimensional wing with chord length c and aileron chord length Ec, constrained by a spring to rotate about an axis at distance ec aft of the aerodynamic center.   In this case we shall say that aileron reversal occurs when the change of lift due to an aileron displacement vanishes.   The lift per unit length of this airfoil is, according to Eq. 1,

$$L' = qc\left(a\alpha + \frac{\partial C_l}{\partial \beta}\beta\right) \tag{5}$$

Hence, at the critical reversal condition,

$$\frac{dL'}{d\beta} = qc\left(a\frac{\partial \alpha}{\partial \beta} + \frac{\partial C_l}{\partial \beta}\right) = 0$$

i.e.,

$$\frac{\partial \alpha}{\partial \beta} = -\frac{1}{a}\frac{\partial C_l}{\partial \beta}$$

$$\tag{6}$$

Now the angle of twist α of the airfoil is related to the angle of deflection of the aileron through the elastic constraint.   The aerodynamic pitching moment per unit length of span about the axis of rotation is, according to Eq. 2,

$$M' = qc^2\left(eC_l + \beta\frac{\partial C_m}{\partial \beta} + C_{m0}\right)$$

This pitching moment is balanced by the elastic restoring moment.   Let K be the stiffness of the torsional restraint per unit span of the airfoil.   The elastic moment induced by a rotation of the airfoil through an angle α is αK.   Hence,

$$\alpha K = qc^2\left(eC_l + \beta\frac{\partial C_m}{\partial \beta} + C_{m0}\right) = qc^2\left(ea\alpha + e\beta\frac{\partial C_l}{\partial \beta} + \beta\frac{\partial C_m}{\partial \beta} + C_{m0}\right)$$

$$\tag{7}$$

Differentiating with respect to β, we obtain (C_{m0} being a constant)

$$K\frac{\partial \alpha}{\partial \beta} = qc^2\left(ea\frac{\partial \alpha}{\partial \beta} + e\frac{\partial C_l}{\partial \beta} + \frac{\partial C_m}{\partial \beta}\right) \tag{7a}$$

Substituting $\partial\alpha/\partial\beta$ from Eq. 6, we obtain the critical dynamic pressure

$$q_{\text{rev}} = \frac{1}{a}\frac{\partial C_l}{\partial \beta}\left(-\frac{\partial C_m}{\partial \beta}\right)^{-1}\frac{K}{c^2} \tag{8}$$

Hence, the reversal speed is given by

$$U_{\text{rev}} = \left(\frac{1}{a}\frac{\partial C_l}{\partial \beta}\right)^{1/2}\left(-\frac{\partial C_m}{\partial \beta}\right)^{-1/2}\left(\frac{2K}{\rho c^2}\right)^{1/2} \tag{9}$$

which shows that the critical reversal speed increases with increasing stiffness K of the airfoil.   Note that the quantity e is absent from this formula; i.e., the position of the torsional axis does not affect the reversal speed. The reason is that the net change of lift due to aileron displacement is zero at the reversal speed.

In order to compare the loss of controlling power of an aileron due to the elastic deformation, let us define, in the two-dimensional case, the *elastic efficiency* of the aileron as the ratio of the lift force produced by a unit deflection of the aileron on an elastic wing to that produced by the same aileron deflection on a fictitious rigid wing of the same chord length. Now, from Eq. 7a, we obtain

$$\frac{d\alpha}{d\beta} = \frac{qc^2 \left(e \frac{\partial C_l}{\partial \beta} + \frac{\partial C_m}{\partial \beta} \right)}{K - qc^2 ea}$$

Hence, the rate of change of lift due to an aileron deflection is, according to Eq. 5,

$$\frac{dL'}{d\beta} = qc \left[\frac{aqc^2 \left(e \frac{\partial C_l}{\partial \beta} + \frac{\partial C_m}{\partial \beta} \right)}{K - qc^2 ea} + \frac{\partial C_l}{\partial \beta} \right]$$

On the other hand, if the wing is perfectly rigid so that $K \to \infty$, then $d\alpha/d\beta = 0$, and the rate of change of lift would be

$$\frac{dL'_0}{d\beta} = qc \frac{\partial C_l}{\partial \beta}$$

The elastic efficiency of the aileron is, therefore,

$$\frac{dL'/d\beta}{dL'_0/d\beta} = 1 + \left[aqc^2 \left(e + \frac{\partial C_m}{\partial \beta} \Big/ \frac{\partial C_l}{\partial \beta} \right) \right] \Big/ [K - qc^2 ea]$$

Introducing the critical-divergence dynamic pressure q_{div} given by Eq. 6 of § 3.1, and the critical aileron-reversal dynamic pressure q_{rev} given by Eq. 8, we can write the above ratio as

$$\text{Elastic efficiency} = \frac{1 - q/q_{\text{rev}}}{1 - q/q_{\text{div}}} \tag{10}$$

When $q_{\text{rev}} < q_{\text{div}}$, the aileron efficiency decreases to zero when $q \to q_{\text{rev}}$, as shown in Fig. 4.3, where the aileron efficiency is plotted against q/q_{rev}, with the ratio $R = q_{\text{div}}/q_{\text{rev}}$ as a parameter. On the other hand, if $q_{\text{rev}} > q_{\text{div}}$, $dL'/d\beta \to \infty$ when $q \to q_{\text{div}}$, as shown in Fig. 4.4; but, for $q_{\text{div}} < q < q_{\text{rev}}$, the aileron efficiency is negative, and the control is already reversed, q_{div} in this case is also a critical reversal speed.   Aileron-control reversal occurs at either q_{rev} or q_{div}, *whichever is smaller*.

From the curves of Fig. 4.3 it is clear that, in order to maintain good aileron efficiency, it is desirable to have q_{rev} and q_{div} *as close to each other as possible*.   Both q_{rev} and q_{div} are proportional to the wing rigidity, and

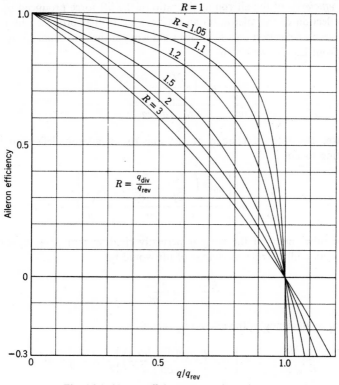

Fig. 4.3.   Aileron efficiency versus dynamic pressure
when $q_{\text{div}} > q_{\text{rev}}$.

can be raised by increasing the rigidity. But $q_{\text{div}} \sim 1/e$, and $q_{\text{rev}} \sim \dfrac{\partial C_l}{\partial \beta} \Big/ \dfrac{\partial C_m}{\partial \beta}$.   Hence, the requirement $q_{\text{rev}} \doteq q_{\text{div}}$ implies an optimum relation between the aileron dimensions and the wing aerodynamic center–elastic-axis eccentricity.   In the two-dimensional incompressible case, an optimum aileron chord ratio as a function of the eccentricity e is given by the condition

$$\frac{q_{\text{rev}}}{q_{\text{div}}} = 1 = e\,\frac{\partial C_l}{\partial \beta} \Big/ \left(-\frac{\partial C_m}{\partial \beta}\right) \tag{11}$$

and Glauert's formulas (Eqs. 3 and 4); i.e.,

$$e = \frac{(1 - E)\sqrt{E(1 - E)}}{\arccos(1 - 2E) + 2\sqrt{E(1 - E)}} \tag{12}$$

For example, if the elastic axis is located at 40 per cent chord behind the leading edge, the optimum aileron chord ratio is 31 per cent.

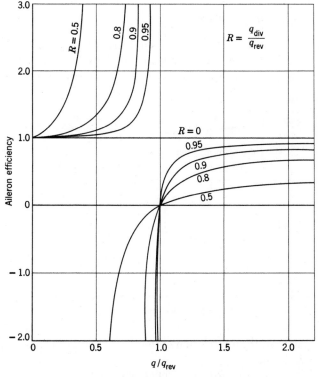

Fig. 4.4. Aileron efficiency versus dynamic pressure when $q_{\mathrm{div}} < q_{\mathrm{rev}}$.

4.2 AILERON REVERSAL—GENERAL CASE

Let us consider a wing without sweep and having a straight elastic axis when the aileron is locked, and again make the strip assumption on aerodynamic forces. The wing is assumed to be built in at the fuselage which is immobile in roll. Let the aileron deflection angle be denoted by $\beta(y)$ and the corresponding wing twisting angle about the elastic axis by

$\theta(y)$.* These functions are linearly related through elastic equilibrium. We shall write

$$\theta(y) = \theta_0 f(y) \qquad (0 \leqslant y \leqslant s)$$

$$\beta(y) = \beta_0 g(y),$$

where $g(y) \neq 0$ in $(\gamma_1 s \leqslant y \leqslant \gamma_2 s)$ (1)

$g(y) = 0$ elsewhere

θ_0 and β_0 are constants, and the functions $f(y)$ and $g(y)$ define the deformation pattern of the wing and the aileron, respectively. The ranges of definition of $f(y)$ and $g(y)$ are indicated in the parentheses, where $\gamma_1 s$ and

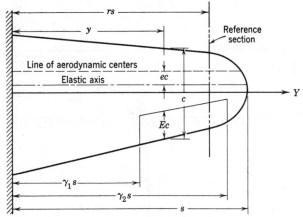

Fig. 4.5.   A wing planform.

$\gamma_2 s$ are locations of the inboard and outboard ends of the aileron, respectively (see Fig. 4.5).

In accordance with the *strip* assumption, the lift acting on a chordwise element of span dy is given by†

$$L'(y)\, dy = q \left[a\, \theta(y) + \frac{\partial C_l}{\partial \beta} \beta(y) \right] c\, dy \qquad (2)$$

---

* $\theta(y)$ and $\beta(y)$ are measured from the steady-flight values. Initially, at the flight speed U, the airplane wing may have some twist, and the aileron may have some deflection for trimming, but the whole airplane is in a steady-state equilibrium. These initial values do not affect our problem.

† The lift-curve slope a should be corrected for finite-aspect ratio (cf. § 1.5 and § 4.3) parent.

The total induced rolling moment about the airplane centerline is therefore

$$M_\phi = \int_0^s y\, L'(y)\, dy$$
$$= q \int_0^s \left[a\, \theta_0 f(y) + \frac{\partial C_l}{\partial \beta} \beta_0\, g(y) \right] cy\, dy \tag{3}$$

At the *critical aileron-reversal speed*, a deflection of the aileron produces no resultant rolling moment, so that

$$\left(\frac{\partial M_\phi}{\partial \beta_0} \right)_{\beta_0 = 0} = 0$$

i.e.,

$$\int_0^s \left[a\, \frac{d\theta_0}{d\beta_0} f(y) + \frac{\partial C_l}{\partial \beta} g(y) \right] cy\, dy = 0 \tag{4}$$

The relation between $f(y)$ and $g(y)$ and the ratio $d\theta_0/d\beta_0$ must be found from the condition of *elastic equilibrium* of the wing. Again, by the strip assumption, the contribution to the aerodynamic twisting moment about the elastic axis from an element dy is

$$M'_a\, dy = L'(y)\, e(y)\, c(y)\, dy + q\, c^2(y) \frac{\partial C_m}{\partial \beta} \beta(y)\, dy \tag{5}$$

Let $G(x, y)$ be the influence function of wing rotation at x due to a unit couple at y; then, the wing being in static equilibrium, the total angle of rotation θ at x is

$$\theta(x) = \int_0^s G(x, y)\, M'_a(y)\, dy \tag{6}$$

According to the strip-assumption equations 2 and 5, the above integral becomes

$$\theta(x) = q \int_0^s G(x,y) \left\{ a\, e(y)\, \theta(y) + \left[\frac{\partial C_l}{\partial \beta} e(y) + \frac{\partial C_m}{\partial \beta} \right] \beta(y) \right\} c^2(y)\, dy \tag{7}$$

or, substituting Eq. 1,

$$\theta_0 f(x) = q \int_0^s G(x, y) \left\{ a\, e(y)\, \theta_0 f(y) + \left[\frac{\partial C_l}{\partial \beta} e(y) + \frac{\partial C_m}{\partial \beta} \right] \beta_0\, g(y) \right\} c^2(y)\, dy \tag{8}$$

A second equation governing the angle of aileron deflection $\beta(y)$ across the aileron span can be derived in a similar manner. Generally, however, the aileron may be regarded as perfectly rigid. Then $\beta(y)$ is constant, and we may assume $g(y) = 1$. This approximation is sufficiently accurate for most ailerons. For more flexible ailerons a semirigid mode of the aileron deflection may be assumed. Thus $g(y)$ may be regarded as a known function of y.

The function $g(y)$ being known, the function $\theta_0 f(x)/\beta_0$ will be given by Eq. 8. The solution depends on q. By substituting $g(y)$ and $\theta_0 f(y)/\beta_0 = (d\theta_0/d\beta_0) f(y)$ into Eq. 4, the critical value of q can be computed.

Semirigid Solution. As a first approximation let us again apply the semirigid theory. We assume that the modes of the aileron deflection and the wing twisting are known; i.e., reasonable forms of the functions $f(y)$ and $g(y)$ are assumed. A process of reasoning entirely analogous to that in § 3.3 may be used to derive the solution corresponding to the assumed semirigid modes. We choose first a reference section at $y = r$, transfer all aerodynamic moments about the elastic axis to the reference section, and consider the rotation of the reference section to obtain the final result. For a greater variety in the ways of reasoning, however, we may proceed slightly differently as follows: Since the wing twisting mode $f(y)$ is assumed not to vary with the load distribution, it also represents the deformation pattern of the wing due to a couple acting at the reference section. Let an angle of rotation at the reference section $\theta_r = \theta_0 f(r)$ correspond to a couple M acting at the reference section, so that

$$M = K\theta_0 f(r) \tag{9}$$

K being the torsional stiffness at the reference section as defined by Eq. 5 of § 3.3. Then a unit moment at the reference section will produce a rotation

$$\frac{\theta_0 f(y)}{M} = \frac{1}{K}\frac{f(y)}{f(r)} \tag{10}$$

across the span. This is, by Maxwell's reciprocal theorem, precisely the influence function $G(r, y)$. Hence, the semirigid assumption implies an approximation

$$G(r, y) = \frac{1}{K}\frac{f(y)}{f(r)} \tag{11}$$

Substituting Eq. 11 into Eq. 8, putting $x = r$, dividing through by $f(r)$ ($\neq 0$), rearranging terms, and introducing the mean aerodynamic chord $\bar{c}$ as a characteristic length, we obtain

$$\theta_0 \left\{ K - q\bar{c}^2 \int_0^s \frac{f^2(y)}{f^2(r)} \frac{c^2(y)}{\bar{c}^2} a\, e(y)\, dy \right\}$$
$$= \beta_0 q\bar{c}^2 \int_0^s \left[\frac{\partial C_l}{\partial \beta} e(y) + \frac{\partial C_m}{\partial \beta} \right] \frac{f(y)\, g(y)}{f^2(r)} \frac{c^2(y)}{\bar{c}^2}\, dy \tag{12}$$

If we define two nondimensional quantities[4.3, 4.12]

$$k_1 = \frac{1}{s} \int_0^s \frac{f^2(y)}{f^2(r)} \frac{c^2(y)}{\bar{c}^2} \, a \, e(y) \, dy$$

$$k_2 = \frac{1}{s} \int_{\gamma_1 s}^{\gamma_2 s} \left[\frac{\partial C_l}{\partial \beta} e(y) + \frac{\partial C_m}{\partial \beta} \right] \frac{f(y) \, g(y)}{f^2(r)} \frac{c^2(y)}{\bar{c}^2} \, dy$$

(13)

then Eq. 12 can be written as

$$\theta_0 = \frac{q s \bar{c}^2 k_2}{K - q s \bar{c}^2 k_1} \beta_0$$

(14)

Substituting this relation into the critical reversal condition (Eq. 4), we obtain the critical reversal speed

$$U_{\text{cr}} = \sqrt{\frac{k_4}{k_1 k_4 - k_2 k_3}} \cdot \sqrt{\frac{2K}{\rho s \bar{c}^2}}$$

(15)

where

$$k_3 = \frac{1}{s^2} \int_0^s a \frac{c}{\bar{c}} f(y) \, y \, dy, \qquad k_4 = \frac{1}{s^2} \int_{\gamma_1 s}^{\gamma_2 s} \frac{\partial C_l}{\partial \beta} \frac{c}{\bar{c}} g(y) \, y \, dy$$

(16)

The U_{cr} so obtained is influenced by the arbitrariness in the choice of the reference section. To remove such arbitrariness, the method of generalized coordinates (see § 3.4) may be used. The details are left to the reader.

Example. Consider a rectangular wing with a rectangular aileron, and assume

$$f(y) = \frac{y}{s}$$

$$g(y) = 1 \quad \text{when} \quad \gamma_1 s \leqslant y \leqslant \gamma_2 s, \quad = 0 \text{ elsewhere}$$

Let the reference section be taken at $y = rs$; then

$$k_1 = \frac{1}{3r^2} ae, \qquad k_2 = \frac{1}{2r^2} \left(\frac{\partial C_l}{\partial \beta} e + \frac{\partial C_m}{\partial \beta} \right) (\gamma_2^2 - \gamma_1^2)$$

$$k_3 = \frac{a}{3}, \qquad k_4 = \frac{1}{2} \frac{\partial C_l}{\partial \beta} (\gamma_2^2 - \gamma_1^2)$$

Hence,

$$U_{\text{cr}} = r \left(\frac{3}{a} \frac{\partial C_l}{\partial \beta} \right)^{1/2} \left(-\frac{\partial C_m}{\partial \beta} \right)^{-1/2} \left(\frac{2K}{\rho s \bar{c}^2} \right)^{1/2}$$

This should be compared with the two-dimensional result (Eq. 9 of § 4.1). In this particular case, the critical reversal speed is independent of the aileron span.

The above example shows that the terms involving the eccentricity e do not appear in the critical-reversal-speed expression for a uniform rectangular wing with a rectangular aileron. This suggests that in the general case the contribution of the terms involving e is small. Pugsley proposes to neglect these terms entirely.[4.12, 4.9] Equation 15 is then simplified into

$$U_{cr} = \sqrt{\frac{k_4}{k_3 k_5}} \cdot \sqrt{\frac{2K}{\rho s \bar{c}^2}} \tag{17}$$

where

$$k_5 = -\frac{1}{s} \int_{\gamma_1 s}^{\gamma_2 s} \frac{\partial C_m}{\partial \beta} \frac{f(y) g(y)}{f^2(r)} \frac{c^2}{\bar{c}^2} dy \tag{18}$$

In an incompressible fluid, the error induced by neglecting e is unlikely to exceed 4 per cent in any conventional tapered wing.[4.9]

Equations 15 and 17 show that the aileron-reversal speed can be raised by increasing the wing torsional stiffness.

Solution by the Method of Successive Approximations. If the aileron is sufficiently stiff so that $g(y) = 1$, or if $g(y)$ can be assumed as a known function of y, we may put

$$h_1 = \frac{\partial C_l}{\partial \beta} \int_0^s g(y) \, c(y) \, y \, dy$$

$$h_2(x) = \int_0^s G(x, y) \left[\frac{\partial C_l}{\partial \beta} e(y) + \frac{\partial C_m}{\partial \beta} \right] c^2(y) \, g(y) \, dy \tag{19}$$

Equations 4 and 8 then become, respectively,

$$a \int_0^s f(y) \, c(y) \, y \, dy = - h_1 \frac{\beta_0}{\theta_0} \tag{20}$$

$$f(x) = qa \int_0^s G(x, y) \, e(y) \, c^2(y) f(y) \, dy + q \, h_2(x) \frac{\beta_0}{\theta_0} \tag{21}$$

Eliminating β_0/θ_0, we have

$$f(x) = q \int_0^s K(x, y) f(y) \, dy \tag{22}$$

where

$$K(x, y) = a \, G(x, y) \, e(y) \, c^2(y) - \frac{a \, h_2(x)}{h_1} c(y) \, y \tag{23}$$

$K(x, y)$ is a continuous function of x, y. Equation 22 is a homogeneous Fredholm's integral equation of the second kind, of the same form as the divergence equation 3 of § 3.2. The method of successive approximations described in § 3.5 can be applied here.   The smallest positive eigenvalue b is the physically significant reversal dynamic pressure.

Equation 21, regarding $h_2(x)$ as a known function, is a nonhomogeneous Fredholm integral equation of the second kind.   If we compare Eq. 21 with Eq. 3 of § 3.2 it is seen that they are identical but for the last term. Hence, when $\dfrac{\beta_0}{\theta_0} q \, h_2(x) = 0$, Eq. 21 reduces to the wing-divergence equation for which nontrivial solution $f(y)$ exists only for the eigenvalues q_{div}. It is well known that a nonhomogeneous equation such as 21 has no solution when q is equal to an eigenvalue,* whereas a unique solution $f(x)$ exists if q is not an eigenvalue.   Physically, if $q = q_{\text{div}}$ there will be no question of aileron reversal since control is impossible.   If $q \neq q_{\text{div}}$, then, for each specified deflection of the aileron, there corresponds a unique deflection curve of the wing.

The method of successive approximation can be applied to Eq. 21 as follows.   Let

$$\phi(x) = \frac{\beta_0}{\theta_0} q \, h_2(x)$$

$$H(x, y) = a \, G\,(x, y)e(y)c^2(y)$$

We compute the following sequence of functions:

$$f_0(x) = \phi(x)$$

$$f_1(x) = q \int_0^s H(x, y) f_0(y) \, dy + \phi(x)$$

$$f_2(x) = q \int_0^s H(x, y) f_1(y) \, dy + \phi(x) \tag{24}$$

$$\cdots \cdots \cdots \cdots \cdots \cdots \cdots \cdots \cdots$$

$$f_n(x) = q \int_0^s H(x, y) f_{n-1}(y) \, dy + \phi(x)$$

The sequence converges to a unique solution $f(x)$ of Eq. 21 for values of q within a finite radius of convergence $r \leq |q_{\text{div}}|$.

4.3   THE LIFT DISTRIBUTION ON A RIGID WING

In previous sections the spanwise-lift distribution over a finite airfoil is generally assumed to be given by the *strip* theory, according to which the local lift coefficient is proportional to the local angle of attack α.   This approximation is a crude one, but is sufficient for certain purpose.

---

* Except possibly when $h_2(y)$ is orthogonal to the eigenfunction.

To account for "induction" properly, more refined theory should be used.

For a wing with very small angle of sweep, Prandtl's *lifting-line* theory may be used.   In this case the local lift coefficient $C_l(y)$ is given by the following integral equation:*

$$\alpha(y) = \frac{C_l(y)}{a_0(y)} + \frac{1}{8\pi} \int_{-s}^{s} \frac{d(cC_l)}{d\eta} \frac{d\eta}{y - \eta} \tag{1}$$

where y is the spanwise location, $a_0(y)$ is the two-dimensional lift-curve slope of the airfoil section at y, and $\alpha(y)$ is the angle-of-attack distribution required to produce the lift-coefficient distribution $C_l(y)$. $\alpha(y)$ is the geometrical angle of attack between the zero-lift line of the section and the flight direction.   Methods of solving Eq. 1 have been proposed by Glauert, Lotz, Hildebrand, Multhopp, and Sears.†

For swept wings and wings of small aspect ratio (say less than 5) Prandtl's lifting-line theory does not apply.   To obtain accurate results it is necessary to use a *lifting-surface* theory in which an airfoil is replaced by a continuous distribution of vorticity over a surface.   Unfortunately the calculation becomes involved and only a small number of solutions exist at present (see refs. 4.39, 40, 41, 44, 45, 48, 49, 50, 56).   However, a modification of the Prandtl theory, known as Weissinger's method, gives a good approximation for swept wings.   In Weissinger's method the bound vorticity of the wing is concentrated into the $1/4$-chord line, and the downwash is calculated at the $3/4$-chord line.   The condition that the downwash angle be equal to the slope of the wing with respect in the direction of flow is then applied at the $3/4$-chord line.   This results in an integral equation governing the lift distribution similar to Eq. 1, but with a continuous function added to the original kernel.‡   Practical methods of solution are given by Mutterperl[4.52] and Weissinger.[4.58]

---

* The integral in Eq. 1 is divergent in the ordinary sense, but can be defined by its Cauchy principal value.   The Cauchy principal value of an integral is defined as follows:

$$\int_a^b f(x)\, dx = \lim_{\varepsilon \to 0} \left[\int_a^{x_0 - \varepsilon} f(x)\, dx + \int_{x_0 + \varepsilon}^b f(x)\, dx \right]$$

where x_0 is a singular point of $f(x)$ in (a, b).

† For Glauert's method, see his book, *Aerofoil and Airscrew Theory* (1926), and R. F. Anderson: *NACA Rept.* **572** (1940).

For Lotz's method, see her paper in *Z. Flugtech. u. Motorlyft.* **22**, 189–195 (1931), and H. A. Pearson, *NACA Rept.* **585** (1937).

For Hildebrand's method (a least-squares procedure), see *NACA Tech. Note* **925** (1944).

For Multhopp's method, see *Luftfahrt-Forsch.* **15**, 153–169 (1938).

For Sears's method, see *Quart. Applied Math.* **6**, 239–255 (1948).

‡ See Eqs. 36, 37 of Ref. 4.58.   See also Ref. 4.57.

A short empirical approximate method for estimating the spanwise lift distribution is given by Schrenk[4.54] for a normal (unswept) wing, and is extended to sweptback wings by Pope and Haney.[4.53]

So far the compressibility effect is neglected. The compressibility of the fluid is expressed in terms of the Mach number M.*

When the airfoil speed is subsonic ($M < 1$), the change in the lift-curve slope is given approximately by Glauert's formula[4.46]:

$$a'_0 = \frac{1}{\sqrt{1 - M^2}} a_0 \quad \text{(2-dimensional)} \tag{2}$$

where a'_0 is the value of $\partial C_l/\partial \alpha$ for a compressible fluid, and a_0 that for an incompressible fluid ($M = 0$), both in a two-dimensional flow.† The effect of finite span is then approximately given by the following formulas:

$$a' = \frac{a'_0}{1 + a'_0 \dfrac{1 + \tau}{\pi R}} = \frac{a_0}{\sqrt{1 - M^2} + a_0 \dfrac{1 + \tau}{\pi R}} \quad \left(\begin{matrix}\text{unswept, subsonic,} \\ \text{symmetric loading,} \\ \text{moderate } R\end{matrix}\right)^{1.44}$$

$$\tag{3}$$

where a' is the lift-curve slope of a wing of finite aspect ratio in a compressible flow, and τ is the same Glauert's correction factor for nonelliptic planform (Fig. 1.13, p. 35). For very small R, we have

$$a' = \frac{a'_0}{\sqrt{1 + \left(\dfrac{a'_0}{\pi R}\right)^2} + \dfrac{a'_0(1 + \tau)}{\pi R}} \quad \left(\begin{matrix}\text{unswept, subsonic, sym-} \\ \text{metric loading, small } R\end{matrix}\right)^{\ddagger} \tag{4}$$

Equation 4 yields the correct limiting value for wings of very low aspect ratio.§

$$\lim_{R \to 0} a = \frac{\pi}{2} R \quad \left(\begin{matrix}\text{unswept, incompressible} \\ \text{symmetric loading, } R \to 0\end{matrix}\right) \tag{5}$$

The effect of sweep angle is incorporated in the following formulas:

$$(a_0)_{\text{swept}} = a_0 \cos \Lambda \quad \left(\begin{matrix}\text{incompressible, infinite} \\ \text{span, swept wing}\end{matrix}\right) \tag{6}$$

---

* Unless stated to the contrary, the Mach number spoken of in this book is the *free-stream* Mach number, which is sometimes denoted by M_∞.

† A more accurate method is given by von Kármán and Tsien. *J. Aeronaut. Sci.* **8**, 337 (1941), and **6**, 399 (1939).

‡ H. B. Helmbold, *Jahrb. deut. Luftfahrt-Forsch.*, I, 111–113 (1942).

§ R. T. Jones, *NACA Rept.* **835** (1946).

where a_0 is the lift-curve slope of the airfoil section normal to the leading edge.

$$(a'_0)_{\text{swept}} = \frac{1}{\sqrt{1 - M^2 \cos^2 \Lambda}} (a_0)_{\text{swept}}$$

$$= \frac{a_0 \cos \Lambda}{\sqrt{1 - M^2 \cos^2 \Lambda}} \quad \text{(subsonic, infinite span)} \tag{7}$$

$$a' = \frac{(a'_0)_{\text{swept}}}{\sqrt{1 + \left(\dfrac{(a'_0)_{\text{swept}}}{\pi R}\right)^2} + \dfrac{(a'_0)_{\text{swept}}(1 + \tau)}{\pi R}} \quad \begin{pmatrix} \text{finite } R, \text{ subsonic,} \\ \text{symmetric loading} \end{pmatrix} \tag{8}$$

where R is the aspect ratio b^2/S ($b =$ the wing-span from tip to tip, $S =$ the wing area).*

If the wing is tapered, the sweep angle Λ should be measured along the $1/4$-chord line. The values of τ are given in Fig. 1.13 (p. 35).

For antisymmetric spanwise lift distribution, the effective aspect ratio is approximately one-half that of the entire wing. Therefore, an approximation is obtained by replacing R in Eq. 8 by

$$R_e = R/2 \quad \text{(antisymmetric loading)} \tag{9}$$

Eqs. 3, 4, 8, and 9 are useful in aeroelasticity. By means of these corrections, the errors of the *strip theory* are greatly reduced.

Because of the difference in the effective aspect ratio for the symmetric and antisymmetric spanwise load distributions, the corrected values of a should be distinguished for the two cases. Let the spanwise angle of attack $\alpha(y)$ be separated into two parts, one symmetrical in y and another antisymmetrical in y,

$$\alpha(y) = \alpha_{\text{sym}}(y) + \alpha_{\text{antisym}}(y) \tag{10}$$

then we may write

$$C_L(y) = a\alpha_{\text{sym}}(y) + \tilde{a}\alpha_{\text{antisym}}(y) \tag{11}$$

For a, Eq. 4 or 8 can be used. For $\tilde{a}$, the aspect ratio in Eqs. 4 and 8 should be replaced by $R_e = R/2$.

The equations quoted above are not valid when the wing approaches stalling, either because of too high an angle of attack or because of too high a speed of flight. The former is *stalling* in the conventional sense; the latter is called *shock stall*, and is due to the formation of shock waves over the wing. Both involve a loss of lift and an increase of drag. Shock stall occurs when the free-stream Mach number exceeds the *Mach number*

---

* T. A. Toll and M. J. Queijo, *NACA Tech. Note* **1581** (1948); F. W. Diederich, *NACA Tech. Note* **2335** (1951).

of *"divergence"* M_{cr} which is defined as the point of inflection of the curve of lift coefficient plotted against the Mach number (see Fig. 4.6 and Fig. 9.2).

In Fig. 4.6 the lift-curve slope is plotted against M for several airfoils. It is seen that Glauert's formula (Eq. 2) gives a fair approximation for

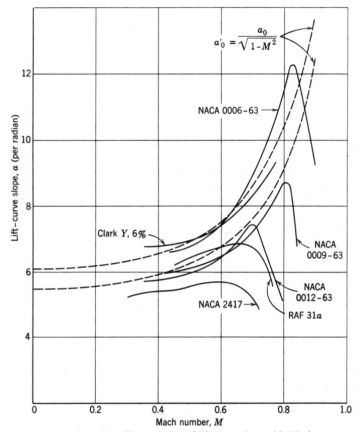

Fig. 4.6.   The variation of lift-curve slope with Mach number.

lower Mach numbers, and that, when M_{cr} is exceeded, the lift-curve slope decreases sharply.[4.37]

When the free-stream Mach number is sufficiently large, both subsonic and supersonic regimes are present in the field of flow. This is the *transonic regime* in which both the force and moment coefficients and the angle of zero lift change substantially. The variations are too complicated to be reviewed here.

When the free-stream Mach number is larger than one, the flow is *supersonic*. When M is sufficiently high, the linearized theory again gives a good approximation.* Let a pressure coefficient be defined by the equation

$$C_p = \frac{p - p_0}{\frac{1}{2}\rho_0 U^2} \tag{12}$$

where p is the local pressure, and p_0, ρ_0, U refer to quantities at a large distance from the airfoil. Then, according to the linearized theory, the pressure coefficient at any point on a two-dimensional airfoil is proportional to the local slope of the surface. If the surface of a two-dimensional airfoil is represented by the function $y = Y(x)$, then†

$$(C_p)_{\text{upper}} = \frac{2}{\sqrt{M^2 - 1}} \frac{dY}{dx} \tag{13}$$

where the subscript "upper" indicates the upper surface of the airfoil. The corresponding formula for the lower surface is obtained by replacing the right-hand side with a negative sign. The total lift coefficient is given by

$$C_L = \frac{4}{\sqrt{M^2 - 1}} \frac{\tau}{c} \quad \text{(two-dimensional, supersonic)} \tag{14}$$

where c is the chord length, τ is the normal distance between the trailing edge and the line parallel to the flow direction passing through the leading edge. τ/c can be identified as the angle of attack. Hence,

$$a_0 = \left(\frac{\partial C_L}{\partial \alpha}\right)_0 = \frac{4}{\sqrt{M^2 - 1}} \quad \text{(supersonic)} \tag{15}$$

For a flat-plate airfoil, the slope dY/dx is constant; Eq. 13 implies that the center of pressure is at the *mid-chord point*.

For wings of finite span a supersonic flow differs from the two-dimensional one only in the tip region. For wings of small aspect ratio, the lifting-surface theory must be used. Many cases of lifting-surface theory have been worked out for a linearized supersonic flow.[4.47]

For a swept wing of large aspect ratio with α measured in the free-stream direction, we obtain, analogous to 7,

$$\frac{\partial C_L}{\partial \alpha} = \frac{4 \cos \Lambda}{\sqrt{M^2 \cos^2 \Lambda - 1}} \quad \left(\begin{array}{c}\text{supersonic leading edge,}\\ M \cos \Lambda > 1\end{array}\right) \tag{16}$$

---

* See for example, R. E. Bolz and J. D. Nicolaides, *J. Aeronaut. Sci.* **17**, 609–621 (1950).

† See Liepmann and Puckett, Ref. 1.46, p. 145.

4.4 THE EFFECT OF ELASTIC DEFORMATION ON THE LIFT DISTRIBUTION

For a wing of small aspect ratio, or arbitrary planform, or with large cutouts, or subject to serious effect of warping restraint at the wing root, it is necessary to consider the entire deflection surface of the wing (instead of assuming rigid chordwise sections), and to compute the lift distribution by a lifting-surface theory.   In this case the problem of determining the effect of the elastic deformation on the lift distribution can be formulated

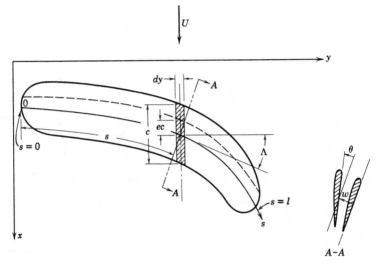

Fig. 4.7. A slender wing.

by means of influence functions into integral equations of two independent variables (involving double integrals).   In most practical cases, however, the warping of the chordwise sections is not of great importance, and reasonable accuracy can be obtained by considering the deformation pattern at each chordwise section as characterized by a deflection at a reference point and a rotation in the chordwise section about that point.

For a wing other than a normal simple beam, the concept of elastic axis loses its simplicity, and it becomes more straightforward to define the wing deformation with respect to an arbitrary reference line. As remarked in § 3.2, when an arbitrary reference line is used, two influence functions are required to specify the wing rotation and two others to specify the wing deflection.   As an example, consider a wing of large aspect ratio whose cross sections normal to a reference line may be assumed rigid.   Let s be the distance measured from an origin along the reference line (see Fig. 4.7).   Let $w(s)$ be the deflection at a point s on the reference line, normal to the plane of the wing (the wing is assumed to be planar, and the displacement in the plane of the wing is assumed to be negligible), and let $\theta(s)$ be the angle of twist of a *normal* cross section at s about the reference line.   In the lift-distribution problem, the slope $\partial w/\partial s$, rather than w itself, is of importance.   The elastic property of the wing can be characterized by the following influence functions:

$F_1(s, \sigma)$,   giving $\dfrac{\partial w}{\partial s}$ at s due to a unit force at σ

$F_2(s, \sigma)$,   giving θ at s due to a unit force at σ

$H_1(s, \sigma)$,   giving $\dfrac{\partial w}{\partial s}$ at s due to a unit external twisting moment at σ

$H_2(s, \sigma)$,   giving θ at s due to a unit external twisting moment at σ

By an external *twisting moment* at the point σ is meant a couple whose vector is *tangent* to the reference line at the point σ.   If there acts a unit couple at σ whose vector is *perpendicular* to the reference line at σ, which will be referred to as a unit external *bending moment*, the deflection surface can be obtained from the following influence functions:*

$\dfrac{\partial F_1(s, \sigma)}{\partial \sigma}$,   giving $\partial w/\partial s$ at s due to a unit external bending moment at σ

$\dfrac{\partial F_2(s, \sigma)}{\partial \sigma}$,   giving θ at s due to a unit external bending moment at σ

In defining these influence functions, the structure is assumed to be rigidly supported in a manner appropriate to the particular problem under consideration (cf. footnote, p. 19).   In order to define the positive senses of w, θ etc., we define a set of local orthogonal coordinates (n, s), with $\mathbf{s}$ tangent to the reference axis and $\mathbf{n}$ normal to it; $\mathbf{n}$ is to $\mathbf{s}$ as $\mathbf{x}$ is to $\mathbf{y}$ axis (see Figs. 4.7 and 4.8).   The senses of the force and deflection agree, and

---

* This can be easily verified by considering a bending couple as the limit of a pair of equal and opposite forces, the magnitude of which increases as the distance between them decreases, in such a way that a unit moment is maintained.   Thus the slope $\partial w/\partial s$ at a point s due to a force $M/\Delta\sigma$ acting at $\sigma + \Delta\sigma/2$ and another force $- M/\Delta\sigma$ at $\sigma - \Delta\sigma/2$ is

$$\frac{\partial w}{\partial s} = \left[F_1\!\left(s, \sigma + \frac{\Delta\sigma}{2}\right) - F_1\!\left(s, \sigma - \frac{\Delta\sigma}{2}\right) \right] \frac{M}{\Delta\sigma}$$

The limit as $\Delta\sigma \to 0$ gives $\partial w/\partial s = M\partial F_1/\partial\sigma$, which verifies the statement.

are positive in the positive direction of the z axis (xyz a right-hand system). The bending moment and the twisting moment are positive when their vectors are in the positive directions of **n** and **s**, respectively.

Let the external loading be resolved into forces acting on the reference line and bending and twisting moments about it. If we denote the normal external load per unit length on the reference line by $f(s)$, the distributed external bending moment per unit length by $m(s)$, and the distributed external twisting moment per unit length by $t(s)$, we have

$$\frac{\partial w}{\partial s}(s) = \int_0^l F_1(s, \sigma) f(\sigma)\, d\sigma + \int_0^l \frac{\partial F_1(s, \sigma)}{\partial \sigma} m(\sigma)\, d\sigma + \int_0^l H_1(s, \sigma) t(\sigma)\, d\sigma \quad (1)$$

$$\theta(s) = \int_0^l F_2(s, \sigma) f(\sigma)\, d\sigma + \int_0^l \frac{\partial F_2(s, \sigma)}{\partial \sigma} m(\sigma)\, d\sigma + \int_0^l H_2(s, \sigma) t(\sigma)\, d\sigma \quad (2)$$

where the integration covers the entire length of the wing.

Consider now the lift problem. The angle of attack at any section s, denoted by $\alpha(s)$, can be expressed as

$$\alpha(s) = \alpha^{(r)} + \alpha^{(e)} \quad (3)$$

where $\alpha^{(r)}$ denotes the angle of attack of the airfoil *if it were perfectly rigid*; and $\alpha^{(e)}$ indicates the change due to elastic deformation. $\alpha(s)$ *is supposed to be measured in the flight direction* and does not include induction effects of the lift.

For a given $\alpha(s)$, the aerodynamic force and moment distribution can be determined by the methods mentioned in § 4.3. For a given lift and moment distribution, the elastic deformation can be computed according to Eqs. 1 and 2. The elastic deformation in turn determines $\alpha(s)$. Thus the nature of an aeroelastic system as a *feedback* system is clearly seen.

It is natural to apply a *process of successive approximation* to find the effect of elastic deformation on the lift distribution, especially when the speed of flight is considerably lower than the divergence speed of the wing. In this process we start from the angle-of-attack distribution $\alpha^{(r)}(s)$ of a rigid wing, find first the lift and moment distribution corresponding to $\alpha^{(r)}(s)$, and then the elastic deformation $\alpha^{(e)}(s)$. Next we determine lift and moment distribution corresponding to $\alpha(s) = \alpha^{(r)} + \alpha^{(e)}$ computed in the first cycle and determine the elastic deformation $\alpha^{(e)}(s)$ again. If, by repeating the process, we arrive at a limiting function $\alpha(s)$, then that limiting function is the equilibrium angle-of-attack distribution of the elastic wing corresponding to $\alpha^{(r)}(s)$.

Because of its importance, we shall examine the problem in greater detail and give a numerical example below.*

* The treatment follows essentially that of Pai and Sears.[4.35]

Equations of Equilibrium. The aerodynamic forces acting on a deformed wing can be expressed in terms of "local" lift and moment coefficients, which are defined by considering the lift and moment acting on an elementary strip of small width dy parallel to the x axis (see Fig. 4.7). The length of this strip is the chord length c measured in the free-stream direction. The lift force on the strip is $qC_l c\, dy$, and the moment about the aerodynamic center of this strip is $qC_m c^2\, dy$. We shall assume that the drag force is negligible and that the force normal to the wing is equal to the lift force. Let the distance between the aerodynamic center and the reference line be ec, also measured in the free-stream direction, and taken as positive if the latter lies behind the former. Then the moment about a point on the reference line due to forces on the elementary strip is $qC_l ec^2\, dy + qC_m c^2\, dy$. The vector of this moment is parallel to the y axis, and can be resolved into an external bending moment $-q(C_l e + C_m)c^2 \sin \Lambda\, dy$ and an external twisting moment $q(C_l e + C_m)c^2\, dy \cos \Lambda$ about the reference line, where Λ denotes the sweepback angle at the point s as shown in Fig. 4.7.

Let w and C_l be positive when the wing tends to deflect upward. Let θ and the twisting moment be positive when the wing tends to increase the angle of attack. Since only the aerodynamic loads need be considered, we have, replacing dy by $\cos \Lambda\, d\sigma$, and using Eqs. 1 and 2:

$$
\begin{aligned}
\frac{\partial w(s)}{\partial s} = q \bigg\{ &\int_0^l F_1(s, \sigma)\, C_l(\sigma)\, c(\sigma) \cos \Lambda(\sigma)\, d\sigma \\
&- \int_0^l \frac{\partial F_1(s, \sigma)}{\partial \sigma} [C_l(\sigma) e(\sigma) + C_m(\sigma)]\, c^2(\sigma) \cos \Lambda(\sigma) \sin \Lambda(\sigma)\, d\sigma \\
&+ \int_0^l H_1(s, \sigma)[C_l(\sigma) e(\sigma) + C_m(\sigma)]c^2(\sigma) \cos^2 \Lambda(\sigma)\, d\sigma \bigg\}
\end{aligned}
\tag{4}
$$

A similar equation for $\theta(s)$ is obtained by replacing F_1, H_1 by F_2 and H_2, respectively, on the right-hand side.

The elastic angle of attack $\alpha^{(e)}$ is given by

$$
\alpha^{(e)} = \theta \cos \Lambda - \frac{\partial w}{\partial s} \sin \Lambda
\tag{5}
$$

which can be derived as follows: Let the wing-deflection surface be denoted by $w(x, y)$; then the slope in the x direction is

$$
\frac{\partial w}{\partial x} = \frac{\partial w}{\partial s} \frac{\partial s}{\partial x} + \frac{\partial w}{\partial n} \frac{\partial n}{\partial x}
\tag{6}
$$

where (s, n) are local orthogonal coordinates in directions parallel and perpendicular to the reference axis (Fig. 4.8). But

$$\frac{\partial s}{\partial x} = \sin \Lambda, \quad \frac{\partial n}{\partial x} = \cos \Lambda, \quad \text{and} \quad \frac{\partial w}{\partial n} = -\theta \tag{7}$$

Hence, Eq. 5, because $\alpha^{(e)}$ is equal to the negative of the slope $\partial w/\partial x$ in the particular coordinate system chosen.

The determination of $C_l(s)$ and $C_m(s)$ from $\alpha(s) = \alpha^{(r)} + \alpha^{(e)}$ is a problem in aerodynamics. When a definite theory is adopted so that a relation

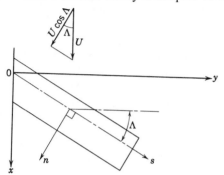

Fig. 4.8.   Relations for a swept wing.

between $\alpha(s)$ and C_l and C_m is known, that relation, in conjunction with Eqs. 4 and 5, determines the functions $\alpha(s)$, $C_l(s)$, and $C_m(s)$.

As an example, assume that Λ is small so that Prandtl's lifting-line theory can be used. α and C_l are therefore related by Eq. 1 of § 4.3. Combining that equation with Eqs. 4 and 5, we obtain the integral equation for C_l:

$$\frac{C_l(s)}{a_0(s)} + \frac{1}{8\pi} \int \frac{d}{d\eta} [C_l(\eta) \, c(\eta)] \frac{d\eta}{y - \eta}$$

$$= \alpha^{(r)}(s) + q \left\{ \int_0^l [F_2(s, \sigma) \cos \Lambda(s) - F_1(s, \sigma) \sin \Lambda(s)] C_l(\sigma) \, c(\sigma) \cos \Lambda(\sigma) \, d\sigma \right.$$

$$+ \int_0^l \left[(H_2(s, \sigma) \cos \Lambda(s) - H_1(s, \sigma) \sin \Lambda(s)) \cos^2 \Lambda(\sigma) \right.$$

$$\left. - \left(\frac{\partial F_2(s, \sigma)}{\partial \sigma} \cos \Lambda(s) - \frac{\partial F_1(s, \sigma)}{\partial \sigma} \sin \Lambda(s) \right) \sin \Lambda(\sigma) \cos \Lambda(\sigma) \right]$$

$$\left. [C_l(\sigma) \, e(\sigma) + C_m(\sigma)] c^2(\sigma) \, d\sigma \right\} \tag{8}$$

where y, η are, respectively, the projections of s, σ on the y axis, $a_0(s)$ is the two-dimensional lift-curve slope at s, and the integrals are taken over the entire span.   The integral on the left hand side of the equation is a Cauchy integral.

Equation 8 can be solved by the method of successive approximation,[3.12] but the most expedient method for practical application is to reduce it into a matrix equation (§ 3.7).   If the approximation of the "strip" theory can be accepted, the second term, which is a singular integral, drops out, whereas a_0 is replaced by the over-all lift-curve slope of the wing.

Reduction to Matrix Equations.   For a numerical solution, α, C_l, C_m at a number of spanwise coordinates are sought.   It is best to approximate the integral equation by a matrix equation.[4.26, 4.35]   The column matrix $\{cC_l\}$ can be regarded, of course, as a product of a square matrix $\mathbf{c}$ with the column matrix $\mathbf{C}_l$, $\mathbf{c}$ being a diagonal matrix with the ith element in the principle diagonal equal to the chord length c_i at $s = s_i$.

Let $\alpha(s)$, $C_l(s)$, and $cC_l(s)$ be represented by column matrices $\boldsymbol{\alpha}$, $\mathbf{C}_l$, and $\mathbf{c}\mathbf{C}_l$ with elements $\alpha_1, \dot{\alpha}_2, \cdot \cdot \cdot, \alpha_n$; $C_{l1}, C_{l2}, \cdot \cdot \cdot, C_{ln}$, and $c_1 C_{l1}, c_2 C_{l2}, \cdot \cdot \cdot, c_n C_{ln}$ specified at $s = s_1, s_2, \cdot \cdot \cdot, s_n$, respectively.   The relation between $\boldsymbol{\alpha}$ and $\mathbf{C}_l$ is as follows:

1. *Strip Theory.*

$$\mathbf{C}_l = a\boldsymbol{\alpha} \tag{9}$$

where a is the lift-curve slope of the wing corrected for the aspect ratio and planform.

2. *Lifting-Line Theory: Glauert's Solution.*   Prandtl's lifting-line equation, in Fourier series form, is (p. 139 of Ref. 1.43)

$$C_l = \frac{4b}{c} \sum_{m=1}^{n} A_m \sin m\psi = a_0 \left\{ \alpha - \sum_{m=1}^{n} m A_m \frac{\sin m\psi}{\sin \psi} \right\} \tag{10}$$

where b denotes the total span, c denotes wing chord, $\psi =$ arc cos $(2y/b)$, and A_m are unknown coefficients.   If we write this equation for n values of y, using the known values of b, c, a_0, and α, we obtain n linear equations for A_m:

$$\sum_{m=1}^{n} A_m \left(\frac{4b}{a_0 c_i} \sin m\psi_i + m \frac{\sin m\psi_i}{\sin \psi_i} \right) = \alpha_i \qquad (i = 1, 2, \cdot \cdot \cdot, n) \tag{11}$$

where

$$\psi_i = \text{arc cos } (2y_i/b), \qquad \alpha_i = (\alpha)_{y=y_i}, \qquad c_i = (c)_{y=y_i}$$

Let

$$\boldsymbol{\Sigma} \equiv \{\Sigma_{mi}\} = \left\{ \frac{4b}{a_0 c_i} \sin m\psi_i + m \frac{\sin m\psi_i}{\sin \psi_i} \right\} \qquad (m, i = 1, 2, \cdot \cdot \cdot, n) \tag{12}$$

then Eq. 11 may be written as

$$\mathbf{A}' \, \boldsymbol{\Sigma} = \boldsymbol{\alpha}' \qquad (13)$$

where $\mathbf{A}'$, $\boldsymbol{\alpha}'$ denotes the transpose* of $\mathbf{A}$ and $\boldsymbol{\alpha}$, respectively. Hence,

$$\mathbf{A}' = \boldsymbol{\alpha}' \, \boldsymbol{\Sigma}^{-1} \qquad (14)$$

Let

$$\boldsymbol{\Theta} = \{\Theta_{mi}\} = 4b \, \{\sin m\psi_i\} \qquad (15)$$

Then, according to Eqs. 10, 14, we obtain

$$c\mathbf{C}_l' = \mathbf{A}'\boldsymbol{\Theta} = \boldsymbol{\alpha}'\boldsymbol{\Sigma}^{-1}\boldsymbol{\Theta} \qquad (16)$$

Hence,

$$c\mathbf{C}_l = \boldsymbol{\Theta}'(\boldsymbol{\Sigma}^{-1})'\boldsymbol{\alpha} \qquad (17)$$

or

$$\boldsymbol{\alpha} = \boldsymbol{\Sigma}'(\boldsymbol{\Theta}^{-1})'c\mathbf{C}_l \qquad (18)$$

Similar matrix representation for other methods of solving the lifting-line equation can be made.[4.38] In all cases, we may write the result as

$$c\mathbf{C}_l = \mathcal{A}\boldsymbol{\alpha} \qquad (19)$$

where $\mathcal{A}$ is a square matrix, to be identified as ac according to Eq. 9, or as $\boldsymbol{\Theta}'(\boldsymbol{\Sigma}^{-1})'$ according to Eq. 17.

The deflection of the control surface may be regarded as an initial twist and can be included in $\alpha^{(r)}$. The equivalent angle of attack corresponding to a control-surface deflection β is

$$\alpha^{(r)} = \frac{1}{a_0} \frac{\partial C_l}{\partial \beta} \beta \qquad (20)$$

where a_0 and $\partial C_l/\partial \beta$ are the two-dimensional values of the airfoil section. Of course, C_m must be properly related to β.

Equation 4 can be represented in matrix form as follows:

$$\frac{\partial \mathbf{w}}{\partial \mathbf{s}} = q\left[\mathbf{F}_1 \cos\boldsymbol{\Lambda}\, c\mathbf{S}\mathbf{C}_l + \left(\mathbf{H}_1 \cos^2\boldsymbol{\Lambda} - \frac{\partial \mathbf{F}_1}{\partial\boldsymbol{\sigma}} \sin\boldsymbol{\Lambda}\cos\boldsymbol{\Lambda} \right) c^2\mathbf{S}(e\mathbf{C}_l + \mathbf{C}_m) \right] \qquad (21)$$

A similar expression for $\boldsymbol{\theta}$ is obtained by replacing $\mathbf{F}_1$, $\mathbf{H}_1$ by $\mathbf{F}_2$, $\mathbf{H}_2$, respectively. In Eq. 21. $\mathbf{e}$, $\sin\boldsymbol{\Lambda}$, $\cos\boldsymbol{\Lambda}$ are diagonal matrices whose elements are, respectively, values of e, $\sin\Lambda$, and $\cos\Lambda$ evaluated at the points $(s_1, s_2, \cdots, s_n)$. $\mathbf{S}$ is a diagonal matrix of "weights" defined by Eq. 14 of § 3.7.

---

* Interchange of rows and columns.

Combining Eq. 21 with Eqs. 3, 5, and 19, we obtain

$$c C_l = \mathcal{A}\alpha = \mathcal{A}[\alpha^{(r)} + \alpha^{(e)}]$$

$$= \mathcal{A}\left[\alpha^{(r)} + \cos\Lambda\theta - \sin\Lambda\frac{\partial w}{\partial s}\right] \tag{22}$$

$$= \mathcal{A}[\alpha^{(r)} + q\mathbf{E}cC_l + q\mathbf{G}c^2 C_m]$$

where

$$\mathbf{E} = (\cos\Lambda F_2 - \sin\Lambda F_1)\cos\Lambda S + \left[(\cos\Lambda H_2 - \sin\Lambda H_1)\cos^2\Lambda\right.$$

$$\left. - \left(\cos\Lambda\frac{\partial F_2}{\partial\sigma} - \sin\Lambda\frac{\partial F_1}{\partial\sigma}\right)\sin\Lambda\cos\Lambda\right]ceS \tag{23}$$

$$\mathbf{G} = \left[(\cos\Lambda H_2 - \sin\Lambda H_1)\cos^2\Lambda - \left(\cos\Lambda\frac{\partial F_2}{\partial\sigma} - \sin\Lambda\frac{\partial F_1}{\partial\sigma}\right)\sin\Lambda\cos\Lambda\right]S$$

For conventional airfoils C_m depends on β, but is independent of α. Equation 22 can be written as

$$(\mathbf{I} - q\mathcal{A}\mathbf{E})cC_l = \mathcal{A}[\alpha^{(r)} + q\mathbf{G}c^2 C_m] \tag{24}$$

Hence,

$$cC_l = (\mathbf{I} - q\mathcal{A}\mathbf{E})^{-1}\mathcal{A}[\alpha^{(r)} + q\mathbf{G}c^2 C_m] \tag{25}$$

The lift per unit span qcC_l can therefore be obtained by matrix operations. For a given value of q, the inverse $(\mathbf{I} - q\mathcal{A}\mathbf{E})^{-1}$ can be calculated by Crout's method.

When the matrix $\mathbf{I} - q\mathcal{A}\mathbf{E}$ is singular, i.e., when the determinant $|\mathbf{I} - q\mathcal{A}\mathbf{E}|$ vanishes, the inverse $(\mathbf{I} - q\mathcal{A}\mathbf{E})^{-1}$ does not exist. The lift distribution corresponding to a change in angle of attack $\alpha^{(r)}$ becomes indeterminate (tending to infinity). The value of q that satisfies the condition

$$|\mathbf{I} - q\mathcal{A}\mathbf{E}| = 0 \tag{26}$$

is the dynamic pressure at divergence.

An eigenvector (i.e., a column matrix) $\mathbf{u}$ satisfying the equation

$$\left(\mathcal{A}\mathbf{E} - \frac{1}{q_{\text{div}}}\mathbf{I}\right)\mathbf{u} = 0 \tag{27}$$

can be determined by a method of iteration as follows. Starting with an arbitrary nonvanishing vector $\mathbf{u}_0$, we compute successively the following sequence:

$$\mathbf{u}_0$$
$$\mathbf{u}_1 = \mathcal{A}\mathbf{E}\mathbf{u}_0$$
$$\mathbf{u}_2 = \mathcal{A}\mathbf{E}\mathbf{u}_1 = (\mathcal{A}\mathbf{E})^2\mathbf{u}_0 \tag{28}$$
$$\cdots\cdots\cdots\cdots\cdots$$
$$\mathbf{u}_n = \mathcal{A}\mathbf{E}\mathbf{u}_{n-1} = (\mathcal{A}\mathbf{E})^n\mathbf{u}_0$$

It can be shown that, if q_{div} is the smallest (in absolute value), simple, real eigenvalue,*

$$\lim_{n \to \infty} \frac{u_{n-1}}{u_n} \to q_{\mathrm{div}} \qquad (29)$$

and, aside from a numerical factor,

$$\lim_{n \to \infty} \mathbf{u}_n \to \mathbf{u} \qquad (30)$$

where $\mathbf{u}$ is an eigenvector satisfying Eq. 27 and the ratio u_{n-1}/u_n means the ratio of the corresponding elements in the column matrices $\mathbf{u}_{n-1}$ and $\mathbf{u}_n$. Equation 29 is useful for calculating the divergence speed.

Let us choose a particular vector $\mathbf{u}_0$:

$$\mathbf{u}_0 = \mathcal{A}[\boldsymbol{\alpha}^{(r)} + q\mathbf{Gc}^2\mathbf{C}_m] \qquad (31)$$

Equation 25 can be written as

$$\begin{aligned}
\mathbf{cC}_l &= (\mathbf{I} - q\mathcal{A}\mathbf{E})^{-1}\mathbf{u}_0 \\
&= \mathbf{u}_0 + \sum_{k=1}^{\infty}(q\mathcal{A}\mathbf{E})^k\mathbf{u}_0 \qquad (32) \\
&= \mathbf{u}_0 + q\mathbf{u}_1 + q^2\mathbf{u}_2 + \cdots + q^k\mathbf{u}_k + \cdots
\end{aligned}$$

Let m be so large that $\mathbf{u}_m$ approximates the eigenvector $\mathbf{u}$ with negligible error. Then, for $k \geqslant m$,

$$\mathbf{u}_{k+l} = (\mathcal{A}\mathbf{E})^l\mathbf{u}_k \doteq \left(\frac{1}{q_{\mathrm{div}}}\right)^l\mathbf{u}_k \qquad (33)$$

The remainder after m terms in the series on the right-hand side of Eq. 32 can be summed as follows:

$$\begin{aligned}
\sum_{l=0}^{\infty}q^{m+l}\mathbf{u}_{m+l} &= q^m\sum_{l=0}^{\infty}q^l\left(\frac{1}{q_{\mathrm{div}}}\right)^l\mathbf{u}_m \\
&= q^m\frac{1}{1 - q/q_{\mathrm{div}}}\mathbf{u}_m \qquad (34)
\end{aligned}$$

Therefore,

$$\mathbf{cC}_l = \mathbf{u}_0 + q\mathbf{u}_1 + \cdots + q^{m-1}\mathbf{u}_{m-1} + q^m(1 - q/q_{\mathrm{div}})^{-1}\mathbf{u}_m \qquad (35)$$

The series expansion can be justified* for $q < |q_{\mathrm{div}}|$. The form of the

---

* See Ref. 3.20, §§4.14–4.17, p. 134. The existence of a real eigenvalue cannot be assured when the matrix **AE** is unsymmetric. For Eq. (35), see Ref. 3.20, §3.9, p. 81.

solution 35 was first given by Pines.[4.30] In practical applications, it is often sufficient to take $m = 2$ or 3.

For certain sweptback wings, q_{div} is negative. Physically the wing will not diverge. Yet the convergence of the series in Eq. 35 fails when $q \geqslant |q_{\text{div}}|$. A method of extending the radius of convergence is discussed by Gaugh and Slap.[4.29]

Aileron Reversal. The rolling moment about the airplane centerline is

$$M_\phi = \int_{-b/2}^{b/2} y c C_l(y)\, dy \doteq \{y_n\}' \mathbf{Sc} \mathbf{C}_l \qquad (36)$$

where $\{y_n\}'$ is a row matrix with y_n as elements.

If we take $\alpha^{(r)}$ and C_M in Eq. 25 as those corresponding to the aileron deflection β, the solution q of the equation

$$M_\phi = 0 \qquad (\beta \neq 0) \qquad (37)$$

gives the aileron reversal pressure.

Stability Derivatives. Knowing $\mathbf{cC}_l$ corresponding to various downwash distributions $\alpha^{(r)}$, we can compute a number of stability derivatives. For example, if the rolling moment M_ϕ due to a steady rolling velocity p is desired, we may set $\alpha^{(r)} = \dfrac{p}{U} y$ and calculate the corresponding cC_l and M_ϕ according to Eqs. 25 and 36. Then the stability derivative $\partial M_\phi / \partial(pb/2U)$ can be easily obtained.

Example (Fig. 4.9). Consider a normal rectangular wing of uniform chord and stiffness. $R = b/c = 5.7$, $a_0 = 5.7$ (two-dimensional). $a = 4.16$ (corrected for R). Let 7 stations be taken across the span:

$$y_n = \frac{b}{2} \cos \frac{n\pi}{8} \qquad (n = 1, 2, \cdots, 7) \qquad (38)$$

i.e.,

$$\mathbf{y}_n = \frac{b}{2} \left\{ \begin{array}{r} 0.92388 \\ 0.70711 \\ 0.38268 \\ 0 \\ -0.38268 \\ -0.70711 \\ -0.92388 \end{array} \right\}$$

Using Prandtl's lifting-line theory, find the symmetrical lift distribution corresponding to a change of angle of attack at the wing root.

Since $\Lambda = 0$, only $F_2(s, \sigma)$ and $H_2(s, \sigma)$ are needed. For a uniform beam having a straight elastic axis, $F_2(s, \sigma) = 0$, and

$$H_2(s, \sigma) = \begin{cases} \int_0^s \dfrac{dy}{G\,J(y)} = \dfrac{1}{GJ}\,s, & \text{for} \quad 0 \leqslant s \leqslant \sigma \\[3mm] \int_0^\sigma \dfrac{dy}{G\,J(y)} = \dfrac{1}{GJ}\,\sigma, & \text{for} \quad \sigma \leqslant s \leqslant \dfrac{b}{2} \end{cases} \tag{39}$$

In Eq. 39, s, σ are measured from the wing root. These expressions are valid for positive s, σ. For the other half of the wing, the sign on the

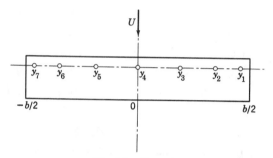

Fig. 4.9. Example.

right-hand side of Eq. 39 should be changed so that $H_2(s, \sigma)$ remain positive.

In expressing $H_2(s, \sigma)$ in the form of a matrix, let the points of loading be $y_1, y_2, \cdots, y_n$ specified by Eq. 38. Then

$$H_2 = \frac{1}{GJ}\frac{b}{2} \begin{matrix} & \sigma \to (y_1) & (y_2) & (y_3) & (y_4) \\ (y_1) & 0.92388 & 0.70711 & 0.38268 & 0 \\ (y_2) & 0.70711 & 0.70711 & 0.38268 & 0 \\ (y_3) & 0.38268 & 0.38268 & 0.38268 & 0 \\ (y_4) & 0 & 0 & 0 & 0 \end{matrix} \tag{40}$$

The auxiliary matrices $\{\cos^2 \Lambda\}$ and $\{ce\}$ are diagonal matrices. Since in this example Λ and ce are constants, we have

$$\{\cos^2 \Lambda\} = \cos^2 \Lambda \, \mathbf{I} = \mathbf{I}, \quad \text{for} \quad \Lambda = 0$$
$$\{ce\} = ce \, \mathbf{I} \tag{41}$$

The matrix $\mathcal{A}$ is to be found by the Glauert's method. The lift distribution being symmetrical, only terms of odd indices in Eq. 10 differ from zero. Hence, let $m = 1, 3, 5, 7$:

$$C_l = \frac{4b}{c} \sum_{M=1,3,5,7} A_m \sin m\theta \tag{42}$$

The matrix $\boldsymbol{\Sigma}$ defined by Eq. 12 depends on the quantity $4b/a_0c$ which is 4 in this example:

$$\frac{4b}{a_0c} = 4 \tag{43}$$

Then $\boldsymbol{\Sigma}$ can be easily calculated:

$$
\boldsymbol{\Sigma} =
\begin{Bmatrix}
m \quad i \rightarrow & (1) & (2) & (3) & (4) \\
(1) & 2.5308 & 3.8284 & 4.6956 & 5 \\
(3) & 10.9382 & 5.8284 & -2.7734 & -7 \\
(5) & 15.7666 & -7.8284 & -3.6018 & 9 \\
(7) & 8.5308 & -9.8284 & 10.6956 & -11
\end{Bmatrix}
\tag{44}
$$

Its inverse is*

$$
\boldsymbol{\Sigma}^{-1} =
\begin{Bmatrix}
0.02314 & 0.03824 & 0.02818 & 0.00924 \\
0.06346 & 0.04752 & -0.02939 & -0.02544 \\
0.09336 & -0.02246 & -0.02086 & 0.03966 \\
0.05202 & -0.03464 & 0.02783 & -0.02245
\end{Bmatrix}
\tag{45}
$$

Moreover,

$$
\{4 \sin m\theta_i\} =
\begin{Bmatrix}
m \quad i \rightarrow (1) & (2) & (3) & (4) \\
(1) \; 1.5308 & 2.8284 & 3.6956 & 4 \\
(3) \; 3.6956 & 2.8284 & -1.5308 & -4 \\
(5) \; 3.6956 & -2.8284 & -1.5308 & 4 \\
(7) \; 1.5308 & -2.8284 & 3.6956 & -4
\end{Bmatrix}
\tag{46}
$$

Hence,

$$\mathcal{A} = \boldsymbol{\Theta}'(\boldsymbol{\Sigma}^{-1})' = b\{4 \sin m\theta_i\}'(\boldsymbol{\Sigma}^{-1})' \tag{47}$$

$$
\mathcal{A} = b
\begin{Bmatrix}
0.29503 & 0.12520 & 0.04353 & 0.02010 \\
0.06777 & 0.46898 & 0.14736 & 0.03394 \\
0.01799 & 0.11275 & 0.55790 & 0.11970 \\
0.01536 & 0.04796 & 0.22120 & 0.54776
\end{Bmatrix}
\tag{48}
$$

---

* Though not tabulated here, 5 significant figures were obtained for $\boldsymbol{\Sigma}^{-1}$ and were used for subsequent calculations.

The matrix S is the weighting factor for the integration. Using Multhopp's formula, (§ 3.7, Eq. 16), we have

$$S = \frac{\pi}{8}\frac{b}{2}\begin{Bmatrix} \sin\theta_1 & 0 & 0 & 0 \\ 0 & \sin\theta_2 & 0 & 0 \\ 0 & 0 & \sin\theta_3 & 0 \\ 0 & 0 & 0 & (^1/_2)\sin\theta_4 \end{Bmatrix}$$

$$= \frac{\pi b}{16}\begin{Bmatrix} 0.38268 & 0 & 0 & 0 \\ 0 & 0.70711 & 0 & 0 \\ 0 & 0 & 0.92388 & 0 \\ 0 & 0 & 0 & 0.5 \end{Bmatrix} \tag{49}$$

Since only the lift distribution corresponding to an angle-of-attack increment

$$\alpha^{(r)} = \text{const} \tag{50}$$

is required, we have

$$\mathbf{E} = ce\,\mathbf{H}_2\cdot\mathbf{S} = \frac{\pi ceb^2}{32GJ}\begin{Bmatrix} 0.35355 & 0.50000 & 0.35355 & 0 \\ 0.27060 & 0.50000 & 0.35355 & 0 \\ 0.14644 & 0.27060 & 0.35355 & 0 \\ 0 & 0 & 0 & 0 \end{Bmatrix} \tag{51}$$

Hence,

$$\mathcal{A}\mathbf{E} = \frac{\pi ceb^3}{32GJ}\begin{Bmatrix} 0.14456 & 0.022189 & 0.16396 & 0 \\ 0.17244 & 0.30825 & 0.24187 & 0 \\ 0.11857 & 0.21634 & 0.24347 & 0 \\ 0.05080 & 0.09152 & 0.10059 & 0 \end{Bmatrix} \tag{52}$$

Writing $\alpha^{(r)}$ as a column matrix of constants, we obtain $\mathbf{u}_0$ from Eq. 31:

$$\mathbf{u}_0 = \mathcal{A}\alpha^{(r)}\begin{Bmatrix} 1 \\ 1 \\ 1 \\ 1 \end{Bmatrix} = b\alpha^{(r)}\begin{Bmatrix} 0.48386 \\ 0.71805 \\ 0.80834 \\ 0.83228 \end{Bmatrix} \tag{53}$$

The iterated vectors according to Eq. 28 are:

$$\mathbf{u}_1 = Kb\alpha^{(r)}\begin{Bmatrix} 0.36181 \\ 0.50029 \\ 0.40952 \\ 0.17161 \end{Bmatrix}, \qquad \mathbf{u}_2 = K^2b\alpha^{(r)}\begin{Bmatrix} 0.23046 \\ 0.31415 \\ 0.25084 \\ 0.10536 \end{Bmatrix} \tag{54}$$

$$\mathbf{u}_3 = K^3 b\alpha^{(r)} \begin{Bmatrix} 0.14415 \\ 0.19725 \\ 0.15636 \\ 0.06569 \end{Bmatrix}, \qquad \mathbf{u}_4 = K^4 b\alpha^{(r)} \begin{Bmatrix} 0.090243 \\ 0.12347 \\ 0.097833 \\ 0.041103 \end{Bmatrix}$$

where

$$K = \frac{\pi c e b^3}{32 GJ}$$

The ratio of the corresponding elements of the successive iterated vectors is:

$$\frac{u_0}{u_1} = \frac{1}{K} \begin{Bmatrix} 1.3373 \\ 1.4353 \\ 1.9739 \\ 4.8498 \end{Bmatrix}, \qquad \frac{u_1}{u_2} = \frac{1}{K} \begin{Bmatrix} 1.5699 \\ 1.5925 \\ 1.6326 \\ 1.6288 \end{Bmatrix},$$

$$\frac{u_2}{u_3} = \frac{1}{K} \begin{Bmatrix} 1.5987 \\ 1.5926 \\ 1.6042 \\ 1.6038 \end{Bmatrix}, \qquad \frac{u_3}{u_4} = \frac{1}{K} \begin{Bmatrix} 1.5973 \\ 1.5975 \\ 1.5982 \\ 1.5981 \end{Bmatrix} \qquad (55)$$

Clearly the convergence of the ratios to q_{div} is approximately established by $n = 3$, and we may take

$$q_{\text{div}} = 1.598 \frac{32 GJ}{\pi c e b^3} \qquad (56)$$

According to Eq. 35, the lift distribution across the span is given by

$$\mathbf{c C}_l \doteq \mathbf{u}_0 + q\mathbf{u}_1 + q^2\mathbf{u}_2 + q^3 \frac{1}{1 - q/q_{\text{div}}} \mathbf{u}_3 \qquad (57)$$

Using Eq. 56, and defining $\bar{\mathbf{u}}_1$, $\bar{\mathbf{u}}_2$, etc. to be the column matrices given by Eq. 54 but with the factors $K^n b\alpha^{(r)}$ removed, we can write Eq. 57 as

$$\mathbf{C}_l = a_0\alpha^{(r)} \left[\mathbf{u}_0 + 1.598 \frac{q}{q_{\text{div}}} \bar{\mathbf{u}}_1 + \left(\frac{1.598q}{q_{\text{div}}}\right)^2 \bar{\mathbf{u}}_2 + \left(\frac{1.598q}{q_{\text{div}}}\right)^3 \frac{1}{1 - q/q_{\text{div}}} \bar{\mathbf{u}}_3 \right]$$

where a_0 is the two-dimensional lift-curve slope which is equal to $\mathbb{R} = b/c$ in this particular example. For a few special values of the ratio q/q_{div}, the results shown on the next page are obtained. This is plotted in Fig. 4.10 for a change of angle of attack $\alpha^{(r)} = 1/a_0$ radian.

If the wing were perfectly rigid, then $\mathbf{H}_2 = \mathbf{E} = 0$, and the lift distribution is simply given by $\mathbf{u}_0$. This is also plotted in Fig. 4.10 as the $q = 0$ line.

q/q_{div}	0	0.5	0.7	0.8	0.9
C_l matrix for	0.4839	1.0671	1.8495	2.8289	5.7691
	0.7180	1.5196	2.7011	3.9316	7.9551
$a_0 \alpha^{(r)} = 1$	0.8083	1.4552	2.3098	3.3752	6.5675
	0.8323	1.1037	1.4626	1.9101	3.2512

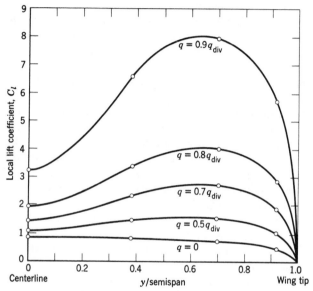

Fig. 4.10. Lift distribution across the span.   ($\alpha^{(r)} = 1/a_0$ radian).

4.5 SWEPT WINGS

For highly swept wings it is necessary to use Weissinger's method to determine the aerodynamic loading. For approximate solutions, however, the modified Schrenk's method[4.53] given by Pope and Haney offers great simplification and fair accuracy. In many occasions, the strip theory is sufficiently accurate.

Besides the change in lift curve slope as shown by Eqs. 6, 7, 8 and 16 of § 4.3, swept wings differ aeroelastically from normal wings in the effective angle of attack due to elastic deformation (Eq. 5, § 4.4):

$$\alpha^{(e)} = \theta \cos \Lambda - \frac{\partial w}{\partial s} \sin \Lambda \qquad (1)$$

where $\partial w/\partial s$ is the slope of the deflection curve in the spanwise direction. Hence, whenever $\Lambda \neq 0$ (Fig. 4.8), the effective angle of attack, and hence the aerodynamic loading, depends on the bending deflection as well as on the torsional deflection. The effect of the bending deflection can be seen qualitatively as follows. For a *sweptback* wing ($\Lambda > 0$), an increase in angle of attack $\alpha^{(r)}$ yields a positive slope $\partial w/\partial s$ and hence reduces $\alpha^{(e)}$. This, together with the reduction of the lift curve slope due to Λ, means that the wing is not so easily twisted as a normal wing. Thus the sweepback increases the torsional stability. Hence, the divergence speed is increased by sweepback. For a *sweptforward* wing ($\Lambda < 0$), the effect of bending tends to decrease the divergence speed, whereas the reduction of lift curve slope tends to increase it. Generally, the divergence speed is decreased by sweepforward.

On the other hand, for a positive (downward) aileron deflection, $\partial w/\partial s$ is again positive and $\alpha^{(e)}$ is again reduced if $\Lambda > 0$, and increased if $\Lambda < 0$. Accordingly, the effect of bending has a tendency to make the aileron less or more effective according to $\Lambda > 0$ or $\Lambda < 0$, respectively. In other words, the effect of bending tends to have the aileron efficiency and the aileron reversal speed reduced by sweepback and increased by sweepforward. Whether a wing's reversal speed would actually be reduced by sweepback or not depends on the balance between several opposing influences, and a careful analysis is required.

For a cantilever wing with the wing root fixed in space, $\partial w/\partial s$ vanishes at the wing root because of the clamping boundary conditions. It increases to a maximum toward the wing tip. Hence, the aforesaid effect of sweep is more seriously felt near the wing tip. It is therefore obvious that, when the pitching angle $\alpha^{(r)}$ of the airplane is increased, the effect of sweepback is to unload the wing tips, and to cause the center of pressure of the wing to move forward. For a sweptforward wing, the load near the wing tip is increased by the elastic bending, and the center of pressure moves forward towards the wing tips. This shift of wing center of pressure has very important effect on the airplane stability.

Detailed presentation of the effects of sweepback and sweepforward can be found in Refs. 4.6, 4.10, 4.26–35.

4.6   TAIL EFFICIENCY

Two aspects of the effect of elastic deformation of the horizontal tail and fuselage will be considered: the elastic efficiency of the horizontal tail and the static longitudinal stability of the airplane. The analysis will be made in an approximate manner. Only unswept wings and tails will be considered.

When an elevator angle changes, the pitching moment about the airplane center of gravity (c.g.) changes. If, originally, the airplane were in the condition of rectilinear symmetric steady flight, then the change of pitching moment induced by the elevator deflection would cause the airplane to pitch, and disturbed motion would ensue. In order to study the efficiency of the tail alone, however, it is more convenient to consider the unbalanced pitching moment, instead of the dynamics of the airplane as a whole. For this purpose, the airplane shall be assumed to be held fixed, against any disturbed motion, at the c.g. of the airplane. The wing and the structures in front of the c.g. shall not be disturbed. In this fictitious condition, the pitching moment (positive if it tends to increase the angle of attack) contributed by the horizontal tail surfaces, about the c.g., is

$$M = -lL_t + M_t \qquad (1)$$

where L_t is the resultant lift of the tail, acting through the tail aerodynamic center; M_t is the pitching moment about the tail aerodynamic center; and l is the distance from the airplane c.g. to the resultant tail lift L_t. Generally, it is sufficiently accurate to assume L_t as normal to the fuselage reference axis, and to neglect the horizontal component of the aerodynamic forces. Then l is a constant independent of the angle of attack.

It is convenient to express the pitching moment about airplane c.g. by a moment coefficient based on wing area and wing chord; and L_t and M_t by coefficients based on tail area and tail chord. Let $(\ )_w$ and $(\ )_t$ denote quantities relating to wing and tail, respectively. Let q denote the dynamic pressure, S the area, c the chord length. We define

$$C_M = \frac{M}{qS_w\bar{c}_w}, \qquad C_{Lt} = \frac{L_t}{q_tS_t}, \qquad C_{Mt} = \frac{M_t}{q_tS_t\bar{c}_t} \qquad (2)$$

The bar over c_w and c_t indicates mean aerodynamic chords. q_t denotes the dynamic pressure at the tail region, which, due to wing-fuselage interference and engine slip stream, may be slightly different from the free-stream dynamic pressure q.

A measure of the tail efficiency is the rate of change of the pitching-moment coefficient with elevator deflection, $\partial C_M/\partial\beta$, β being the (symmetric) elevator deflection angle (positive if deflected downward). Owing to the elastic deformation of the tail and the fuselage, the derivative $\partial C_M/\partial\beta$ for a real airplane is smaller than that of a rigid airplane. The ratio

$$\frac{\partial C_M}{\partial\beta} \bigg/ \left(\frac{\partial C_M}{\partial\beta}\right)_{\text{rigid}} \qquad (3)$$

is called the *elastic efficiency of the elevator*. From Eq. 1, we have

$$\frac{\partial C_M}{\partial \beta} = \frac{q_t S_t}{q S_w} \left(-\frac{l}{\bar{c}_w} \frac{\partial C_{Lt}}{\partial \beta} + \frac{\bar{c}_t}{\bar{c}_w} \frac{\partial C_{Mt}}{\partial \beta} \right) \tag{4}$$

To compute the derivatives $\partial C_{Lt}/\partial \beta$, $\partial C_{Mt}/\partial \beta$, it is necessary to consider the elastic deformation of the airplane. The airplane structure can be represented by a skeleton system of beams as shown in Fig. 4.11. The tail angles of attack shall be represented by a characteristic number θ_t measured at a reference section located at a spanwise coordinate η_t. If

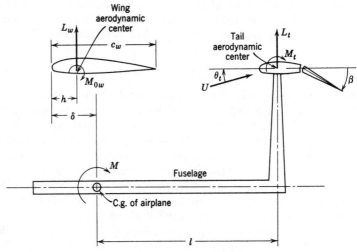

Fig. 4.11. A skeleton airplane.

the tail angle-of-attack distribution is described by $\theta_0 f(y)$, y being the spanwise coordinate, then η_t may be defined according to the following equation:

$$\theta_t = \theta_0 f(\eta_t) = \frac{\theta_0}{S_t} \int_{-s_t}^{s_t} f(y)\, c_t(y)\, dy \tag{5}$$

where s_t is the tail semispan, and S_t is the tail area. Hence θ_t is a weighted average of the tail angle of attack. If a semirigid mode of the tail twisting $f(y)$ is assumed, η_t can be evaluated at once. Generally it lies at 2/3 to 3/4 semispan outboard from the fuselage.

The elastic property of the tail and fuselage may be described by two stiffness-influence coefficients K_1 and K_2 defined as follows. Let K_1 be the total lift force (with $K_1/2$ acting at each of the reference sections on the two halves of the horizontal tail) that is required to act at the tail aerodynamic center to produce a rotation of 1 radian at the tail reference section,* with the fuselage assumed clamped at the airplane c.g. Let K_2 be the total pitching moment (with $K_2/2$ acting at each reference section) that is required to act at the tail reference sections to produce the same rotation. Then the total change of the tail angle of attack is

$$\Delta\theta_t = \frac{L_t}{K_1} + \frac{M_t}{K_2} \tag{6}$$

To evaluate the derivatives involved in Eq. 4, it is only necessary to consider a small deflection angle $\Delta\beta$. In the following, the quantities β, θ_t, L_t, and M_t will denote the changes corresponding to $\Delta\beta$, and the symbol Δ will be omitted. Let us assume that the strip assumption may be used and that the lift-curve slope a_t and the derivatives $\partial C_{Lt}/\partial\beta \equiv a_2$, $\partial C_{Mt}/\partial\beta \equiv -m$ of the tail airfoil section be evaluated at the steady flight conditions. Then, if θ_t is defined as in Eq. 5, we have

$$\begin{aligned} L_t &= q_t S_t (a_t \theta_t + a_2 \beta) \\ M_t &= - m\beta q_t S_t c_t \end{aligned} \tag{7}$$

Combining Eqs. 6 and 7, we obtain

$$\frac{\partial\theta_t}{\partial\beta} = \frac{A}{B} \tag{8}$$

where**

$$A = q_t S_t \left(\frac{a_2}{K_1} - \frac{mc_t}{K_2} \right), \qquad B = 1 - \frac{1}{K_1} q_t S_t a_t \tag{9}$$

From Eqs. 7 and 8, and according to the definitions 2, we derive

$$\frac{\partial C_{Lt}}{\partial\beta} = a_t \frac{A}{B} + a_2 \qquad \frac{\partial C_{Mt}}{\partial\beta} = - m \tag{10}$$

Hence, from Eq. 4,

$$\frac{\partial C_M}{\partial\beta} = - \frac{q_t S_t}{q S_w} \left[\frac{l}{\bar{c}_w} \left(a_t \frac{A}{B} + a_2 \right) + \frac{\bar{c}_t}{\bar{c}_w} m \right] \tag{11}$$

If the airplane were perfectly rigid, then $K_1 = K_2 = \infty$, $A = 0$, and $B = 1$, so

$$\left(\frac{\partial C_M}{\partial\beta} \right)_{\text{rigid}} = - \frac{q_t S_t}{q S_w} \left[\frac{l}{\bar{c}_w} a_2 + \frac{m\bar{c}_t}{\bar{c}_w} \right] \tag{12}$$

---

* Actually we are interested only in small deflections, so the linearity of the structural property can be assumed.

**If the aerodynamic moment of the elevator is resisted by a control stick, the quantity m in A, Eq. 9 and Eq. 16 *infra*, should be replaced by $\partial C_{Mt}/\partial\beta - \partial C_{Ht}/\partial\beta$ where C_{Ht} is the hinge moment coefficient of the elevator based on the tail area and tail chord. Elsewhere no change is needed.

The elastic efficiency of the elevator is therefore

$$\frac{\partial \dot{C}_M}{\partial \beta} \bigg/ \left(\frac{\partial C_M}{\partial \beta}\right)_{\text{rigid}} = 1 + a_t \frac{Al}{B\bar{c}_w}\left(\frac{l}{\bar{c}_w}a_2 + \frac{m\bar{c}_t}{\bar{c}_w}\right)^{-1} \tag{13}$$

$$\doteq 1 + \frac{A}{B}\frac{a_t}{a_2} \quad (\text{if } \bar{c}_t \ll l) \tag{14}$$

We may define, in analogy with the wing divergence, the *critical divergence speed of the horizontal tail* at which an infinitesimal change in β induces a large tail twist. According to Eq. 8 this occurs when $B = 0$, i.e., when

$$q_{t\,\text{div}} = \frac{K_1}{S_t a_t} \tag{15}$$

If K_1 is negative, then $q_{t\,\text{div}}$ given by Eq. 15 has no physical meaning, but is merely a parameter showing that the tail is stable.

We may also define a critical *horizontal tail-control reversal speed* at which a change of elevator angle produces no change in the pitching moment about the airplane c.g. This occurs when the tail efficiency becomes zero. Using the approximate formula 14, we obtain, at the reversal speed,

$$1 + \frac{A}{B}\frac{a_t}{a_2} = 0$$

which gives, according to Eq. 9,

$$q_{t\,\text{rev}} = \frac{K_2}{S_t \bar{c}_t}\frac{a_2}{a_t m} \tag{16}$$

Note that $q_{t\,\text{rev}}$ is independent of K_1, because at the reversal speed the tail lift due to elevator deflection vanishes. Using Eqs. 15 and 16, we may write

$$\text{Elastic efficiency of elevator} = \frac{1 - q_t/q_{t\,\text{rev}}}{1 - q_t/q_{t\,\text{div}}} \tag{17}$$

4.7 THE EFFECT OF ELASTIC DEFORMATION ON THE STATIC LONGITUDINAL STABILITY OF AN AIRPLANE

The static longitudinal stability is measured by the derivative $\partial C_M/\partial C_L$ at the symmetrical level flight condition, where C_L is the total lift coefficient and C_M is the coefficient of the pitching moment about the airplane center of gravity. An airplane is statically stable when $\partial C_M/\partial C_L$ is negative. In power-off condition and for the rearmost center of gravity location, a value of $-\partial C_M/\partial C_L$ from 0.10 to 0.15 usually leads to satisfactory results. A study of the pitching moment about the center of gravity in steady flight is a very complex one when the effects of power, component interference, and free controls are taken into account. However, a large amount of experimental data has been gathered in the past, and the determination of the $\partial C_M/\partial C_L$ derivative for a *rigid airplane* offers no serious difficulty, even in the most complicated cases.*

If the airplane is not rigid, the static longitudinal stability derivative will be affected by the elastic bending of the fuselage and the twisting of the wing and tail surfaces. Let us consider, as an example, an airplane with unswept wing and tail and having the control stick fixed (i.e., elevator locked). To account for the effect of elastic deformation approximately, we shall take a reference section of the wing and a reference section of the tail determined in the same manner as in § 4.6.† Let the angles of attack of the wing and the tail be α_{w0} and α_{t0} respectively, when the total lift coefficient C_L vanishes. If the fuselage is now rotated through an angle α about the airplane center of gravity, the new angle of attack of the wing will be $(\alpha_{w0} + \alpha + \theta_w)$, and that of the tail, $(\alpha_{t0} + \alpha + \theta_t)$; θ_w, θ_t being the elastic twisting angles of the wing and tail, respectively. θ_t can be found from the equations of § 4.6. In the case of a locked elevator, we may regard the tail as a full-chord elevator and put

$$m = 0, \qquad a_2 = a_t, \qquad \beta = \alpha + \alpha_{t0}$$

from which

$$A = q_t/q_{t\,\text{div}}, \qquad B = 1 - A \tag{1}$$

To derive an expression that gives the total angle of twist of the tail as a result of the rotation α, let us add a term $C_{M0t}q_tS_tc_t$ to the right-hand side of the expression for M_t in Eq. 7 of § 4.6, to represent the moment about the aerodynamic center at zero lift. Substituting L_t and M_t from Eqs. 7 of § 4.6 into the relation

$$\theta_t = \frac{L_t}{K_1} + \frac{M_t}{K_2} \tag{2}$$

and solving for θ_t, we obtain

$$\theta_t = \frac{q_t/q_{t\,\text{div}}}{1 - q_t/q_{t\,\text{div}}}(\alpha + \alpha_{t0}) + \frac{q_tS_t}{1 - q_t/q_{t\,\text{div}}}\frac{\bar{c}_tC_{M0t}}{K_2} \tag{3}$$

A similar argument applied to the wing leads to

$$\theta_w = \frac{q/q_{w\,\text{div}}}{1 - q/q_{w\,\text{div}}}(\alpha + \alpha_{w0}) + \frac{qS_w}{1 - q/q_{w\,\text{div}}}\frac{\bar{c}_wC_{M0w}}{K_3} \tag{4}$$

* See for example, C. B. Millikan, *Aerodynamics of the Airplane*, Chapter 4, John Wiley & Sons (1941). C. D. Perkins, and R. E. Hage. *Airplane Performance, Stability and Control*, Chapters 5 and 6, John Wiley & Sons (1949).

† The characteristic wing-twisting angle θ_w and the spanwise location of the wing reference section η_w will be defined by an equation analogous with Eq. 5 of § 4.6.

where K_3 denotes the spring constant for wing twisting with respect to airplane c.g., K_3 is the torque required to act at the wing-reference section to produce a change of angle of attack of 1 radian at the wing-reference section with respect to a fuselage that is clamped at the c.g. Equation 4 is a slightly more general expression than Eq. 7 of § 3.1.

Neglecting the tail lift in comparison with the wing lift, we obtain the total lift coefficient of the airplane from Eq. 4:

$$C_L = a_w(\alpha_{w0} + \alpha + \theta_w)$$

$$= \frac{a_w}{1 - q/q_{w\,\text{div}}}\left\{\alpha + \alpha_{w0} + \frac{qS_w\bar{c}_w C_{M0w}}{K_3}\right\} \tag{5}$$

Solving Eq. 5 for α, and substituting into

$$C_{Lt} = a_t(\alpha_{t0} + \alpha + \theta_t)$$

we obtain

$$C_{Lt} = \frac{a_t}{1 - q_t/q_{t\,\text{div}}}\left\{\left(1 - \frac{q}{q_{w\,\text{div}}}\right)\frac{C_L}{a_w} + \alpha_{t0} - \alpha_{w0}\right.$$
$$\left. - \frac{qS_w\bar{c}_w C_{M0w}}{K_3} + \frac{q_t S_t c_t C_{M0t}}{K_2}\right\} \tag{6}$$

$$C_{Mt} = C_{M0t}$$

If the airplane were perfectly rigid, then $q_{w\,\text{div}}$ and $q_{t\,\text{div}}$ are infinitely large and C_{Lt} and C_{Mt} are obtained from Eqs. 6 by putting all terms involving $q_{w\,\text{div}}$, $q_{t\,\text{div}}$, K_2, and K_3 to be zero. Therefore the change of C_{Lt} and C_{Mt} due to the elastic deformation of the airplane can be easily calculated. The total pitching moment about the airplane c.g. is, clearly from Fig. 4.11,

$$M = L_w(\delta - h) + M_{0w} - L_t l + M_t \tag{7}$$

or, in coefficients form,

$$C_M = C_L\left(\frac{\delta}{c} - \frac{h}{c}\right) + C_{M0w} - \frac{l}{\bar{c}_w}\frac{q_t S_t}{qS_w}C_{Lt} + \frac{\bar{c}_t q_t S_t}{\bar{c}_w qS_w}C_{Mt} \tag{8}$$

where δ is the distance of the airplane c.g. behind the wing leading edge, h is the location of wing aerodynamic center behind the wing leading edge, and M_{0w} is the aerodynamic moment of the wing about its aerodynamic center. Hence, the change of the static longitudinal stability derivative due to the elastic deformation of the airplane is

$$\Delta\left(\frac{\partial C_M}{\partial C_L}\right) = -\frac{l}{\bar{c}_w}\frac{q_t S_t}{qS_w}\Delta\left(\frac{\partial C_{Lt}}{\partial C_L}\right) + \frac{\bar{c}_t q_t S_t}{\bar{c}_w qS_w}\Delta\left(\frac{\partial C_{Mt}}{\partial C_L}\right) \tag{9}$$

where Δ indicates the difference of the values of the derivatives from those of a hypothetical rigid airplane. From Eqs. 6, we have

$$\Delta \left(\frac{\partial C_{Lt}}{\partial C_L} \right) = \frac{a_t}{a_w} \left[\frac{1 - q/q_{w\,\text{div}}}{1 - q_t/q_{t\,\text{div}}} - 1 \right] \tag{10}$$

$$\Delta \left(\frac{\partial C_{Mt}}{\partial C_L} \right) = 0 \tag{11}$$

Hence,

$$\Delta \left(\frac{\partial C_M}{\partial C_L} \right) = \frac{q_t}{q} \frac{l S_t q_t}{\bar{c}_w S_w q_w} \left[\frac{q/q_{w\,\text{div}} - q_t/q_{t\,\text{div}}}{1 - q_t/q_{t\,\text{div}}} \right] \tag{12}$$

The static longitudinal-stability derivative of an elastic airplane is therefore

$$\left(\frac{\partial C_M}{\partial C_L} \right)_{\text{elastic}} = \left(\frac{\partial C_M}{\partial C_L} \right)_{\text{rigid}} + \Delta \left(\frac{\partial C_M}{\partial C_L} \right) \tag{13}$$

From Eq. 12 it is seen that, for an aeroelastically stable airplane, the effect of elastic deformation on the static longitudinal-stability derivative depends on the relative magnitude of the divergence dynamic pressures of the wing and tail. For instance, if the wing is so rigid that $|q_{w\,\text{div}}| \gg |q_{t\,\text{div}}|$, while the fuselage and tail are such that $q_{t\,\text{div}}$ is positive, then a strong stabilizing effect will occur as q approaches $q_{t\,\text{div}}$. On the other hand, if $q_{w\,\text{div}}$ is nearly equal to $q_{t\,\text{div}}$ or, more precisely, if $q_{w\,\text{div}}/q_{t\,\text{div}} = q/q_t$ then the effect of the elastic deformation on the static longitudinal stability disappears.

The same method can be used to study the change of stabilizer "trim" due to elastic deformation of the airplane. The trim problem is, for high-speed airplanes, one of the most important factors in tail design.

4.8 TWISTING OF PROPELLER BLADES

Divergence has an important bearing on many phases of the aeroelastic stability of propellers, not only because the steady-state instability in itself must be avoided, but also because of its close relationship to flutter. It was shown by Theodorsen and Regier[4.36] that in some cases the blade simply twists to the stalling angle and initiates stall flutter when the speed is too close to the divergence speed. Thus the problem of predicting propeller flutter is resolved primarily into the calculation of the speed at which the propeller will stall due to its aeroelastic twist, and the critical-divergence speed becomes a convenient criterion.

The divergence of propellers can be analyzed by the same methods as those exposed before, except that the effect of the centrifugal force must be added.

In a good propeller design, large bending stresses are avoided by making the normal component of the centrifugal force balance approximately the aerodynamic lift.   Hence, it is a fair approximation to assume that the twisting moment acting on the blade consists in a couple formed by a lift force and a centrifugal-force component which is assumed equal to the lift.

Consider a representative section of the propeller (Fig. 4.12).   Let c be

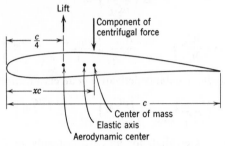

Fig. 4.12.   A propeller section.

the length of the chord, and xc be the distance from the center of mass to the leading edge.   Then the twisting moment is

$$(\text{Lift}) \cdot c(x - \tfrac{1}{4})$$

This is balanced by the elastic moment.   Let K be the torsional stiffness of the representative section defined as in Eq. 5 of § 3.3; then the equilibrium condition is

$$K\theta = sc^2(x - \tfrac{1}{4})qa(\alpha_r + \theta - \alpha_0) \tag{1}$$

where α_r is the angle of attack of the section if the blade is perfectly rigid, θ the angle of twist due to elasticity, α_0 the angle of attack for which there is no twist (to be explained later), q the dynamic pressure of the relative airstream, s a representative length of propeller blade, and a the lift-curve slope of the section.

The condition of divergence of the blade is that $\theta \to \infty$; i.e.,

$$\frac{\theta}{\alpha_r + \theta - \alpha_0} \to 1$$

The critical-divergence speed is then given by

$$q_{\text{div}} = \frac{K}{sc^2a(x - \tfrac{1}{4})} \tag{2}$$

Combining Eqs. 1 and 2, we obtain

$$\theta = (\alpha_r - \alpha_0)\frac{q/q_{\text{div}}}{1 - q/q_{\text{div}}} \tag{3}$$

From this equation, we see that the angle of twist increases rapidly when $q \to q_{\text{div}}$.

The conventional propeller design is based on the assumption of perfectly rigid blades. The design condition is stated in terms of α_r or the corresponding lift coefficient C_{Lr}. The true lift coefficient C_L of the elastic propeller is related to C_{Lr} by the relation

$$C_L = C_{Lr} + a\theta \tag{4}$$

The angle of attack for zero twist α_0 can be calculated from the equilibrium of aerodynamic force and the centrifugal force. In the no-twist condition, the couple of the centrifugal force and the lift just balances the aerodynamic moment about the aerodynamic center, for which the moment coefficient is C_{M0}. Hence, the equilibrium condition is

$$C_{M0} + a\alpha_0(x - \tfrac{1}{4}) = 0 \tag{5}$$

provided that the angles α_0 and α_r are measured from the zero-lift line. Combining Eqs. 3 and 5 with 4, we obtain

$$C_L = C_{Lr}\left(\frac{1}{1 - q/q_{\text{div}}}\right) + \frac{C_{M0}}{x - \tfrac{1}{4}}\frac{q/q_{\text{div}}}{1 - q/q_{\text{div}}} \tag{6}$$

i.e.,

$$C_{Lr} = C_L - \frac{q}{q_{\text{div}}}\left(C_L + \frac{C_{M0}}{x - \tfrac{1}{4}}\right) \tag{7}$$

The second term of the right-hand side of Eq. 7 is the increase in lift coefficient due to the elastic twist of the blade. This increase will be zero, and the blade will not be twisted, if the design lift coefficient is

$$C_{LI} = -\frac{C_{M0}}{x - \tfrac{1}{4}} \tag{8}$$

This is the *ideal design lift coefficient* given by Theodorsen and Regier. For the Clark-Y airfoil with center of gravity at 44 per cent and $C_{M0} = -0.07$, C_{LI} is 0.37. In this case the angle of attack at zero elastic twist is not very far from the optimum angle of attack of the Clark-Y airfoil, which is at $C_L = 0.40$. Since operating a blade at C_{LI} delays the stall, and thus causes an increase in the flutter speed, it is desirable to operate the propeller at the ideal angle of attack.

In using Eq. 2 to determine the divergence speed, the choice of the radius of the representative sections is open to question. Usually the 80 per cent radius section is taken as the representative section, at which L, c, x, and the corresponding stiffness K are measured. A more rational basis is to use Lagrange's equations or a method of successive approximation, similar to those of §§ 3.4, and 3.5. The arbitrariness of choosing a representative section can thus be avoided.

The closeness of the elastic axis, the inertia axis, and the line of aerodynamic centers of practical propeller or helicopter blades makes such blades inherently strong against flutter. An exact flutter analysis, however, must consider the nonuniform flow condition across the span, the centrifugal force, and the Coriolis force, and is not simple. See papers by Morris, Rosenberg, Duncan, Turner, etc., listed in the bibliography of Chapter 7.

BIBLIOGRAPHY

For a general survey, see Refs. 3.1–3.3.
The first paper written on the subject of steady-state aeroelasticity perhaps is

4.1   Reissner, H.: Neuere Probleme aus der Flugzeugstatik. *Z. Flugtech. u. Motorluftschif.* **17**, 137–146 (Apr. 1926).

For lateral control effectiveness and aileron reversal, see

4.2   Collar, A. R., and E. G. Broadbent: The Rolling Power of an Elastic Wing. *Aeronaut. Research Council R. & M.* **2186** (1945).

4.3   Cox, H. Roxbee, and A. G. Pugsley: Theory of Loss of Lateral Control Due to Wing Twisting. *Aeronaut. Research Com. R. & M.* **1506** (1932).

4.4   Cox, H. Roxbee: Problems involving the Stiffness of Aeroplane Wings. *J. Roy. Aeronaut. Soc.* **38**, 73–107 (Feb. 1934).

4.5   Duncan, W. J., and G. A. McMillan: Reversal of Aileron Control Due to Wing Twist. *Aeronaut. Research Com. R. & M.* **1499** (1932).

4.6   Foss, K. A., and F. W. Diederich: Charts and Approximate Formulas for the Estimation of Aeroelastic Effects on the Lateral Control of Swept and Unswept Wings. *NACA Tech. Note* **2747** (1952); *NACA Rept.* **1139** (1953). See also Diederich, *NACA RM* L8H24á (1948).

4.7   Harmon, S. M.: Determination of the Effect of Wing Flexibility on Lateral Maneuverability and a Comparison of Calculated Rolling Effectiveness with Flight Results. *NACA Wartime Rept. ARR* **4A28**, L-525 (1944).

4.8   Hedgepeth, J. M., and R. J. Kell: Rolling Effectiveness and Aileron Reversal of Rectangular Wings at Supersonic Speeds. *NACA Tech. Note* **3067** (1954).

4.9   Hirst, D. M.: Calculation of Critical Reversal Speeds of Wings. *Aeronaut. Research Com. R. & M.* **1568** (1933).

4.10 Johnson, H. C., and G. Fotieo: Rolling Effectiveness and Aileron Reversal Characteristics of Straight and Sweptback Wings (comparison between theory and experiment). *U.S. Air Force Tech. Rept.* **6198** (Aug. 1950); part 2 (Feb. 1951). Wright Air Development Center.

4.11 Pearson, H. A., and W. S. Aiken Jr.: Charts for the Determination of Wing Torsional Stiffness Required for Specified Rolling Characteristics or Aileron Reversal Speed. *NACA Rept.* **799** (1944).

4.12 Pugsley, A. G.: The Aerodynamic Characteristics of a Semirigid Wing Relevant to the Problem of Loss of Lateral Control Due to Wing Twisting. *Aeronaut. Research Com. R. & M.* **1490** (1932). See also Pugsley and Brooke, *R. & M.* **1508** (1932); Pugsley and Roxbee Cox, *R. & M.* **1640** (1934).

4.13 Pugsley, A. G.: Control Surface and Wing Stability Problems. *Aircraft Engg.* 268 (Oct. 1937). Also *J. Roy. Aeronaut. Soc.* **41**, 975–996 (1937).

4.14 Rosenberg, R.: Loss in Aileron Effectiveness because of Wing Twist and Considerations regarding the Internal-Pressure Balanced Aileron. *J. Aeronaut. Sci.* **11**, 41–47 (1944).

4.15 Thomson, W. T.: Aileron Reversal Speed by Influence Coefficients and Matrix Iteration. *J. Aeronaut. Sci.* **13**, 192–194 (Apr. 1946).

4.16 Tucker, W. A., and R. L. Nelson: The Flexible Rectangular Wing in Roll at Supersonic Flight Speeds. *NACA Tech. Note* **1769** (1948).

4.17 Tucker, W. A., and R. L. Nelson: The Effect of Torsional Flexibility on the Rolling Characteristics at Supersonic Speeds of Tapered Unswept Wings. *NACA Rept.* **972** (1950). Supersedes *NACA Tech. Note* **1890**.

4.18 Victory, M.: The Calculation of Aileron Reversal Speed. *Aeronaut. Research Council R. & M.* **2059** (1944).

For longitudinal stability and control of airplanes, see

4.19 Bisgood, P. L., and D. J. Lyons: Preliminary Report on the Flight Measurement of Aero-Elastic Distortion in Relation to Its Effects on the Control and Longitudinal Stability of a Mosquito Aircraft. *Aeronaut. Research Council R. & M.* **2371** (1946).

4.20 Collar, A. R., and F. Grinsted: The Effects of Structural Flexibility of Tailplane, Elevator, and Fuselage on Longitudinal Control and Stability. *Aeronaut. Research Council R. & M.* **2010** (1942).

4.21 Epstein, A.: Some Effects of Structural Deformation in Airplane Design. *Aero Digest* (Feb. 1949).

4.22 Frick, C. W., and R. S. Chubb: The Longitudinal Stability of Elastic Swept Wings at Supersonic Speed. *NACA Rept.* **965** (1950); supersedes *NACA Tech. Note* **1811**. *J. Aeronaut. Sci.* **17**, 691–700 (1950).

4.23 Lyon, H. M., and J. Ripley: A General Survey of the Effects of Flexibility of the Fuselage, Tail Unit, and Control Systems on Longitudinal Stability and Control. *Aeronaut. Research Council R. & M.* **2415** (July 1945).

4.24 Pengelley, C. D., and D. Benun: Aeroelastic Studies on a High-Performance, Swept-Wing Airplane. *Proc. 1st U.S. Natl. Congr. Applied Mech.*, 903–906 (June, 1951). ASME, New York.

4.25 Pugsley, A. G.: The Influence of Wing Elasticity on the Longitudinal Stability of an Aeroplane. *Aeronaut. Research Com. R. & M.* **1548** (1933).

For the fundamental problem of lift distribution over a flexible wing, see

4.26 Diederich, F. W.: Calculation of the Aerodynamic Loading of Swept and Unswept Flexible Wings of Arbitrary Stiffness. *NACA Rept.* **1000** (1950). Supersedes *NACA Tech. Note* **1876**.

4.27 Diederich, F. W., and K. A. Foss: Charts and Approximate Formulas for the Estimation of Aeroelastic Effects on the Loading of Swept and Unswept Wings. *NACA Tech. Note* **2608** (1952). *NACA Rept.* **1140** (1953).

4.28 Flax, A. H.: The Influence of Structural Deformation on Airplane Characteristics. *J. Aeronaut. Sci.* **12**, 94–102 (1945).

4.29 Gaugh, W. J., and J. K. Slap: Determination of Elastic Wing Aerodynamic Characteristics. *J. Aeronaut. Sci.* **19**, 173–182 (1952).

4.30 Pines, S.: A Unit Solution for the Load Distribution of a Nonrigid Wing by Matrix Methods. *J. Aeronaut. Sci.* **16**, 470–476 (1949).

4.31   Skoog, R. B., and H. H. Brown: A Method for the Determination of the Span-
       wise Load Distribution of a Flexible Swept Wing at Supersonic Speeds.
       NACA Tech. Note **2222** (1951).

For steady-state aeroelastic problems of swept wings, see

4.32   Brown, R. B., K. F. Holtby, and H. C. Martin: A Superposition Method for
       Calculating the Aeroelastic Behavior of Swept Wings. *J. Aeronaut. Sci.* **18**,
       531–542 (Aug. 1951).
4.33   Lyon, H. M.: A Method of Estimating the Effect of Aero-elastic Distortion of
       a Sweptback Wing on Stability and Control Derivatives. *Aeronaut. Research
       Council R. & M.* **2331** (1946).
4.34   Miles, J. W.: A Formulation of the Aeroelastic Problem for a Swept Wing.
       J. Aeronaut. Sci. **16**, 477–490 (1949).
4.35   Pai, S. I., and W. R. Sears: Some Aeroelastic Properties of Swept Wings. *J.
       Aeronaut. Sci.* **16**, 105–115 (1949).

For the twisting of propeller blades, see

4.36   Theodorsen, Th., and A. A. Regier: Effect of the Lift Coefficient on Propeller
       Flutter. *NACA Wartime Rept. ACR* **L5F30**, L-161 (1945).

For the aerodynamic problem of lift distribution over rigid lifting surfaces, see the
following papers.   Further references can be obtained from Refs. 4.43 and 4.45.

4.37   Abbott, I. H., and A. E. von Doenhoff: *Theory of Wing Sections.* McGraw-
       Hill, New York (1949).
4.38   Benscoter, S. U.: Matrix Development of Multhopp's Equations for Spanwise
       Air-Load Distribution. *J. Aeronaut. Sci.* **15**, 113–120 (Feb. 1948).
4.39   Blenk, H.: The Monoplane Wing as Lifting Vortex Sheet. *NACA Tech. Memo.*
       1111 (1947); *Z. angew. Math. u. Mech.* **5**, 36–47 (1925).
4.40   Bollay, W.: A Non-Linear Wing Theory and Its Application to Rectangular
       Wings of Small Aspect Ratio. *Z. angew. Math. u. Mech.* **19**, 21–35 (1939).
       Also, *J. Aeronaut. Sci.* **4**, 294–296 (1937).
4.41   De Young, J.: Theoretical Anti-Symmetrical Span Loading for Wings of
       Arbitrary Plan Form at Subsonic Speeds. *NACA Tech. Note* **2140** (1950).
       See also De Young and C. W. Harper, *NACA Rept.* **921** (1948).
4.42   Diederich, F. W., and M. Zlotnick: Calculated Spanwise Lift Distribution and
       Aerodynamic Influence Coefficients for Unswept Wings in Subsonic Flow.
       NACA Tech. Note **3014** (1953).
4.43   Fage, A.: Some Aerodynamic Advances. *3d Anglo-Am. Aeronaut. Conf.
       Brighton*, 329–362 (1951).   Royal Aeronautical Society, London.
4.44   Falkner, V. M.: The Calculation of Aerodynamic Loading on Surfaces of Any
       Shape. *Aeronaut. Research Council R. & M.* **1910** (1943).   See also Falkner,
       R. & M. **2591** (1953); Falkner and W. P. Jones, *R. & M.* **2685** (1953).
4.45   Flax, A. H. and H. R. Lawrence: The Aerodynamics of Low-Aspect-Ratio
       Wings and Wing-Body Combinations. *3d Anglo-Am. Aeronaut. Conf.
       Brighton* 363–398 (1951).   Royal Aeronautical Society, London.
4.46   Glauert, H.: The Effect of Compressibility on the Lift of an Aerofoil. *Aeronaut.
       Research Com. R. & M.* **1135** (1927); *Proc. Roy. Soc. A*, **118**, 113–119 (1928).
4.47   Gunn, J. C.: Linearized Supersonic Aerofoil Theory, parts I and II. *Phil.
       Trans. Roy. Soc. London A.* **240**, 327–373 (1947).
4.48   Jones, R. T.: Theoretical Correction for the Lift of Elliptic Wings. *J. Aeronaut
       Sci.* **9**, 8–10 (1941).

4.49 Krienes, K.: The Elliptical Wing Based on the Potential Theory. *NACA Tech. Memo.* **971**. *Z. angew. Math. u. Mech.* **20**, 65–88 (1940).

4.50 Lawrence, H. R.: The Lift Distribution on Low Aspect Ratio Wings at Subsonic Speeds. *J. Aeronaut. Sci.* **18**, 683–695 (1951).

4.51 Multhopp, H.: Die Berechnung der Auftriebsverteilung von Tragflügeln. *Luftfahrt-Forsch.* **15**, 153–169 (1938); RTP translation no. 2392.

4.52 Mutterperl, W.: The Calculation of Span Load Distributions on Swept-Back Wings. *NACA Tech. Note* **834** (1941).

4.53 Pope, A., and W. R. Haney: Spanwise Lift Distribution for Sweptback Wings. *J. Aeronaut. Sci.* **16**, 505 (1949).

4.54 Schrenk, O.: A Simple Approximation Method for Obtaining the Spanwise Lift Distribution. *NACA Tech. Memo.* **948**.

4.55 Stanton, J. R.: An Empirical Method for Rapidly Determining the Loading Distributions on Swept Back Wings. *College Aeronaut. Cranfield Rept.* **32** (Jan. 1950).

4.56 Stewart, H. J.: The Aerodynamics of a Ring Airfoil. *Quart. Applied Math.* **2**, 136–141 (1944).

4.57 Van Dorn, H. N., and J. De Young: A Comparison of Three Theoretical Methods of Calculating Span Load Distribution on Swept Wings. *NACA Tech. Note* **1476** (1947).

4.58 Weissinger, J.: The Lift Distribution of Swept-back Wings. *Forsch Ber.* **1553** (1942); *NACA Tech. Memo* **1120** (1947).

Chapter 5

FLUTTER PHENOMENON

5.1 THE PHENOMENON OF FLUTTER

A type of oscillation of airplane wings and control surfaces has been observed since the early days of flight. To describe the physical phenomenon, let us consider a cantilever wing, without sweepback and without aileron, mounted in a wind tunnel at a small angle of attack and with a rigid support at the root. When there is no flow in the wind tunnel, and the model is disturbed, say, by a poke with a rod, oscillation sets in, which is damped gradually. When the speed of flow in the wind tunnel gradually increases, the rate of damping of the oscillation of the disturbed airfoil first increases. With further increase of the speed of flow, however, a point is reached at which the damping rapidly decreases. At the *critical flutter speed*, an oscillation can just maintain itself with steady amplitude. At speeds of flow somewhat above the critical, a small accidental disturbance of the airfoil can serve as a trigger to initiate an oscillation of great violence. In such circumstances the airfoil suffers from oscillatory instability and is said to *flutter*.*

Experiments on wing flutter show that the oscillation is self-sustained; i.e., no external oscillator or forcing agency is required. The motion can maintain itself or grow for a range of wind speed which is more or less wide according to the design of the wing and the conditions of the test. For a simple cantilever wing, flutter occurs at any wind speed above the critical. In other instances, for example, in flutter involving aileron motion, there may be one or more ranges of speed for which flutter occurs, and these are bounded at both ends by critical speeds at which an oscillation of constant amplitude can just maintain itself.

The oscillatory motion of a fluttering cantilever wing has both flexural and torsional components. A rigid airfoil so constrained as to have only the flexural degree of freedom does not flutter. A rigid airfoil with only the torsional degree of freedom can flutter only if the angle of attack is at or near the stalling angle ("stall flutter," Chapter 9), or for some special mass distributions and elastic-axis locations. In ordinary circumstances, oscillations of a control surface (aileron, flap, etc.), in a single degree of

---

* In the following text, the terms *flutter speed* and *flutter frequency* refer to the critical flutter speed and the frequency at the critical condition.

freedom, are also damped at all speeds unless a flow separation is involved. *Let us restrict the term "flutter" to the oscillatory instability in a potential flow, in which neither separation nor strong shocks are involved.* Then, in general, the coupling of several degrees of freedom is an essential feature for flutter. The steady oscillation that occurs at the critical speed is harmonic. Experiments on cantilever wings show that the flexural movements at all points across the span are approximately in phase with one another, and likewise the torsional movements are all approximately in phase,† but the flexure is considerably out of phase from the torsional movement. It will be seen later that mainly it is this phase difference that is responsible for the occurrence of flutter.

The importance of phase shift between motions in various degrees of freedom suggests at once the importance of the number of degrees of freedom on flutter. An airplane wing, as an elastic body, has infinitely many degrees of freedom. But owing to its particular construction, its elastic deformation in any chordwise section can usually be described with sufficient accuracy by two quantities: the deflection at a reference point, and the angle of rotation about that point, i.e., the flexural and the torsional deformations, respectively. Similarly, for a control surface, such as a flap or an aileron, its freedom to turn about the hinge line is so much more important than its elastic deformation, that ordinarily it is possible to describe the deflection of a control surface simply by the angle of rotation about its hinge line. In general, then, it is sufficient to consider three variables in wing flutter: the flexure, the torsion, and the control-surface rotation. A flutter mode consisting of all three elements is called a *ternary* flutter. In special cases, however, two of the variables predominate, and the corresponding flutter modes are called *binary* flutter modes. Similar consideration applies to airplane tail surfaces. In fact, most airplanes can be replaced by a substitutional system of simple beams, so that the elastic deformation can be described by the deflection and torsion of the elastic axes of the substitutional beams, in addition to the rotation of control surfaces about their hinge lines.

These degrees of freedom, together with the freedom of the airplane to move as a rigid body, offer a large number of possible combinations of binary, ternary, and higher modes of flutter. Since it is not clear which of these modes correspond to the actual critical speeds, it is necessary

* In order to distinguish "flutter" from "stall flutter," some authors use the term "classical flutter."

† From the results of § 1.9, it is natural to expect that spanwise shift in phase angle exists at the critical flutter condition, because the aerodynamic force contributes a term proportional to the velocity of movement, but such spanwise shift in phase is relatively small.

either to resort to experiments or, in a theoretical approach, to analyze all cases. This is why a successful flutter analysis depends so much on the analyst's experience. He must be able to choose, among all possible modes, those that are likely to be critical for a given structure.

Since flutter analysis is a rather extensive subject, we shall divide our discussions into three chapters. In the present chapter only some general considerations based on dimensional arguments will be given. The role of the elastic stiffness and the mass balancing in flutter prevention are explained on the empirical basis. The origin of flutter from the aero-dynamic point of view is then considered in § 5.4. It will be shown that flutter occurs because the speed of flow affects the amplitude ratios and phase shifts between motion in various degrees of freedom in such a way that energy can be absorbed by the airfoil from the airstream passing by. Some remarks on the experimental approach to the flutter problem are given in § 5.6, and the dynamic similarity rules are discussed in § 5.7. A brief historical review of the earlier developments is included in § 5.8. The details of the dynamical process, however, are left for the next two chapters.

5.2 NONDIMENSIONAL PARAMETERS

Two mechanical systems are said to be *similar* when they are similar in geometry and in the distribution of mass and elasticity. In flutter analysis, let us consider two similar systems and assume that the motion is dependent on the following fundamental variables:

Symbol	Significance	Physical Dimensions
l	Typical linear dimension	L
U	Air speed	LT^{-1}
ρ	Air density	ML^{-3}
σ	Typical density of structural material	ML^{-3}
K	Typical torsional stiffness constant (ft-lb per rad)	ML^2T^{-2}

These five variables can be combined into two independent nondimensional parameters, such as

$$\frac{\rho}{\sigma}, \qquad \frac{K}{\sigma l^3 U^2} \tag{1}$$

Any nondimensional quantity relating to the motion can be expressed as a function of these parameters. Thus if, in a free oscillation, the deflection at a point is described by an expression $e^{-\varepsilon t} \cos \omega t$, the damping factor ε, of dimension (T^{-1}), can be combined with U and l

to form a nondimensional parameter $\varepsilon l/U$, and hence satisfies a functional relation:

$$\frac{\varepsilon l}{U} = F\left(\frac{\rho}{\sigma}, \frac{K}{\sigma l^3 U^2}\right) \tag{2}$$

where F is some function of the arguments (1). Therefore, a sufficient condition for two similar systems to have the same value of $\varepsilon l/U$ (in particular, to have $\varepsilon = 0$, which corresponds to the critical flutter condition) is that they have the same values of the parameters ρ/σ and $K/(\sigma l^3 U^2)$.

The frequency of oscillation ω (radians per second), with dimension (T^{-1}), can be expressed nondimensionally in the parameter

$$k = \frac{\omega l}{U} \tag{3}$$

which is called the *reduced frequency* or *Strouhal number* (§ 1.5). Hence, there exists a functional relation

$$k = f\left(\frac{\rho}{\sigma}, \frac{K}{\sigma l^3 U^2}\right) \tag{4}$$

Combining Eqs. 4 and 2, we see that two similar systems having the same values of ρ/σ and $K/(\sigma l^3 U^2)$ flutter at the same reduced frequency.

Since all derived concepts relating to the motion can be expressed in functional relations as above, it is clear that the equality of the values of the parameters ρ/σ and $K/(\sigma l^3 U^2)$ is sufficient to guarantee dynamic similarity of the two systems.

In a more careful consideration, the energy dissipation of the structure and the viscosity and compressibility of the fluid must be added to the list of fundamental variables. These can be incorporated non-dimensionally as the material damping coefficient g, the Reynolds number R, and the Mach number M. Dynamic similarity requires the equality of g, R, and M in addition to parameters in expression (1). In general, g is important in control-surface flutter, R is important in stall flutter, and M is important in high-speed flight; otherwise their effects are small.

The Strouhal number, or the reduced frequency k, is the most natural parameter in the consideration of unsteady aerodynamic forces. Whenever convenient, the parameters ρ/σ and k, instead of those in (1), may be taken as the fundamental parameters for dynamic similarity of flutter models. It should be noted that, if the Strouhal number is calculated for the fundamental oscillation frequency of the structure, it can be identified with the second parameter in (1): The factor $\sqrt{K/\sigma l^5}$, of dimension T^{-1}, represents a frequency of oscillation.   Hence, the parameter $K/(\sigma l^3 U^2)$ can be identified with the square of the Strouhal number.

The Strouhal number, or reduced frequency, characterizes the variation of the flow with time.   Its inverse, $U/(\omega l)$, is called the *reduced speed*. An interesting interpretation of the reduced frequency is given by von Kármán as follows.   Consider that a disturbance occurs at a point on a body and oscillates together with the body.   The fluid influenced by the disturbance moves downstream with a mean velocity U.   Let the frequency of oscillation of the body and the disturbance be ω.   Then the spacing, or "wave length" of the disturbance, is $2\pi U/\omega$.   Therefore, the ratio

$$l \div \frac{2\pi U}{\omega} = \frac{l\omega}{2\pi U}$$

which is proportional to the reduced frequency shows that k represents a ratio of the characteristic length l of the body to the wave length of the disturbance.   In other words, the reduced frequency characterizes the way a disturbance is felt at other points of the body.   Since every point of an oscillating body disturbs the flow, one may say that the reduced frequency characterizes the mutual influence between the motion at various points of the body.

5.3   STIFFNESS CRITERIA

According to an analysis of the data on several airplanes in which wing flutter has been observed, and on others that showed no tendency to flutter, Küssner found in 1929 that, for airplane wings with mass-unbalanced ailerons, of the type of construction prevalent at that time, wing-aileron flutter occurs when the reduced frequency is lower than the following critical value:

$$k_{cr} = \frac{\omega c}{2U} = 0.9 \pm 0.12 \tag{1}$$

where $U = $ the mean speed of flow, feet per second; $\omega = $ the fundamental frequency of the wing in torsional oscillation in still air, radians per second; and $c = $ the chord length of the vibrating portion of the wing, feet.

For safety against flutter, the reduced frequency should be higher than k_{cr}.   In other words, the design speed of the airplane must be lower than

$$U_{cr} = \frac{\omega c}{2k_{cr}} \tag{2}$$

This is Küssner's well-known formula.   The frequency ω may be determined by ground vibration experiments in still air or computed by a

theoretical analysis.   It increases with increasing stiffness of the structure.
Therefore, the critical speed can be raised by increasing the wing stiffness.
From this point of view we see that when the lower limit of U_{cr} is prescribed (e.g., the maximum speed of flight) a minimum value of ω, and
hence a minimum value of the torsional rigidity is also prescribed.

Thus a stiffness criterion can be specified for the purpose of flutter
prevention.   If we return to the parameters used at the beginning of the
last section, Küssner's formula can be expressed in the form

$$\frac{K}{\sigma l^3 U^2} \geqslant \text{const} \tag{3}$$

If the inequality sign is satisfied, flutter will not occur.

The stiffness criterion, regarded as a specification of the wing stiffness
with regard to flutter prevention, is convenient for a designer.

Such stiffness criteria arise also in the consideration of steady-state
instabilities (cf. Chapters 3 and 4).   Most types of aeroelastic instabilities
can be avoided by sufficiently high structural stiffness.   For the safety
against each type of instability, inequalities of the following form are to
be satisfied.

1. Torsional-stiffness criterion:

$$\frac{K_\theta}{\rho U^2 s \bar{c}^2} \geqslant \text{const} \tag{4}$$

2. Flexural-stiffness criterion:

$$\frac{K_h}{\rho U^2 s^3} \geqslant \text{const} \tag{5}$$

where

K_θ = torsional stiffness of the wing = $\dfrac{\text{torque applied at effective tip}}{\text{angle of twist at effective tip}}$

K_h = flexural stiffness of the wing

$\quad = \dfrac{\text{maximum bending moment} \times \text{semispan}}{\text{linear deflection of tip}}$

s = semispan

$\bar{c}$ = mean chord of the wing

The constants in the above equations depend on many parameters, such
as the type of structural construction, the locations of the elastic axis and
the center of mass of the wing, the moment of inertia of the sections, the
amount of mass balancing.   The final form of the stiffness criteria can be
obtained for a particular type of structure only after consideration has
been given to all the possible instabilities.

As an example, let us quote the case of flutter of a thin-walled circular cylindrical shell of uniform thickness in a supersonic flow along the axis of the cylinder. Such a shell is used often in large liquid-fueled rockets, particularly at the interstage area. Experiments in a wind tunnel revealed that such a shell may have several types of oscillations: the small amplitude random oscillations, the sinusoidal flutter oscillations, and the sinusoidal flutter motion whose amplitude varies periodically over the circumference of the cylinder and moves as a traveling wave. Tested at a Mach number of 2.49, the amplitude of the last two types of oscillations would suddenly rise at the critical condition

$$\left(\frac{q}{\bar{\beta}E}\right)^{\frac{1}{3}}\frac{R}{h} = 7 \tag{6}$$

in which $q = \frac{1}{2}\rho U^2$ is the dynamic pressure of the main flow, R is the radius of the middle surface of the circular cylinder, h is the thickness of the shell wall, E is the Young's modulus of the shell material, and $\bar{\beta} = \sqrt{M^2 - 1}$ is a function of the Mach number. In a limited range of experiments this critical condition is independent of the internal pressure as long as it is positive. The critical condition (6) may be regarded as a stiffness criterion. Details of the experiments are given in the author's paper, Ref. 5.73.

5.4 THE EXTRACTION OF ENERGY FROM AIRSTREAM

Since flutter is an oscillation induced by the aerodynamic forces without any external source of energy other than the airstream, it is possible only if the oscillating body, the mean position of which is assumed stationary, can extract energy from the airstream.* Hence, the possibility of flutter can be discussed by considering the energy relation.

An oscillation will be called aerodynamically unstable if the oscillating body gains energy from the airstream in completing a cycle. If the oscillating body has neither external excitation nor internal friction, then the aerodynamic instability can be identified with flutter. Internal friction dissipates energy, external excitation imposes a source of energy exchange; both modify the kinematic relations (the amplitude ratio and phase relationship between various degrees of freedom) of the oscillation of an aeroelastic system. Hence when there is external excitation or internal

---

* The inertia force and the elastic force are both conservative and do not contribute any net gain or loss of energy in each complete cycle of motion.

friction, the aerodynamic instability alone cannot be directly identified with flutter.

Consider an airfoil performing a vertical translatory oscillation with a constant amplitude h_0. Let the vertical displacement be described by the expression

$$h = h_0 e^{i\omega t} \tag{1}$$

We shall define h as positive *downward*. The speed of downward motion is therefore

$$\dot{h} = i\omega h_0 e^{i\omega t} \tag{2}$$

where a dot indicates a differentiation with respect to time. If $\dot{h}$ were a constant, the downward motion will induce a lift force L_0 on the airfoil:

$$L_0 = \frac{1}{2}\rho U^2 S \frac{dC_L}{d\alpha} \frac{\dot{h}}{U} \tag{3}$$

where $\frac{1}{2}\rho U^2$ is the dynamic pressure and S is the wing area. The lift is defined as positive *upward* in the usual sense. When the airfoil is oscillating, the true instantaneous lift acting on the airfoil differs from L_0 both in magnitude and in phase. Let us call L_0 the *quasi-steady lift* and write the true instantaneous lift as

$$L = L_0 r e^{i\psi} \tag{4}$$

Then r represents the ratio of the absolute value of the instantaneous lift to that of the quasi-steady lift, and ψ the phase angle by which the actual lift leads the quasi-steady value. The quantities r and ψ depend on the reduced frequency k, the Mach number M, and the Reynolds number R. For a nonviscous incompressible fluid, r and ψ are functions of k alone. The ratio $L/L_0 = re^{i\psi}$ can be plotted as vector with length r and angle ψ. Such a vector diagram, for a flow of an incompressible fluid ($M = 0$), with k as a parameter, is given in Fig. 5.1. The theoretical derivation of this diagram will be given in Chapter 13.

When the airfoil moves through a distance dh, the work done by the lift is, in real variables,

$$dW = -L\, dh = -L\dot{h}\, dt \tag{5}$$

It must be recognized that, when L and h are expressed in the complex forms 1 and 4, the physical quantities are represented only by the *real parts* of the complex representations. For example, the physical displacement h represented by Eq. 1 is $h_0 \cos \omega t$. The proper form of the work, in the complex representation, is therefore

$$dW = -Rl\,[L] \cdot Rl\,[\dot{h}]\, dt \tag{5a}$$

Integrating through a cycle of oscillation, we obtain the total work done *by* the air *on* the airfoil:

$$W = -\int_0^{2\pi/\omega} Rl\,[L] \cdot Rl\,[h]\,dt$$

$$= -\frac{1}{2}\rho US \frac{dC_L}{d\alpha}(\omega h_0)^2 r \int_0^{2\pi/\omega} \sin(\omega t + \psi)\sin \omega t\,dt$$

$$= -\frac{\pi}{2}\rho US \frac{dC_L}{d\alpha}\omega h_0^2 r \cos \psi \qquad (6)$$

Hence, the gain of energy W by the airfoil from the airstream is proportional to $(-\cos \psi)$. If $-\pi/2 < \psi < \pi/2$, W is negative; i.e., the oscillating airfoil will lose energy to the airstream. The oscillation is therefore stable. If we refer back to Fig. 5.1, it is seen that the condition

$$-\frac{\pi}{2} < \psi < \frac{\pi}{2}$$

is satisfied. Hence, in a nonviscous incompressible fluid, the vertical translation oscillation is aerodynamically stable.

This example shows the importance of the phase angle between the aerodynamic force and the oscillatory motion. Although purely translational flutter is impossible, it is conceivable that, when several degrees of freedom are involved, a certain combination of the phase relations will render the energy input to the airfoil positive.

Thus the fundamental cause of flutter is quite clear. The airfoil, by adjusting its phase shift, extracts energy from the airstream. In fact, the airfoil can be regarded as a flutter engine.[5.2] The fact that the phase shift and amplitude ratio of the flexural and torsional wing motions which follow an imposed disturbance depend largely on the speed of flow over the wing is of fundamental importance. It is this dependence that makes flutter occur at certain critical speed of flow.

By calculating the energy input from the airstream the stability of more complicated motions can be determined. The bending-torsion case, in an incompressible fluid, has been calculated by J. H. Greidanus[5.43] whose result, with a slight addition, is reproduced in Fig. 5.2. The airfoil is assumed rigid. The vertical translation, called bending, is denoted by h and is positive downward. The rotation about the $\frac{1}{4}$-chord point, called torsion, is denoted by α and is positive nose up. The fluid is assumed nonviscous and incompressible, and the linearized two-dimensional aerodynamic theory is used. Let $\bar{E}$ be the mean work per unit time done by the aerodynamic force per unit span in a harmonic oscillation of

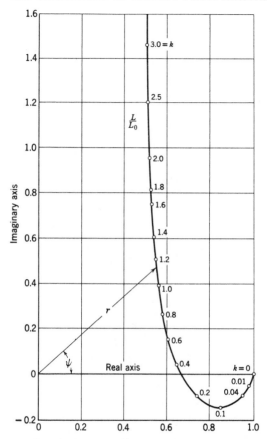

Fig. 5.1.   Vector diagram of lift in vertical translation oscillation.   A vector drawn from the origin to an appropriate point on the curve gives a complex number (real part, abscissa; imaginary part, ordinate) that is the value of L/L_0 given by the linearized theory.   Curve given by von Kármán and Sears, Ref. 13.21.   (Courtesy of the Institute of the Aeronautical Sciences.)

frequency ω (radians per second). The expressions given by Greidanus are

$$\bar{E} = \tfrac{1}{4}\pi\rho c^4\omega^3\alpha_0{}^2 C_E$$
$$C_E = f_1(k)\xi^2 - [f_2(k)\sin\phi + f_3(k)\cos\phi]\xi - \tfrac{1}{4} \tag{7}$$

where

$$\xi = \text{dimensionless ratio } h_0/(\alpha_0 c)$$
$$h = h_0 e^{i\omega t} \tag{8}$$
$$\alpha = \alpha_0 e^{i(\omega t - \phi)}$$

Here ϕ represents the *phase lag* of the torsion behind the bending. The functions f_1, f_2, f_3 are functions of the reduced frequency k derived from the theory of oscillating airfoils which will be presented in Chapter 13.

All critical oscillations are given by the equation

$$C_E = 0 \qquad (9)$$

The solutions of this equation are plotted in Fig. 5.2. Inside each loop, C_E is positive and the oscillation is unstable.

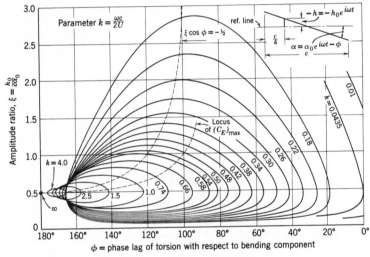

Fig. 5.2. Energy coefficient in bending-torsion oscillations according to Greidanus, Ref. 5.43. (Courtesy of the Institute of the Aeronautical Sciences.)

For $U \to \infty$, $k \to 0$, the entire half-strip $\xi \geqslant 0$, $0 \leqslant \phi \leqslant \pi$ becomes unstable. $k \to 0$ also when $\omega \to 0$ if U remain finite, then $\bar{E}$ tends to zero as fast as ω^3.

For $U \to 0$, $k \to \infty$, only one limiting point reaches the critical condition of neutral stability. This is

$$\xi = \tfrac{1}{2}, \qquad \phi = \pi$$

which implies that the wing is oscillating about the $^3/_4$-chord point, because the downward displacement at a point located at a distance xc *behind* the $^1/_4$-chord point is

$$z = h_0 e^{i\omega t} + \alpha_0 x c e^{i(\omega t - \phi)} \qquad (10)$$

which vanishes when $x = \xi$ and $\phi = \pi$. When $\xi = ^1/_2$, the $^3/_4$-chord point is stationary. Thus, in the absence of structural damping, a wing reaches the critical flutter condition at zero airspeed if the oscillation node is located at the $^3/_4$-chord point. In practice, structural damping always exists and flutter in this case does not occur; but Biot and Arnold[5.42] have shown that, if the nodal line of oscillation of a wing is located close to the $^3/_4$-chord line, flutter at low airspeed is likely to occur.

For intermediate values of U, k is finite. The loops of instability become smaller as k increases.

From Eq. 10, the location x where the amplitude of $|z|$ becomes a minimum can be calculated. The result is

$$x = -\xi \cos \phi \qquad (11)$$

It coincides with the $^3/_4$-chord point if $\xi \cos \phi = -^1/_2$, a relation represented by a dotted curve in Fig. 5.2, which appears to pass right through the dangerous area. This indicates again that flutter is liable to occur if the node of oscillation is located near the $^3/_4$-chord point.

It is particularly interesting to consider the possibility of one-degree-of-freedom flutter. A purely translation motion corresponds to $\alpha_0 = 0$ or $\xi = \infty$, which by Fig. 5.2 is stable for all $k > 0$, confirming a result obtained previously. A purely rotational motion exists if there is a node. From Eq. 10 we see that z vanishes at all time t if and only if

$$h_0 e^{i\omega t} + \alpha_0 c x e^{i(\omega t - \phi)} = 0 \qquad (12)$$

Eq. 12 is satisfied by pitching about the $^1/_4$-chord point, corresponding to $x = h_0 = 0$, which, for $k > 0$, is seen from Fig. 5.2 to be stable for all values of ϕ between 0 and π. For nonvanishing x, Eq. 12 is satisfied if

(a) $\phi = \pi$, $x = \xi$ (13)

(b) $\phi = 0$, $x = -\xi$

Case a is the oscillation about the $^3/_4$-chord point discussed above. Case b leads to flutter, according to Fig. 5.2, only if $k < 0.0435$, and for axis of rotation located forward of the $^1/_4$-chord point, yet not too far forward of the airfoil leading edge. This torsional flutter was first found by Glauert in 1929. It is discussed in detail by Smilg.[5.50] Recently, several types of single-degree-of-freedom flutter involving control surfaces at both subsonic and supersonic speeds have been found,[5.45-5.49] all requiring the fulfillment of certain special conditions on the rotational-axis locations, the reduced frequency, and the mass moment of inertia. Pure-bending flutter is possible for a cantilever swept wing if it is heavy enough relative to the surrounding air and has a sufficiently large sweep angle.[5.46]

The preceding analysis is purely kinematical. It does not tell how the phase shift ϕ and the reduced frequency k will vary with the flow speed U for a given structure. The latter information must be obtained from a consideration of the balance of the inertia and elastic forces with the aerodynamic forces. Thus a dynamic analysis is necessary to determine which point on a figure such as Fig. 5.2 corresponds to a given structure at a given airspeed U. Such a dynamic analysis is the subject matter of the next two chapters.

5.5   DYNAMIC MASS BALANCING*

Since purely translational, purely rotational, and purely control-surface oscillations are stable in most cases, it is clear that a key to flutter prevention is to break up any coupling between the various degrees of freedom.

Consider a two-dimensional wing of infinite torsional rigidity fitted with an aileron whose center of mass lies behind the hinge line. If this wing is initially at rest and is suddenly given an upward motion, the aileron will tend to lag behind the wing to produce a relationship as shown in Fig. 5.3.

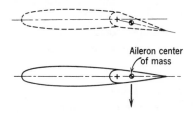

Fig. 5.3.   Bending-aileron coupling.

Thus an aileron motion is induced by a vertical motion of the wing by inertia force. This is called an *inertia coupling*. If the aileron center of mass lies on the hinge line, the inertia coupling will vanish, but the aileron motion may still be excited by the vertical motion because a nonvanishing aerodynamic moment about the hinge line may exist. This is called an *aerodynamic coupling*. Finally, some elastic linkage may exist so that a vertical deflection causes the aileron to rotate, thus forming an *elastic coupling*. Similar terms are used for the interconnections between other degrees of freedom.†

An airplane designer has only a limited control over the elastic and aerodynamic couplings, but inertia coupling is more controllable. Since the critical flutter speed is often very sensitive to inertia coupling, a careful consideration of "dynamic mass balancing" can be very rewarding.

---

* Much of the concept of dynamic mass balancing is due to von Baumhauer and Koning,[5.6] who, in 1923, gave results of wind-tunnel tests in which flutter had been eliminated by adding weights to the paddle balance then used for aerodynamic balance.

† A coupling between the wing and aileron can be broken by using a rigid, irreversible control system, i.e., one in which the aileron is not free to rotate.

Consider first the *flexure-aileron* coupling of a cantilever wing. The inertia coupling can be broken if it is possible to arrange the mass distribution in such a way that the inertia force due to bending induces no rotational moment about the aileron hinge line. An aileron with mass distribution so arranged is said to be dynamically mass-balanced with respect to flexural motion. Otherwise it is mass-unbalanced. A measure of the mass unbalance is the product of inertia of the aileron about two perpendicular axes, one coinciding with the aileron hinge line, the other corresponding approximately to the wing root.[5.20] Let an aileron (Fig. 5.4) be mounted on a hinge line Y, which in turn is mounted rigidly on a wing oscillating about an axis X (assumed perpendicular to Y). The

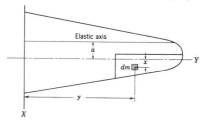

Fig. 5.4.   Notations for Aileron mass balance.

inertia force acting on any element of the aileron is proportional to the amplitude of oscillation of that element. If the aileron is perfectly rigid, and the wing oscillates as a rigid body as if it were hinged about the X axis, the acceleration will be proportional to the distance y, and the inertia moment about the aileron hinge line due to an element of mass dm will be proportional to $x\,y\,dm$, where x is the distance of the element dm behind the aileron hinge line. Hence, the total moment is proportional to the product of inertia of the mass of the aileron:

$$J_{xy} = \int xy\, dm$$

where the integral is taken over the entire aileron. If $J_{xy} = 0$, the inertia coupling is eliminated for the mode of oscillation assumed. In practice, J_{xy} seldom vanishes. As a measure of the mass unbalance, a nondimensional "dynamic-balance coefficient"

$$\frac{J_{xy}}{\text{Mass} \times \text{area of aileron}}$$

is introduced by Roche[5.35].

In a modern sense, mass balancing means the best arrangement of masses. For an airplane the location of engines, fuel tanks, radar equipment, and so on has a profound influence on the critical flutter condition.

As an example, Figure 5.5 shows the results of a systematic study of the effect of engine locations on the flutter speed of airplanes with unswept and swept wings by means of an electric analog computation. A point on the contour curves of constant flutter speed represents the actual location of the center of gravity of the added mass, and the numbers shown refer to unity based on the flutter speed of the bare wing, without any added mass.

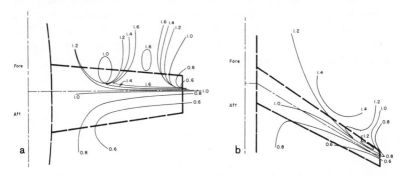

Fig. 5.5. Contours of constant flutter speed for (a) unswept and (b) swept tapered wing showing effect of concentrated mass location. Symmetric Flutter. (From C. H. Wilts, "Incompressible Flutter Characteristics of Representative Aircraft Wings." *NACA Tech. Note 3780.*)

Thus an engine whose center of gravity is located at a point on a contour labeled 1.4 will improve the flutter characteristics of the bare wing by raising its flutter speed by a factor of 1.4. A knowledge of such trend curves is of great value to the engineer; but such accurate information certainly cannot be given by a simple criterion. It is a general practice in airplane design to start flutter analysis at an early stage, so that certain decisions such as the location of engines and fuel tanks, etc., can be made.

Mass balancing is extremely critical for rockets and satellites. For artificial satellites the mass distribution reacts with gravity gradient and influences the guidance and control. Moving masses are employed to control the damping of a satellite. Active motion of masses makes the system nonconservative. For example, in Ref. 5.74 the author considered the stability of a spinning space station due to periodic motions of the crew. Instability may occur if the period of an astronaut's motion bears certain ratios to the half-period of the spin of the satellite. If he moves back and forth along the radius of a circular, planar satellite, instability will occur when the period of his motion is approximately an integral multiple of the half-period of the satellite spin. A similar conclusion holds if the astronauts move with constant speed or oscillate periodically in circumferential direction. The heavier the moving masses or the larger their amplitude of motion, the wider is the region of instability.

5.6 MODEL EXPERIMENTS

Much valuable information can be derived from wind-tunnel model tests. Sometimes the behavior of a specific airplane is so complex that the accuracy of simplified theoretical analysis becomes doubtful; then model testing is almost indispensable in arriving at a sound design. Model testing is often used to determine the optimum location of engines or fuel tanks, and other design parameters.

Because of the requirements on geometrical, kinematical, and dynamic similarity, wind-tunnel flutter models are often quite expensive and difficult to construct. Elaborate techniques of model construction and test instrumentation have been evolved in the past decades.[5.65–5.72] In recent years attention is called to the method of support of the model in the wind tunnel. For example, the rigid-body degrees of freedom (translation and rotation of the airplane as a whole), may have important effects on the flutter of swept wings and tails. Yet it is impractical in model tests to allow all the degrees of freedom corresponding to free flight conditions. Some simplification is achieved by separating the constituent oscillations of the airplane into symmetric and antisymmetric types and examining them separately.

The critical condition can be found by observing either the free oscillation of the structure following an initial disturbance, or the response of the structure to an external periodic excitation. In the former method, the airspeed is increased until there results a maintained oscillation of a specific amplitude in a chosen degree of freedom. In the second method, one or several exciters (e.g., eccentric rotating masses, air pulse exciter, etc.) are used to excite the oscillation. At each airspeed, the amplitude response is recorded for varying exciter frequencies. The critical flutter condition is specified as the extrapolated airspeed at which the amplification becomes very large.

5.7 DIMENSIONAL SIMILARITY

From a mere knowledge of the number of significant physical parameters, dimensional analysis can be made to determine the characteristic dimensionless parameters that govern the dynamic similarity (see § 5.2). Without further information, a flutter model and its prototype must be geometrically similar, have similar mass and stiffness distribution, and have the same geometrical attitude relative to the flow. The scale factors must be such that the density ratio σ/ρ, the reduced frequency k, the Mach number M, and the Reynolds number R be the same for the model as for the prototype. These requirements are, of course, exceedingly severe, and hard to be met.

When more specific information about a physical phenomenon is known, certain conditions of geometric, mass, and elastic similarity may be dispensed with, without loss of exactness. Consider the simple example of the bending deflection of a beam. In this case only the flexural stiffness EI is of significance. Hence, in constructing a beam model for the purpose of deflection measurements, only the EI distribution needs to be simulated; the cross-sectional shape can be distorted if desired. Such freedom in distorting a model greatly simplifies the model design and testing, and will be discussed in greater detail below.

If the differential or integral equation, or equations, governing a physical phenomenon are known, they provide a deeper insight into the laws of similarity than a mere knowledge of the variables that enter the problem, and offer ways in which a distorted model may be used.   Illustrations of this point can be found in many problems.[5.62] As a classical example, let us quote the problem of George Stokes, who, in 1850, introduced the term "dynamic similarity" into the literature. Stokes' problem is the motion of a pendulum in a viscous fluid. The basic equation is the Navier-Stokes equation which, for a two-dimensional flow of an incompressible viscous fluid with fixed boundaries, may be written as

$$\frac{\partial \omega}{\partial t} + u \frac{\partial \omega}{\partial x} + v \frac{\partial \omega}{\partial y} = \nu \left(\frac{\partial^2 \omega}{\partial x^2} + \frac{\partial^2 \omega}{\partial y^2} \right) \tag{1}$$

where u and v are velocity components in the x and y directions, ν is the kinematic viscosity, and $\omega = (\partial v/\partial x) - (\partial u/\partial y)$ is the vorticity of the fluid. The same differential equation applies to the model:

$$\frac{\partial \omega'}{\partial t'} + u' \frac{\partial \omega'}{\partial x'} + v' \frac{\partial \omega'}{\partial y'} = \nu' \left(\frac{\partial^2 \omega'}{\partial x'^2} + \frac{\partial^2 \omega'}{\partial y'^2} \right) \tag{2}$$

where the primes refer to model. Let us introduce the *scale factors* K_L, K_t, etc., between dimensions of the model and the prototype, so that

$$x' = K_L x, \qquad y' = K_L y, \qquad t' = K_t t, \qquad v' = K_\nu \nu \tag{3}$$
$$\omega' = K_\omega \omega, \qquad u' = K_V u, \qquad v' = K_V v$$

These scale factors are subject to the *kinematic similarity* imposed by the relations

$$u = \frac{dx}{dt}, \qquad u' = \frac{dx'}{dt'}, \qquad \omega = \frac{\partial v}{\partial x} - \frac{\partial u}{\partial y}, \qquad \omega' = \frac{\partial v'}{\partial x'} - \frac{\partial u'}{\partial y'}$$

so that

$$K_V = K_L/K_t, \qquad K_\omega = K_V/K_L \tag{4}$$

When Eqs. 3 are introduced into Eq. 1, we obtain

$$\frac{\partial \omega'}{\partial t'} + \frac{K_L}{K_t K_V}\left(u'\frac{\partial \omega'}{\partial x'} + v'\frac{\partial \omega'}{\partial y'}\right) = \nu' \frac{K_L{}^2}{K_t K_\nu}\left(\frac{\partial^2 \omega'}{\partial x'^2} + \frac{\partial^2 \omega'}{\partial y'^2}\right) \tag{5}$$

If Eq. 5 is identified with Eq. 2, we must have

$$\frac{K_L}{K_t K_V} = 1, \qquad \frac{K_L{}^2}{K_t K_\nu} = 1 \tag{6}$$

The first of Eq. 6 is a kinematic similarity equation 4. The second of Eq. 6 gives the Stokes' rule for similarity for flows *with similar boundary conditions*. Using the first of Eqs. 6, the second equation may be written

$$\frac{K_V K_L}{K_\nu} = 1 \tag{7}$$

which means, of course, in present terminology, that both the prototype and the model must have the same Reynolds number.*

An alternate procedure, which differs from the previous one only in form and not in basic reasoning, is also commonly used. The idea is to express the differential equations in dimensionless form. Introduce a characteristic length L, a characteristic time T, a characteristic velocity V, a characteristic vorticity Ω, and a characteristic number for the kinematic viscosity N. Let $\bar{x}$, $\bar{y}$, $\bar{t}$, etc., be dimensionless quantities so that

$$x = \bar{x}L, \qquad y = \bar{y}L, \qquad u = \bar{u}L/T, \qquad v = \bar{v}L/T \tag{8}$$

$$t = \bar{t}T, \qquad \Omega = \bar{\omega}/T, \qquad \nu = N\bar{\nu}$$

Then Eq. 1 may be written

$$\frac{\partial \bar{\omega}}{\partial \bar{t}} + \frac{VT}{L}\left(\bar{u}\frac{\partial \bar{\omega}}{\partial \bar{x}} + \bar{v}\frac{\partial \bar{\omega}}{\partial \bar{y}}\right) = \frac{NT}{L^2}\bar{\nu}\left(\frac{\partial^2 \bar{\omega}}{\partial \bar{x}^2} + \frac{\partial^2 \bar{\omega}}{\partial \bar{y}^2}\right) \tag{9}$$

Now, for dynamically similar systems, the dimensionless values $\bar{x}$, $\bar{y}$, $\bar{t}$, etc., have the same value for the model as for the prototype. Hence the coefficients VT/L and NT/L^2 must be the same for the two systems; i.e.,

$$\frac{VT}{L} = \frac{V'T'}{L'}, \qquad \frac{NT}{L^2} = \frac{N'T'}{L'^2} \tag{10}$$

where primes refer to model.

The results obtained in Eqs. 6 and 10 are of course identical. The first method has the advantage of requiring fewer notations.

---

* Stokes anticipated Osborne Reynolds' results by thirty years. See Langhaar, Ref. 5.62.

As a second example, consider the flexural oscillation of a beam. The governing equation is

$$\frac{\partial^2}{\partial x^2}\left(EI\frac{\partial^2 w}{\partial x^2}\right) + \rho A\frac{\partial^2 w}{\partial t^2} = 0 \tag{11}$$

where ρ is the density of the beam material, A is the area of beam cross section, and EI is the flexural rigidity.   Let

$$I = A_0 r_0^2 f_1\left(\frac{x}{l}\right), \qquad A = A_0 f_2\left(\frac{x}{l}\right) \tag{12}$$

where A_0, r_0 are the cross-section area and radius of gyration at a reference section, respectively, and l is the length of the beam, f_1, f_2 being dimensionless functions involving only the dimensionless parameter x/l.   Then

$$\frac{\partial^2}{\partial x^2}\left[f_1\left(\frac{x}{l}\right)\frac{\partial^2 w}{\partial x^2}\right] + \frac{\rho}{E r_0^2}f_2\left(\frac{x}{l}\right)\frac{\partial^2 w}{\partial t^2} = 0 \tag{13}$$

Introducing scale factors

$$x' = K_L x, \qquad l' = K_L l, \qquad w' = K_L w, \qquad r_0' = K_{r0} r_0$$
$$\rho' = K_\rho \rho, \qquad E' = K_E E, \qquad t' = K_t t \tag{14}$$

where primes refer to model, we have

$$\frac{\partial^2}{\partial x'^2}\left[f_1\left(\frac{x'}{l'}\right)\frac{\partial^2 w'}{\partial x'^2}\right] + \frac{K_t^2 K_E K_{r0}^2}{K_\rho K_L^4}\frac{\rho'}{E' r_0'^2}f_2\left(\frac{x'}{l'}\right)\frac{\partial^2 w'}{\partial t'^2} = 0 \tag{15}$$

Since Eq. 13 applies as well to the model, we must have

$$\frac{K_t^2 K_E K_{r0}^2}{K_\rho K_L^4} = 1 \tag{16}$$

This result can be expressed in more familiar form if we write ω as the frequency of oscillation and notice that, according to the definition $\omega' = K_\omega \omega$, the scale factor K_ω must be the reciprocal of K_t.   Then Eq. 16 may be written as

$$\frac{K_\omega K_L^2}{K_{r0}}\sqrt{\frac{K_\rho}{K_E}} = 1 \tag{17}$$

i.e., the scale factor of the dimensionless parameter

$$\frac{\omega l^2}{r_0}\sqrt{\frac{\rho}{E}} \tag{18}$$

must be equal to unity.   This remains true for all systems expressible by means of l, r_0, ρ, E, $f_1(x/l)$, and $f_2(x/l)$.   No restriction to any particular shape of the cross sections is imposed. Thus the design of a model for the purpose of measuring oscillation modes needs only to simulate $f_1(x/l)$, $f_2(x/l)$, and the parameter (18) above, and is left with complete freedom in selecting cross-sectional shape. Without the auxiliary information contained in the differential equation, other dimensionless parameters such as l/r_0 would have appeared, and the conclusions would have been more restrictive.

The same reasoning applies to aeroelastic models. As an example, consider the torsion-bending flutter of a cantilever wing in an incompressible fluid. If one is satisfied with Theodorsen's approximation of characterizing flutter condition by the conditions at a "typical" section, the equations describing flutter are given by Eqs. 11 of § 6.9, which are already written in dimensionless form, and hence must apply equally well to the model as to the prototype. An examination of these equations shows that at the flutter condition $P_0 = Q_0 = 0$ the following ten dimensionless ratios are involved:

$$\mu, \quad x_\alpha, \quad \omega_h/\omega_\alpha, \quad a_h, \quad r_\alpha, \quad \omega_h/\omega, \quad \omega_\alpha/\omega, \quad k, \quad h_0/b, \quad \alpha_0 \quad (19)$$

k, ω_α/ω, and $h_0/b\alpha_0$ are determined by the condition of existence of flutter: the vanishing of the flutter determinant. Hence a model must simulate the first five parameters which are to be evaluated at a typical section of the wing. Any wing having these same dimensionless ratios can be considered as a flutter model of the prototype. In particular, the flutter of a cantilever wing may be simulated by a two-dimensional wing model.

Clearly, model testing intended to conform with more comprehensive theories would require more restrictive similarity laws.

5.8 HISTORICAL REMARKS

The earliest study of flutter seems to have been made by Lanchester,[5.29] Bairstow, and Fage[5.4] in 1916 in connection with the antisymmetrical (fuselage torsion-elevator torsion) flutter of a Handley Page bomber. Blasius,[5.8] in 1918, made some calculations after the failure of the lower wing of the Albatross D3 biplane. But the real development of flutter analysis had to wait for the development of the nonstationary airfoil theory, the foundation of which was laid by Kutta and Joukowsky in the period 1902 to 1906. The first numerical calculation of the aerodynamic force on a harmonically oscillating flat plate in a two-dimensional flow was given twenty years later in 1922 by Birnbaum in his thesis at Göttingen. It is well known that Prandtl's theory of bound vortices was completed in 1918, and was applied by Ackermann to compute the lift of a stationary

airfoil.   At the suggestion of Prandtl, Birnbaum extended Ackermann's
concept to nonstationary airfoils.[13.2, 13.3]   He obtained numerical results
up to a reduced frequency $k = 0.12$.

About the same time, Wagner[15.30] investigated the aerodynamic forces
acting on a body that moves suddenly from a stationary configuration to
a constant velocity U.   The sudden change of angle of attack was also
treated.

The next landmark was recorded in 1929.   In that year, Glauert[15.23, 13.14]
published data on the force and moment acting on a cylindrical body due
to an arbitrary motion, and aerodynamic coefficients of an oscillating
wing up to $k = 0.5$.   The calculation was based on Wagner's method.
In the same year, Küssner[5.28] extended the method of Birnbaum to obtain
the aerodynamic coefficients up to $k = 1.5$.

In 1934, Theodorsen's exact solution of a harmonically oscillating wing
with a flap was published;[13.32] the range of k is then unlimited.   Much
additional work on aerodynamics appeared since then.   It will be reviewed in Chapters 12–15.

Up to 1934, only a few cases of flutter were recorded.   In those days,
only airplane wings showed flutter.   Aileron mass unbalance and low
torsional stiffness of the wing were responsible for most of these accidents.

As early as 1929, the theory of flutter was clarified by Küssner[5.28] with
respect to many fundamental details—elimination of the time coordinate,
substitution of the wing structure by a simple beam, iterative solution of
the resulting system of differential equations, representation of the internal
damping by a phase lag in the elastic restoring force, etc.   On the other
hand, Duncan and Frazer measured[5.21, 5.25] (1928) flutter derivatives in a
wind tunnel and introduced the concept of semirigidity and the methods
of matrices.   Simple rules of flutter prevention were derived from statistical studies both in Germany (by Küssner) and in England (by Roxbee
Cox).[5.12]

From 1934 to 1937, the development of new types of airplane was lively,
owing to the arms race of the great powers.   Numerous cases of flutter
occurred, not only with wings, but also with tail surfaces.   The experience
of accidents demonstrated the decisive effect of the mass unbalance of the
control surfaces, and dynamic mass-balance requirements were generally
incorporated into design specifications.   In this period, intensive research
on flutter was reflected by numerous publications.   Many methods of
analysis were discussed, and details of aerodynamic forces for control
surfaces were published.   The two-dimensional problem of airfoil flutter
with two degrees of freedom no longer involved any difficulty.   Quick
solutions (for example, Kassner and Fingado's graphical method[6.15])
became available.   Two-dimensional problems with three degrees of

freedom—airfoils with flap—were treated satisfactorily. For a three-dimensional wing, Galerkin's method was applied together with the "strip theory" of aerodynamics. Above all, the theory was confirmed by flutter model tests in wind tunnels.

On the engineering side, ground-vibration tests of an airplane became a routine. The stiffness criteria were generally accepted and proved satisfactory from the point of view of safety.

It was supposed before 1938 that the solution of the flutter problem could be found in flight testing. Unfortunately, in February 1938, during a carefully planned flight test, a four-engined Junkers plane Ju 90 VI crashed, killing all scientists aboard. Since this accident it has been recognized that the inherent difficulties and uncertainties of flight-flutter testing are great. It is only one of many means of investigation, and is justified only if the flutter characteristics are investigated before the test and the dangerous points to be observed are approximately known.

This picture led to an emphasis on theoretical research. With the development of multi-engined wings, twin rudders, auxiliary control surfaces, etc., the two-dimensional analysis had to concede to more complicated three-dimensional analyses. Flutter analysis became more and more a specialized field.

In the period 1937 to 1939, the most frequent cause of flutter accidents was the control-surface tabs. Investigation of the aerodynamically balanced flaps became a central problem. Wind-tunnel tests in this period indicated that, aerodynamically, the "strip theory" gives reasonable accuracy for calculating the critical speed, at least in the incompressible range and for wings of moderate aspect ratio.

In the early part of World War II, most wing flutter cases were due to insufficient aileron mass balance and most tail-surface flutter cases were due to control-surface tabs. Toward the latter part of World War II, airplane speed increased toward the transonic range, and supersonic missiles appeared. Sweptback wings and delta wings attracted the attention of research workers. Steady-state instabilities, especially the control-surface effectiveness of large airplanes, became a real problem. Buffeting, another aeroelastic phenomenon, emerged with new threats because of the shock stall. On the other hand, airplane dynamics, which so far was regarded as a distant relative to aeroelasticity, now strengthened its tie to flutter and other aeroelastic problems.

At present, transonic flight is a daily event, and supersonic flight is a reality. Aeroelastic analysis becomes an organic part of the design. Many problems still await the solution.

BIBLIOGRAPHY

For an excellent introduction, see

5.1   Duncan, W. J.: Flutter and Stability. *J. Roy. Aeronaut. Soc.* **53**, 529–549 (1949).
5.2   Duncan, W. J.: The Fundamentals of Flutter. *Aeronaut. Research Council R. & M.* **2417** (1951); also *Aircraft Engg.* (Jan. and Feb. 1945).

The following are classical papers of historical interest:

5.3   Ackeret, J. and H. L. Studer: Bemerkungen über Tragflügelschwingungen. *Helv. Phys. Acta*, 501 (1934).
5.4   Bairstow, L., and A. Fage: Oscillations of the Tail Plane and Body of an Aeroplane in Flight. *Aeronaut. Research Com. R. & M.* **276**, part ii (July 1916).
5.5   Bairstow, L.: The Theory of Wing Flutter. *Aeronaut. Research Com. R. & M.* **1041** (1925).
5.6   Baumhauer, A. G. von, and C. Koning: On the Stability of Oscillations of an Aeroplane Wing. *Intern. Air Congr. London* **221** (1923). Royal Aeronautical Society, London. *NACA Tech. Memo.* **223** (1923).
5.7   Birnbaum, W.: Der Schlagflügelpropeller und die kleinen Schwingungen elastischer befestigter Tragflügel. *Z Flugtech. u. Motorluftschif.* **15**, 128–134 (1924).
5.8   Blasius, H.: Über Schwingungserscheinungen an Einholmigen Unterflügeln. *Z. Flugtech. u. Motorluftschif.* **16**, 39–42 (1925).
5.9   Blenk, H., and F. Liebers: Gekoppelte Torsions- und Biegungsschwingungen von Tragflügeln. *Z. Flugtech. u. Motorluftschif.* **16**, 479–486 (1925).
5.10 Blenk, H., and F. Liebers: Flügelschwingungen von freitragenden Eindeckern. *Luftfahrt-Forsch.* **1**, 1–17 (1928).
5.11 Blenk, H., and F. Liebers: Gekoppelte Biegungs- Torsions- und Querruderschwingungen von Freitragenden und halbfreitragenden Flügeln. *Luftfahrt-Forsch.* **4**, 69–93 (1929).
5.12 Cox, H. Roxbee: A Statistical Method of Investigating the Relations between the Elastic Stiffness of Aeroplane Wings and Wing-Aileron Flutter. *Aeronaut. Research Com. R. & M.* **1505** (1932).
5.13 Duncan, W. J.: The Wing Flutter of Biplanes. *Aeronaut. Research Com. R. & M.* **1227** (1930).
5.14 Duncan, W. J.: Models for the Determination of Critical Flutter Speeds. *Aeronaut. Research Com. R. & M.* **1425** (1931).
5.15 Duncan, W. J., and A. R. Collar: A Theory of Binary Servo-Rudder Flutter, with Applications to a Particular Aircraft. *Aeronaut. Research Com. R. & M.* **1527** (1933).
5.16 Duncan, W. J., and A. R. Collar: The Present Position of the Investigation of Airscrew Flutter. *Aeronaut. Research Com. R. & M.* **1518** (1933).
5.17 Duncan, W. J., D. L. Ellis, and A. G. Gadd: Experiments on Servo-Rudder Flutter. *Aeronaut. Research Com. R. & M.* **1652** (1934).
5.18 Essers, J.: Untersuchung von Flügelschwingungen im Windkanal. *Luftfahrt-Forsch.* **4**, 107–132 (1929).
5.19 Frazer, R. A.: An Investigation of Wing Flutter. *Aeronaut. Research Com. R. & M.* **1042** (1926).
5.20 Frazer, R. A., and W. J. Duncan: The Flutter of Aeroplane Wings. *Aeronaut. Research Com. R. & M.* **1155**, monograph (1928).

5.21 Frazer, R. A., and W. J. Duncan: A Brief Survey of Wing Flutter with an Abstract of Design Recommendations. *Aeronaut. Research Com. R. & M.* **1177** (1928).

5.22 Frazer, R. A., and W. J. Duncan: Wing Flutter as Influenced by the Mobility of the Fuselage. *Aeronaut. Research Com. R. & M.* **1207** (1929).

5.23 Frazer, R. A.: The Flutter of Aeroplane Wings. *J. Roy. Aeronaut. Soc.* **33**, 407–454 (1929).

5.24 Frazer, R. A., and W. J. Duncan: Conditions for the Prevention of Flexural-Torsional Flutter of an Elastic Wing. *Aeronaut. Research Com. R. & M.* **1217** (1930).

5.25 Frazer, R. A., and W. J. Duncan: The Flutter of Monoplanes, Biplanes, and Tail Units. *Aeronaut. Research Com. R. & M.* **1255** (1931).

5.26 Hesselbach, B.: Über die gekoppelten Schwingungen von Tragflügel und Verwindungsklappe. *Z. Flugtech. u. Motorluftschif.* **18**, 465–470 (1927).

5.27 Koning, C.: Einige Bemerkungen über nichtstationäre Strömungen an Tragflügeln. *Proc. 1st Intern. Congr. Applied Mech. Delft*, 414–417 (1924).

5.28 Küssner, H. G.: Schwingungen von Flugzeugflügeln. *Luftfahrt-Forsch.* **4**, 2, 41–62 (1929).

5.29 Lanchester, F. W.: Torsional Vibration of the Tail of an Aeroplane. *Aeronaut. Research Com. R. & M.* **276**, part i (July 1916).

5.30 Lockspeiser, B., and C. Callen: Wind Tunnel Tests of Recommendations for Prevention of Wing Flutter. *Aeronaut. Research Com. R. & M.* **1464** (1932).

5.31 Nagel, F.: Flügelschwingungen im stationären Luftströmungen. *Luftfahrt-Forsch.* **3**, 111–134 (1929).

5.32 Perring, W. G. A.: Wing Flutter Experiments upon a Model of a Single Seater Biplane. *Aeronaut. Research Com. R. & M.* **1197** (1928).

5.33 Raab, A.: Flügelschwingungen an freitragenden Eindeckern. *Z Flugtech. u. Motorluftschif.* **17**, 146–147 (1926).

5.34 Rauscher, M.: Über die Schwingungen freitragender Flügel. *Luftfahrt-Forsch.* **4**, 94–106 (1924).

5.35 Roche, J. A.: Airplane Vibrations and Flutter Controllable by Design. *SAE J.* **33**, 305–312 (Sept. 1933).

5.36 Scheubel, N.: Über das Leitwerkflattern und die Mittel zu seiner Verhütung. *Ber. u. Abh. Wiss. ges. Luftfahrt*, 103 (1926).

5.37 Younger, J. E.: Wing Flutter Investigation on Brady's Wind Tunnel Model. *Air Corps Informat. Circ.* **608**, 1–3 (1928).

5.38 Zahm, A. F., and R. M. Bear: A Study of Wing Flutter. *NACA Rept.* **285** (1928).

For stiffness criteria, see

5.39 Collar, A. R., E. G. Broadbent, and E. B. Puttick: An Elaboration of the Criterion for Wing Torsional Stiffness. *Aeronaut. Research Council R. & M.* **2154** (1946).

5.40 Hanson, J.: Critical Speeds of Monoplanes. *J. Roy. Aeronaut. Soc.* **41**, 703–726 (1937).

The extraction of energy from a flow by an oscillating airfoil with specific modes of motion is discussed in the following papers. See also Garrick, *NACA Rept.* **567** (1936), and Cicala, Ref. 15.1.

5.41 Barton, M. V.: Stability of an Oscillating Airfoil in Supersonic Airflow. *J. Aeronaut. Sci.* **15**, 371–376 (1948).

184 FLUTTER PHENOMENON

5.42 Biot, M. A., and L. Arnold: Low Speed Flutter and Its Physical Interpretation. *J. Aeronaut. Sci.* **15**, 232–236 (Apr. 1948).

5.43 Greidanus, J. H.: Low-Speed Flutter. *J. Aeronaut. Sci.* **16**, 127–128 (1949).

5.44 Rott, N.: Flügelschwingungsformen in ebener Kompressibler Potentialströmung. *Z. angew. Math. u. Physik.* **1**, 380–410 (1950).

For flutter in single degree of freedom at various Mach numbers, see

5.45 Cheilik, H., and H. Frissel: Theoretical Criteria for Single Degree of Freedom Flutter at Supersonic Speeds. *Cornell Aeronaut. Lab. Rept.* CAL-7A (May 1947).

5.46 Cunningham, H. J.: Analysis of Pure-Bending Flutter of a Cantilever Swept Wing and Its Relation to Bending-Torsion Flutter. *NACA Tech. Note* **2461** (1951).

5.47 Runyan, H. L.: Single-Degree-of-Freedom Flutter Calculations for a Wing in Subsonic Potential Flow and Comparison with an Experiment. *NACA Tech. Note* **2396** (1951).

5.48 Runyan, H. L.: Effect of Various Parameters including Mach Number on the Single-Degree-of-Freedom Flutter of a Control Surface in Potential Flow. *NACA Tech. Note* **2551** (1951).

5.49 Runyan, H. L., H. J. Cunningham, and C. E. Watkins: Theoretical Investigation of Several Types of Single-Degree-of-Freedom Flutter. *J. Aeronaut. Sci.* **19**, 101–110, 126 (1952). Comments by K. P. Abichandani and R. M. Rosenberg and author's reply, 215–216; 503–504.

5.50 Smilg, B.: The Instability of Pitching Oscillations of an Airfoil in Subsonic Incompressible Potential Flow. *J. Aeronaut. Sci.* **16**, 691–696 (Nov. 1949).

Dynamic mass balancing:

5.51 Collar, A. R.: Note on Virtual Inertia and Control Surface Flutter. *Brit. RAE TN SME* **218** (Feb. 1944); *ARC* **7469**, *RAE SME* 5/22, 92/104.

5.52 Collar, A. R.: A Proposal for a Control System Operated by an Aerodynamic Servo and Avoiding Large Mass-Balance Weights. *Brit. RAE TN SME* **293** (Dec. 1944).

5.53 Frazer, R. A.: Graphical Treatment of Binary Mass-Balancing Problems. *Aeronaut. Research Council R. & M.* **2551** (1942).

5.54 Williams, D.: Elevator Mass-Balance in Relation to Flexibility of the Arm and Position of Fuselage Nodal Point. *Aeronaut. Research Council R. & M.* **2062** (Aug. 1944).

5.55 Wolff, R. von: Dynamischer Ruderausgleich. *Forsch. Gebiete Ing-Wes.* **8**, 184–191 (1937).

Dimensional analysis and model theory. Refs. 5.64, 5.57, and 5.58 are of historical interest. Ref. 5.60 contains many examples of aeronautical applications. Rigorous mathematical theories are developed in Refs. 5.56 and 5.62.

5.56 Bridgman, P. W.: *Dimensional Analysis.* Yale Univ. Press (1931).

5.57 Buckingham, E.: On Physically Similar Systems; Illustrations of the Use of Dimensional Equations. *Phys. Rev. Ser. 2*, **4**, 345–376 (1914).

5.58 Buckingham, E.: Model Experiments and the Forms of Empirical Equations. *Trans. ASME* **37**, 263–96 (1915).

5.59 Duncan, W. J.: *Physical Similarity and Dimensional Analysis; An Elementary Treatise.* Edward Arnold & Co., London (1953).

5.60   Goodier, J. N., and W. T. Thomson: Applicability of Similarity Principles to Structural Models. *NACA Tech. Note* **933** (1944).

5.61   Kaufmann, W.: Über die Ähnlichkeitsbedingungen für Flatterschwingungen von Tragflügeln. *Luftfahrt-Forsch.* **16**, 21–25 (1939).

5.62   Langhaar, H. L.: *Dimensional Analysis and Theory of Models.* John Wiley & Sons, New York (1951).

5.63   Murphy, G.: *Similitude in Engineering.* Ronald Press Co., New York (1950).

5.64   Stokes, G. G.: On the Effect of the Internal Friction of Fluids on the Motion of Pendulums. *Trans. Cambridge Phil. Soc.* **9**, part 2 (1856). (Read Dec. 9, 1850.)

For model construction, see

5.65   Beckley, L. E., and M. Rauscher: The Design of Sectional Type Flutter and Dynamic Models of Aircraft Structures. *MIT Aeroelastic Research Lab. Rept.* to Bureau of Aeronautics, U.S. Navy, Contract Noa (s)7493 (1948).

5.66   Rauscher, M.: Report on the Suitability of Various Materials and Methods of Construction of Wind Tunnel Flutter Models. *MIT Aeroelastic Research Lab. Rept.* **1.0**, to Bureau of Aeronautics, U.S. Navy (June 1942). See also, *Repts.* **1.1, 1.2, 1.3, 1.4, 4.0, 9.0, 10.0, 17.0.**

Flight flutter tests are discussed in the following:

5.67   Cunningham, H. J., and R. R. Lundstrom: Description and Analysis of Rocket-Vehicle Experiment on Flutter involving Wing Deformation and Body Motions. *NACA RM* **L50I29** (1950).

5.68   Rosenbaum, R., and R. H. Scanlan: A Note on Flight Flutter Testing. *J. Aeronaut. Sci.* **15**, 366–370 (1948).

5.69   Schlippe, B. von: The Question of Spontaneous Wing Oscillations. Determination of critical velocity through flight vibration tests. *NACA Tech. Memo.* **806** (1936). Translated from *Luftfahrt-Forsch.* **13**, 41–45 (1936).

5.70   Thompson, F. L.: Flight Research at Transonic and Supersonic Speeds with Free-Falling and Rocket-Propelled Models. *2d Intern. Aeronaut. Conf. N.Y.,* 582–597 (1949). Institute of Aeronautical Sciences, New York.

The following papers describe some of the difficulties encountered in wind-tunnel testing. References to procedures and results of wind-tunnel and flight tests are listed in Chapters 7 and 9.

5.71   Kinnaman, E. B.: Flutter Analysis of Complex Airplanes by Experimental Methods. *J. Aeronaut. Sci.* **19**, 577–584 (1952).

5.72   Lambourne, N. C.: An Experimental Investigation on the Flutter Characteristics of a Model Flying Wing. *Aeronaut. Research Council R. & M.* **2626** (1952).

The panel flutter result quoted in Sec. 5.3 is given in

5.73   Fung, Y. C.: Some Recent Contributions to Panel Flutter Research. *AIAA Journal* **1**, 4, 898–909 (1963).

5.74   Fung, Y. C. and Thomson, W. T.: Instability of Spinning Space Stations Due to Crew Motion. *AIAA Journal* **3**, 1082–1087 (1965).

For mass balancing, see

5.75   Wilts, C. H.: Incompressible Flutter Characteristics of Representative Aircraft Wings. *NACA Tech. Note* **3780** (1957).

Chapter 6

FUNDAMENTALS OF FLUTTER ANALYSIS

This chapter is essentially divided into two parts. In §§ 6.1–6.6 the flutter of a cantilever wing having a straight elastic axis is treated under the assumption of quasi-steady aerodynamic derivatives. In §§ 6.7–12, the quasi-steady assumption is removed, and the analysis is based on the linearized thin-airfoil theory. This division is necessary because the results of the unsteady airfoil theory are so complicated that a great deal of numerical work is required in the solution, and the main analytical features are masked by the calculative complications. On the other hand, the quasi-steady assumption introduces such great simplifications that one will have no difficulty in carrying through a detailed analysis explicitly. Furthermore, the unsteady airfoil theory, as outlined in Chapters 13–15, is quite elaborate. Therefore it seems advantageous to use the quasi-steady theory as an introduction.

Although the quasi-steady assumption is used here chiefly as an introduction, the results so obtained may find practical applications for low-speed airplanes. To present a greater variety of methods of attack, Galerkin's method, as applied to flutter analysis by Grossman,[6.13] is used in the first part, whereas the method of generalized coordinates is used in the second part. The advantage of Galerkin's method is its direct relationship with the partial differential equations of motion. It is a natural extension of the approach used in § 1.10 for torsion-flexure oscillation of a cantilever beam. The main mathematical feature of the method, particularly in the first approximation, is very similar to the method of generalized coordinates to be explained later in § 7.1. It is easy to see how the analysis can be improved by successive approximation, or generalized so that the results of more accurate aerodynamic theory can be incorporated. However, these refinements will not be discussed. Here, after a review of the thin-airfoil theory in a two-dimensional steady incompressible flow in § 6.1, the quasi-steady aerodynamic coefficients are derived in § 6.2. Then, in § 6.3, the partial differential equations of motion of a straight cantilever wing in a flow are derived, and the possibility of flutter and divergence is discussed. In § 6.4 Galerkin's method for calculating the critical flutter speed is given. This is followed by § 6.5 treating the stability of the wing as the speed of flow varies. In § 6.6 some conclusions regarding the effect of changing several structural parameters on the critical torsion-flexure flutter speed are drawn.*

To obtain more accurate answers, the unsteady airfoil theory may be used. To simplify the calculation, we introduce not only the linearization of the hydrodynamic equations, but also the "strip" assumption regarding the finite-span effect: that the aerodynamic force at any chordwise section is the same as if that section were situated in a two-dimensional flow. Information regarding the finite-span effect is still incomplete, and the inclusion of known results in the analysis will introduce tremendous complications in the numerical work. The linearity and strip assumptions will be discussed further in Chapter 7.

The unsteady aerodynamic forces in an incompressible fluid are summarized in § 6.7 and are used to study the forced oscillation of a two-dimensional wing in a flow (§ 6.9). The existence of a critical flutter speed is again demonstrated, thus confirming a result obtained earlier in § 6.3. Using the linearized airfoil theory with compressibility effects properly included, the flutter of a two-dimensional wing is analyzed in § 6.10. Methods of solving the flutter determinant are discussed in § 6.11. The general problem of determining the critical speed is then discussed in § 6.12. Further remarks about the practical engineering flutter analysis are reserved for the next chapter.

6.1 TWO-DIMENSIONAL THIN AIRFOILS IN A STEADY FLOW OF AN INCOMPRESSIBLE FLUID

Let us consider a strip of unit span of a two-dimensional thin airfoil as shown in Fig. 6.1. Let the origin of a right-handed rectangular coordinate system be taken at the leading edge of the airfoil profile. Let

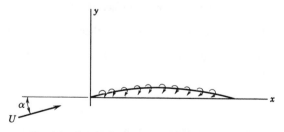

Fig. 6.1.   Steady flow over a two-dimensional airfoil.

---

* It should be pointed out that the method used in these sections can be applied whenever the aerodynamic forces are expressed in the form of the so-called "classical derivative coefficients" (see p. 228), and are not restricted to quasi-steady coefficients.

the x axis be taken along the chord line and the y axis be perpendicular to it. The airfoil camber line is described by the equation

$$y = Y(x) \qquad (0 \leqslant x \leqslant c) \tag{1}$$

If a steady two-dimensional flow with speed U and an angle of attack α (at large distance from the airfoil) streams past the airfoil, disturbances are introduced into the flow by the airfoil in such a manner that the resultant flow is tangent to the airfoil. According to the aerodynamic theory, the thin airfoil can be replaced by a continuous distribution of vorticity (a vortex sheet). Let the strength of the vorticity over an element of unit length in the spanwise direction and dx in the chordwise direction be $\gamma(x)\, dx$. The lift force contributed by the element dx is, according to Joukowsky's theorem,

$$dL = \rho U\, \gamma(x)\, dx \tag{2}$$

The total lift per unit span is therefore

$$L = \rho U \int_0^c \gamma(x)\, dx \tag{3}$$

where c is the chord length of the airfoil.

For a thin airfoil of small camber ($y(x) \leq c$), the surface of the airfoil differs only infinitesimally from a flat plate. The induced velocity over the airfoil surface, to the first order of approximation, can be calculated by assuming the vortices to be situated on the x axis. The y component of the induced velocity at a point x on the x axis is*

$$v_i(x) = \int_0^c \frac{\gamma(\xi)\, d\xi}{2\pi(\xi - x)} \tag{4}$$

which, to the first order of approximation, is the same as the component of velocity normal to and on the airfoil surface at the chordwise location x. The slope of the fluid stream on the airfoil is then $\alpha + \dfrac{v_i}{U}$. This must be equal to the slope of the airfoil surface dY/dx. Hence, the boundary condition on the airfoil is

$$\alpha + \frac{v_i}{U} = \frac{dY}{dx} \tag{5}$$

The vorticity distribution $\gamma(x)$ must be determined from Eq. 5. In addition, the Kutta condition that $\gamma(c) = 0$ must be satisfied, which is equivalent to the statement that the fluid must leave the trailing edge smoothly.

* This is a Cauchy integral. See footnote on p. 126.

In place of x let a new independent variable ψ be introduced so that

$$x = \frac{c}{2}(1 - \cos \psi) \tag{6}$$

When x varies from 0 to c along the chord, ψ varies from 0 to π. The vorticity distribution can be written as*

$$\gamma = 2U \left(A_0 \cot \frac{\psi}{2} + \sum_1^\infty A_n \sin n\psi \right) \tag{7}$$

Substituting (7) into (4), we obtain (see Ref. 1.43 for details)

$$v_i = U \left(-A_0 + \sum_1^\infty A_n \cos n\psi \right) \tag{8}$$

Equation 5 then implies

$$\alpha - A_0 + \sum_1^\infty A_n \cos n\psi = \frac{dY}{dx} \tag{9}$$

The left-hand side is a Fourier series. The coefficients can therefore be determined by the usual method. Multiplying Eq. 9 by $\cos n\psi$ ($n = 0, 1, 2, \cdots$), and integrating from 0 to π, we obtain

$$\alpha - A_0 = \frac{1}{\pi} \int_0^\pi \frac{dY}{dx} d\psi$$
$$A_n = \frac{2}{\pi} \int_0^\pi \frac{dY}{dx} \cos n\psi \, d\psi \tag{10}$$

From Eqs. 3 and 7, the total lift can be obtained. The result, expressed as the lift coefficient, is

$$C_L = \pi(2A_0 + A_1) \tag{11}$$

Similarly, the moment about the leading edge, expressed as the moment coefficient, is

$$(C_M)_{\text{l.e.}} = -\frac{\pi}{2} \left(A_0 + A_1 - \frac{1}{2} A_2 \right) = -\frac{\pi}{4}(A_1 - A_2) - \frac{1}{4} C_L \tag{12}$$

A substitution of Eq. 10 into Eqs. 11 and 12 gives

$$C_L = 2\pi(\alpha + \varepsilon_0) \tag{13}$$

$$(C_M)_{\text{l.e.}} = \left(\mu_0 - \frac{\pi}{2} \varepsilon_0 \right) - \frac{1}{4} C_L \tag{14}$$

* A form assumed by H. Glauert.[1.43]

where

$$\varepsilon_0 = -\frac{1}{\pi} \int_0^\pi \frac{dY}{dx} (1 - \cos \psi)\, d\psi$$

$$\mu_0 = -\frac{1}{2} \int_0^\pi \frac{dY}{dx} (1 - \cos 2\psi)\, d\psi \qquad (15)$$

From Eq. 13 it is seen that ε_0 is the negative of the angle of zero lift and that the theoretical value of $dC_L/d\alpha$ is 2π. The experimental value of $dC_L/d\alpha$ is somewhat smaller than 2π. It is convenient to write Eq. 13 as

$$C_L = \frac{dC_L}{d\alpha} (\alpha + \varepsilon_0) \qquad (16)$$

From Eq. 14 it is seen that $\mu_0 - \frac{\pi}{2} \varepsilon_0$ is the moment coefficient at zero lift and that the lift force acts through the $1/_4$-chord point (the aerodynamic center). The coefficient of moment about a point at a distance x from the leading edge is

$$(C_M)_x = (C_M)_{\text{l.e.}} + \frac{x}{c} C_L \qquad (17)$$

Differentiating, we obtain

$$\frac{\partial (C_M)_x}{\partial \alpha} = \frac{dC_L}{d\alpha} \left[\frac{x}{c} - \frac{1}{4} \right] \qquad (18)$$

Equation 15 shows that, for an airfoil with parabolic or circular camber, the lift is determined by the slope of the camber line at the $3/_4$-chord point. For such an airfoil can be described by the equation

$$y = Y(x) = ax(c - x) \qquad (19)$$

where $ac^2/4$ is the maximum camber at the mid-chord point. Hence,

$$\frac{dY}{dx} = ac - 2ax = ac \cos \psi \qquad (20)$$

So, when $x = 3c/4$, $dY/dx = -ac/2$. Now, from Eqs. 15,

$$-\varepsilon_0 = \frac{1}{\pi} \int_0^\pi ac \cos \psi (1 - \cos \psi)\, d\psi = -\frac{ac}{2}$$

which is exactly the value of dY/dx at the $3/_4$-chord point. Therefore the line of zero lift is parallel to the tangent to the airfoil at the $3/_4$-chord point.

Since the resultant lift force acts through the $1/_4$-chord point, the vorticity can be regarded as concentrated there in calculating the lift.

The downwash due to a concentrated vortex of strength Γ located at the $^1/_4$-chord point is

$$v_i(x) = \frac{\Gamma}{2\pi(c/4 - x)} = \frac{qcC_L}{2\pi(c/4 - x)\rho U} = \frac{U(\alpha + \varepsilon_0)}{2(1/4 - x/c)} \qquad (21)$$

At the $^3/_4$-chord point, $x/c = {}^3/_4$, we have

$$-\frac{1}{U}(v_i)_{x/c = {}^3/_4} = \alpha + \varepsilon_0 = \alpha - \left(\frac{dY}{dx}\right)_{x/c = {}^3/_4} \qquad (22)$$

In other words, if the vorticity is considered as concentrated at the aerodynamic center ($^1/_4$-chord point), the boundary condition for downwash is satisfied at the $^3/_4$-chord point.

We have quoted in § 4.3 that, for a swept or unswept wing of finite span, if the lifting line (the vortex line) is located at the $^1/_4$-chord line and the downwash condition is satisfied at the $^3/_4$-chord line, the result will give a good approximation to the spanwise lift distribution (Weissinger's method). Thus the $^3/_4$-chord line seems to have a unique importance. We shall see later that this line is also uniquely significant for oscillating airfoils.

6.2   QUASI-STEADY AERODYNAMIC DERIVATIVES

In an unsteady flow, the fundamental equation 2 of § 6.1 does not hold because $\gamma(x)$ now consists of both free and bound vortices. Moreover, since the lift, and hence the vorticity strength, varies with time, vortices

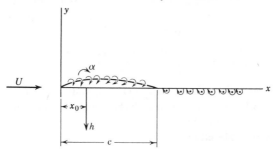

Fig. 6.2.   Unsteady flow over a two-dimensional airfoil.

must be shed at the trailing edge of the wing and carried downstream by the flow. The reason for this is that the total circulation in a contour enclosing all the singularities (Fig. 6.2) must remain zero in a nonviscous fluid. Therefore every vortex element on the wing must be balanced by another in the wake. The vortices in the wake, having no force to support them, cannot have relative velocity with the flow and hence move downstream along the streamlines. These wake vortices, however, induce vertical components of velocity on the wing, and therefore the second fundamental equation (Eq. 4 of § 6.1) also becomes invalid.

In order to make a simplified analysis, let us introduce the following *quasi-steady assumption*: The aerodynamic characteristics of an airfoil whose motion consists of variable linear and angular motions are equal, at any instant of time, to the characteristics of the same airfoil moving with constant linear and angular velocities equal to the actual instantaneous values. The inclination of the flow-velocity vector to the profile is also taken to be constant and equal to the actual instantaneous inclinations. Thus, at any instant of time, we assume that Eqs. 2 and 4 of § 6.1 hold, in spite of the objections named above.

Under the quasi-steady assumption, the results of the last section can be applied directly. Let us consider a flat plate in a stream whose velocity at infinity is U in the x-axis direction. Let the plate have two degrees of freedom: a vertical translation h and a rotation α about an axis located at x_0 behind the leading edge, h being positive downward and α positive nose up, both measured from the x axis. Let the coordinate system be as shown in Fig. 6.2. As before, the plate will be replaced by a vortex sheet, and the camber line, when $h = \alpha = 0$, is a line $Y(x) = 0$.

At a point x on the airfoil, the vertical velocity component is

$$- \frac{dh}{dt} + (x_0 - x) \frac{d\alpha}{dt}$$

The instantaneous slope of the airfoil at x is $- \alpha$. As the fluid velocity must be tangential to the airfoil, the vertical component of fluid velocity v_i on the airfoil must satisfy the equation

$$\frac{v_i}{U} = - \alpha - \frac{1}{U} \frac{dh}{dt} + \frac{(x_0 - x)}{U} \frac{d\alpha}{dt} \qquad (1)$$

v_i is the velocity induced by the vorticity $\gamma(x)$. Equation 1 is the equivalent to Eq. 5 of § 6.1. A comparison with the fundamental equations of § 6.1 shows that everything is the same except that the term dY/dx in that section must now be replaced by

$$- \frac{1}{U} \frac{dh}{dt} + \frac{(x_0 - x)}{U} \frac{d\alpha}{dt}$$

Making this replacement in Eqs. 15 of § 6.1, we obtain, from Eqs. 16 and 14 of § 6.1

$$C_L = \frac{dC_L}{d\alpha} \left[\alpha + \frac{1}{U} \frac{dh}{dt} + \frac{1}{U} \left(\frac{3}{4} c - x_0 \right) \frac{d\alpha}{dt} \right] \qquad (2)$$

$$(C_M)_{\text{l.e.}} = -\frac{c\pi}{8U}\frac{d\alpha}{dt} - \frac{1}{4}C_L \qquad (3)$$

Equations 2 and 3 give the lift and moment for variable h and α. Equation 3 shows that the resultant lift acts through the $1/4$-chord point, but, in addition, there is a damping couple $-\dfrac{c\pi}{8U}\dfrac{d\alpha}{dt}$ proportional to the angular velocity.

It is interesting to note that the expression inside [ ] in Eq. 2 is the downwash angle of the flow at the $3/4$-chord point induced by the motion of the airfoil. The vertical velocity at the $3/4$-chord due to translation is dh/dt: that due to rotation about x_0 is $(\frac{3}{4}c - x_0)d\alpha/dt$, both positive downward. The resultant vertical velocity downward is $dh/dt + (\frac{3}{4}c - x_0)d\alpha/dt$, which, divided by the velocity of flow U, gives an induced angle of attack as shown in Eq. 2.

Thus, in the unsteady flow, as well as in the steady flow, a correct total vortex strength is obtained by regarding the vortices as concentrated at the $1/4$-chord point, computing the induced velocity at the $3/4$-chord point, and satisfying there the boundary condition of tangency to the airfoil. It will be shown in Chapter 13 that this fact remains true, even when the quasi-steady assumption is removed.

The lift and moment discussed above arise from vorticity or circulation. However, it is well known that, for unsteady motions of a body in a flow, there are "apparent mass" forces whose origin is not associated with the creation of vorticity. These apparent mass forces are of minor importance in the bending-torsion flutter of a cantilever wing to be considered below, especially when the reduced frequency is low, and will be neglected when the quasi-steady assumption is used. It should be remarked, however, that, for oscillations of control surfaces at a higher reduced frequency, the force and moment acting on the control surfaces are principally of the apparent mass origin.

6.3 FLUTTER OF A CANTILEVER WING

Consider an unswept cantilever wing having a straight elastic axis perpendicular to the fuselage which is assumed to be fixed in space. The wing deformation can be measured by a deflection h and a rotation α about the elastic axis, h being positive downward and α positive leading edge up. The chordwise distortion will be neglected. Let a frame of reference be chosen as shown in Fig. 6.3, with the y axis coinciding with the elastic axis. Let x_α be the distance between the center of mass and the elastic axis at any section, positive if the former lies behind the latter.

Let c be the chord length and x_0 be the distance of the elastic axis aft the leading edge. In a steady flow of speed U, the wing will have some elastic deformation, which is, however, of no concern to the problem of flutter. In the following, *the free motion of the wing following an initial disturbance is considered.* Thus let h and α be the deviations from the equilibrium state, and let the inertia, elastic, and aerodynamic forces

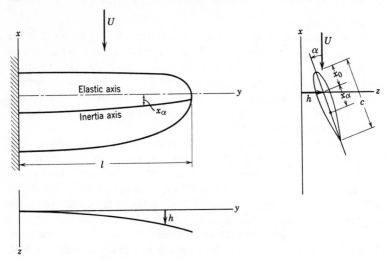

Fig. 6.3. Notations for a cantilever wing.

correspond also to the deviations from the steady-state values; then, for small disturbances, the principle of superposition holds, and we have the following equations of motion (see § 1.10):

$$\frac{\partial^2}{\partial y^2}\left(EI\frac{\partial^2 h}{\partial y^2}\right) + m\frac{\partial^2 h}{\partial t^2} + mx_\alpha\frac{\partial^2 \alpha}{\partial t^2} + L = 0$$

$$\frac{\partial}{\partial y}\left(GJ\frac{\partial \alpha}{\partial y}\right) - I_\alpha\frac{\partial^2 \alpha}{\partial t^2} - mx_\alpha\frac{\partial^2 h}{\partial t^2} + M = 0 \tag{1}$$

where EI and GJ are the bending and torsional rigidity of the wing, m and I_α are the mass and mass moment of inertia (about the elastic axis) of the wing section at y, per unit length along the span, and L and M (about the elastic axis) are the aerodynamic lift and moment per unit span, respectively. Now

$$L = \frac{\rho U^2}{2}cC_L, \qquad M = \frac{\rho U^2}{2}c^2 C_M = \frac{\rho U^2}{2}c^2\left[(C_M)_{\text{l.e.}} + \frac{x_0}{c}C_L\right] \tag{2}$$

Hence, from Eqs. 2 and 3 of § 6.2, we obtain

$$\frac{\partial^2}{\partial y^2}\left(EI\frac{\partial^2 h}{\partial y^2}\right) + m\frac{\partial^2 h}{\partial t^2} + mx_\alpha\frac{\partial^2 \alpha}{\partial t^2}$$

$$+ \frac{\rho U^2}{2}c\frac{dC_L}{d\alpha}\left[\alpha + \frac{1}{U}\frac{\partial h}{\partial t} + \frac{1}{U}\left(\frac{3}{4}c - x_0\right)\frac{\partial \alpha}{\partial t}\right] = 0$$

$$\frac{\partial}{\partial y}\left(GJ\frac{\partial \alpha}{\partial y}\right) - I_\alpha\frac{\partial^2 \alpha}{\partial t^2} - mx_\alpha\frac{\partial^2 h}{\partial t^2} + \frac{\rho U^2}{2}c^2\left\{-\frac{c\pi}{8U}\frac{\partial \alpha}{\partial t}\right.$$

$$\left. + \left(\frac{x_0}{c} - \frac{1}{4}\right)\frac{dC_L}{d\alpha}\left[\alpha + \frac{1}{U}\frac{\partial h}{\partial t} + \frac{1}{U}\left(\frac{3}{4}c - x_0\right)\frac{\partial \alpha}{\partial t}\right]\right\} = 0$$

$$(3)$$

The boundary conditions are:

$$h = \frac{\partial h}{\partial y} = \alpha = 0 \quad \text{at} \quad y = 0$$

$$\frac{\partial^2 h}{\partial y^2} = \frac{\partial^3 h}{\partial y^3} = \frac{\partial \alpha}{\partial y} = 0 \quad \text{at} \quad y = l$$

$$(4)$$

where l is the semispan.

Should x_α and U be zero, Eqs. 3 would be reduced to two independent equations, one for h and one for α. The terms involving x_α and U indicate inertia and aerodynamic couplings.

Since Eqs. 3 are linear equations with constant coefficients, the solution is a sum of linearly independent solutions of the form

$$h = A f(y)e^{\lambda t}, \qquad \alpha = B \phi(y)e^{\lambda t} \qquad (5)$$

where $\lambda, f(y), \phi(y)$ and A, B are to be determined from Eqs. 3 and 4, and the initial conditions. Since the solution of a differential equation depends continuously on the coefficients of the equation, and since the coefficients of Eq. 3 vary continuously with U, the constant λ will vary continuously with U. In general, λ is a complex number. Let

$$\lambda = p + iq \qquad (6)$$

When p is positive, the amplitude of the motion will increase with increasing time. When p is negative, the opposite is true. If p is negative at U_1, and positive at U_2, $(U_2 > U_1)$, then there exists at least one value of U, say U_0, between U_1 and U_2, at which p vanishes. At U_0, p is purely imaginary, corresponding physically to a simple-harmonic motion. Such a speed will separate the speed range in its neighborhood into two regions, in one of which $p < 0$, where the motion is damped and stable; in the other $p > 0$, where the amplitude increases with time and is unstable.

Beginning with $U = 0$, let us gradually increase the speed of flow and consider the variation of p with U. When $U = 0$, the motion is simple harmonic, and so $p = 0$, as is shown in § 1.10 (material and air damping neglected). Next let U be a very small positive number. We shall show that p is small and negative. Instead of solving the differential system directly, let us recall that $p < 0$ implies that the motion is damped. For a *free* system this imples that the wing is losing energy to the surrounding air. Hence, it is sufficient to consider the energy relations. When U is small, the solutions h and α differ only slightly from those obtained for $U = 0$. Free oscillation in still air has been investigated in § 1.10, where it is shown that the solution can be written as

$$h = A\,f(y) \sin qt, \qquad \alpha = B\,\phi(y) \sin qt \qquad (7)$$

where A, B, q are real numbers and f, ϕ are real functions. h and α are in phase with each other, so the initial phase angles may be omitted from Eqs. 7 by suitably choosing the origin of time. The oscillation modes h and α at very small U will be assumed to be given by Eqs. 7. Substituting Eqs. 7 into Eqs. 2 to obtain L and M, we can compute the energy *gain* per unit span of the wing at a spanwise station y in each complete cycle of oscillation:

$$\text{Gain of energy} = -\int_0^{2\pi/q} L \cdot \frac{\partial h}{\partial t}\,dt + \int_0^{2\pi/q} M \cdot \frac{\partial \alpha}{\partial t}\,dt \qquad (8)$$

The result can be written as

$$\text{Gain of energy} = -\frac{dC_L}{d\alpha}\frac{\rho U}{4}\,q^2 c\,(\alpha_1 A^2 + \alpha_2 B^2 + \alpha_3 AB) \qquad (9)$$

where

$$\alpha_1 = f^2$$

$$\alpha_2 = \left[\left(\frac{x_0}{c} - \frac{1}{4}\right)\left(\frac{x_0}{c} - \frac{3}{4}\right) + \frac{\pi}{8\,dC_L/d\alpha}\right] c^2\phi^2 \qquad (10)$$

$$\alpha_3 = \left(\frac{1}{2} - \frac{x_0}{c}\right) 2cf\phi$$

Evidently $\alpha_1 > 0$. α_2 reaches a minimum at $x_0/c = {}^1\!/_2$ where it becomes zero if $dC_L/d\alpha = 2\pi$, and is positive if $dC_L/d\alpha < 2\pi$. Furthermore, if $dC_L/d\alpha < 2\pi$, the discriminant of Eq. 9 is negative; i.e.,

$$\frac{\alpha_3^2}{4} - \alpha_1\alpha_2 < 0 \qquad (11)$$

For we may reduce the discriminant into the following form:

$$\frac{\alpha_3^2}{4} - \alpha_1\alpha_2 = \left(1 - \frac{2\pi}{dC_L/d\alpha}\right)\frac{c^2 f^2\phi^2}{16}$$

which is zero when $dC_L/d\alpha = 2\pi$ and is negative when $dC_L/d\alpha < 2\pi$. Therefore the quadratic form of A and B in the parenthesis of Eq. 9 is nonnegative,* and Eq. 9 shows that the airfoil cannot gain energy from the flow. In other words, when the speed of flow U is infinitesimal, the flow is stable and p is negative.† Since $p = 0$ when $U = 0$, it is evident that p will remain negative until it becomes zero again at certain higher value of U, say U_{cr} (see Fig. 6.4). For speeds $U > U_{cr}$, p may become positive, corresponding to an unstable motion.

We shall call the speed at which $p = 0$ a *critical speed*. We have shown that between $U = 0$ and U_{cr} the torsion-flexure motion of the cantilever wing is stable when $dC_L/d\alpha < 2\pi$. At the critical speed, two cases are possible: Either q (the imaginary part of λ) vanishes, or it does not vanish.

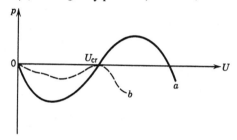

Fig. 6.4.   The variation of the sign of p with U.

If $q = 0$, then the displacements h and α are independent of time, but the structure has lost its power to recover its original form when disturbed. The wing is said to be in critical *divergent* condition. If $q \neq 0$, the motion is harmonic with an indefinite amplitude. It is said to be in the critical *flutter* condition. Hence,

$$p = 0, \quad q = 0 \quad \text{implies divergence}$$
$$p = 0, \quad q \neq 0 \quad \text{implies flutter} \tag{12}$$

In both cases the aeroelastic system may be said to be neutrally stable.

The continuity argument cannot establish definitely the sign of p for $U > U_{cr}$. As shown in Fig. 6.4, at U_{cr} there are two possibilities. The curve of p vs. U may cross the U axis to $p > 0$ for $U > U_{cr}$ (curve a), or it may have a horizontal tangent at U_{cr} and then turn back to the lower side (curve b). More careful study in the neighborhood of U_{cr} is necessary

* A quadratic form $ax^2 + bxy + cy^2$ can never change its sign for all real values of (x, y), if
$$b^2 - 4ac < 0$$

† This result has no universal validity since the case $x_0/c = \frac{1}{2}$, $dC_L/d\alpha = 2\pi$ must be excepted. Nor is it true for all other types of oscillations. For example, in tab flutter problems, it may happen that a disturbed motion is actually unstable for zero airspeed upwards for bad values of tab frequency and mass balance. In general, a test of stability is needed. Such a test will be discussed in § 6.5 and in § 10.6.

to establish the tendency definitely. In practice, the first case is what generally occurs: The motion at supercritical speeds is unstable.

When we continue curve a or b further for larger values of U, it may cross the U axis again. Hence, higher critical speeds may exist. But, since aircraft flight starts from $U = 0$ and increases continuously, and since generally no instability can be tolerated, the flight speed cannot be permitted to be larger than the first critical value. Thus the finding of the first critical speed is the main object of flutter analysis.

That only the critical speed is of interest implies the following important facts:

1. Only undamped harmonic oscillations are of interest. Hence the aerodynamic coefficients need only be evaluated for harmonic motions.

2. At the critical condition, the amplitude of oscillation may be considered as infinitesimal. There is no need to discuss finite deformations. The linearized aerodynamical theory and the linearized equations in elasticity can thus be justified in most cases.

6.4 APPROXIMATE DETERMINATION OF THE CRITICAL SPEED BY GALERKIN'S METHOD

We shall consider the motion at the critical flutter condition and determine the critical speed of flight. In the following, therefore, U represents the U_{cr} of the preceding section. The motion is harmonic and representable as

$$h = A f(y)e^{i\omega t}, \qquad \alpha = B \phi(y)e^{i\omega t} \qquad (1)$$

where ω is real and A, B are complex constants.

We assume A and B to be of the same dimensions as h and α, respectively, and $f(y)$, $\phi(y)$ to be dimensionless. Substituting Eqs. 1 into the differential equation 3 of § 6.3, and canceling the factor $e^{i\omega t}$ throughout, we obtain

$$A \left[\frac{d^2}{dy^2} \left(EI \frac{d^2 f}{dy^2} \right) - m\omega^2 f + \frac{dC_L}{d\alpha} \frac{\rho c}{2} Ui\omega f \right] + B \left[- mx_\alpha \omega^2 \phi \right.$$

$$\left. + \frac{dC_L}{d\alpha} \frac{\rho c}{2} U^2 \phi + \frac{dC_L}{d\alpha} \frac{\rho c^2}{2} \left(\frac{3}{4} - \frac{x_0}{c} \right) Ui\omega \phi \right] = 0$$

$$A \left[mx_\alpha \omega^2 f + \frac{dC_L}{d\alpha} \frac{\rho c^2}{2} U \left(\frac{x_0}{c} - \frac{1}{4} \right) i\omega f \right] \qquad (2)$$

$$+ B \left\{ \frac{d}{dy} \left(GJ \frac{d\phi}{dy} \right) + I_\alpha \omega^2 \phi + \frac{\rho U^2 c^2}{2} \left(\frac{x_0}{c} - \frac{1}{4} \right) \frac{dC_L}{d\alpha} \phi \right.$$

$$\left. + \frac{\rho U c^3}{2} \left[-\frac{\pi}{8} + \left(\frac{x_0}{c} - \frac{1}{4} \right) \left(\frac{3}{4} - \frac{x_0}{c} \right) \frac{dC_L}{d\alpha} \right] i\omega \phi \right\} = 0$$

These are simultaneous differential equations with complex coefficients. The boundary conditions are given by Eqs. 4 of § 6.3, provided that the h and α functions were replaced by f and ϕ.

As a first approximation to the critical speed and frequency, the method of Galerkin may be used.[11.8] Assume that the functions $f(y)$ and $\phi(y)$ are known and real valued. Multiplying the first of Eqs. 2 by $f(y)\,dy$ and the second by $\phi(y)\,dy$, and integrating from 0 to l, we obtain

$$A(a_{11} - c_{11}\omega^2 + i\omega U d_{11}) + B(c_{12}\omega^2 - b_{12}U^2 - i\omega U d_{12}) = 0$$
$$A(c_{21}\omega^2 - i\omega U d_{21}) + B(a_{22} - c_{22}\omega^2 + b_{22}U^2 + i\omega U d_{22}) = 0 \tag{3}$$

where*

$$a_{11} = \int_0^l \frac{d^2}{dy^2}\left(EI\frac{d^2f}{dy^2}\right)f\,dy = \int_0^l EI\left(\frac{d^2f}{dy^2}\right)^2 dy$$

$$a_{22} = -\int_0^l \frac{d}{dy}\left(GJ\frac{d\phi}{dy}\right)\phi\,dy = \int_0^l GJ\left(\frac{d\phi}{dy}\right)^2 dy$$

$$b_{12} = -\frac{dC_L}{d\alpha}\frac{\rho}{2}\int_0^l cf\phi\,dy$$

$$b_{22} = -\frac{\rho}{2}\frac{dC_L}{d\alpha}\int_0^l \left(\frac{x_0}{c} - \frac{1}{4}\right)c^2\phi^2\,dy$$

$$c_{11} = \int_0^l mf^2\,dy$$

$$c_{12} = c_{21} = -\int_0^l mx_\alpha f\phi\,dy \tag{4}$$

$$c_{22} = \int_0^l I_\alpha\phi^2\,dy$$

$$d_{11} = -\frac{dC_L}{d\alpha}\frac{\rho}{2}\int_0^l cf^2\,dy$$

$$d_{12} = -\frac{dC_L}{d\alpha}\frac{\rho}{2}\int_0^l c^2\left(\frac{3}{4} - \frac{x_0}{c}\right)f\phi\,dy$$

$$d_{21} = \frac{dC_L}{d\alpha}\frac{\rho}{2}\int_0^l c^2\left(\frac{x_0}{c} - \frac{1}{4}\right)f\phi\,dy$$

$$d_{22} = -\frac{\rho}{2}\int_0^l \left[\left(\frac{3}{4} - \frac{x_0}{c}\right)\left(\frac{x_0}{c} - \frac{1}{4}\right)\frac{dC_L}{d\alpha} - \frac{\pi}{8}\right]c^3\phi^2\,dy$$

* Note that, if $f(y)$ and $\phi(y)$ represent the uncoupled flexural and torsional modes of the wing in a vacuum, respectively, the ratio a_{11}/c_{11} represents the square of the frequency of the flexural oscillation, and a_{22}/c_{22} that of the torsional oscillation of the wing.

The homogeneous equations 3 admit nonvanishing solutions A, B only if the determinant of their coefficients vanishes. This determinant being complex, both the real and imaginary parts must vanish. On setting the determinant to zero and separating the real and imaginary parts, we obtain two equations:

$$A_1\omega^4 - (C_1 + C_2 U^2)\omega^2 + (E_1 + E_2 U^2) = 0$$
$$- B_1\omega^2 + (D_1 + D_2 U^2) = 0 \tag{5}$$

where

$$A_1 = c_{11}c_{22} - c_{12}c_{21} \tag{6}$$
$$B_1 = d_{11}c_{22} + c_{11}d_{22} - c_{12}d_{21} - c_{21}d_{12}$$
$$C_1 = c_{11}a_{22} + a_{11}c_{22}, \qquad C_2 = c_{11}b_{22} - b_{12}c_{21} + d_{11}d_{22} - d_{12}d_{21}$$
$$D_1 = d_{11}a_{22} + a_{11}d_{22}, \qquad D_2 = d_{11}b_{22} - b_{12}d_{21}$$
$$E_1 = a_{11}a_{22}, \qquad E_2 = a_{11}b_{22}$$

The second equation of 5 gives

$$\omega^2 = \frac{D_1 + D_2 U^2}{B_1} \tag{7}$$

Substitution of this expression into the first equation of 5 gives

$$LU^4 + MU^2 + N = 0 \tag{8}$$

where

$$L = D_2(B_1 C_2 - D_2 A_1)$$
$$M = B_1 C_2 D_1 + B_1 C_1 D_2 - B_1^2 E_2 - 2D_1 D_2 A_1 \tag{9}$$
$$N = B_1 C_1 D_1 - B_1^2 E_1 - D_1^2 A_1$$

From Eq. 8, we obtain the critical speed

$$U^2 = \frac{-M \pm \sqrt{M^2 - 4LN}}{2L} \tag{10}$$

The sign in front of the radical must be chosen in such a way as to give the smallest positive value of U^2.

Thus a solution can be obtained if the flutter modes f and ϕ are known. But they are not known at the beginning. Hence, it is necessary to approximate them on the basis of empirical information.* As a simple approach, f and ϕ may be assumed to be the fundamental modes of purely

---

* The assumed functions $f(y)$ and $\phi(y)$ must satisfy the "rigid" boundary conditions at $y = 0$ (Eqs. 4, § 6.3). See footnote on p. 54.

flexural and purely torsional oscillations in still air, of a cantilever beam of uniform cross section:

$$f(y) = \cosh \kappa y - \cos \kappa y - 0.734 (\sinh \kappa y - \sin \kappa y) \tag{11}$$

where $\kappa = 1.875/l$, and

$$\phi(y) = \sin \frac{\pi y}{2l} \tag{12}$$

For these functions, tables for the integrals of $f^2(y)$, $\phi^2(y)$, $f(y)\phi(y)$, etc., are available.[6.25]

When the distances from the centers of mass to the elastic axis are large, or the wing planform is such that the oscillation modes differ considerably from those of a uniform beam, better results can be obtained by assuming $f(y)$ and $\phi(y)$ as the coupled flexure-torsional oscillation modes of a wing. The predominantly torsional mode should usually be used. Sometimes it is advantageous to use the uncoupled flexure and torsion modes of the actual wing for f and ϕ.

Corresponding to the two solutions of U^2 from Eq. 10, there are two values of ω^2 from Eq. 7. Usually the smaller U^2 is associated with the higher ω^2; for, in Eq. 7, the coefficients B_1 and D_1 are always positive, whereas D_2 is negative if the elastic axis lies behind the $1/_4$-chord point, as is usually so. Since, for conventional wings the torsional frequency (in still air) is higher than the flexural, the above conclusion indicates that the flutter mode is generally predominantly torsional.

It has been shown by a simple example in § 1.9 that the oscillations at different spanwise coordinates of a beam of variable cross section with damping are out of phase. Since at the critical flutter condition the aerodynamic force has a component proportional to $\partial h/\partial t$ and $\partial \alpha/\partial t$, (of the nature of a "damping" force), it is natural to expect that the flutter mode is also generally out of phase in the spanwise direction. Mathematically, this means that the functions $f(y)$ and $\phi(y)$ must have complex coefficients (see § 1.9). This is actually true, but the change in phase angle along the span is generally small. By assuming $f(y)$, $\phi(y)$ to be real functions, little error will result. On the other hand, the ratio A/B, solved from Eqs. 3, when the critical value of ω and U are used, is generally complex, indicating a shift of phase between the flexural and torsional motion. This phase shift is of fundamental importance in governing the energy exchange between the wing and the flow, as we have shown in § 5.4.

6.5 THE STABILITY OF A DISTURBED MOTION

Let us now return to the general problem, and consider the stability of the motion of the cantilever wing of § 6.3 at any speed of flight U, following an initial disturbance. The equation of motion is given by Eqs. 3 of § 6.3 and the boundary conditions by Eqs. 4 of that section. The solution can be expressed in the form

$$h = A f(y)e^{\lambda t}, \qquad \alpha = B \phi(y)e^{\lambda t} \tag{1}$$

where A, B, λ are constants; λ, $f(y)$, $\phi(y)$ are to be determined by the differential system, and A, B by the initial conditions. To obtain an approximate solution, Galerkin's method may be used. We shall assume approximate forms of $f(y)$, $\phi(y)$ to be known (as discussed in § 6.4). Let us substitute Eqs. 1 into Eqs. 3 of § 6.3, multiply the first equation by $f(y)$ and the second by $\phi(y)$, integrate both with respect to y from 0 to l, to obtain two linear homogeneous equations for A and B, which are exactly the same as Eqs. 3 of § 6.4, except that the factor $i\omega$ in those equations must be replaced by λ. For a nontrivial solution corresponding to a disturbed motion, the determinant of the coefficients of A and B must vanish. This determinantal equation, known as the characteristic equation, may be written as

$$A_0\lambda^4 + B_0\lambda^3 + C_0\lambda^2 + D_0\lambda + E_0 = 0 \tag{2}$$

where

$$
\begin{aligned}
A_0 &= A_1, & B_0 &= B_1 U \\
C_0 &= C_1 + C_2 U^2, & D_0 &= D_1 U + D_2 U^3 \\
E_0 &= E_1 + E_2 U^2,
\end{aligned} \tag{3}
$$

and A_1, B_1, etc., are given by Eqs. 6 of § 6.4.

From the discussion in § 6.3, it is seen that the condition for stability of the disturbed motion is that the real parts of all the roots of the characteristic equation be negative. The necessary and sufficient condition for this is that the coefficients A_0, B_0, C_0, D_0, E_0 and the Routh discriminant

$$R = B_0 C_0 D_0 - B_0{}^2 E_0 - D_0{}^2 A_0 \tag{4}$$

have the same sign.* It can be shown on the basis of Schwarz inequality†

---

* See Appendix 2.

† For arbitrary functions $f_1(x)$ and $f_2(x)$, we have

$$\int [\alpha f_1(x) + \beta f_2(x)]^2 \, dx \geqslant 0$$

i.e.

$$\alpha^2 \int f_1{}^2(x) \, dx + 2\alpha\beta \int f_1(x) f_2(x) \, dx + \beta^2 \int f_2{}^2(x) \, dx \geqslant 0$$

Since this is a positive definite function of α, β, the discriminant must be negative, which leads to Schwarz inequality

$$[\int f_1{}^2(x) \, dx][\int f_2{}^2(x) \, dx] > [\int f_1(x) f_2(x) \, dx]^2$$

that the coefficient A_0 is always positive.   Hence the conditions of stability are

$$B_0 > 0, \quad C_0 > 0, \quad D_0 > 0, \quad E_0 > 0, \quad R > 0 \qquad (5)$$

These inequalities are satisfied when U is very small, because it is shown in § 6.3 that the disturbed motion is stable.   As U increases, the coefficients B_0, C_0, $\cdots$, R vary.   Since $B_0 = B_1 U$ and B_1 does not depend on U, B_0 is positive at all values of $U > 0$.   As for C_0 and D_0, we shall show that E_0 and R become zero before C_0 and D_0.   For, if $D_0 = 0$, then

$$R = - B_0{}^2 E_0 \qquad (6)$$

which shows that either E_0 or R must be negative.   Similarly, if $C_0 = 0$, then

$$R = - B_0{}^2 E_0 - D_0{}^2 A_0 \qquad (7)$$

which again shows that either E_0 or R must be negative.   Hence, as long as E_0 and R remain positive, C_0 and D_0 must be positive.

The stability is then determined by the signs of E_0 and R.   If E_0 and R_0 remain positive, the motion is stable.   If either one becomes negative, the motion is unstable.

The physical meaning of the critical condition $E_0 = 0$ or $R = 0$ can be clarified as follows.   When $E_0 = 0$, Eq. 2 has a root $\lambda = 0$.   Hence, according to Eqs. 12 of § 6.3, a condition of wing divergence is reached. The *divergence speed* is given by

$$E_0 = a_{11}(a_{22} + b_{22}U^2) = 0 \qquad (8)$$

i.e.,

$$U_{\text{div}}{}^2 = - \frac{a_{22}}{b_{22}} = \frac{\displaystyle\int_0^l GJ \left(\frac{d\phi}{dy}\right)^2 dy}{\dfrac{\rho}{2} \displaystyle\int_0^l \frac{\partial C_M}{\partial \alpha} c^2 \phi^2 dy} \qquad (9)$$

which agrees with the result of § 3.4, C_M being the coefficient of aerodynamic moment about the elastic axis.

The condition $R = 0$, on the other hand, corresponds to the critical *flutter* condition.   For, a substitution of A_0, B_0, etc., from Eqs. 3 leads to

$$R \equiv U^2(LU^4 + MU^2 + N) = 0 \qquad (10)$$

which is exactly the critical flutter equation 8 of § 6.4, except for the factor U^2.   That Eq. 10 has a root $U = 0$ simply reflects the fact that the structure may oscillate in still air, which is a natural consequence of our assumption that the viscosity of the air and the internal damping of the structure are negligible.

Combining the above results with the discussion at the end of § 6.3, we see that, at U slightly larger than the smallest critical flutter speed, the motion will be unstable if $R < 0$. The very special case that R becomes zero at U_{cr} but becomes positive again at supercritical speeds occurs only when R reaches a relative minimum at U_{cr}. This is possible if and only if*

$$R = \frac{dR}{dU} = 0, \qquad \frac{d^2R}{dU^2} > 0 \tag{11}$$

Using Eq. 10, we see that the dimensions and the aerodynamic characteristics of the wing must be such as to satisfy the conditions

$$M^2 = 4LN \quad \text{and} \quad M < 0 \tag{12}$$

If Eq. 12 is satisfied, then the critical nature of the motion at the speed U_{cr} is only superficial. For, at speeds differing slightly from U_{cr} (either larger or smaller), the disturbed motion is stable. If Eq. 12 is not satisfied, then the speed U_{cr} is truly critical; for supercritical speeds the motion diverges. The exacting condition 12 has little chance of being satisfied in practice.

This method of stability investigation can be applied to other aeroelastic systems whenever the time variable appears in the governing equation as derivatives of finite order (see footnote, p. 228). The general method consists of deriving a characteristic equation by a substitution such as Eq. 1, and then check the signs of Routh or Hurwitz discriminants (Appendix 2).

6.6 THE EFFECT OF STRUCTURAL PARAMETERS OF THE WING ON THE CRITICAL SPEED OF THE TORSION-FLEXURE FLUTTER

Under the assumptions of § 6.3 and using the method of § 6.4, Grossman proved a number of relations expressing the effects of structural parameters of the wing on the critical torsion-flexure flutter speed:†

1. A simultaneous change of the flexural and torsional rigidities by a factor n changes both the critical flutter speed and the flutter frequency by a factor $\sqrt{n}$, and leaves the critical reduced frequency unchanged.

For, other things remaining equal, a multiplication of EI and GJ

---

* Note that it is impossible to have $R = \dfrac{dR}{dU} = \dfrac{d^2R}{dU^2} = \dfrac{d^3R}{dU^3} = 0$ while $\dfrac{d^4R}{dU^4} \neq 0$.

† The validity of these relations is restricted by the quasi-steady assumption made in their derivations. The fluid is incompressible. The wing is unswept, and the aileron is locked. For a more exact survey, see, for example, Ref. 6.23.

by a factor n changes the values of the coefficients a_{11}, a_{22}, etc., by factors listed below, as can be easily verified according to Eqs. 4, 6, 9 of § 6.4:

Coefficients changed by a factor n: a_{11}, a_{22}, C_1, D_1, E_2, M.

Coefficients changed by a factor n^2: E_1, N.

Coefficients unaffected: b_{12}, b_{22}, d_{11}, d_{12}, d_{21}, d_{22}, c_{11}, $c_{12} = c_{21}$, c_{22}, A_1, B_1, C_2, D_2, L.

Equations 7 and 10 of § 6.4 then show the conclusion of the theorem at once. The constancy of the reduced frequency follows by definition.

The effect of the individual changes of the flexural and torsional rigidities cannot be stated in such general terms. Examples of flutter analysis generally show that, when the torsional rigidity alone is increased, the flutter speed is also increased, but, when the flexural rigidity alone is varied, the change in critical flutter speed is small. The flutter speed reaches a minimum when the flexural rigidity becomes so high that the frequency of (uncoupled) flexural oscillation is equal to that of the (uncoupled) torsional oscillation. Further increase of flexural rigidity increases the flutter speed.

An important consequence of the above result concerns the accuracy required in determining the rigidity constants: It is permissible to admit considerable error in the flexural rigidity of a wing without causing serious error in the calculated critical flutter speed.

2. A similar investigation gives the following: A change in all the geometric dimensions of a wing by a factor n without a change in the elastic constants (E and G) has no effect on the magnitude of the critical speed, but changes the flutter frequency by a factor $1/n$. The critical reduced frequency remains unchanged.

As for the effect of variation of individual geometric parameters, the results of sample calculations can be stated most concisely in the following form, which, however, cannot be proved without introducing further assumptions in addition to those stated in § 6.3:

3. When the characteristic geometric dimensions of a wing are varied individually, with the mass density and elasticity distributions remaining unchanged (while the absolute values of the mass density and the torsional stiffness may vary), the "apparent" reduced frequency for torsion-flexure flutter

$$k'_{\text{cr}} = \frac{\omega_\alpha c}{2U_{\text{cr}}} \tag{1}$$

remains approximately unchanged. In the formula above, c represents the chord length at a reference section, and ω_α the fundamental frequency of the torsional oscillation of the wing.*

Equation 1 is an approximate empirical rule whose validity must be questioned when unconventional wing designs are considered.

4. The effects of the relative positions of the elastic, inertia, and aerodynamic axes are so important that each particular case should be computed separately. Generally speaking, the closer the inertia and elastic axes are to the line of aerodynamic centers, the higher is the critical flutter speed.

6.7 UNSTEADY AERODYNAMIC FORCES ON AN AIRFOIL IN AN INCOMPRESSIBLE FLUID

The simplified analysis of the preceding sections are based on the quasi-steady aerodynamic derivatives. We shall now show how the quasi-steady assumptions can be removed. In this and later sections, the results of the linearized aerodynamic theory, as presented in Chapters 12–15, will be used.

In order to show the existence of the critical flutter conditions, we shall first summarize the aerodynamic forces acting on a two-dimensional

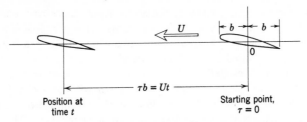

Fig. 6.5.   Impulsive motion of an airfoil.

airfoil in unsteady motion in an incompressible fluid. The equations of motion of the airfoil are derived in § 6.8. The forced oscillation of the airfoil due to a periodic excitation is then considered in § 6.9. Following this, in § 6.10, the flutter of a two-dimensional airfoil is discussed in general terms, without restricting to the incompressible fluid. The solution of the flutter determinant and a summary of the methods of determining the critical speeds are presented in the last two sections.

The unsteady aerodynamic force acting on a thin airfoil in unsteady

---

* In the definition of the critical reduced frequency of flutter, ω is the frequency at the critical flutter speed, which, in general, differs somewhat from the fundamental frequency of the torsional oscillation of the wing. But, since the flexural rigidity plays only a minor part in the flexure-torsional flutter, the "apparent" reduced frequency as defined above has a physical significance.

motion in a two-dimensional incompressible fluid was obtained by Wagner, Küssner, von Kármán and Sears, and others.   Let the chord of the airfoil be $2b$, and the angle of attack (assumed infinitesimal) α.   Consider the *growth of circulation about the airfoil which starts impulsively from rest to a uniform velocity U*.   Let the impulsive motion take place at the origin when $\tau = 0$ (Fig. 6.5).   The vertical velocity component of the fluid, the so-called downwash, is $w = U \sin \alpha \doteq U\alpha$ on the airfoil, since the flow must be tangent to the airfoil.   Then, on the physical assumption that the velocity at the trailing edge must be finite, one derives the lift due to circulation on a strip of unit span as a function of time:

$$L_1 = 2\pi b\rho Uw \, \Phi(\tau), \qquad \Phi(\tau) = 0 \quad \text{if} \quad \tau < 0 \tag{1}$$

where

$$\tau = Ut/b \tag{2}$$

is a nondimensional quantity proportional to time.   The function $\Phi(\tau)$, called *Wagner's function*, is illustrated in Fig. 6.6.   An approximate

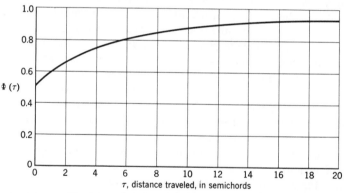

$\Phi(\tau)$

τ, distance traveled, in semichords

Fig. 6.6.   Wagner's function for an incompressible fluid.

expression which agrees within 2 per cent of the exact value in the entire range $0 < \tau < \infty$ is given by Garrick[15.9, 15.17]:

$$\Phi(\tau) \doteq 1 - \frac{2}{4 + \tau} \qquad (\tau > 0) \tag{3}$$

Another approximate expression is given by R. T. Jones[15.43]:

$$\Phi(\tau) \doteq 1 - 0.165e^{-0.0455\tau} - 0.335e^{-0.300\tau} \qquad (\tau > 0) \tag{4}$$

whereas W. P. Jones gives[15.80]

$$\Phi(\tau) \doteq 1 - 0.165e^{-0.041\tau} - 0.335e^{-0.32\tau} \tag{4a}$$

The expression 4a gives slightly better approximation than Eq. 4 for $\tau < 2.2$.

The exact form of $\Phi(\tau)$ is the following (see § 15.1)

$$\Phi(\tau) = 1 - \int_0^\infty \{(K_0 - K_1)^2 + \pi^2(I_0 + I_1)^2\}^{-1} e^{-x\tau} x^{-2} \, dx \tag{5}$$

where K_0, K_1; I_0, I_1 are modified Bessel functions of the second and first kind, respectively, with argument x implied. It is seen that half of the final lift is assumed at once and that the lift approaches asymptotically its steady-state value $2\pi b\rho Uw$ when $\tau \to \infty$. The center of pressure of this lift (due to circulation) is at the $1/4$-chord point behind the leading edge.

Let us now consider a more general type of motion. Let the airfoil have two degrees of freedom: a vertical translation h, called *bending*, positive downward, and a rotation a, called *pitching*, positive nose up, about an axis located at a distance $a_h b$ from the mid-chord point, a_h being positive toward the trailing edge (Fig. 6.7).* The flow is assumed to be two-dimensional, h and α are infinitesimal, and the mean flow speed U is a constant. In this case part of the lift arises from circulation, and part from noncirculatory origin—the so-called "apparent mass" forces.

Wagner's function gives the growth of circulation about the airfoil due to a sudden increase of downwash which is uniform over the airfoil. For a general motion having two degrees of freedom h and α, the downwash over the airfoil is not uniform. Now, in the theory of oscillating airfoils, it can be shown that for bending and pitching oscillations the circulation about the airfoil is determined by the downwash velocity at the $3/4$-chord point from the leading edge of the airfoil (§ 13.4). By a reciprocal relation between the harmonic oscillations and the response to unit-step functions (§ 15.1), and the principle of superposition, this result holds also for arbitrary bending and pitching motions. Hence, if we replace w in Eq. 1 by the increment of downwash at the $3/4$-chord point, the circulatory lift can be obtained.

The downwash at the $3/4$-chord point due to the h and α degrees of freedom consists of the following: (1) a uniform downwash corresponding to a pitching angle α, $w = U \sin \alpha \doteq U\alpha$ (α being infinitesimal), (2) a uniform downwash due to vertical translation h, which may also be written as $\dfrac{dh}{d\tau}\dfrac{d\tau}{dt} = \dfrac{U}{b} h'$, where *a prime denotes a differentiation with respect to the nondimensional time τ, and a dot denotes a differentiation with respect*

* These notations are different from those of the preceding sections. In the airfoil theory (Chapters 12–15), the origin of reference axes is usually taken at the mid-chord point, and the semichord length b is taken as the characteristic length. In keeping with these standard notations, the symbols are redefined in Fig. 6.7.

to the physical time t, (3) a nonuniform downwash due to $\dot{\alpha}$, its value at the $^3/_4$-chord point being $(^1/_2 - a_h)b\dfrac{d\alpha}{dt}$ or $(^1/_2 - a_h)U\alpha'$.

Summing up, we find

$$w(\tau) = U\alpha(\tau) + \frac{U}{b}h'(\tau) + (\tfrac{1}{2} - a_h)U\alpha'(\tau) \tag{6}$$

In the time interval $(\tau_0, \tau_0 + d\tau_0)$, the downwash $w(\tau_0)$ increases by an amount $\dfrac{dw(\tau_0)}{d\tau_0}d\tau_0$. When $d\tau_0$ is sufficiently small, this may be regarded as an impulsive increment and the corresponding circulatory lift per unit span is

$$dL_1 = 2\pi b\rho U\Phi(\tau - \tau_0)\frac{dw(\tau_0)}{d\tau_0}d\tau_0 \quad \text{for} \quad \tau \geqslant \tau_0$$

By the principle of superposition which holds when w remains small, we have the circulatory lift per unit span for arbitrary time history of w:*

$$L_1(\tau) = 2\pi b\rho U\int_{-\infty}^{\tau}\Phi(\tau - \tau_0)\frac{dw}{d\tau_0}(\tau_0)\,d\tau_0 \tag{7}$$

where the lower limit is taken as $-\infty$, meaning before the very beginning of motion. If the motion starts at time $\tau = 0$, $w = 0$ for $\tau < 0$, Eq. 7 reduces to

$$L_1(\tau) = 2\pi b\rho U\left[w_0\Phi(\tau) + \int_0^{\tau}\Phi(\tau - \tau_0)\frac{dw(\tau_0)}{d\tau_0}d\tau_0\right] \tag{7a}$$

where w_0 is the limiting value of $w(\tau)$ when $\tau \to 0$ from the positive side.†
It is, of course, to be remembered that $\Phi(\tau - \tau_0) = 0$ if $\tau < \tau_0$. Combining Eq. 7 with Eq. 6, we obtain:

$$L_1(\tau) = 2\pi b\rho U^2\int_{-\infty}^{\tau}\Phi(\tau - \tau_0)\left[\alpha'(\tau_0) + \frac{1}{b}h''(\tau_0) + \left(\frac{1}{2} - a_h\right)\alpha''(\tau_0)\right]d\tau_0 \tag{8}$$

An expression corresponding to Eq. 7a can be written down if the motion starts at time $\tau = 0$.

When the airfoil has a general motion, the lift and moment of the noncirculatory origin (the apparent mass forces) must be added. The resultant lift and moment include the following terms:

1. A *lift* force with center of pressure at the mid-chord, of amount equal to the apparent mass $\rho\pi b^2$ times the vertical acceleration at the mid-chord point:

$$L_2 = \rho\pi b^2(\ddot{h} - a_h b\ddot{\alpha}) = \rho\pi U^2(h'' - a_h b\alpha'') \tag{9}$$

---

* Duhamel's integral. See § 8.1.

† The first term in Eq. 7a gives the effect of initial disturbance. It can be obtained formally from Eq. 7 by noticing the jump of $w(\tau_0)$ at $\tau_0 = 0$.

2. A *lift* force with center of pressure at the $^3/_4$-chord point, of the nature of a centrifugal force, of amount equal to the apparent mass $\rho \pi b^2$ times $U\dot{\alpha}$. (This term is, however, circulatory.)

$$L_3 = \rho \pi b^2 U\dot{\alpha} = \rho \pi b U^2 \alpha' \tag{10}$$

3. A nose-down *couple* equal to the apparent moment of inertia* $\rho \pi b^2 (b^2/8)$ times the angular acceleration $\ddot{\alpha}$:

$$M_a = -\frac{\rho \pi b^4}{8} \ddot{\alpha} = -\frac{\rho \pi b^2 U^2}{8} \alpha'' \tag{11}$$

The total lift per unit span is then

$$L = L_1 + L_2 + L_3 \tag{12}$$

The total moment per unit span about the elastic axis is

$$M = (\tfrac{1}{2} + a_h)bL_1 + a_h bL_2 - (\tfrac{1}{2} - a_h)bL_3 + M_a \tag{13}$$

All these are valid for a two-dimensional flow of an incompressible fluid.

6.8 EQUATIONS OF MOTION OF A TWO-DIMENSIONAL AIRFOIL

Let us consider a strip of unit width of a two-dimensional flat-plate airfoil having two degrees of freedom: a bending h (positive downward measured at the elastic axis) and a pitching α (positive nose-up) about the

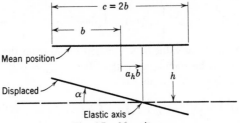

Fig. 6.7. Notations.

elastic axis (Fig. 6.7). Let the airfoil be situated in a flow of an incompressible fluid at speed U. The equations of motion of the airfoil will be derived by considering the balance of the inertia, elastic, aerodynamic, and exciting forces.

For an element of mass dm situated at a distance r (positive toward the trailing edge) from the elastic axis, the inertia force is

$$- dm(\ddot{h} + r\ddot{\alpha})$$

---

* It is interesting to note that, whereas the apparent mass of a flat plate is equal to the mass of a cylinder of air with diameter equal to the chord of the plate, the apparent moment of inertia is only one fourth of the mass moment of inertia of that cylinder of air if that cylinder were solid.

The total inertia force per unit span of the wing is therefore

$$- \int dm(\ddot{h} + r\ddot{\alpha}) = - (m\ddot{h} + S\ddot{\alpha}) \tag{1}$$

where $m = \int dm$ = the total mass of the wing per unit span, slugs, and $S = \int r\, dm$ = wing static moment about the elastic axis, slug-feet. The integrals are taken over the entire wing chord.

The inertia force exerts a moment per unit span about the elastic axis of amount

$$- \int r(\ddot{h} + r\ddot{\alpha})dm = - (I_\alpha\ddot{\alpha} + S\ddot{h}) \tag{2}$$

where $I_\alpha = \int r^2 dm$ = wing mass moment of inertia about the elastic axis, slug-ft^2.

Let the bending and pitching displacements be resisted by a pair of springs at the elastic axis with spring constants K_h (pounds per foot) and K_α (foot-pounds per radian), respectively. The elastic restoring force corresponding to a displacement h is $- h K_h$, in the direction opposing h. That against α is $- \alpha K_\alpha$.

The equations of motion can be written according to the condition that the sum of the inertia and elastic forces and moments must balance the externally applied force and moment. Let the latter be denoted by Q_h and Q_α, which include the aerodynamic forces and other mechanical excitations. Hence,

$$m\ddot{h} + S\ddot{\alpha} + hK_h = Q_h, \qquad S\ddot{h} + I_\alpha\ddot{\alpha} + \alpha K_\alpha = Q_\alpha \tag{3}$$

These equations can be written in a slightly different way by expressing the spring constants in terms of certain frequencies. Consider the airfoil to be so restrained that only one degree of freedom, say h, is permitted. Assume further that no external force is acting. The equation of motion of the airfoil is then

$$m\ddot{h} + hK_h = 0 \qquad (\alpha \equiv 0) \tag{4}$$

This represents an independent (uncoupled) harmonic oscillation of frequency

$$\omega_h = \sqrt{K_h/m} \tag{5}$$

Hence, we may write

$$K_h = m\omega_h{}^2 \tag{6}$$

Similarly,

$$K_\alpha = I_\alpha\omega_\alpha{}^2 \tag{7}$$

where ω_α (radians per second) is the uncoupled natural frequency in torsion.

The part of the external forces induced by the motion of the airfoil can be obtained from the results of the preceding section. The aerodynamic lift and moment about the elastic axis induced by h and α are:

$$L(\tau) = 2\pi b\rho U^2 \int_{-\infty}^{\tau} \Phi(\tau - \tau_0) \left[\alpha'(\tau_0) + \frac{1}{b} h''(\tau_0) \right.$$

$$\left. + \left(\frac{1}{2} - a_h \right) \alpha''(\tau_0) \right] d\tau_0 + \rho\pi U^2(h'' - a_h b\alpha'') + \rho\pi b U^2 \alpha' \qquad (8)$$

$$M(\tau) = \left(\frac{1}{2} + a_h \right) 2\pi b^2 \rho U^2 \int_{-\infty}^{\tau} \Phi(\tau - \tau_0) \left[\alpha'(\tau_0) + \frac{1}{b} h''(\tau_0) \right.$$

$$\left. + \left(\frac{1}{2} - a_h \right) \alpha''(\tau_0) \right] d\tau_0 + a_h b\rho\pi U^2(h'' - a_h b\alpha'')$$

$$- \left(\frac{1}{2} - a_h \right) \rho\pi b^2 U^2 \alpha' - \frac{\rho\pi b^2 U^2}{8} \alpha'' \qquad (9)$$

If we let $P(\tau)$ and $Q(\tau)$ denote, respectively, the external applied force (positive downward) and moment (positive nose up) other than the aerodynamic lift and moment, and introduce the dimensionless time $\tau = Ut/b$, Eqs. 3 become*

$$m \frac{U^2}{b^2} h'' + S \frac{U^2}{b^2} \alpha'' + m\omega_h^2 h = -L(\tau) + P(\tau)$$

$$S \frac{U^2}{b^2} h'' + I_\alpha \frac{U^2}{b^2} \alpha'' + I_\alpha \omega_\alpha^2 \alpha = M(\tau) + Q(\tau) \qquad (10)$$

Equations 10 can be solved by the method of Laplace transformation. But a special case is of interest: the steady-state forced oscillation due to a periodic excitation, in which the integrals in Eqs. 9 can be integrated explicitly, thus simplifying the calculations.

6.9 FORCED OSCILLATION DUE TO A PERIODIC EXCITATION

Consider the forced oscillation of a two-dimensional airfoil in a flow, under a harmonic exciting force. We shall have, then, the external force and moment per unit span

$$P = P_0 e^{i\omega t}, \qquad Q = Q_0 e^{i\omega t} \qquad (1)$$

where P_0 and Q_0 are complex constants (see § 1.8).

---

* The lift $L(\tau)$ is defined as positive *upward* according to the usual sign convention in aerodynamics. The bending displacement h is positive downward. The forces on the right-hand side of Eqs. 3 are positive downward. Hence, the negative sign in front of $L(\tau)$ in Eqs. 10.

Let us assume that the excitation has been operative for a long time and that the response of the system has reached a steady state.* The motion of the airfoil must be periodic and of the same period as the exciting force. Furthermore, because the system is linear, the response must be harmonic if the exciting force is. Hence, we may write

$$h = h_0 e^{i\omega t}, \qquad \alpha = \alpha_0 e^{i\omega t} \qquad (2)$$

where h_0 and α_0 are complex constants, the absolute values of which represent the amplitudes, and the arguments, the phase angles.

It is convenient to use the nondimensional time τ. Then, since

$$\tau = \frac{U}{b} t \qquad (3)$$

we may write

$$P(\tau) = P_0 e^{ik\tau}, \qquad Q(\tau) = Q_0 e^{ik\tau}$$
$$h(\tau) = h_0 e^{ik\tau}, \qquad \alpha(\tau) = \alpha_0 e^{ik\tau} \qquad (4)$$

where k is the *reduced frequency*

$$k = \frac{\omega b}{U} \qquad (5)$$

When $h(\tau)$ and $\alpha(\tau)$ are represented by Eqs. 4, the integrals involved in Eqs. 8 and 9 of § 6.8 can be evaluated. In order to avoid the mathematical difficulty of oscillatory divergence at the lower integration limit $(-\infty)$, we shall introduce a convergence factor $e^{\varepsilon\tau_0}$, $(\varepsilon > 0)$, into the integrands and pass to the limit $\varepsilon \to 0$ through real, positive values after the integrals are evaluated. In other words, a divergent motion is considered first, but the degree of divergence is reduced to zero afterwards. The physical problem remains to be the finding of aerodynamic forces for harmonic oscillations. Let us write

$$\lim_{\varepsilon \to 0+} ik \int_{-\infty}^{\tau} \Phi(\tau - \tau_0) e^{ik\tau_0 + \varepsilon\tau_0} \, d\tau_0 = C(k) e^{ik\tau} \qquad (6)$$

and therefore, on substituting Eq. 4,

$$\lim_{\varepsilon \to 0+} \int_{-\infty}^{\tau} \Phi(\tau - \tau_0) \left[\alpha'(\tau_0) + \frac{1}{b} h''(\tau_0) + \left(\frac{1}{2} - a_h \right) \alpha''(\tau_0) \right] e^{\varepsilon\tau_0} \, d\tau_0$$
$$= C(k) \left[\alpha_0 + \frac{i}{b} k h_0 + \left(\frac{1}{2} - a_h \right) ik\alpha_0 \right] e^{ik\tau} \qquad (7)$$

* If the system is *stable*, the effect of the initial disturbance will die out as time increases. If the system is *unstable*, the effect of the initial disturbance will actually be magnified as time increases. Nevertheless, in both cases, the response of the system can be separated into two parts: (1) the steady-state response to the periodic force, and (2) the transient response to the initial disturbance. The first part is studied in this section.

The function $C(k)$ is called *Theodorsen's function,** the exact expression of which, corresponding to Eq. 5 of § 6.7, is

$$C(k) = F(k) + i\, G(k) = \frac{H_1^{(2)}(k)}{H_1^{(2)}(k) + i\, H_0^{(2)}(k)} = \frac{K_1(ik)}{K_0(ik) + K_1(ik)} \quad (8)$$

Table 6.1   The Function $C(k) = F + iG$, and Related Quantities

k	$1/k$	F	$-G$	$-2G/k$	$2F/k^2$
∞	0.000	0.5000	0	0	$0^\bullet$
10.00	0.100	0.5006	0.0124	0.00248	0.010012
6.00	0.16667	0.5017	0.0206	0.00686	0.02787
4.00	0.250	0.5037	0.0305	0.01525	0.06296
3.00	0.33333	0.5063	0.0400	0.02667	0.1125
2.00	0.500	0.5129	0.0577	0.0577	0.2565
1.50	0.66667	0.5210	0.0736	0.0948	0.4631
1.20	0.83333	0.5300	0.0877	0.1462	0.7361
1.00	1.000	0.5394	0.1003	0.2006	1.0788
0.80	1.250	0.5541	0.1165	0.2912	1.7316
0.66	1.51516	0.5699	0.1308	0.3964	2.6166
0.60	1.66667	0.5788	0.1378	0.4593	3.2156
0.56	1.78572	0.5857	0.1428	0.5100	3.7353
0.50	2.000	0.5979	0.1507	0.6028	4.7832
0.44	2.27273	0.6130	0.1592	0.7236	6.3326
0.40	2.500	0.6250	0.1650	0.8250	7.8125
0.34	2.94118	0.6469	0.1738	1.022	11.192
0.30	3.33333	0.6650	0.1793	1.195	14.778
0.24	4.16667	0.6989	0.1862	1.552	24.267
0.20	5.000	0.7276	0.1886	1.886	36.380
0.16	6.250	0.7628	0.1876	2.345	59.592
0.12	8.33333	0.8063	0.1801	3.002	111.99
0.10	10.000	0.8320	0.1723	3.446	166.4
0.08	12.500	0.8604	0.1604	4.010	268.9
0.06	16.66667	0.8920	0.1426	4.753	495.6
0.05	20.000	0.9090	0.1305	5.220	727.2
0.04	25.000	0.9267	0.1160	5.800	1158.3
0.025	40.000	0.9545	0.0872	6.976	3054.4
0.01	100.000	0.9824	0.0482	9.640	19648
0	∞	1.000	0	∞	∞

where H and K are the Hänkel functions and the modified Bessel functions, respectively.   The standard notations for the real and imaginary parts of $C(k)$ are F and G, which are tabulated in Table 6.1,† and shown in Fig. 6.8.

* See § 13.4.

† From Ref. 13.33. More extensive numerical tables of $C(k)$ are given by Luke and Dengler, and Brower and Lassen, in Refs. 13.25 and 13.6, respectively.

Approximate expressions of $C(k)$ corresponding to Eqs. 4 (R. T. Jones) and 4a (W. P. Jones) of § 6.7 are, respectively:*

$$C(k) \doteq 1 - \frac{0.165}{1 - \dfrac{0.0455}{k} i} - \frac{0.335}{1 - \dfrac{0.3}{k} i} \qquad (8a)$$

$$C(k) \doteq \cdot 1 - \frac{0.165}{1 - \dfrac{0.041}{k} i} - \frac{0.335}{1 - \dfrac{0.32}{k} i} \qquad (8b)$$

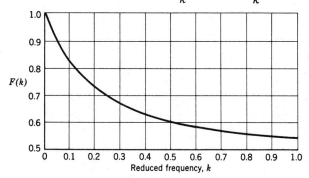

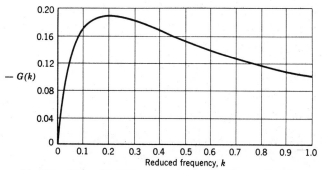

Fig. 6.8. The real and imaginary parts of Theodorsen's function $F(k)$ and $G(k)$. Note the difference in vertical scale in these two figures. $F(k)$ tends to 1/2 and $G(k)$ tends to zero as k tends to infinity.

* The expression 8a gives somewhat better approximation for $k < 0.5$, whereas 8b is better for $k > 0.5$. The real part F is well approximated by both 8a and 8b, the maximum percentage errors based on the exact values are $+ 2.6$ per cent, $- 2.1$ per cent throughout the range $(0, \infty)$. The error of the absolute value of the imaginary part, $- G$, is much larger. The maximum percentage errors based on the exact values are as follows: For 8a (R. T. Jones), $+ 8.5$ per cent, $- 13.5$ per cent. For 8b (W. P. Jones), $+ 10$ per cent, $- 11$ per cent. A detailed numerical comparison can be found in Ref. 15.80.

Substituting Eqs. 4 and 7 into Eqs. 8 and 9, § 6.8, we obtain

$$\frac{L(\tau)e^{-ik\tau}}{\rho\pi b U^2} = -k^2\left(\frac{h_0}{b} - a_h\,\alpha_0\right) + ik\alpha_0$$

$$+ 2C(k)\left[\alpha_0 + \frac{i}{b}kh_0 + \left(\frac{1}{2} - a_h\right)ik\alpha_0\right]$$

$$\tag{9}$$

$$\frac{M(\tau)e^{-ik\tau}}{\rho\pi b^2 U^2} = \left(\frac{1}{2} + a_h\right)2C(k)\left[\alpha_0 + \frac{i}{b}kh_0 + \left(\frac{1}{2} - a_h\right)ik\alpha_0\right]$$

$$- k^2 a_h\left(\frac{h_0}{b} - a_h\alpha_0\right) - \left(\frac{1}{2} - a_h\right)ik\alpha_0 + \frac{k^2}{8}\alpha_0$$

These expressions may be compared with the quasi-steady lift and moment given by Eqs. 2 and 3 of § 6.2. Specializing to harmonic oscillations, we obtain, from Eq. 2 of § 6.2,

$$\frac{L(\tau)e^{-ik\tau}}{\rho\pi b U^2} = 2\left[\alpha_0 + \frac{i}{b}kh_0 + \left(\frac{1}{2} - a_h\right)ik\alpha_0\right]$$

Comparing this with the last term of the first of Eq. 9, we see that the lift due to circulation can be obtained by multiplying the quasi-steady lift by Theodorsen's function $C(k)$. An investigation of the expression for the moment M shows that the resultant of the circulatory lift acts through the 1/4-chord point. The remaining terms in Eqs. 9 are of noncirculatory origin.

Let us introduce the dimensionless coefficients:

$$\mu = \frac{m}{\pi\rho b^2} \quad = \text{the mass ratio}$$

$$x_\alpha = \frac{S}{mb} \quad = \begin{array}{l}\text{the distance of wing center of mass aft of}\\ \text{the elastic axis in semichords}\end{array} \tag{10}$$

$$r_\alpha = \sqrt{\frac{I_\alpha}{mb^2}} \quad = \begin{array}{l}\text{the radius of gyration (about the elastic axis)}\\ \text{in semichords}\end{array}$$

The equations of motion (Eqs. 10 of § 6.8) then become, on dividing throughout the first by $\pi\rho b^3\omega^2$ and the second by $\pi\rho b^4\omega^2$, and omitting the time factor $e^{ik\tau}$,

$$- \mu \frac{h_0}{b} - \mu x_\alpha \alpha_0 + \mu \frac{\omega_h^2}{\omega^2} \frac{h_0}{b} - \frac{h_0}{b} + a_h \alpha_0 + \frac{i}{k} \alpha_0$$

$$+ 2 \frac{C(k)}{k^2} \left[\alpha_0 + ik \frac{h_0}{b} + \left(\frac{1}{2} - a_h \right) ik\alpha_0 \right] = \frac{P_0}{\rho \pi b^3 \omega^2}$$

$$- \mu x_\alpha \frac{h_0}{b} - \mu r_\alpha^2 \alpha_0 + \mu r_\alpha^2 \frac{\omega_\alpha^2}{\omega^2} \alpha_0 + a \left(\frac{h_0}{b} - a_h \alpha_0 \right) \qquad (11)$$

$$+ i \left(\frac{1}{2} - a_h \right) \frac{1}{k} \alpha_0 - \frac{1}{8} \alpha_0 - \left(\frac{1}{2} + a_h \right) 2 \frac{C(k)}{k^2} \left[\alpha_0 + ik \frac{h_0}{b} \right.$$

$$\left. + \left(\frac{1}{2} - a_h \right) ik\alpha_0 \right] = \frac{Q_0}{\rho \pi b^4 \omega^2}$$

For the forced-vibration problem, the effect of the structural damping (so far neglected) must be considered, particularly in resonant conditions. The structural damping in aircraft structures is generally small. To account for the damping approximately, let us assume that the actual form of the hysteresis curve is unimportant and that the hysteresis loop can be replaced by an ellipse whose area is the same as the actual one. In other words, we assume that the effect of structural damping is revealed through the energy it dissipates per cycle. Furthermore, following Theodorsen, let us assume that the energy dissipation varies with the square of the amplitude of oscillation. Under these assumptions the effect of damping can be represented by a shift of the phase angle of the elastic restoring force. For aircraft structures the amount of this phase shift is very small, and so the structural damping may be described by a force in phase with the velocity, against the direction of motion, and of a magnitude proportional to the elastic restoring force. Thus in association with the elastic restoring force $- K_h h_0 e^{ik\tau}$ which acts against the vertical translation, there is a damping force $- i g_h K_h h_0 e^{ik\tau}$. g_h is called the *damping coefficient*. The net result is simply that the restoring force terms $h K_h$, αK_α be replaced by terms of the form $h K_h (1 + i g_h)$, $\alpha K_\alpha (1 + i g_\alpha)$, or, equivalently, ω_h^2, ω_α^2 be replaced by $\omega_h^2 (1 + i g_h)$, $\omega_\alpha^2 (1 + i g_\alpha)$, respectively. Cf. Appendix 3.

Equations 11, with the above-named modifications for the structural damping, take the form:

$$A \frac{h_0}{b} + B\alpha_0 = p_0, \qquad D \frac{h_0}{b} + E\alpha_0 = q_0 \qquad (12)$$

where

$$p_0 = \frac{P_0}{\rho \pi b^3 \omega^2}, \qquad q_0 = \frac{Q_0}{\rho \pi b^4 \omega^2} \qquad (13)$$

The coefficients A, B, D, E are functions of ω, U, g_h, g_α, and k. Let us take ω and k as fundamental parameters and let

$$\frac{\omega_\alpha^2}{\omega^2} = X \tag{14}$$

then the coefficients may be written as

$$A = A_R + iA_I + (1 + ig_h)\mu \frac{\omega_h^2}{\omega_\alpha^2} X$$

$$B = B_R + iB_I \tag{15}$$

$$D = D_R + iD_I$$

$$E = E_R + iE_I + (1 + ig_\alpha)\mu r_\alpha^2 X$$

where

$$A_R = -(\mu + 1) - \frac{2G}{k}$$

$$A_I = \frac{2F}{k}$$

$$B_R = -(\mu x_\alpha - a_h) + \frac{2F}{k^2} - \left(\frac{1}{2} - a_h\right)\frac{2G}{k}$$

$$B_I = \frac{1}{k}\left[1 + \frac{2G}{k} + \left(\frac{1}{2} - a_h\right)2F\right]$$

$$D_R = -(\mu x_\alpha - a_h) + \left(\frac{1}{2} + a_h\right)\frac{2G}{k} \tag{16}$$

$$D_I = -\left(\frac{1}{2} + a_h\right)\frac{2F}{k}$$

$$E_R = -\left(\mu r_\alpha^2 + a_h^2 + \frac{1}{8}\right) + \left(\frac{1}{4} - a_h^2\right)\frac{2G}{k} - \left(\frac{1}{2} + a_h\right)\frac{2F}{k^2}$$

$$E_I = \frac{1}{k}\left[\left(\frac{1}{2} - a_h\right) - \left(\frac{1}{2} + a_h\right)\frac{2G}{k} - \left(\frac{1}{4} - a_h^2\right)2F\right]$$

The solution of Eqs. 12 can be written as

$$\frac{h_0}{b} = \frac{1}{\Delta}\begin{vmatrix} p_0 & B \\ q_0 & E \end{vmatrix}, \qquad \alpha_0 = \frac{1}{\Delta}\begin{vmatrix} A & p_0 \\ D & q_0 \end{vmatrix} \tag{17}$$

where

$$\Delta = \begin{vmatrix} A & B \\ D & E \end{vmatrix} \equiv \Delta_R + i\Delta_I \qquad (18)$$

with Δ_R, Δ_I denoting the real and imaginary parts of the determinant Δ. From Eqs. 15, we obtain

$$\Delta_R = (1 - g_h g_\alpha)\mu^2 r_\alpha^2 \frac{\omega_h^2}{\omega_\alpha^2} X^2 + \left[\mu \frac{\omega_h^2}{\omega_\alpha^2} (E_R - g_h E_I) \right.$$

$$\left. + \mu r_\alpha^2 (A_R - g_\alpha A_I) \right] X + A_R E_R - B_R D_R - A_I E_I + B_I D_I$$

$$\Delta_I = (g_h + g_\alpha)\mu^2 r_\alpha^2 \frac{\omega_h^2}{\omega_\alpha^2} X^2 + \left[\mu \frac{\omega_h^2}{\omega_\alpha^2} (g_h E_R + E_I) \right.$$

$$\left. + \mu r_\alpha^2 (A_I + g_\alpha A_R) \right] X + A_I E_R - B_R D_I + A_R E_I - B_I D_R$$

(19)

The responses h_0/b and α_0 depend on the magnitude and phase relations of the excitations p_0, q_0, the frequency ω, the speed of flow U, and the wing's geometric, elastic, and damping characteristics. Owing to the complicated expressions, the nature of the response can be best seen by examining individual examples.

Let us consider the following particular two-dimensional wing model:*

$$\mu = 76, \qquad a_h = -0.15, \qquad x_\alpha = 0.25, \qquad r_\alpha^2 = 0.388 \qquad (20)$$

$$b = 5 \text{ inches},$$

$$\omega_\alpha = 64.1 \text{ radians per second},$$

$$\omega_h = 55.9 \text{ radians per second}.$$

Assume the wing to be excited by a periodic force acting on the elastic axis so that

$$P_0 = \text{const}, \qquad Q_0 = 0 \qquad (21)$$

It is convenient to define certain static deflection h_{st} corresponding to P_0:

$$h_{\text{st}} = \frac{P_0}{K_h} = \frac{P_0}{m\omega_h^2} = \frac{\rho\pi b^3 \omega^2 p_0}{m\omega_h^2} = \frac{b}{\mu} \left(\frac{\omega}{\omega_h} \right)^2 p_0 \qquad (22)$$

---

* The dimensions are not very different from those given by Bollay and Brown.[9.20] In comparing the following theoretical results with the experimental results of Ref. 9.20, we must note that the values of r_α^2 and g_h, g_α, as well as Q_0, were not given in that reference. The general agreement between theory and experiment seems satisfactory.

Then, according to Eqs. 16, 17, and 21,

$$\left|\frac{h_0}{h_{st}}\right| = \mu \frac{\omega_h^2}{\omega^2}\left|\frac{E}{\Delta}\right| \tag{23}$$

where

$$|E| = \sqrt{(E_R + \mu r_\alpha^2 X)^2 + (E_I + g_\alpha \mu r_\alpha^2 X)^2}, \qquad |\Delta| = \sqrt{\Delta_R^2 + \Delta_I^2}$$

The phase angle ψ_h between the response h and the exciting force is given by the equation

$$\tan \psi_h = \frac{\Delta_R(E_I + g_\alpha \mu r_\alpha^2 X) - \Delta_I(E_R + \mu r_\alpha^2 X)}{\Delta_R(E_R + \mu r_\alpha^2 X) + \Delta_I(E_I + g_\alpha \mu r_\alpha^2 X)} \tag{24}$$

The response h leads the exciting force if ψ_h is positive.

At a given speed of flow U, the response ratio $|h_0/h_{st}|$ can be computed as follows: (1) Assume a value of k. From Table 6.1 find the corresponding values of F and G. (2) Calculate the coefficients A_I, A_R, etc., with the assumed parameters. (3) Since $\omega = kU/b$ and k and U are specified, ω and $X = (\omega_\alpha/\omega)^2$ can be obtained. (4) A substitution into Eqs. 19 gives Δ_R and Δ_I. (5) Obtain the response ratio $|h_0/h_{st}|$ and the phase shift ψ_h from Eqs. 23 and 24.

The results of such a calculation are shown in Fig. 6.9 where the response ratios of a wing, characterized by the dimensions and frequencies listed in Eqs. 20, at three speeds of flow $U = 7.42$, 29.7, and 86.4 ft per sec, are plotted. It is seen that, as the frequency increases from zero, two peaks of response are reached, one at a frequency close to the natural frequency in bending, and the other at that close to the natural frequency in torsion. These frequencies in still air, $U = 0$, are 7.92 and 12.37 cycles per second respectively. Two sets of curves are presented: The solid curves are referred to an airfoil without structural damping, $g_h = g_\alpha = 0$; and the dotted curves are referred to an airfoil with $g_h = g_\alpha = 0.05$, a value probably high for most metal aircraft structures. It is seen that the effect of structural damping on the response is small, except near the peaks.* At zero airspeed, the resonance peaks tend to infinity at the natural frequencies of vibration if the structural damping is zero. With damping g_h and g_α, the peak responses vary approximately as $1/g_h$ and $1/g_\alpha$, respectively.

Such response calculations can be repeated for other speeds of flow U. A relief map of the response as a function of U and ω may be plotted.

---

* This is not true in all cases. For flutter involving ailerons, for instance, the effect of structural damping can be exceedingly large. See *NACA Rept.* **741**.

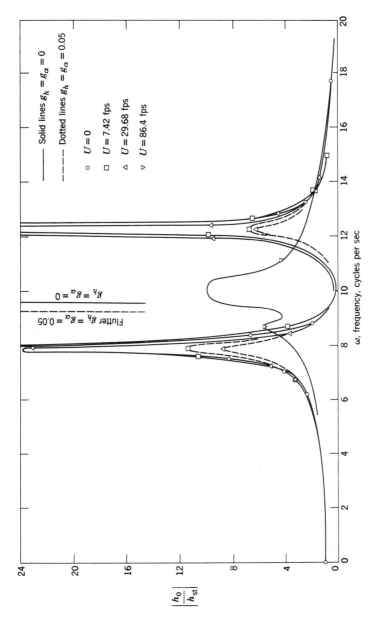

Fig. 6.9. Amplitude response of the vertical displacement of a wing subjected to a periodic exciting force acting on the elastic axis.

221

Figure 6.10 is such a relief map of the response ratio $|h_0/h_{st}|$ for the above numerical example, at $g = 0.05$, for which a critical flutter speed exists. The curves of Fig. 6.9 are the intersections of the relief map with particular planes perpendicular to the U axis. It is seen that, as U increases, the peak response diminishes, until, along one frequency branch, the response becomes negligibly small. Along the other frequency branch, however, a

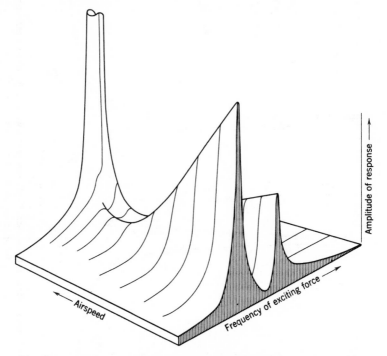

Fig. 6.10. Relief map of the amplitude response of the vertical displacement of a wing subjected to a periodic exciting force acting on the elastic axis. Damping factor $g_h = g_\alpha = 0.05$.

minimum response is first reached, after which the response increases (usually rapidly) until the flutter speed is reached.

The critical flutter condition is reached when, at certain combinations of U and ω, the determinant Δ (Eq. 18) vanishes. Then the response ratio $|h_0/h_{st}|$ becomes infinity. (Note that the response tends to infinity at the flutter condition, no matter whether structural damping is present or not.) The critical flutter frequency usually lies between the two natural frequencies that exist at zero airspeed.

To find the critical flutter speed and frequency, the *characteristic* equation

$$\Delta = 0 \tag{25}$$

must be solved for the pair of real variables U and ω. Since Δ is complex-valued, both the real and imaginary parts of Δ must vanish. Equation 25 is actually equivalent to two real equations,

$$\Delta_R = 0, \qquad \Delta_I = 0 \tag{26}$$

The solution can be obtained as follows. Let a series of values of k be assumed and the corresponding coefficients A, B, D, E calculated. By

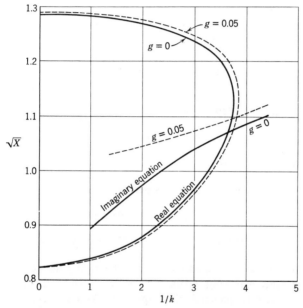

Fig. 6.11. Solution of the flutter determinant.

equating Δ_R and Δ_I given by Eqs. 19 to zero, two quadratic equations in X are obtained. These equations can be solved for X. Since X by definition (Eq. 14) is a positive quantity, only the real positive roots have a physical meaning. Curves of $\sqrt{X}$ vs. $1/k$ can then be plotted, such as those shown in Fig. 6.11. If the curves corresponding to $\Delta_R = 0$ and $\Delta_I = 0$ intersect, the points of intersection determine the values of $\sqrt{X}$ and $1/k$ at which the determinant Δ vanishes. Now

$$\sqrt{X} = \frac{\omega_\alpha}{\omega}, \qquad \frac{1}{k} = \frac{U}{\omega b} \tag{27}$$

Hence, the critical flutter frequency and speed are given by

$$\omega = \frac{\omega_\alpha}{\sqrt{X}}, \qquad U = \frac{\omega_\alpha b}{k\sqrt{X}} \tag{28}$$

If the curves $\Delta_R = 0$ and $\Delta_I = 0$ do not intersect, there will be no flutter. If they intersect at more than one point, then each intersection represents a critical condition. The one having the lowest value of U is the most important. In general, it represents a transition speed below which the wing is stable and above which the wing becomes unstable. In case of doubt, however, a test of stability by the methods of §§ 6.5 and 10.6 should be made.

For the given example, Fig. 6.11 shows the existence of a critical flutter condition for $g_h = g_\alpha = 0$, with

$$\frac{1}{k} = 3.62, \qquad \sqrt{X} = 1.072$$

Hence, the critical flutter speed is

$$U_{\text{cr}} = 90.1 \text{ ft per sec}$$

and the flutter frequency is

$$\omega_{\text{cr}} = 9.52 \text{ cycles per second}$$

The corresponding values for $g_h = g_\alpha = 0.05$ are

$$U_{\text{cr}} = 93.0 \text{ ft per sec}, \qquad \omega_{\text{cr}} = 9.27 \text{ cycles per second}$$

The main feature of the response is given by the peak values on the two ridges of the relief map (Fig. 6.10). It can be shown that, for a fixed value of the reduced frequency k, the response peaks are reached *approximately* where Δ_R vanishes. Hence, if the values of Δ_I are calculated for various selected points $(1/k, \sqrt{X})$ along $\Delta_R = 0$, these values may be used to determine Δ. In fact,

$$|\Delta| = |\Delta_I| \tag{29}$$

along the $\Delta_R = 0$ line. The approximate values of the response peaks can then be determined from Eq. 23, the corresponding ω and U being given by Eqs. 28.

Figure 6.12 gives the peak response ridges of the example cited above. The relief map of Fig. 6.10 is constructed on the basis of Fig. 6.12.

If the exciting force is generated by a rotating eccentric weight, the exciting force is proportional to ω^2; the response can be obtained from the above by appropriately multiplying the $|h_0/h_{st}|$ curve by a factor ω^2.

Note that the coefficients of X in Eqs. 26 are nondimensional quantities, depending on the ratios μ, a_h, x_α, $r_\alpha{}^2$, and ω_h/ω_α. Hence, the solution X is nondimensional and does not depend on the absolute values (or physical units) of the semichord b and the frequency ω_α. For example, if in the example specified by Eqs. 20 the semichord is changed to 2 ft, and the

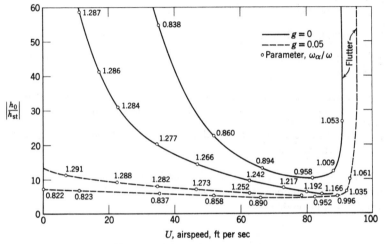

Fig. 6.12.   The peak response ridges.

frequency ω_α to 48 radians per second, while other quantities (μ, a_h, x_α, $r_\alpha{}^2$, ω_h/ω_α) remain the same, then the critical reduced frequency remains the same, but the critical flutter frequency and speed become, respectively, when $g = 0.05$,

$$\omega_{cr} = 43.6 \text{ radians per second,} \qquad U_{cr} = 334 \text{ ft per sec}$$

6.10 FLUTTER OF A TWO-DIMENSIONAL AIRFOIL

The example of the last section shows that for certain wing there exists an airspeed at which the wing response to a periodic forcing function tends to infinity at some particular frequency. At this combination of airspeed and frequency, the determinant of the coefficients of Eqs. 12 of § 6.9 vanishes and a nontrivial solution exists even when the forcing function vanishes: the aeroelastic system will oscillate harmonically without further excitation after an initial disturbance. The aeroelastic system is then said to be in the critical flutter condition.

In order to indicate the form of available aerodynamic tables and their application to flutter calculations, we shall consider in this section the critical flutter condition of a two-dimensional airfoil with an aileron and a control tab. Extensions to three-dimensional wings, tail surfaces, and other structures will be made in the next chapter.

The Airfoil. Let us consider a two-dimensional airfoil of unit length in the spanwise direction, having four degrees of freedom h, α, β, and δ, as shown schematically in Fig. 6.13, where

$h =$ bending deflection of the elastic axis, positive downward, feet

$\alpha =$ pitching about the elastic axis, relative to the direction of flow, positive nose up, radians

$\beta =$ angular deflection of aileron about aileron hinge, relative to wing chord, positive for aileron trailing edge down, radians

$\delta =$ angular deflection of tab relative to aileron, positive trailing edge down, radians

We assume that h, α, β, and δ are infinitesimal, so that the flow remains potential and unseparated, and the linearization of the aerodynamic equations is justifiable.

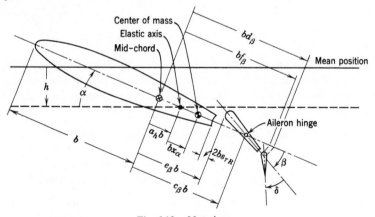

Fig. 6.13. Notations.

The notations are shown in the figure. Note that the semichord is denoted by b. Other dimensions are referred *nondimensionally* to the semichord. Distances are measured with the *mid-chord* point as origin.

Thus,

$c_\beta b$ = distance between mid-chord and aileron hinge, positive if aileron hinge is aft of mid-chord

$e_\beta b$ = distance between mid-chord and aileron leading edge, positive if aileron leading edge is aft of mid-chord

$a_h b$ = distance between elastic axis and mid-chord, positive if elastic axis is aft of mid-chord

Similarly, the dimensions $d_\beta b$, $f_\beta b$, $x_\alpha b$, etc., can be identified.

Elastic Restraints. The displacements of the airfoil are restrained by elastic springs. A linear spring located at the elastic axis restrains the bending h. A torsional spring located at the same axis restrains the pitching α. The aileron rotation is restrained by a torsional spring at the aileron hinge line, and the tab by a spring at its own hinge. The spring constants will be denoted by K_h, K_α, K_β, and K_δ. The elastic force due to a displacement h is $- hK_h$, acting at the elastic axis, and in the direction opposing h. The moment about the elastic axis due to α is $- \alpha K_\alpha$. Similarly, the moments about the aileron and tab hinges are $- \beta K_\beta$ and $- \delta K_\delta$, respectively.

It is desirable to express the spring constants in terms of uncoupled natural frequencies as in § 6.8 (Eq. 6), and write

$$K_h = m\omega_h{}^2, \qquad K_\alpha = I_\alpha \omega_\alpha{}^2, \qquad K_\beta = I_\beta \omega_\beta{}^2, \qquad K_\delta = I_\delta \omega_\delta{}^2 \qquad (1)$$

where m is the mass per unit span of the wing–aileron–tab combination, I_α is the moment of inertia per unit span of the wing–aileron–tab about the elastic axis, I_β is the aileron–tab moment of inertia about the aileron hinge, and I_δ is the tab moment of inertia about the tab hinge. The frequencies ω_h, ω_α, ω_β, ω_δ are the uncoupled frequencies in radians per second, obtainable approximately by experiments, if necessary.

Structural Damping. The small structural damping of metal aircraft may be approximated by a force that opposes the motion and is in phase with the velocity. For simplicity of analysis, and also from lack of more accurate knowledge, we shall assume that the magnitude of the damping force is proportional to the elastic restoring force (cf. § 11.4). Since the motion of the airfoil is harmonic at the critical flutter condition, the condition is the same as in § 6.9, and the effect of the structural damping can be accounted for simply by replacing the terms hK_h, αK_α, etc., with terms $hK_h(1 + ig_h)$, $\alpha K_\alpha(1 + ig_\alpha)$, etc. The constants g_h, g_α, g_β, g_δ are the damping coefficients.*

---

* Measurement by Fearnow on a full-scale airplane wing (C-46D) shows that g depends on the amplitude of motion and varies from 0.002 at an amplitude of vibration of 0.05 in. to approximately 0.012 at an amplitude of $\pm$ 5 in. (at wing tips). See Ref. 11.1.

*Aerodynamic Forces.** The theory of oscillating airfoils, to be presented in Chapters 12 through 15, gives the necessary information about the aerodynamic forces acting on a fluttering wing. The theory is based on thin airfoils oscillating at infinitesimal amplitudes. At the flutter condition, we shall assume that

$$\frac{h}{b} = \frac{h_0}{b} e^{i\omega t}, \qquad \alpha = \alpha_0 e^{i(\omega t + \theta_1)}, \qquad \beta = \beta_0 e^{i(\omega t + \theta_2)}, \qquad \delta = \delta_0 e^{i(\omega t + \theta_3)} \quad (2)$$

where h_0, α_0, β_0, δ_0 are real numbers (small compared to 1), θ_1, θ_2, θ_3 are the phase angles by which α, β, δ lead the wing bending displacement, and ω is the flutter frequency in radians per second. For a two-dimensional airfoil having these four degrees of freedom, in a flow of speed U, the aerodynamic forces are functions of the dimensionless numbers:

$$M = \text{Mach number} = U/c \quad (3)$$

$$k = \text{reduced frequency or Strouhal number} = \omega b/U \quad (4)$$

where c is the speed of sound propagation in the undisturbed flow. The aerodynamic expressions are simpler if the motion, as well as the forces and moments, are referred to the aerodynamic center. For a subsonic flow the aerodynamic center is located at the $1/4$-chord point aft of the wing leading edge. Hence, let us introduce the following nondimensional generalized displacements:

$$q_1 = \frac{(h)_{c/4}}{b}, \qquad q_2 = \alpha \quad (5)$$

---

* In most British papers, the so-called *classical derivative theory* is used. Following Frazer and Duncan, the aerodynamic forces are assumed to be linear functions of the generalized displacements, velocities, and accelerations. For example, let z denote the linear *downward* displacement of the *leading edge* and α the increase in angle of attack from a mean position at a local chordwise section of the wing. Then it is assumed that the corresponding local aerodynamic lift L (positive upward) and the moment M (referred to the leading edge, positive nose up), per unit span, may be expressed in the form

$$L/(\rho U^2 c) = (l_{\ddot{z}}\ddot{z} + l_{\dot{z}}\dot{z} + l_z z)/c + (l_{\ddot{\alpha}}\ddot{\alpha} + l_{\dot{\alpha}}\dot{\alpha} + l_\alpha \alpha)$$

$$-M/(\rho U^2 c^2) = (m_{\ddot{z}}\ddot{z} + m_{\dot{z}}\dot{z} + m_z z)/c + (m_{\ddot{\alpha}}\ddot{\alpha} + m_{\dot{\alpha}}\dot{\alpha} + m_\alpha \alpha)$$

where c is the local chord and a superposed dot denotes differentiation with respect to a dimensionless time $\tau \equiv Ut/c$. The coefficients l_z, $l_{\dot{z}}$, etc., are assumed to be independent of the reduced frequency, aspect ratio, planform, and modes of motion of the wing. In an incompressible fluid, a set of "mean experimental values" given by J. Williams (Ref. 6.24) is the following:

$$l_{\dot{z}} = 1.6, \qquad l_{\dot{\alpha}} = 1.6, \qquad l_\alpha = 1.8, \qquad l_z = 0$$
$$m_{\dot{z}} = 0.4, \qquad m_{\dot{\alpha}} = 0.7, \qquad m_\alpha = 0.45, \qquad m_z = 0$$
$$l_{\ddot{z}} = \pi/4, \qquad l_{\ddot{\alpha}} = m_{\ddot{z}} = \pi/8, \qquad m_{\ddot{\alpha}} = 9\pi/128$$

where

$$(h)_{c/4} = h - b(\tfrac{1}{2} + a_h)\alpha$$

$$= \text{bending displacement of the } 1/_4\text{-chord point} \tag{6}$$

$$\alpha = \text{pitching displacement about the } 1/_4\text{-chord point}$$

Similarly, for an aileron with aerodynamic balance, it is convenient to resolve the aileron motion about its hinge line into two components, a rotation about the aileron leading edge and a vertical translation of the

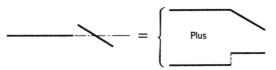

Fig. 6.14.   Resolution of aileron motion.

aileron in the direction perpendicular to the wing chord (positive downward) (Fig. 6.14).   The rotation about the aileron leading edge is β.   The vertical translation must have an amplitude

$$z = - (c_\beta - e_\beta)b\beta \tag{7}$$

so that the resultant of the rotation about the leading edge β and the translation z may leave the aileron hinge line fixed with respect to the wing chord.

The tab rotation can be similarly treated.   But for simplicity let us assume the tab to be hinged at its leading edge.

Thus in addition to q_1, q_2 defined by Eqs. 5, we have the following nondimensional displacements:

$$q_3 = \beta, \qquad q_4 = z/b, \qquad q_5 = \delta \tag{8}$$

These five displacements are illustrated in Fig. 6.15.

When h, α, β, z, δ all vary sinusoidally as given in Eqs. 2, the aerodynamic lift per unit span, $L_{c/4}$, acting at the $1/_4$-chord point,* positive upward in the usual sense, can be written as:

$$L_{c/4} = - \pi\rho b^3\omega^2 \left[\left(\frac{h}{b}\right)_{c/4} L_h + \alpha L_\alpha + \beta L_\beta + \frac{z}{b} L_z + \delta L_\delta \right] \tag{9}$$

---

* In supersonic flow, it is more convenient to resolve lift and moment about the mid-chord point.   Note here that a negative sign is attached to the right-hand side of Eqs. 9, 10 and 12.   The lift forces $L_{c/4}$ and $P_{1.e.}$ are defined as positive when their vector directions are *upward, as in the usual sign convention* for a lift force.   But, in Smilg and Wasserman's paper,[6.20] from which extensive tables for L_h, L_α, etc., are obtained, the opposite sign conventions are used.   The negatives signs are chosen for Eqs. 9, 10 and 12 so that the existing tables for L_h, L_α, etc., can be used.

where L_h, L_α, etc., are nondimensional coefficients. To shorten the writing, let us write the sum in the bracket [ ] as $\sum q_i L_i$. Thus

$$L_{c/4} = -\pi\rho b^3 \omega^2 \sum q_i L_i \tag{10}$$

In a similar way, the aerodynamic moment per unit span about the $1/4$-chord point, $M_{c/4}$, *positive in the nose-up sense*, can be written as

$$M_{c/4} = \pi\rho b^4 \omega^2 \sum q_i M_i \tag{11}$$

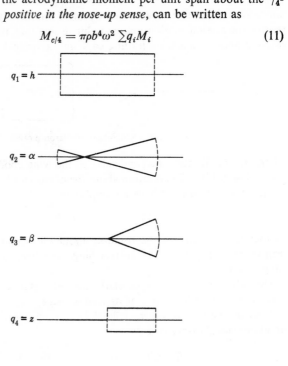

Fig. 6.15.   Degrees of freedom considered.

Analogously, the force per unit span on the aileron, $P_{\text{l.e.}}$ *positive up*, is

$$P_{\text{l.e.}} = -\pi\rho b^3 \omega^2 \sum q_i P_i \tag{12}$$

The moment per unit span about the leading edge of the aileron, $T_{\text{l.e.}}$ *positive trailing edge down*, is

$$T_{\text{l.e.}} = \pi\rho b^4 \omega^2 \sum q_i T_i \tag{13}$$

The aerodynamic coefficients L_i, M_i, P_i, T_i, ($i = 1, 2, \cdots, 5$) are completely defined by the above equations.   Linearity is assumed in order that these equations may be valid.   In the theoretical derivation of these coefficients (Chapters 13 and 14), the fluid is assumed to be nonviscous so that L_i, M_i, etc., are functions of the Mach number and the reduced frequency.   For a real fluid, they are also functions of the Reynolds number.   Figure 6.16 show the theoretical curves of the complex numbers

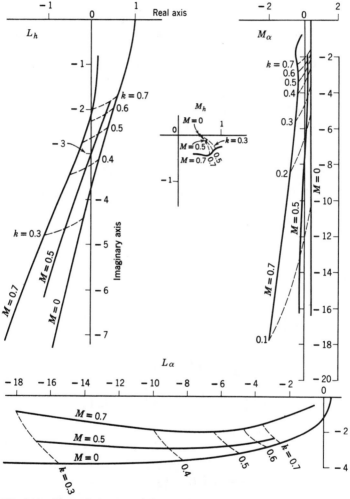

Fig. 6.16.   Theoretical values of the complex numbers L_h, L_α, M_h, and M_α for various Mach numbers and reduced frequencies.   A vector drawn from the origin to an appropriate point on a curve gives a complex number (real part, abscissa; imaginary part, ordinate) which is the value of L_h, etc., given by the linearized theory.   Note the difference in scales between L_h, M_h and L_α, M_α.

L_h, L_α, M_h, M_α at several Mach numbers with the reduced frequency k as a parameter.

The source references from which the aerodynamic coefficients L_i, M_i, etc., can be obtained are reviewed in §§ 13.6 and 14.7. Throughout the literature various notations and forms of the coefficients have been used. A comparison between the notations of different authors is given in Tables 13.1 and 14.2. The notations used in Eqs. 10 through 13 are introduced in *Army Air Force Technical Report* 4798, (1942), by Smilg and Wasserman.

The aerodynamic forces and moments about the *elastic axis* of the wing and the *hinge line* of the aileron can be easily computed from $L_{c/4}$, $M_{c/4}$, $P_{l.e.}$, and $T_{l.e.}$. Let the force and moment at the elastic axis be $L_{e.a.}$ and $M_{e.a.}$, respectively, and the moment about the aileron hinge line T_{hinge}. From Fig. 6.17 is it clear that

$$M_{e.a.} = M_{c/4} + L_{c/4} \left(\tfrac{1}{2} + a_h\right)b$$
$$T_{hinge} = T_{l.e.} + P_{l.e.}(c_\beta - e_\beta)b \tag{14}$$

If we use Eqs. 6 and 7, a little calculation leads to the following expressions:

$$
L_{e.a.} = -\pi\rho b^3 \omega^2 \left\{ \frac{h}{b} L_h + \alpha \left[L_\alpha - \left(\frac{1}{2} + a_h\right) L_h \right] \right.
$$
$$
\left. + \beta[L_\beta - (c_\beta - e_\beta)L_z] + \delta L_\delta \right\}
$$

$$
M_{e.a.} = \pi\rho b^4 \omega^2 \left\{ \frac{h}{b} \left[M_h - \left(\frac{1}{2} + a_h\right) L_h \right] \right.
$$
$$
+ \alpha \left[M_\alpha - \left(\frac{1}{2} + a_h\right)(L_\alpha + M_h) + \left(\frac{1}{2} + a_h\right)^2 L_h \right]
$$
$$
+ \beta \left[M_\beta - \left(\frac{1}{2} + a_h\right) L_\beta - (c_\beta - e_\beta)M_z \right.
$$
$$
\left. + (c_\beta - e_\beta)\left(\frac{1}{2} + a_h\right) L_z \right] + \delta \left[M_\delta - L_\delta \left(\frac{1}{2} + a_h\right) \right] \Big] \Big\}
\tag{15}
$$

$$
T_{hinge} = \pi\rho b^4 \omega^2 \left\{ \frac{h}{b}[T_h - (c_\beta - e_\beta)P_h] + \alpha \left[T_\alpha - (c_\beta - e_\beta)P_\alpha \right. \right.
$$
$$
\left. - \left(\frac{1}{2} + a_h\right) T_h + \left(\frac{1}{2} + a_h\right)(c_\beta - e_\beta)P_h \right]
$$
$$
+ \beta[T_\beta - (c_\beta - e_\beta)(P_\beta + T_z) + (c_\beta - e_\beta)^2 P_z]
$$
$$
+ \delta[T_\delta - P_\delta(c_\beta - e_\beta)] \Big\}
$$

The Equations of Motion. By a summation of the inertia, elastic, damping, and aerodynamic forces, we obtain the equations of motion in a manner described in § 6.8. In the following discussions, we shall assume that the tab is geared to the aileron so that

$$\delta = n\beta \tag{16}$$

where n is a proportional constant. The rigidity of the tab constraint about its hinge is considered as infinite. The equation of motion for the tab can therefore be omitted.

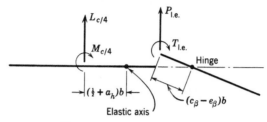

Fig. 6.17.   Change of moment axes.

Denoting the time derivatives by dots, we obtain the following equations of motion:

$$m\ddot{h} + S_\alpha\ddot{\alpha} + (S_\beta + nS_\delta)\ddot{\beta} + (1 + ig_h)K_h h + L_{\text{e.a.}} = 0$$

$$S_\alpha\ddot{h} + I_\alpha\ddot{\alpha} + [(c_\beta - a_h)bS_\beta + I_\beta + nI_\delta + b(d_\beta - a_h)nS_\delta]\ddot{\beta}$$
$$+ (1 + ig_\alpha)K_\alpha\alpha - M_{\text{e.a.}} = 0 \tag{17}$$

$$S_\beta\ddot{h} + [I_\beta + b(c_\beta - a_h)S_\beta]\ddot{\alpha} + [I_\beta + nI_\delta + b(d_\beta - c_\beta)nS_\delta]\ddot{\beta}$$
$$+ (1 + ig_\beta)K_\beta\beta - T_{\text{hinge}} = 0$$

where

m = mass of wing–aileron–tab (per unit span)

S_α = static moment of wing–aileron–tab about wing elastic axis (per unit span)

S_β = aileron–tab static moment (per unit span) about aileron hinge line

S_δ = tab static moment (per unit span) about tab hinge line

I_α = mass moment of inertia of wing–aileron–tab about wing elastic axis

I_β = mass moment of inertia of aileron–tab about aileron hinge line

I_δ = mass moment of inertia of tab about tab hinge line

Let us assume that flutter exists. The wing thus can oscillate harmonically as given by Eqs. 2 so that

$$\ddot{h} = -\omega^2 h, \qquad \ddot{\alpha} = -\omega^2 \alpha, \qquad \ddot{\beta} = -\omega^2 \beta \qquad (18)$$

Substituting Eqs. 15 and 18 into 17, and dividing the first equation by $\pi \rho b^3 \omega^2$ and the other two by $\pi \rho b^4 \omega^2$, we obtain

$$A\frac{h}{b} + B\alpha + C\beta = 0$$

$$D\frac{h}{b} + E\alpha + F\beta = 0 \qquad (19)$$

$$G\frac{h}{b} + H\alpha + I\beta = 0$$

where

$$A = \mu\left[1 - \left(\frac{\omega_\alpha}{\omega}\right)^2\left(\frac{\omega_h}{\omega_\alpha}\right)^2(1 + ig_h)\right] + L_h$$

$$B = \mu x_\alpha + L_\alpha - L_h(\tfrac{1}{2} + a_h)$$

$$C = \mu(x_\beta + nx_\delta) + L_\beta - L_z(c_\beta - e_\beta) + nL_\delta$$

$$D = \mu x_\alpha + M_h - L_h(\tfrac{1}{2} + a_h)$$

$$E = \mu r_\alpha^2\left[1 - \left(\frac{\omega_\alpha}{\omega}\right)^2(1 + ig_\alpha)\right] - \frac{1}{2}\left(\frac{1}{2} + a_h\right)$$

$$+ M_\alpha - L_\alpha\left(\frac{1}{2} + a_h\right) + L_h\left(\frac{1}{2} + a_h\right)^2$$

$$F = \mu[r_\beta^2 + (c_\beta - a_h)x_\beta + nr_\delta^2 + n(d_\beta - a_h)x_\delta]$$

$$+ M_\beta - L_\beta(\tfrac{1}{2} + a_h) - M_z(c_\beta - e_\beta) \qquad (20)$$

$$+ L_z(\tfrac{1}{2} + a_h)(c_\beta - e_\beta) + nM_\delta - nL_\delta(\tfrac{1}{2} + a_h)$$

$$G = \mu x_\beta + T_h - P_h(c_\beta - e_\beta)$$

$$H = \mu[r_\beta^2 + (c_\beta - a_h)x_\beta] + T_\alpha - T_h(\tfrac{1}{2} + a_h) - P_\alpha(c_\beta - e_\beta)$$

$$+ P_h(c_\beta - e_\beta)(\tfrac{1}{2} + a_h)$$

$$I = \mu r_\beta^2\left[1 - \left(\frac{\omega_\alpha}{\omega}\right)^2\left(\frac{\omega_\beta}{\omega_\alpha}\right)^2(1 + ig_\beta)\right] + n\mu[r_\delta^2 + (d_\beta - c_\beta)x_\delta]$$

$$+ T_\beta + nT_\delta - (P_\beta + T_z + nP_\delta)(c_\beta - e_\beta) + P_z(c_\beta - e_\beta)^2$$

and

$$S_\alpha = mbx_\alpha, \qquad S_\beta = mbx_\beta, \qquad S_\delta = mbx_\delta$$

$$I_\beta = mb^2 r_\beta^2, \qquad I_\delta = mb^2 r_\delta^2$$

If flutter exists, h, α, β do not vanish identically. Such a nontrivial solution exists when and only when the determinant of coefficients in Eqs. 19 vanishes:

$$\begin{vmatrix} A & B & C \\ D & E & F \\ G & H & I \end{vmatrix} = 0 \qquad (21)$$

This characteristic equation involves the following real variables, U, ω, k, ρ, ω_h, ω_α, ω_β, g_h, g_α, g_β, and parameters defining the geometrical and mass distributions of the airfoil. Since Eq. 21 is an equation with complex coefficients, and since the real and imaginary parts of the determinant must vanish separately, it is actually equivalent to two real equations for these real variables. It can be used to determine any two of these variables, while others must be specified. The choice of particular variables to be considered as the unknowns depends on the information desired and the expediency of calculation.

When Eq. 21 is satisfied. Eqs. 19 may be solved for the ratios $h/b : \alpha : \beta$, which are complex numbers showing both the amplitude ratios and phase relationship.

Application of the Two-Dimensional Analysis. Regarding the two-dimensional airfoil as a *typical* section of a three-dimensional wing, and adjusting the spring constants in such a way that the frequencies ω_h, ω_α, etc., coincide with the actual uncoupled free vibration frequencies of the wing, while the mass and geometric properties are taken as those of a typical section, one may expect that the critical flutter speed calculated for the two-dimensional case approximates that of the actual wing. This was shown to be true by Theodorsen and Garrick[6.22] for wings without appreciable sweep angle, without large concentrated mass, with more or less uniform distribution of structural properties across the span, with straight elastic and inertial axes, and with high chordwise rigidity. The location of the *typical* section is of some importance. Generally it is taken in the neighborhood of 0.7 span from the root, or at the mid-span of the aileron.

6.11   SOLUTION OF THE FLUTTER DETERMINANT

The method used in § 6.9, of finding the flutter speed by plotting the solutions of the real and imaginary parts of the flutter determinant, is often referred to in U.S. literature as *Theodorsen's method*. This method can be used to solve the determinantal equation 21 of § 6.10.   Let

$$\left(\frac{\omega_\alpha}{\omega}\right)^2 = X \qquad (1)$$

Only the diagonal terms A, E, and I in Eq. 21 of § 6.10 involve X. For a given value of k, all the terms of the determinant are known (complex numbers). If the real and imaginary parts are separated, it can then be written as two cubic equations with real coefficients involving X as an unknown. The roots of these equations, when plotted against $1/k$, yield two curves, the intersection of which determines the reduced frequency k and frequency ratio ω_α/ω at flutter. The critical speed and frequency are then given by Eqs. 28 of § 6.9.

For a compressible fluid, the available aerodynamic data are such that it is necessary to assume both a Mach number and a reduced frequency in order to obtain numerical values of the determinant. Let the assumed value of the Mach number be M. For this M, let the flutter-speed calculated by the procedure outlined above be U, which corresponds to a Mach number M'. Varying M will lead to other values of M'. The true solution is obtained when $M = M'$. Hence, a process of trial and error is necessary. In § 10.6, a process based on Nyquist's criterion of stability will be described, which makes it possible to state whether flutter will or will not occur at a given Mach number. Although the Nyquist approach does not give the flutter speed directly, it does make a definite statement about the aeroelastic stability.

Example 1. Consider the torsion-bending flutter of the following two-dimensional section (of a suspension bridge):

$$b = 30 \text{ ft,} \qquad\qquad r_\alpha{}^2 = 0.6222$$
$$m = 269 \text{ slugs per foot,} \qquad\qquad\qquad (2)$$
$$\omega_h{}^2 = 0.775, \qquad\qquad \omega_\alpha{}^2 = 2.41$$

The elastic axis lies on the mid-chord line, and the mass distribution is symmetrical. Hence,

$$a_h = 0, \qquad x_\alpha = 0 \qquad\qquad (3)$$

In this case

$$\mu = \frac{m}{\pi \rho b^2} = 40 \qquad\qquad (4)$$

$$\rho = 0.002378 \text{ slug per cubic foot}$$

Consider first $g_h = g_\alpha = 0$. Then

$$
\begin{aligned}
A &= 40(1 - 0.3216X) + L_h \\
B &= L_\alpha - 0.5L_h \\
D &= 0.5 - 0.5L_h \\
E &= 24.8928(1 - X) - 0.25 + M_\alpha - 0.5L_\alpha + 0.25L_h
\end{aligned} \qquad (5)
$$

Assuming $1/k = 2.00$, we find from the tables of L_h, L_α, etc. (Smilg and

Wasserman[6.20], *A.F.Tech. Report* 4798; or p. 409, Rosenberg and Scanlan,[6.19]:

$$L_h = 0.3972 - 2.3916i, \qquad L_\alpha = -4.8860 - 3.1860i$$
$$M_\alpha = 0.3750 - 2.0000i \tag{6}$$

Hence,

$$A = 40.4049 - 2.3916i - 12.8665X$$
$$B = -5.0846 - 1.9902i$$
$$D = 0.3014 + 1.1958i \tag{7}$$
$$E = 27.5601 - 1.0049i - 24.8928X$$

The determinant

$$\begin{vmatrix} A & B \\ D & E \end{vmatrix} = 0$$

becomes

$$320.283X^2 + 74.463iX - 1360.393X - 99.836i + 1110.312 = 0 \tag{8}$$

Separating the real and imaginary parts, we get

1. Real Equation:

$$320.283X^2 - 1360.393X + 1110.312 = 0$$

The roots are

$$X = 1.1022 \quad \text{and} \quad X = 3.150$$

or

$$\sqrt{X} = 1.0499 \quad \text{and} \quad \sqrt{X} = 1.466 \tag{9}$$

2. Imaginary Equation:

$$74.463X - 99.836 = 0$$

The root is

$$X = 1.3777 \quad \text{or} \quad \sqrt{X} = 1.1738 \tag{10}$$

By repeating the process for other assumed values of $1/k$, a table of the roots $\sqrt{X}$ as a function $1/k$ can be prepared (Table 6.2).

Table 6.2

$\dfrac{1}{k}$	$\sqrt{X}$	
	From Real Eq.	From Imaginary Eq.
2	1.0499	1.1738
2.94	1.1097	1.2043
3.33	1.1420	1.2155
4.17	1.2241	1.2364
5.00	1.3236	1.2538

Figure 6.18 shows a plot of these roots vs. $1/k$. It is seen that the real and imaginary equations are both satisfied at the intersection of the curves:

$$\frac{1}{k} = 4.31, \qquad \sqrt{X} = 1.239 \tag{11}$$

Hence, the corresponding critical flutter speed

$$U_{cr} = \frac{b\omega_\alpha}{k\sqrt{X}} = 162 \text{ ft per sec} \tag{12}$$

Only the smaller of the two roots $\sqrt{X}$ of the real equation is listed in Table 6.2 and plotted in Fig. 6.18. The larger of the two roots of the

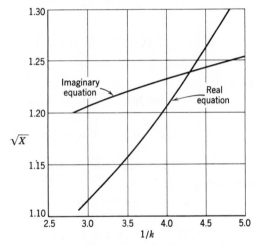

Fig. 6.18. Example 1.

real equation forms a branch that does not intersect the curve for the imaginary equation. Hence, the above intersection is the only intersection, and the critical flutter speed given by Eq. 12 is the only critical speed.

A different point of view is to consider the flutter determinant not as two real equations but as a single equation for the variable X. For each assumed value of k, this equation may be solved for X. Since the coefficients are complex functions of k, the roots are, in general, complex. But X, by definition (Eq. 1), must be real to have a physical meaning. Hence, we must find the particular value of k that renders the root X real. A plot of the real and complex parts of the roots X with k as a parameter will indicate whether such a solution exists. If the fluid is compressible, the calculations must be repeated for different values of the Mach number M. The correct M's are those for which the assumed values and the calculated values agree.

In practical application, it is more convenient to use a different variable than X. Note that ω^2 and g always appear together in the flutter determinant as $(\omega_\alpha/\omega)^2(1 + ig_\alpha)$, etc. For small values of g_h, g_α, g_β, we may write

$$1 + ig_\alpha = 1 + ig_h + i(g_\alpha - g_h) \doteq (1 + ig_h)[1 + i(g_\alpha - g_h)] \quad (13)$$

provided that $g_h(g_\alpha - g_h)$ is negligibly small in comparison to g_h. Hence, let

$$Z = \left(\frac{\omega_\alpha}{\omega}\right)^2 (1 + ig_h) \quad (14)$$

Then the flutter determinant (Eq. 21 of § 6.10) can be written as a cubic equation in Z. For a given k, the (complex) roots Z can be found. If we plot the roots Z with their real parts $(\omega_\alpha/\omega)^2$ as the abscissa and the ratios of the imaginary part to the real part, g_h, as the ordinate, we obtain a parametric curve showing the variation of Z with k. The correct values of k at which flutter is possible are those that correspond to the value of g_h of the specific structure. The compressibility effects, if any, must be catered for as before.

Example 2. For the torsion-bending flutter of the numerical example 1 considered above, the determinantal equation for $1/k = 2$ is given by Eq. 8. It is a quadratic equation because only two degrees of freedom are allowed. Equation 8 is derived for $g = 0$. When $g \neq 0$, we must replace X by Z as defined by Eq. 14. Hence for $1/k = 2$, and assuming $g_h = g_\alpha = g$, we have

$$320.283Z^2 + 74.463iZ - 1360.393Z - 99.836i + 1110.312 = 0 \quad (15)$$

This is of the form

$$aZ^2 + bZ + c = 0 \quad (16)$$

and the roots are

$$Z_{(+)} = \frac{-b + \sqrt{b^2 - 4ac}}{2a}, \qquad Z_{(-)} = \frac{-b - \sqrt{b^2 - 4ac}}{2a} \quad (17)$$

Denoting the two roots of Eq. 15 by $Z_{(+)}$ and $Z_{(-)}$ as in Eqs. 17, we obtain

$$Z_{(-)} = 1.1051 - 0.0303i, \qquad Z_{(+)} = 3.1424 - 0.1960i$$

By repeating this process for several other assumed values of $1/k$, Table 6.3 is obtained.

Table 6.3

$\mu = 40, \qquad b = 30, \text{ft}, \qquad \omega_\alpha^2 = 2.41$

$1/k$	$Z_{(-)}$	$g = \dfrac{\text{Im}(Z_{(-)})}{\text{Rl}(Z_{(-)})}$	$Z_{(+)}$
2	$1.1051 - 0.0303i$	-0.0274	$3.1424 - 0.1960i$
2.5	$1.1842 - 0.0384i$	-0.0324	$3.1249 - 0.2647i$
2.94	$1.2390 - 0.0426i$	-0.0344	$3.1088 - 0.3344i$
3.33	$1.3134 - 0.0411i$	-0.0313	$3.0947 - 0.4059i$
4.17	$1.5023 - 0.0102i$	-0.0078	$3.0723 - 0.5975i$
5.00	$1.7042 + 0.0745i$	$+0.0437$	$3.0911 - 0.8568i$

Clearly the imaginary part of $Z_{(+)}$ remains negative for all values of k. On the other hand, the sign of the imaginary part of $Z_{(-)}$ changes at certain value of k. According to the definition of Z (Eq. 14), the real part of Z is $(\omega_\alpha/\omega)^2$. The imaginary part is $(\omega_\alpha/\omega)^2 g$. Hence,

$$g = \frac{\text{Im}(Z)}{\text{Rl}(Z)} \tag{18}$$

A plot of g vs. $(\omega_\alpha/\omega)^2$ with $1/k$ as a parameter is shown in Fig. 6.19. If

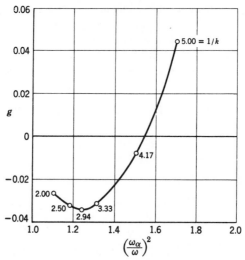

Fig. 6.19.   Plot of the damping factor g versus the square of the frequency ratio $(\omega_\alpha/\omega)^2$.

the value of g for the structure is zero, the critical condition is reached when

$$\left(\frac{\omega_\alpha}{\omega}\right)^2 = 1.545, \qquad \frac{1}{k} = 4.31$$

The flutter speed is therefore

$$U_{cr} = \frac{b}{k} \frac{\omega}{\omega_\alpha} \omega_\alpha = 162 \text{ ft per sec}$$

By the relation

$$U = \frac{1}{k} \left(\frac{\omega}{\omega_\alpha}\right) b\omega_\alpha \tag{19}$$

the function g can be plotted against U.   The result is shown in Fig. 6.20.

This method was used by Smilg and Wasserman,[6.20] and is sometimes known as "AMC" (Air Materiel Command) method.   A tabular adaption for routine calculations has been published by Scanlan and Rosenbaum.[6.19]

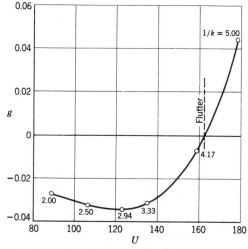

Fig. 6.20.   Plot of the damping factor g versus the
speed of flow U.

The values of ω^2 and g so obtained satisfy the flutter determinant, and determine a mode of flutter motion (i.e., the amplitude ratios and phase relations of h/b, α, β, δ).   Now suppose that, at a specific critical condition, the value of g is increased by imposing certain dampers on the wing; would this stabilize the wing?   The answer is uncertain, because by the change in g the phase relationship between various components of motion changes.   A new energy balance is set up, and it is not obvious whether this results in a stable or unstable motion.   In other words, an increase in damping does not necessarily raise the flutter speed.

Although the numerical example given above shows an increase in U_{cr} by an increase of g, examples to the contrary can be constructed. Collar (see Chapter 8 of Ref. 3.20) has shown, for the case of pitching oscillations of a frictionally constrained airfoil, and Frazer[6.8], for the case of wing bending-aileron rotation, that it is possible to adjust the wing mass distribution in such a way that the flutter speed decreases by imposing several types of damping: dry friction, viscous damping proportional to amplitude, and damping force proportional to the square of the amplitude.*

It may be pointed out that the destabilizing effect of damping is not entirely unfamiliar in aerodynamics.   For example, the origin of turbulence in a flow lies in the viscosity of the fluid.

6.12   METHODS OF DETERMINING THE CRITICAL SPEEDS

The methods discussed in the last two sections can be extended to systems having any finite number (say n) of degrees of freedom.   Several different points of view may be used to determine the critical flutter speeds. In the forced-oscillation method (§ 6.9), we may proceed to write down the equations of motion, with an (undetermined) external exciting force acting in one of the degrees of freedom (say the first equation), and with no excitation term in the remaining $(n - 1)$ equations.   The amplitude and phase angle of one of the remaining degrees of freedom (say the nth) may be arbitrarily assumed.   For a given speed of flow and given frequency, these $(n - 1)$ homogeneous equations can be used to determine the amplitude and phase angle of various degrees of freedom in proportion to the nth, and the first equation can then be used to determine the magnitude and phase of the exciting force required to act in the first degree of freedom to produce the assumed motion.   A variation in the speed of flow and the oscillation frequency will produce a corresponding change in the required exciting forces.   By repeating the calculations for different combinations of speed and frequency, the trend of changes in the exciting force can be determined, and the speed and frequency at which the exciting force vanishes can be obtained.   These are the flutter speed

---

* Ref. 6.8 also gives examples to show that the stability of a wing *cannot* be determined by measuring the work done by an exciting force in a steady forced oscillation.   The reason is very simple: The amplitude and phase relationship between various components of motion vary with the frequency and the point of application of the exciting force. In particular, the amplitude and phase relationship in forced oscillation are in general different from those in a free oscillation.   From § 5.4 it is clear that such variations will cause an important change in the energy exchange between the wing and the airstream. Frazer shows that in certain cases it is possible to provide a mechanism which extracts energy from a wing, yet causing an otherwise stable wing to oscillate sinusoidally.

and frequency.* This method has been applied by Duncan[6.6, 6.7] and Myklestad[6.17] to systems having many degrees of freedom.

It is also possible to study the free oscillations following an initial disturbance and define flutter as a condition at which an oscillation of nondecreasing amplitude is obtained. A method of iteration for such calculation has been used by Goland and Luke.[6.11]

When the existence of flutter is presupposed, as in § 6.10, the airspeed and oscillation frequency are sought. The similarity of this formulation to the classical vibration problem of a mechanical system in still air indicates that flutter is a complex eigenvalue problem.* Besides Theodorsen's method, this problem may be solved by matrix methods as given by Duncan, Frazer, and Collar,[6.5] or by the method of iteration, as expounded by Jordan,[6.14] Küssner,[5.28] Wielandt,[11.24] Greidanus, and van de Vooren.[6.12]

A method proposed by R. A. Frazer[6.10] regards the flutter determinant as defining an eigenvalue problem involving two real-valued eigenvalues, one related to the reduced frequency and the other to the stiffness of the structure.

The most straightforward method is to write the flutter equation as a matrix equation, form the characteristic equation, and solve the complex eigenvalues. The quickest way, however, is to use an analog computer. For analog approach we refer to the book by McNeal.[6.28] The aeroelastic system is simulated by an electric network. An example of such an analog sets up the following correspondence:

Capacitors—concentrated or lumped inertia properties
Inductors—lumped flexibility properties
Transformers—geometric properties
Voltage—velocities
Current—forces

Such an electric analog can be regarded as a model of aircraft whose properties can be altered easily. A parametric study of the aircraft can be done with great rapidity. A particular advantage of the electric analog method is that tuned pulses may be used to separate two or more nearly unstable modes of motion.

Whereas the most convenient form of aerodynamic information used in the influence-functions formulation is the frequency response, in the analog method indicial responses are used: e.g., the lift due to a sudden change of angle of attack, the moment due to a sudden aileron deflection, etc. They can be approximated by a finite sum of exponential functions. The electric analog for the aerodynamic system can then be recognized

---

* Further references regarding the mathematical problem of complex eigenvalues can be found in Chapter 11.

relatively easily by an examination of the Laplace transforms of the indicial responses.

In all the methods mentioned above, one tries to find the actual airspeed at which flutter may occur, but often this is more than necessary from the point of view of airplane design. In fact, an airplane may be regarded as nonexistent for airspeeds exceeding the design speeds. The immediate question regarding the safety against flutter is this: Will the airplane flutter within the "design envelope" of Mach number and altitude? This question has a different content from the question of finding the flutter speeds in terms of miles per hour. We are concerned only with the stability of the aeroelastic system in a given speed range. Mathematically, we are dealing with the problem of the *existence* of an eigenvalue within a specified range, instead of finding out its numerical value. Certain methods applicable to this problem will be discussed in § 10.6.

BIBLIOGRAPHY

For aerodynamic tables, see Chapters 13, 14, and 15.

6.1   Arnold, L.: A Vector Solution of the Three Degree Case of Wing Bending, Wing Torsion, Aileron Flutter. *J. Aeronaut. Sci.* **9**, 497–500 (Nov. 1942).

6.2   Bergen, W. B., and L. Arnold: Graphical Solution of the Bending-Aileron Case of Flutter. *J. Aeronaut. Sci.* **7**, 495–508 (Oct. 1940).

6.3   Bleakney, W. M., and J. D. Hamm: Vector Methods of Flutter Analysis. *J. Aeronaut. Sci.* 439–451 (Oct. 1942).

6.4   Cicala, P.: Le oscillazioni flesso-torsionali di un'ala in corrente uniforme. *Aerotecnica* **16**, 785–801 (1936).

6.5   Duncan, W. J., and A. R. Collar: Matrices Applied to the Motions of Damped Systems. *Phil. Mag. Ser.* 7, **19**, 197–219 (1935). See also, *Elementary Matrices*, by Frazer Duncan, and Collar, Cambridge Univ. Press, London (1938).

6.6   Duncan, W. J., A. R. Collar, and H. M. Lyon: Oscillations of Elastic Blades and Wings in an Airstream. *Aeronaut. Research Com. R. & M.* **1716** (1936).

6.7   Duncan, W. J.: Flutter of Systems with Many Freedoms. *College Aeronaut. Cranfield Rept.* **19** (1948).

6.8   Frazer, R. A.: On the Power Input Required to Maintain Forced Oscillations of an Aeroplane Wing in Flight. *Aeronaut. Research Com. R. & M.* **1872** (1939).

6.9   Frazer, R. A., and W. P. Jones: Wing-Aileron-Tab Flutter, parts I and II. *Aeronaut. Research Council Rept. ARC* **5668** (Mar. 1942).

6.10   Frazer, R. A.: Bi-variate Partial Fractions and Their Applications to Flutter and Stability Problems. *Proc. Roy. Soc. London A* **185**, 465–484 (1946).

6.11   Goland, M., and Y. L. Luke: A Study of the Bending-Torsion Aeroelastic Modes for Aircraft Wings. *J. Aeronaut. Sci.* **16**, 389–396 (1949). See also discussions in *J. Aeronaut. Sci.* **19**, 213–214 (1952).

6.12   Greidanus, J. H., and A. I. van de Vooren: Mathematical Principles of Flutter Analysis. *Natl. Luchtvaartlab. Amsterdam Rept.* F **43** (1949).

6.13   Grossman, E. P.: Flutter. *Joukowsky Mem. Central Aero-Hydrodynamic Inst. Rept.* **186** (1935). In Russian. *Air Force Translation* **F-TS-1225-IA**. (GDAM-A9-T-44). Air Material Command, Wright Field, Dayton, Ohio.

6.14   Jordan, P.: The Wing Flutter as a Complex Characteristic Value Problem. *ForschBer.* **1719**, AVA Göttingen. *Air Force Translation* **F-TS-990-RE**. Wright Air Development Center (1947).

6.15 Kassner, R., and H. Fingado: The Two-Dimensional Problem of Wing Vibra-
      tion. *J. Roy. Aeronaut. Soc.* **41**, 921–944 (1937). Translated from *Luftfahrt-
      Forsch.* **13**, 374–387 (1936).

6.16 Kassner, R.: The Consideration of Internal Damping in the Two-Dimensional
      Problem of Wing Vibration. *J. Roy. Aeronaut. Soc.* **41**, 945–952 (1937).
      Translated from *Luftfahrt-Forsch.* **13**, 388–390 (1936).

6.17 Myklestad, N. O.: *Vibration Analysis.* McGraw-Hill, New York (1944).

6.18 Price, H. L.: Note on Frazer's Proposed Method for the Numerical Solution of
      Flutter and Stability Problems. *Aeronaut. Research Council R. & M.* **2393**
      (1944).

6.19 Scanlan, R. H. and R. Rosenbaum: *Introduction to the Study of Aircraft
      Vibration and Flutter.* Macmillan Co., New York (1951).

6.20 Smilg, B., and L. S. Wasserman: Application of Three-Dimensional Flutter
      Theory to Aircraft Structures. *U.S. Army Air Force Tech. Rept.* **4798** (1942).

6.21 Theodorsen, T.: General Theory of Aerodynamic Instability and the Mechanism
      of Flutter. *NACA Rept.* **496** (1935).

6.22 Theodorsen, T., and I. E. Garrick: Mechanism of Flutter. A Theoretical and
      Experimental Investigation of the Flutter Problem. *NACA Rept.* **685** (1940).

6.23 Theodorsen, T., and I. E. Garrick: Flutter Calculations in Three Degrees of
      Freedom. *NACA Rept.* **741** (1942).

6.24 Williams, J.: Methods of Predicting Flexure-Torsion Flutter of Cantilever
      Wings. *Aeronaut. Research Council R. & M.* **1990** (1943).

6.25 Young, D., and R. P. Felgar: Tables of Characteristic Functions Representing
      Normal Modes of Vibration of a Beam. *Univ. Texas Rept.* **4913** (1949). See
      also Felgar, Formulas for Integrals Containing Characteristic Functions of a
      Vibrating Beam. *Univ. Texas Circ.* **14** (1950).

For Galerkin's method, and other methods of generalized coordinates, see

6.26 Duncan, W. J.: Galerkin's Method in Mechanics and Differential Equations.
      Aeronaut. Research Com. R. & M. **1798** (1937), *R. & M.* **1848** (1938).

6.27 Fung, Y. C.: *Foundations of Solid Mechanics.* Prentice-Hall, New Jersey (1965)
      p. 333–339, 498.

6.28 MacNeal, R. H.: *Electric Circuit Analogies for Elastic Structures.* John Wiley &
      Sons, New York (1962).

Chapter 7

ENGINEERING FLUTTER ANALYSIS AND STRUCTURAL DESIGN

Extension of flutter analysis of the preceding chapter to three-dimensional structure is straightforward. In § 7.1, a practical approach, based on generalized coordinates, will be outlined in greater detail. The underlying principles for the selection of generalized coordinates is discussed in § 7.2. The limitations of the theory are then pointed out in § 7.3. Some general remarks on the control and prevention of flutter are presented in § 7.4.

Flutter analysis should be comprehensive so that no factor is overlooked. We know many instances in which the crucial factors that caused commercial disaster or loss of lives were so hidden that they were forgotten. One fine turboprop plane had several disastrous flutter failures because of an ignorance of the propeller yawing effect (oscillatory aerodynamic force acting on the propeller in a direction perpendicular to the axis of the propeller). One high performance fighter development lost many millions of dollars because of the use of a rubbing block (for dry friction) in the control system. A famous company which made very extensive flutter analysis in the course of development of an airplane was met with flutter failure because the engineers employed the *same simplifying assumptions* in building the mathematical model for analysis and the wind tunnel model for testing; in which case the confidence derived from the agreement between theory and experiment was meaningless with regard to the prototype.

On the other hand, the varied conditions in the service of the aircraft must be considered. Icing, fuel displacement, minor structural damages, may cause flutter. One small private airplane crashed because of accumulation of dust in the wing.

Safety can be purchased only with exhaustive care!

7.1 FLUTTER OF A THREE-DIMENSIONAL WING

As a concrete example, let us consider an unswept cantilever wing with its root fixed in space while an airstream flows over it. We shall assume that the wing has an elastic axis that is a straight line, which is then chosen as the reference line for measuring the wing deflections. Assume that the wing deformation of each chordwise section can be described by three quantities: the bending (positive downward) of the elastic axis, the pitching (positive nose up) about the elastic axis, and the rotation (positive trailing edge down) of the aileron about the aileron hinge line. Let these be denoted, for a section located at the spanwise coordinate y, and at time t, by $h(y, t)$, $\alpha(y, t)$, $\beta(y, t)$, respectively. A complete description of the arbitrary functions $h(y, t)$, $\alpha(y, t)$, and $\beta(y, t)$ requires a complete set of generalized coordinates. As is usual in the vibration analysis, however, let us choose a few, say three, generalized coordinates as representative of the fluttering wing. As an example, let

$$h(y, t) = \bar{h}(t) f(y)$$
$$\alpha(y, t) = \bar{\alpha}(t) \phi(y) \tag{1}$$
$$\beta(y, t) = \bar{\beta}(t) g(y)$$

where $\bar{h}(t)$, $\bar{\alpha}(t)$, and $\bar{\beta}(t)$ are unknown functions of time, whereas $f(y)$, $\phi(y)$, $g(y)$ are three assumed functions of y. By fixing $f(y)$, $\phi(y)$, $g(y)$, the deformation pattern of the wing is restricted.*

If $f(y)$, $\phi(y)$, $g(y)$ were suitably chosen, a good approximation of the true flutter speed can be obtained. These functions are termed *semirigid modes* of deformation, or simply *modes*, on the basis of which the generalized coordinates $\bar{h}(t)$, $\bar{\alpha}(t)$, $\bar{\beta}(t)$ are defined. We shall assume that the functions $\bar{h}(t)$, $\bar{\alpha}(t)$, $\bar{\beta}(t)$ have the dimensions of h, α, and β, respectively, so that $f(y)$, $\phi(y)$, $g(y)$ are nondimensional functions.

Generally, the functions $f(y)$, $\phi(y)$, and $g(y)$ may be chosen as the uncoupled modes of free vibrations of the wing.† For ease of computation, they may also be chosen as polynomials or other elementary functions approximating the uncoupled vibration modes. The boundary conditions specifying the end constraints must be satisfied. As examples, the

---

* Usually the control surface may be regarded as a rigid body, and $g(y)$ can be taken as a constant. A sufficient condition for this is that the fundamental torsional frequency of the surface, with the control lever rigidly attached to the wing, shall be well above the flutter frequency (see Duncan[7.25]). This is one reason why the control lever should be located at, or near, the middle of the span of the control surface. For the same reason the balancing masses should be located near the control lever, or at least not far away from it.

† Better results can be obtained by using the *normal modes* of free vibration of the wing. For example, if the first three normal modes are represented by the columns

$$\begin{Bmatrix} h \\ \alpha \\ \beta \end{Bmatrix} = \begin{Bmatrix} f_1(y) \\ \phi_1(y) \\ g_1(y) \end{Bmatrix}, \quad \begin{Bmatrix} f_2(y) \\ \phi_2(y) \\ g_2(y) \end{Bmatrix}, \quad \begin{Bmatrix} f_3(y) \\ \phi_3(y) \\ g_3(y) \end{Bmatrix}$$

we may write

$$h(y, t) = q_1(t) f_1(y) + q_2(t) f_2(y) + q_3(t) f_3(y)$$
$$\alpha(y, t) = q_1(t) \phi_1(y) + q_2(t) \phi_2(y) + q_3(t) \phi_3(y)$$
$$\beta(y, t) = q_1(t) g_1(y) + q_2(t) g_2(y) + q_3(t) g_3(y)$$

where $q_1(t)$, $q_2(t)$, $q_3(t)$ are generalized coordinates.

following sets of functions may give a fair representation of a cantilever wing with the root section fixed in space:

$$f(y) = \frac{1}{l^2} y^2, \qquad \phi(y) = \frac{1}{l} y, \qquad g(y) = 1 \tag{2}$$

$$f(y), \phi(y) \text{ given by Eqs. 11 and 12 of } \S 6.4, \qquad g(y) = 1 \tag{3}$$

If more than three generalized coordinates were taken, we have lengthier expressions. Thus, instead of Eqs. 1 we may write

$$h(y, t) = h_1(t) f_1(y) + h_2(t) f_2(y) + \cdots + h_n(t) f_n(y) \tag{4}$$

etc., where $f_1(y), \cdots, f_n(y)$ are modes, on the basis of which the generalized coordinates $h_1(t), \cdots, h_n(t)$ are defined. A comparison of the flutter speed obtained by using n generalized coordinates with that obtained by using $n + 1$ generalized coordinates will give some indication of the degree of accuracy achieved. A rigorous proof of convergence by successively increasing the number of degrees of freedom, however, is rarely undertaken.

Let us take Eqs. 1 as an explicit example for the application of the method of generalized coordinates. Let $m(y)$, $I_\alpha(y)$, $S_\alpha(y)$, $I_\beta(y)$, $S_\beta(y)$, $b(y)$, $c_\beta(y)$, and $a_h(y)$ be the mass per unit length, moment of inertia per unit length, etc., at the spanwise station y, corresponding to the same symbols used in § 6.10 (p. 223). We can write the kinetic energy per unit span for a section at y as

$$T'(y) = \tfrac{1}{2} m(y) \, h^2(y) + \tfrac{1}{2} I_\alpha(y) \, \dot\alpha^2(y) + \tfrac{1}{2} I_\beta(y) \, \dot\beta^2(y) + S_\beta(y) \, h(y) \, \dot\beta(y)$$
$$+ S_\alpha(y) \, h(y) \, \dot\alpha(y) + \{ S_\beta(y) [c_\beta(y) - a_h(y)] b(y) + I_\beta(y) \} \dot\beta(y) \, \dot\alpha(y) \tag{5}$$

Substituting h, α, β from Eqs. 1 and integrating with respect to y over the wing span, we obtain the total kinetic energy of the wing:

$$T = \tfrac{1}{2} \bar{m} \dot{h}^2 + \tfrac{1}{2} I_\alpha \dot\alpha^2 + \tfrac{1}{2} I_\beta \dot\beta^2 + \bar{S}_\beta \dot{h} \dot\beta + \bar{S}_\alpha \dot{h} \dot\alpha + \bar{P}_{\alpha\beta} \dot\alpha \dot\beta \tag{6}$$

where

$$\bar{m} = \int_0^l m(y) f^2(y) \, dy$$

$$I_\alpha = \int_0^l I_\alpha(y) \, \phi^2(y) \, dy$$

$$I_\beta = \int_{l_1}^{l_2} I_\beta(y) \, g^2(y) \, dy$$

$$\bar{S}_\alpha = \int_0^l S_\alpha(y) f(y) \, \phi(y) \, dy \tag{7}$$

$$\bar{S}_\beta = \int_{l_1}^{l_2} S_\beta(y) f(y) \, g(y) \, dy$$

$$\bar{P}_{\alpha\beta} = \int_{l_1}^{l_2} \{ S_\beta(y) [c_\beta(y) - a_h(y)] \, b(y) + I_\beta(y) \} \, \phi(y) \, g(y) \, dy$$

The symbol l denotes the semispan of the wing, and l_1, l_2 denote, respectively, the spanwise locations of the inner and outer ends of the aileron. $\bar{m}$, $\bar{I}_\alpha$, etc., are called the *generalized mass, generalized mass moment of inertia*, etc.

The strain energy stored in the wing is

$$V = \frac{1}{2} \int_0^l E\, I(y) \left[\frac{\partial^2 h(y, t)}{\partial y^2}\right]^2 dy + \frac{1}{2} \int_0^l G\, J(y) \left[\frac{\partial \alpha(y, t)}{\partial y}\right]^2 dy$$

$$+ \text{ a quadratic function of } \beta(y, t) \qquad (8)$$

Substituting Eqs. 1 into Eq. 8, we may write the result as

$$V = \tfrac{1}{2}K_h \bar{h}^2 + \tfrac{1}{2}K_\alpha \bar{\alpha}^2 + \tfrac{1}{2}K_\beta \bar{\beta}^2 \qquad (9)$$

where

$$
\begin{aligned}
K_h &= \int_0^l E\, I(y) \left[\frac{d^2}{dy^2} f(y)\right]^2 dy \\
K_\alpha &= \int_0^l G\, J(y) \left[\frac{d}{dy} \phi(y)\right]^2 dy
\end{aligned}
\qquad (10)
$$

and K_β is the torsional "spring constant" of the aileron control system. As in § 6.9, the generalized spring constants K_h, K_α, K_β may be replaced by the corresponding "uncoupled" free-vibration frequencies. Thus, if ω_h is the uncoupled free vibration frequency in the h degree of freedom, we have

$$K_h = \bar{m}\omega_h{}^2 \qquad (11)$$

Similarly,

$$K_\alpha = I_\alpha \omega_\alpha{}^2, \qquad K_\beta = I_\beta \omega_\beta{}^2$$

Hence, Eq. 9 may be written as

$$V = \tfrac{1}{2}\bar{m}\omega_h{}^2 \bar{h}^2 + \tfrac{1}{2}I_\alpha \omega_\alpha{}^2 \bar{\alpha}^2 + \tfrac{1}{2}I_\beta \omega_\beta{}^2 \bar{\beta}^2 \qquad (13)$$

The Lagrange's equations of motion are

$$\frac{d}{dt}\left(\frac{\partial T}{\partial \dot{q}_i}\right) + \frac{\partial V}{\partial q_i} = Q_i \qquad (q_i = \bar{h}, \bar{\alpha}, \bar{\beta})$$

Using Eqs. 6 and 13, we obtain the equations of motion:

$$
\begin{aligned}
\bar{m}\,\ddot{\bar{h}} + \bar{S}_\alpha \ddot{\bar{\alpha}} + \bar{S}_\beta \ddot{\bar{\beta}} + \bar{m}\omega_h{}^2 \bar{h} &= Q_h \\
\bar{S}_\alpha \ddot{\bar{h}} + I_\alpha \ddot{\bar{\alpha}} + \bar{P}_{\alpha\beta}\ddot{\bar{\beta}} + I_\alpha \omega_\alpha{}^2 \bar{\alpha} &= Q_\alpha \\
\bar{S}_\beta \ddot{\bar{h}} + \bar{P}_{\alpha\beta}\ddot{\bar{\alpha}} + I_\beta \ddot{\bar{\beta}} + I_\beta \omega_\beta{}^2 \bar{\beta} &= Q_\beta
\end{aligned}
\qquad (14)
$$

At the critical flutter condition, the wing motion is simple harmonic. We may write, in complex representation, (§ 1.8)

$$\bar{h}(t) = h_0 e^{i\omega t}, \qquad \bar{\alpha}(t) = \alpha_0 e^{i\omega t}, \qquad \bar{\beta}(t) = \beta_0 e^{i\omega t} \qquad (15)$$

Such a simple-harmonic solution is possible if we assume that the amplitude of oscillation is very small, because then the generalized forces Q_h, Q_α, Q_β are linear functions of $\bar{h}$, $\bar{\alpha}$, $\bar{\beta}$, and Eqs. 14 become a system of linear equations. A solution of the form of Eqs. 15, with h_0, α_0, β_0 complex valued, is admissible for such a system of equations. When we substitute Eqs. 15 into Eqs. 14 and canceling the factor $e^{i\omega t}$, the time-free equations of motion are obtained. If, in these equations, the elastic-restoring-force terms are multiplied by factors of the form $(1 + ig)$ to represent the structural damping forces, in phase with the velocity and opposing the motion (see § 6.9), we obtain

$$- \omega^2(\bar{m}\bar{h} + \bar{S}_\alpha\bar{\alpha} + \bar{S}_\beta\bar{\beta}) + (1 + ig_h)\bar{m}\omega_h{}^2\bar{h} = Q_h$$

$$- \omega^2(\bar{S}_\alpha\bar{h} + \bar{I}_\alpha\bar{\alpha} + \bar{P}_{\alpha\beta}\bar{\beta}) + (1 + ig_\alpha)\bar{I}_\alpha\omega_\alpha{}^2\bar{\alpha} = Q_\alpha \qquad (16)$$

$$- \omega^2(\bar{S}_\beta\bar{h} + \bar{P}_{\alpha\beta}\bar{\alpha} + \bar{I}_\beta\bar{\beta}) + (1 + ig_\beta)\bar{I}_\beta\omega_\beta{}^2\bar{\beta} = Q_\beta$$

The generalized forces Q_h, Q_α, Q_β must be computed. The aerodynamic forces and moments at a section at y are given by Eqs. 15 of § 6.10. For a three-dimensional wing, however, the coefficients L_h, L_α, M_h, M_α, etc., are not only functions of the Mach number M and the reduced frequency k, but also of the aspect ratio of the wing, the wing geometry (taper ratio, sweep angle, etc.), the wing deflection mode (the nodal-line location and the shape of the deflection curve), as well as the spanwise location y. To simplify the calculation, the *strip assumption* is often introduced, which states that the aerodynamic force acting at any section is the same as if that section were a part of a two-dimensional wing. Furthermore, the induced downwash due to trailing vortices is neglected. Therefore, the coefficients L_h, L_α, etc., are given by the same tables as those for the two-dimensional case.

Since $k = \omega b / U$ is proportional to the semichord b, it generally varies across the span. If a particular cross section of the wing is defined as a *reference cross section*, of which the semichord and the corresponding "reference" reduced frequency are denoted by b_r and $k_r = \omega b_r / U$, respectively, then any other cross section with a chord $b(y)$ will have the reduced frequency

$$k = \frac{b(y)}{b_r} k_r$$

The coefficients L_h, L_α, etc., being functions of k, can thereafter be comprehended as functions of y, with the reference reduced frequency k_r as a parameter.

The generalized force Q_h for the oscillating bending motion is obtained by considering the virtual work due to a variation of the generalized coordinate $\delta \bar{h}$, which corresponds to a virtual displacement $\delta \bar{h} f(y)$. This gives*

$$Q_h(t)\delta \bar{h} = - \int_0^l L_{e.a.}(y, t)\delta \bar{h}f(y)\,dy$$

i.e.,

$$Q_h(t) = - \int_0^l L_{e.a.}(y, t)f(y)\,dy \tag{17}$$

Similarly,

$$Q_\alpha(t) = \int_0^l M_{e.a.}(y, t)\,\phi(y)\,dy, \qquad Q_\beta(t) = \int_{l_1}^{l_2} T_{hinge}(y, t)\,g(y)\,dy \tag{18}$$

Substituting Eqs. 15 of § 6.10 into the above equations, we obtain

$$Q_h = \pi\rho b_r{}^3\omega^2 \left(A_{hh}\frac{h}{b_r} + A_{h\alpha}\bar{\alpha} + A_{h\beta}\bar{\beta}\right)$$

$$Q_\alpha = \pi\rho b_r{}^4\omega^2 \left(A_{\alpha h}\frac{h}{b_r} + A_{\alpha\alpha}\bar{\alpha} + A_{\alpha\beta}\bar{\beta}\right) \tag{19}$$

$$Q_\beta = \pi\rho b_r{}^4\omega^2 \left(A_{\beta h}\frac{h}{b_r} + A_{\beta\alpha}\bar{\alpha} + A_{\beta\beta}\bar{\beta}\right)$$

---

* In Eq. 7 of § 1.6, the generalized coordinate q_n and the generalized force Q_n are real valued; then the virtual work δW is simply $\Sigma Q_n \delta q_n$. If the motion q_n is sinusoidal, Q_n is also sinusoidal, and q_n, Q_n can be represented by complex representations. As is shown in Eq. 5a of § 5.4, a simple multiplication $Q_n \delta q_n$ does not represent the virtual work. However, it is readily shown (Nilsson and Langefors[7.50]) that the sum of the product of Q_n with the complex conjugate of δq_n

$$\delta \overline{W} = \Sigma\, Q_n \cdot \delta q^*{}_n$$

represents a complex invariant with respect to coordinate transformation from the physical (real) coordinates to the generalized coordinates. The real and imaginary parts of $\delta \overline{W}$ can be interpreted physically as the "active" and "reactive" energy. Let $q_n = q_{n0}e^{i\omega t}$, where q_{n0} is a real number, and $Q_n = (Q_{nR} + iQ_{nI})e^{i\omega t}$; then

$$\delta \overline{W} = (Q_{nR} + iQ_{nI}) \cdot \delta q_{n0}$$

This equation can be used to compute the proper form of the generalized force Q_n, since the complex invariant $\delta \overline{W}$ can be calculated in the rectangular Cartesian coordinates. Equations 17 and 18 are based on this argument.

where b_r is the semichord at a reference section, and

$$A_{hh} = \int_0^l \left(\frac{b}{b_r}\right)^2 f^2(y)\, L_h(y)\, dy$$

$$A_{h\alpha} = \int_0^l \left(\frac{b}{b_r}\right)^3 f(y)\, \phi(y) \left[L_\alpha - \left(\frac{1}{2} + a_h\right) L_h \right] dy$$

$$A_{h\beta} = \int_{l_1}^{l_2} \left(\frac{b}{b_r}\right)^3 f(y)\, g(y)\, [L_\beta - (c_\beta - e_\beta) L_z]\, dy$$

$$A_{\alpha h} = \int_0^l \left(\frac{b}{b_r}\right)^3 f(y)\, \phi(y) \left[M_h - \left(\frac{1}{2} + a_h\right) L_h \right] dy$$

$$A_{\alpha\alpha} = \int_0^l \left(\frac{b}{b_r}\right)^4 \phi^2(y) \left[M_\alpha - \left(\frac{1}{2} + a_h\right)(L_\alpha + M_h) \right.$$
$$\left. + \left(\frac{1}{2} + a_h\right)^2 L_h \right] dy \qquad (20)$$

$$A_{\alpha\beta} = \int_{l_1}^{l_2} \left(\frac{b}{b_r}\right)^4 \phi(y)\, g(y) \left[M_\beta - \left(\frac{1}{2} + a_h\right) L_\beta - (c_\beta - e_\beta) M_z \right.$$
$$\left. + (c_\beta - e_\beta)\left(\frac{1}{2} + a_h\right) L_z \right] dy$$

$$A_{\beta h} = \int_{l_1}^{l_2} \left(\frac{b}{b_r}\right)^3 f(y)\, g(y)[T_h - (c_\beta - e_\beta) P_h]\, dy$$

$$A_{\beta\alpha} = \int_{l_1}^{l_2} \left(\frac{b}{b_r}\right)^4 \phi(y)\, g(y) \left[T_\alpha - (c_\beta - e_\beta) P_\alpha - \left(\frac{1}{2} + a_h\right) T_h \right.$$
$$\left. + \left(\frac{1}{2} + a_h\right)(c_\beta - e_\beta) P_h \right] dy$$

$$A_{\beta\beta} = \int_{l_1}^{l_2} \left(\frac{b}{b_r}\right)^4 g^2(y)\, [T_\beta - (c_\beta - e_\beta)(P_\beta + T_z) + (c_\beta - e_\beta)^2 P_z]\, dy$$

Combining Eqs. 16 and 19, we obtain the equations of motion at the critical flutter condition. Comparing these equations with Eqs. 19 of § 6.10, we see that they are of the same form, except that the mass, moment of inertia, and aerodynamic coefficients must all be interpreted as the generalized quantities. The solution of the final determinantal equation can be obtained by the same methods used in § 6.11.

We have assumed in the above analysis (1) the existence of an unswept straight elastic axis, and (2) the strip assumption for aerodynamic action. Let us consider now cases in which these assumptions must be relaxed.

If the structure does not behave like a simple beam (in the sense that the formulas listed in § 1.1 hold), but its chordwise sections can be regarded as rigid, then the simplest procedure is to choose any convenient straight line as a reference line; the vertical deflection of a point on this line may be denoted by h, and the rotations of a cross section normal to the reference line in its own plane may be denoted by α. The expressions of the kinetic energy T and the generalized forces Q_h, Q_α, etc., take the same form as those presented above, except that the word "elastic axis" must now be interpreted as the "reference line." The strain energy, however, cannot be obtained from Eq. 8 or 13, whose validity depends on the concepts of shear center and elastic axis. Nevertheless, from the general principles of elasticity, we can conclude that the strain energy must be expressible as a quadratic function of elastic displacements. (By Castigliano's theorem the elastic displacements are linearly proportional to the loading only if the strain energy expression is quadratic.) In the present case, the strain energy is a quadratic function of h, α, and β. A practical way of deriving such a strain energy expression is to use the stiffness influence coefficients, as shown in Eq. 7 of § 1.4.

For a monocoque wing with cutouts or with restrained cross-sectional warping, it is necessary to solve the shear lag problem in order to obtain the "flexibility" influence coefficients, from which the stiffness influence coefficients can be derived by a matrix inversion. If a structure is available for testing, the influence coefficients can be measured directly. The use of stiffness influence coefficients avoids differentiation of the assumed semirigid modes in computing the strain energy, and thus is favorable for numerical accuracy.

If the chordwise sections cannot be regarded as rigid, the deflection surface must be described by a continuous function in more than one dimension. For example, if a wing occupies a region on the (x, y) plane, the deflection $w(x, y; t)$ may be expressed as

$$w(x, y; t) = q_1(t)f_1(x, y) + q_2(t)f_2(x, y) + \cdots + q_n(t)f_n(x, y)$$

where $f_1(x, y)$, $f_2(x, y)$, etc., are the modes on which the generalized coordinates q_1, q_2, $\cdots$, etc., are defined. The calculation of the kinetic energy presents no difficulty. The strain energy can be expressed in terms of stiffness-influence coefficients. The aerodynamic force, however, would have to be calculated from the general theory of oscillating airfoils.

When the strip assumption for aerodynamic action is relaxed, it is necessary to solve the three-dimensional oscillating-airfoil problem for the oscillation mode of each generalized coordinate. This can be done in principle, but so far no convenient table of aerodynamic coefficients exists. For low-aspect ratio wings, the wing-fuselage interference effect can also become important.

It is evident that in all cases Lagrange's equations reduce to a set of linear equations in the generalized coordinates, and the methods of § 6.11 are sufficient to derive the flutter conditions.

7.2 CHOICE OF THE GENERALIZED COORDINATES

Representing the deformation of a continuous elastic structure by a finite number of generalized coordinates is equivalent to imposing certain *constraints* on the elastic body. The structure is no longer *elastic*, but *"semirigid."*

A choice of the generalized coordinates in describing the elastic displacements is a choice of the semirigid modes. When the generalized coordinates are chosen, the equations of motion can be derived according to Lagrange's equations. Explicit use of the differential equations of the theory of elasticity is avoided. The elastic properties of the structure are summed up completely in the expression of the elastic strain energy.

The crucial question remains whether such a simplified model can yield sufficiently accurate results for engineering purposes. The answer is affirmative provided that the semirigid modes were properly chosen. Guidance to this choice has to be sought from model tests and mathematical experiments.

Some of the consequences of the semirigid approximation can be clarified at once. To be concrete, consider the particular representation of a cantilever wing as adopted in Eqs. 1 of § 7.1. This representation implies (1) that the wing-bending deformation can only assume the form $f(y)$, the torsion $\phi(y)$, the aileron $g(y)$, for all airstream speed, density, and direction of flow; (2) if f, ϕ, g are real-valued functions, no phase shift occurs across the span in bending, torsion, or aileron deflection. Now it is known that a wing is capable of oscillating in many different modes when immersed in a flow of a given speed and density. But most of these multitudes of modes are irrelevant to the flutter problem, since at a critical speed the flutter mode assumes a definite form. At, or near, the critical flutter speed, the other oscillation modes, if accidentally excited, are relatively heavily damped and quickly die out. Thus, if the assumed semirigid mode closely approximates the actual displacements which occur in flutter at the critical speed, little error will result from neglecting other possible modes.

As for the phase shift across the span, examples show that generally considerable phase shift exists in the bending displacement, but not in the torsional motion. In many examples the bending phase shift across the span is unimportant. Duncan[7.25] offers the following explanation: In flutter the motion is much larger near the wing tip than inboard. Thus, at half the wing span from the wing root, the amplitude is of the order of one quarter of that at the tip, which implies that the dynamical importance of the motion at half-span is only of the order of one sixteenth of that at the tip. Thus, phase differences are only of importance when they occur in the region near the tip, and within this region they are in fact very small for both bending and torsion. This argument, of course, fails when a wing carries large isolated masses, for then a motion of small amplitude may be dynamically important.

Duncan and his associates[6.6, 5.24, 5.25] have shown that, for a cantilever wing of uniform rectangular planform and uniform cross sections along the span, of either (1) a two-spar construction which derives its torsional stiffness entirely from the differential bending, or (2) a monocoque structure with bending stiffness negligibly small in comparison with its torsional stiffness, the semirigid assumption is exact—meaning that

(a) The distribution of the bending and torsion displacements along the span are independent of air speed for the mode of oscillation which develops into flutter at the critical speed, and

(b) For this mode, the bending oscillations at all parts of the span are in phase, and likewise for the torsional oscillations.

Generally speaking, the application of the semirigid concept has been extremely successful. Disagreement between the calculated flutter speed and the experimental value can usually be attributed to other causes than the semirigidity approximations. In general, the calculated flutter speed is remarkably insensitive to the exact form of the semirigid modes, but it is rather sensitive to the aerodynamic assumptions. Thus, in the past, efforts have been made repeatedly to replace one or a few of the aerodynamic coefficients according to the experimental evidence. Such partial tampering with aerodynamics often leads to inferior results. Better results are obtained by using either the whole set of coefficients exactly as given by the linearized aerodynamic theory or the whole set of experimental derivatives.

There are two ways of representing the deformation of a structure: (1) by a series expansion in terms of a set of continuous functions, (2) by recording displacements at a number of points on the structure. The first type leads to generalized coordinates such as those used in § 7.1, whereas the second type leads to the so-called *lumped-mass* method. In the lumped-mass method, a wing is divided into a number of strips, each of which is supposed to move as a unit. For certain types of electrical analog computors, this is the most convenient method. A full description of the lumped-mass method can be found in Myklested's book[6.17].

If the strip assumption for aerodynamic force is relaxed and the finite-span effect is to be estimated, the method of generalized coordinates is preferred, because the deformation pattern of the whole wing is fixed for each coordinate, and the aerodynamic problem can be solved. On the other hand, in the lumped-mass method the deformation pattern across the span is not known until the displacements at all sections are determined; thus finite-span effect cannot be easily accounted for.

The idea of Rauscher's "station-function" method[7.53] stems from a desire to adopt the simplicity of the lumped-mass method to the calculation of the aerodynamic finite-span effect. Consider the deflection of a wing as an example. The deflections $Z_1, Z_2, \cdots, Z_n$ at a series of stations $y_1, y_2, \cdots, y_n$ are chosen to describe the wing deformation. The deflection curve of the entire wing, expressed as a continuous function $f(y)$, may be approximated by

$$f(y) = Z_1 f_1(y) + \cdots + Z_n f_n(y)$$

where $f_1(y), \cdots, f_n(y)$ are certain functions of y, independent of $Z_1, \cdots, Z_n$. Each of the reference deflections $Z_1, \cdots, Z_n$ thus makes its own contribution to $f(y)$: As $f(y)$ must satisfy the boundary conditions regardless of the values of $Z_1, \cdots, Z_n$, the component functions $f_1(y), \cdots, f_n(y)$ must individually satisfy all boundary conditions. In order that $f(y)$ may have the value Z_1 at the station y_1, irrespective of the values $Z_2, \cdots, Z_n$, it is necessary that $f_1(y_1) = 1$, and $f_2(y_1) = \cdots = f_n(y_1) = 0$. In a like manner, for all $i, j = 1, 2, \cdots, n$; we must have $f_i(y_i) = 1$, and $f_i(y_j) = 0$ if $i \neq j$. The functions $f_i(y)$ so defined are called "station functions." Their application to the flutter problem is similar to the generalized coordinates used in § 7.1.

A question of paramount importance is the minimum number of degrees of freedom (i.e., the number of generalized coordinates) that should be allowed in each particular problem to insure reasonable accuracy of the final result. For example, for a multi-engined airplane wing, or for a wing having a large fuel tank attached to the outer span, a simple semirigid representation, with one pure bending and one pure torsion mode of deflection, is, in general, inadequate. Better results can be obtained by using the first few normal modes of free oscillation of the wing as generalized coordinates, or by numerical integrations of the differential equations of motion (see Runyan and Watkins[7.97]). In some cases the flexibility of the engine mount and the fore and aft motion of the wing are important, and must be accounted for.

The most complicated case is probably the tail flutter. A large number of distinct kinds of deformation of the tail unit and the fuselage are possible, and the effect of the freedom of the airplane as a whole is serious.

To reduce the complexity of the analysis, one may first distinguish the *symmetrical* and *antisymmetrical* types which are independent of each other. In the former, the motions occur symmetrically about the fuselage centerline; in the latter, antisymmetrically. The number of degrees of freedom for the general case of each type is so great that it is impractical to make routine calculations of critical speeds without introducing simplifying assumptions, which must be guided by experimental results (see Duncan[7.25]).

Unless there are reasons to believe otherwise, at least the following degrees of freedom should be considered.[7.25]

For symmetrical tail flutter:

1. Elevator rotation about the hinge line (elevators treated as a single rigid unit).
2. Vertical bending of the fuselage.
3. Bending of the horizontal tail surface.
4. Pitching of the airplane as a whole.

For antisymmetrical tail flutter:

(a) Elevator flutter:

1. Rotation of elevators about their hinge line, in opposition.
2. Torsion of the fuselage.
3. Bending of the horizontal tail surface.

(b) Rudder flutter:

1. Rotation of the rudder about its hinge line.
2. Lateral bending of fuselage.
3. Fuselage torsion or bending of the fin.

7.3 LIMITATIONS OF THE THEORY

It is of basic importance to remember that all we discussed above is based on the linearized theory. Since real physical phenomena are not linear, the question always arises how good the linearized theory is as an approximation to the real case, and to what order of magnitude of the variables concerned is the linearized theory valid.

Unfortunately, so little is known about the nonlinear case that the questions so raised cannot be answered. At present, it can only be said that experimental evidences show that the linearized theory of flutter represents fairly closely the real situation in the neighborhood of the critical flutter speed, provided that the amplitude of motion remains neither too small nor too large.

Phenomena that disagree with the linearized theory are often attributed to the nonlinearity of the system. For example, it is often observed that it is possible to exceed the critical flutter speed without encountering flutter, whereas a sufficiently large disturbance may at once initiate flutter with great violence. This is often regarded as a consequence of the nonlinear characteristics of the structural damping, particularly of dry friction. Küssner[7.42] also points out that at very low amplitudes the laws of potential flow do not hold, because of the effect of viscosity of the air. If the amplitude is of the order of the thickness of the boundary layer, the aerodynamic forces induced by the oscillations are probably smaller than would be expected from the potential theory. It seems plausible to assume that a disturbance of certain minimum value is required to initiate flutter.

On the other hand, violent flutter motions cannot be treated by the linearized theory. Thus it is impossible, within the scope of the linearized theory, to trace the divergent flutter motion. Flutter has been observed whose amplitude does not increase indefinitely at super critical speeds. On the contrary, definite maximum amplitudes are often recorded. The prediction of the violence of the fluttering motion can be made only if the nonlinear characteristics of the structures, as well as those of the aerodynamic forces, are allowed.

For control surfaces, the natural frequencies often depend on the amplitude of oscillation, because of dry friction. This may cause some peculiar behavior in control-surface flutter.

Even within the framework of a linearized theory, flutter analyses are not generally made to their full logical extent. Additional assumptions are introduced to simplify the calculation. Examples of these are: (1) A three-dimensional body is replaced by a system of simple beams. (2) The elastic model is replaced by a mechanical substitutional system having only a finite number of degrees of freedom. (3) The "strip" assumption is used to simplify the aerodynamic expressions. (4) The compressibility effect of the air is sometimes neglected. (5) The aerodynamic coefficients are computed for flat-plate airfoils at zero mean angle of attack.

Of these additional simplifying assumptions, the first and the second have been discussed in § 7.2. The effects of the remaining assumptions vary with the wing planform, Mach number, and flutter mode. They are subjects of current research.

The Effect of Finite Span. The effects of finite span on flutter are complex. The trailing vortices shed from the wing as a result of the spanwise variation of circulation, induce a downwash distribution that is not always negligible. This induced downwash field, however, depends not only on the geometric aspect ratio but also on the mode of deformation. Its theoretical prediction is naturally complicated. Owing to the complexity, flutter analysis based on the finite-span theory is rarely made.

A few examples, incorporating three-dimensional wing theory, seem to show that an increase of the critical flutter speed of the order of 10 to 15 per cent above that computed by the strip theory might be expected as a result of the finite-span effect. The effect is more pronounced for wings of small aspect ratio and low reduced frequency, but tends to be negligible for high-frequency oscillations.

The Effect of the Compressibility of the Fluid. For a flow of sufficiently high Mach number, say $M > 0.5$, the aerodynamic coefficients differ considerably from those for an incompressible fluid. As was shown in § 6.10, however, the formal flutter analysis can be made in the same way for all Mach numbers.

In a subsonic flow, for Mach numbers below the critical Mach number Garrick[7.110] concluded that, for an ordinary wing of normal density and low ratio of bending-to-torsion frequency, the compressibility correction to the flutter speed is of the order of a few per cent. There are experimental indications that this is true through the transonic speed range, provided that the wing is not stalled.

The character of a supersonic flow differs entirely from that of a subsonic flow, and the effect of the compressibility of air must be taken into account. Large effect of compressibility at supersonic speeds is expected, not only because of the larger effect of finite span on the transient lift distribution, but also because the flutter modes of a supersonic wing can be very different from those of subsonic wings. For example, a delta wing may exhibit a flapping motion at the wing tip, which cannot be described adequately by bending and pitching about an elastic axis. Whereas under static loading condition it may be a good approximation to assume that the streamwise cross sections remain rigid, such an assumption is in general not very good in the flutter analysis of a delta wing. In other words, the location of the nodal line becomes very important in the flutter problem. The use of normal modes of free vibration of the wing is very helpful in such cases. The flutter mode usually approaches one of the higher modes of free vibration.

In the transonic-speed range, the effect of shock waves on the aeroelastic properties of a wing are not yet entirely clarified. The legitimacy of treating transonic-flow problems by means of linearized aerodynamic equations is often questioned, although it has been shown that the familiar singularity (of infinite lift) at $M = 1$, which occurs in a steady flow, disappears if the flow is unsteady; thus there exists no fundamental contradiction within the linear theory itself (see references 14.26–14.31).

Airfoil Angle of Attack, Camber, and Thickness. Since in flutter calculation only the *deviation* from the wing's steady-state configuration need be considered, it is implied by the linearized theory that the actual angle of attack, camber, and thickness of the airfoil section have no effect. However, experiments do show the effect of finite angle of attack, camber, and thickness of the wing. The effect of angle of attack can often be detected at angles considerably below the static stalling angle. This is particularly evident for thin wings at transonic speeds. Finite angle of attack usually results in a reduction of the critical flutter speed. As the angle of attack approaches the static stalling angle, very severe drop in critical flutter speed occurs, accompanied by other important changes in the fluttering motion. This is the stall flutter to be discussed in Chapter 9.

When great accuracy is desired, the linearized strip theory alone can hardly be trusted. Experimental model investigations, flight tests, etc., must be performed in conjunction with the theoretical analysis.

7.4 OTHER FLUTTER PROBLEMS

In the flutter problems discussed so far, the chordwise deformation (distortion of the airfoil cross section as distinguished from its rigid body motion) is of minor importance, and can be neglected entirely except for wings of small aspect ratio. However, there is another type of flutter in

Fig. 7.1. Deformation of a two-dimensional panel.

which the chordwise deformation is of paramount importance. A typical example is a flat plate which spans two rigidly supported edges, as shown in Fig. 7.1. Let the air flow over one side of the plate and remain stagnant on the other side. In a supersonic flow, a type of self-excited oscillation may occur in certain ranges of critical dynamic pressure, whose value depends on the initial curvature and the stiffness of the plate, the ratio of the density of air to that of the plate, the dimensions of the plate, and the thrust exerted by the supports at the edges of the plate. This is called *panel flutter*.

Aircraft wings of high-speed aircraft are often of multi-spar multi-rib construction. The spars and ribs support the skin against flexural deflection. In some cases the resistance of the spars and ribs to displacements in the plane of the skin is small, then panel flutter can be treated by a linearized theory.[7.120–7.122] In other cases the spars and ribs are of such rigid construction that they resist displacements in the plane of the skin; then the thrust offered by the spars and ribs is a nonlinear function of the deflection of the skin. This nonlinearity may induce a "relaxation" type of oscillation which is associated with the "oil canning" or the *Durchschlag* of the skin.[7.119]

One of the practical causes of panel flutter is the thermal stress induced in the skin due to aerodynamic heating in flight at high speed.   If the skin is hotter than the supporting structures, compressive stress may be induced in the skin.   If the temperature difference between the supporting structures and the skin is sufficiently large, the skin may become buckled (one can easily verify that often it takes only a few degrees of temperature difference to buckle a plate).   A buckled skin has a much lower critical dynamic pressure than an unbuckled one.

The most practical method of preventing panel flutter is to introduce tension into the skin, for example, by internally pressurizing the wing or the fuselage.

7.5   FLUTTER PREVENTION OR FLUTTER CONTROL

The following measures may be used to secure stability in the design airspeed range:

1. Provide sufficient stiffness, so that the critical speeds of aeroelastic instabilities are inherently high.

2. Furnish good aerodynamic design, so that the flow remains unseparated in service conditions.   If, on the other hand, the aerodynamic force is undesirable, as in a suspension bridge, attempt should be made to render the structure aerodynamically ineffective, to reduce the lift and drag.   Drag reduction is especially beneficial in the case of stall flutter.

3. Break the inertia and aerodynamic couplings:

   (*a*) By a suitable arrangement of mass and elasticity distribution so that the elastic axis, the inertia axis, and the line of aerodynamic centers are as close to each other as possible.

   (*b*) By addition of masses to achieve dynamic mass balancing.

   (*c*) By arrangement of mass and elasticity so that the lower modes of free oscillation of the structure do not have a nodal line close to the $^3/_4$-chord point.

4. Provide servomechanisms to control the phase relationship between various components of motion.

To put any of these measures on a quantitative basis, a detailed analysis of the special type of structure under consideration is necessary.

A successful aeroelastic design secures stability without adding much material to the structure in excess of what is required to carry the live and

dead loads for which the structure is intended.   Proper mass distribution
is of supreme importance.   For example, in a particular transport design,
it is found that the mass of the fuel in the outer panel of the wing (near
the wing tip) has a very strong destabilizing effect; and a more economical
design results if the fuselage or the inner panel of the wing is made larger
to carry the excess fuel so that the wing tip region will be relieved of heavy
masses.   Considerations of this nature indicate clearly that a comprehensive flutter analysis which is made at the early stages of an airplane design, and which takes into account a wide range of variations of structural
parameters, can be of vital importance.

Aircraft design has advanced to a stage where the safeguarding against
aeroelastic instabilities demands as much attention as providing sufficient
strength for flight and ground loads.   One wants an airplane of minimum
weight that has structural integrity for prescribed design requirements
(loads, geometry, etc.).   Although in the early stages of aviation history
the structural integrity against flutter could be achieved by minor changes
in the design, with little cost in weight, the new trend toward optimum
design of high-performance airplanes creates a very different picture.

We have seen that, other things remaining equal, the rigidity of the
structure is the ultimate safeguard against flutter.   Although the rigidity
depends on the manner in which the structure is fabricated, the ultimate
limitation always lies with the materials of construction.

In comparing different materials for aircraft construction, an interesting
criterion is the speed of propagation of sound in the material.   To realize
this let us consider two airplanes identical in geometry and construction,
but differing in material.   Let the density of the two materials be σ and σ'
and their Young's moduli be E and E', respectively.   For dynamic
similarity the dimensionless parameters σ/ρ, $E/(\rho U^2)$ must have the same
values for both airplanes.   Hence,

$$\frac{\sigma'}{\sigma} = \frac{\rho'}{\rho} = \frac{E'}{E}\left(\frac{U}{U'}\right)^2 \tag{1}$$

or

$$\left(\frac{U'}{U}\right)^2 = \left(\frac{E'}{\sigma'}\right)\Big/\left(\frac{E}{\sigma}\right) \tag{2}$$

In other words, the higher the E/σ ratio of a material, the higher will
be the critical speed.   Since the sound speed in a material is proportional
to $\sqrt{E/\sigma}$, we may say that the critical speed is directly proportional to the
speed of sound in the material of construction.

The speeds of sound $\sqrt{E/\sigma}$ at room temperature in several materials are
approximately as given in Table 7.1.   The speeds of sound in most

structural metals and wood are surprisingly close to each other.   Plastics
have lower E/σ ratios.

Table 7.1

	E, Tension, 10^6 psi	$E/\sigma g$, 10^6 in.	Speed of Sound, 10^3 ft per sec
Metals			
Steels	30	100	16.3
Aluminum alloys	10	102	16.5
Magnesium alloys	6.5	103	16.6
Titanium	16	100	16.3
Molybdenum	46	125	18.2
Tungsten carbide	96	177	21.8
Titanium carbide	55	280	27.4
Beryllium	42	640	41.4
Wood (veneer)			
Spruce	1.3	75	14.1
Mahogany	1.5	59	12.5
Balsa (at 8.8 lbs per ft^3)	0.5	98	16.2
Plastics			
Cellulose acetate	0.22	4.8	3.6
Vinylchloride acetate	0.46	0.5	5.1
Phenolic laminates	1.23	25.6	8.2

BIBLIOGRAPHY

In addition to the references listed in the two preceding chapters, the following
references can be quoted.

Extensive bibliographies to published and unpublished papers are found in the
following general reviews:

7.1   Flax, A. H.: Aeroelastic Problems at Supersonic Speed. *2d Intern. Aeronaut.
      Conf. N.Y.*, 322–360 (1949).   Institute of Aeronautical Sciences, New York.

7.2   Garrick, I. E.: Some Research on High-Speed Flutter. *3d Anglo-Am. Aeronaut.
      Conf. Brighton*, 419–446 (1951).   Royal Aeronautical Society, London.

7.3   Küssner, H. G., and P. Jordan, et al.: Aeroplane Flutter. Section G of *Monographs
      on German Aeronautical Research since* 1939.   AVA Göttingen for Ministry of
      Supply.   Brit. *Ministry of Aircraft Production Translations* MAP-VG 24, 156,
      172, 187, 311, 312, 313, 314. MAP Völkenrode. See also, *FIAT* (Field
      Information Agency, Technical) *Review of German Sciences, 1939–1946, Review
      of Aerodynamics*, Chapter VI, by H. G. Küssner and H. Billing.   Office of
      Military Government for Germany (1948).

7.4   Williams, J.: Aircraft Flutter. *Aeronaut. Research Council R. & M.* 2492 (1951).

For damping and oscillation tests, see

7.5 Coleman, R. P.: Damping Formulas and Experimental Values of Damping in Flutter Models. *NACA Tech. Note* **751** (1940).

7.6 Cooper, D. H. D.: A Suggested Method of Increasing the Damping of Aircraft Structures. *Aeronaut. Research Council R. & M.* **2398** (1946).

7.7 Kennedy, C. C., and C. D. P. Pancu: Vector Methods in Vibration Measurements and Analysis. *J. Aeronaut. Sci.* **14**, 603–625 (Nov. 1947).

7.8 Lyman, C. B.: Airplane Vibration Tests as Related to the Flutter Problem. *J. Aeronaut. Sci.* **9**, 24–30 (1941).

7.9 Plantin, C. P.: The Technique of Resonance Testing and Flutter Calculations as Applied to Fighter Aircraft Design. *J. Roy. Aeronaut. Soc.* **56**, 117–137 (1952).

7.10 Plunkett, R.: Vibration Damping. *Applied Mech. Rev.*, 313 (July 1953), 56 refs. Survey of literature.

7.11 Robertson, J. M., and A. J. Yorgiadis: Internal Friction in Engineering Materials. *J. Applied Mech.* **13**, A-173–182 (1946).

7.12 Schlippe, B. von: Die innere Dampfung, Berechnungsansatze. *Ing. Arch.* **6**, 127 (1935).

7.13 Soroka, W. W.: Note on the Relations between Viscous and Structural Damping Coefficients. *J. Aeronaut. Sci.* **16**, 409–410 (1949). Comments by W. Pinsker, p. 699.

7.14 de Vriess, G.: Contribution to the Determination of Oscillation Qualities of Aeroplanes by Ground Tests, Especially Considering the New Method of Measuring the Phases. *Forsch. Ber.* **1882**. See also *Tech. Ber.* **11**, 165, 441 (1944). *Zentr. Wiss. Ber. Luftfahrt-Forsch. des General luftzeugmeisters*, Berlin.

The "complementary energy" method is discussed in the following:

7.15 Reissner, E.: Complementary Energy Procedure for Flutter Calculations. *J. Aeronaut. Sci.* **16**, Readers' Forum, 316 (1949). Comments by Greidanus and van de Vooren, **17**, 454 (1950), and by Libby and Sauer, **16**, 700 (1949).

Wing-flutter analysis:

7.16 Barnes, R. H.: An Analysis of the Effect of a Power Boost System on Wing-Torsion Control Surface Flutter. *Inst. Aeronaut. Sci. Preprint* **345** (1952).

7.17 Bergh, H., and J. Ijff: Application of Experimental Aerodynamic Coefficients to Flutter Calculations. *Natl. Luchtvaartlab., Amsterdam Rept.* **F122** (1952).

7.18 Biot, M. A.: Flutter Analysis of a Wing Carrying Large Concentrated Weights. *GALCIT Flutter Rept.* **1A** (1941). California Institute of Technology.

7.19 Bleakney, W. M.: Three-Dimensional Flutter Analysis. *J. Aeronaut. Sci.* **9**, 56–63 (Dec. 1941).

7.20 Buxton, G. H. L., and I. T. Minhinnick: Expressions for the Rates of Change of Critical Flutter Speeds and Frequencies with Inertial, Aerodynamic, and Elastic Coefficients. *Aeronaut. Research Council R. & M.* **2444** (1945).

7.21 Chopin, S.: Influence de Divers Paramètres sur les Vitesses Critiques des Avions. *Recherche Aéronaut.* **4**, 41–46 (Mar.–Apr. 1952).

7.22 Cicala, P.: Comparison of Theory with Experiment in the Phenomenon of Wing Flutter. *NACA Tech. Memo.* **887** (1939). Translated from *Aerotecnica*, **18**, 412–433 (Apr. 1938).

7.23 Duncan, W. J., and H. M. Lyon: Calculated Flexural-Torsional Flutter of some Typical Cantilever Wings. *Aeronaut. Research Com. R. & M.* **1782** (1937).

7.24 Duncan, W. J., and C. L. T. Griffith: The Influence of Wing Taper on the Flutter of Cantilever Wings. *Aeronaut. Research Com. R. & M.* **1869** (1939).
7.25 Duncan, W. J.: The Representation of Aircraft Wings, Tails, and Fuselages by Semi-rigid Structures in Dynamic and Static Problems. *Aeronaut. Research Council R. & M.* **1904** (1943).
7.26 Ellenberger, G.: Berechnung der kritischen Geswindigkeit für das ebene Problem eines Tragflügels mit Querruder. *Luftfahrt-Forsch.* **15**, 395–405 (1938).
7.27 Falkner, V. M.: The Effect of Variation of Aileron Inertia and Damping on the Flexure-Aileron Flutter of a Typical Cantilever Wings. *Aeronaut. Research Com. R. & M.* **1685** (1935).
7.28 Flax, A. H.: Three-Dimensional Wing Flutter Analysis. *J. Aeronaut. Sci.* **10**, 41–47 (1943).
7.29 Frazer, R. A.: The Influence of Differential Aileron Control on Wing Flutter. *Aeronaut. Research Com. R. & M.* **1723** (1936).
7.30 Frazer, R. A., and W. P. Jones: Forced Oscillations of Aeroplanes with Special Reference to von Schlippe's Method of Predicting Critical Speeds for Flutter. *Aeronaut. Research Com. R. & M.* **1795** (1937).
7.31 Frazer, R. A., and S. W. Skan: A Comparison of the Observed and Predicted Flexure-Torsion Flutter Characteristics of a Tapered Model Wing. *Aeronaut. Research Council R. & M.* **1943** (1941).
7.32 Frazer, R. A.: On the Influence of Chordwise Flexibility on Wing Flutter. *Aeronaut. Research Council R. & M.* **1954** (1940).
7.33 Frazer, R. A.: The Prevention of Binary Flutter by Artificial Damping. *Aeronaut. Research Council R. & M.* **2552** (1944).
7.34 Frazer, R. A., W. P. Jones, C. Scruton, D. V. Dunsdon, and P. M. Ray: Influence of Tuned Dampers on Flexure-Aileron Flutter. I, Theoretical Investigation. II, Flutter Calculations. III, Experiments. *Aeronaut. Research Council R. & M.* **2559** (1946).
7.35 Goland, M.: The Flutter of a Uniform Cantilever Wing. *J. Applied Mech.* **12** (Dec. 1945). *Trans. ASME* **67**, A197–208 (1945).
7.36 Goland, M., and Y. L. Luke: The Flutter of a Uniform Wing with Tip Weights. *J. Applied Mech.* **15** (Mar. 1948). *Trans. ASME* **70**, A13–20 (1948).
7.37 Goland, M., and M. A. Dengler: Comparison between Calculated and Observed Flutter Speeds. *U.S. Air Force Tech. Rept.* **6184** (1950).
7.38 Graham, M. E. K.: Estimation of Wing Flutter Speeds from the Curves of R. & M. 1869. *Aeronaut. Research Council R. & M.* **2608** (1942).
7.39 Jones, W. P.: Effect of Flexurally-Geared Aileron Control on Binary Flutter of a Wing-Aileron System. *Aeronaut. Research Council R. & M.* **2362** (1944).
7.40 Jones, W. P.: Antisymmetrical Flutter of a Large Transport Aeroplane. *Aeronaut. Research Council R. & M.* **2363** (1944).
7.41 Jones, W. P.: Wing Fuselage Flutter of Large Aeroplanes. *Aeronaut. Research Council R. & M.* **2656** (1953).
7.42 Küssner, H. G.: Status of Wing Flutter. *NACA Tech. Memo.* **782** (1936). Translated from *Luftfahrt-Forsch.* **12**, 193–209 (Oct. 1935).
7.43 Laasonen, P.: On the Theory of Flutter and an Iterative Method of Calculating the Critical Speed of a Wing. *Tech. Note KTH-Aero.* **TN 11**, *Kgl. Tek. Högskol. Sweden* (1950).
7.44 Leiss, K.: Einfluss der einzelnen Baugrössen auf das Flattern und das aperiodische Auskippen von Tragflächen mit und ohne Ruder. *Jahrb. 1938 deut. Versuchsanstalt Luftfahrt*, 276–296.

7.45 Lombard, A. E., Jr.: An Investigation of the Conditions of Occurence of Flutter in Aircraft, and the Development of Criteria for the Prediction and Elimination of Such Flutter. Ph.D. thesis, California Institute of Technology (1938).

7.46 Loring, S. J.: General Approach to the Flutter Problem. *SAE J.* **49**, 345–356 (1941).

7.47 Loring, S. J.: Use of Generalized Coordinates in Flutter Analysis. *SAE J.* **52**, 113–132 (1944).

7.48 Mazet, R.: Application de la méthode de "l'effet d'accompagnement" à la détermination de la vitesse critique de vol et du degré d'explosiveté du flutter. *Proc. 7th Intern. Congr. Applied Mech.* **4**, 96–108 (1948).

7.49 Naylor, G. A.: An Approximation Simplifying Wing Flutter Calculations. *Aeronaut. Research Council R. & M.* **2605** (1942).

7.50 Nilsson, J. E. V., and N. B. Langefors: Generalized Complex Forces in Flutter Calculations. *J. Aeronaut. Sci.* **18**, *Readers' Forum* 139 (1951).

7.51 Pugsley, A. G., J. Morris, and G. A. Naylor: The Effect of Fuselage Mobility in Roll upon Wing Flutter. *Aeronaut. Research Council R. & M.* **2009** (1939).

7.52 Pugley, A. G.: A Simplified Theory of Wing Flutter. *Aeronaut. Research Council R. & M.* **1839** (1938).

7.53 Rauscher, M.: Station Functions and Air Density Variations in Flutter Analysis. *J. Aeronaut. Sci.* **16**, 345–353 (1949). See also *Theoretical and Experimental Methods of Flutter Analysis*, Vol. 1, Phase 1, *Comparison of Theoretical Methods of Calculating Vibration Modes and Flutter Speeds*. MIT Rept. to Bur. Aeronaut. (Nov. 15, 1948). Comments by Greidanus and van de Vooren: *J. Aeronaut. Sci.* **17**, 178–179 (1950).

7.54 Reismann, H., and G. C. Best: Two-Dimensional Transient Motion and Flutter of a Wing Having Four Degrees of Freedom. *J. Aeronaut. Sci.* **19**, 540–542 (1952).

7.55 Ruggiero, R. J.: Investigation of Three Methods for Solving the Flutter Equation and Their Relative Merits. *J. Aeronaut. Sci.* **13**, 3–22 (1946).

7.56 Rühl, K. H.: Forschungaufgaben über der Sicherheitsgrad und die Schwingungssicherhiet von Flugzeugen. *Lillienthal Ges. Jahrb.* (1936).

7.57 Schaefer, H.: Liestungsbetrachtungen bei Flatteruntersuchung. *Jahrb.* 1939 *deut. Luftfahrt-Forsch*, I 526–534.

7.58 Schallenkamp, A.: Flutter Calculations for Profiles of Small Chord. *Luftfahrt-Forsch*. **19**, 11–12 (1942). R.T.P. Translation 1556.

7.59 Teichmann, A.: State and Development for Flutter Calculation. *NACA Tech. Memo.* **1297** (1951). Translated from *Lillienthal Ges. Ber.* **135** (1941).

7.60 van de Vooren, A. I., and J. H. Greidanus: Diagrams of Critical Flutter Speed for Wings of a Certain Standard Type. *Natl. Luchtvaartlab. Amsterdam Rept.* **V 1297** (1946).

7.61 von de Vooren, A. I.: Diagrams of Flutter, Divergence, and Aileron Reversal Speeds for Wings of a Certain Standard Type. *Natl. Luchtvaartlab. Amsterdam Rept.* **V 1397** (1947).

7.62 Williams, J.: Theory of Wing Flexure-Torsion Flutter. *Aeronaut. Research Council R. & M.* **2274** (1945).

7.63 Woolston, D. S., and H. L. Runyan: Appraisal of Method of Flutter Analysis Based on Chosen Modes by Comparison with Experiment for Cases of Large Mass Coupling. *NACA Tech. Note* **1902** (1949).

7.64 Woolston, D. S., and H. L. Runyan: On the Use of Coupled Modal Functions in Flutter Analysis. *NACA Tech. Note* **2375** (1951).

Wing-flutter experiments (see also Chapter 9):

7.65   Falkner, V. M., W. P. Jones, and C. Scruton: Flutter Experiments on a Model Wing Fitted with a Dead-Centre Aileron Control. *Aeronaut. Research Com. R. & M.* **1686** (1935).

7.66   Falkner, V. M., W. P. Jones, and C. Scruton: The Effect of a Reduction of Aileron Torsional Stiffness on the Flutter of a Model Wing. *Aeronaut. Research Com. R. & M.* **1722** (1935).

7.67   Gayman, W. H.: An Investigation of the Effect of a Varying Tip-Weight Distribution on the Flutter Characteristics of a Straight Wing. *J. Aeronaut. Sci.* **19**, 289–301 (1952).

7.68   Jones, W. P., and N. C. Lambourne: Derivative Measurements and Flutter Tests on a Model Tapered Wing. *Aeronaut. Research Council R. & M.* **1945** (1941).

7.69   Hall, A. H.: Wing Flutter Experiments with Variations in Stiffness and Distorsion Form. *Natl. Research Council Canada Aeronaut. Rept.* **AR-6** (1948).

7.70   Scruton, C.: Interim Report of Flexure-Aileron Flutter Tests on a Model of B.A.C. Wing Type 167. The Effect of Artificial Aileron Damping. *Aeronaut. Research Council R. & M.* **2480** (1944).

7.71   Sezawa, K.: The Nature of Wing Flutter as Revealed through Its Vibrational Frequencies. *J. Aeronaut. Sci.* **4**, 30–34 (1936).

7.72   Voigt, H.: Wing-Tunnel Investigations on Flexural-Torsional Wing Flutter. *NACA Tech. Memo.* **877** (1938). Translated from *Luftfahrt-Forsch.* **14**, 427–433 (1937).

7.73   Voigt, H.: Weitere Versuche über Tragflügelschwingungen. *Jahrb. 1938 deut. Versuchsanstalt Luftfahrt*, 71–80.

7.74   Williams, J.: An Examination of Experimental Data Relating to Flexural-Torsional Wing Derivatives. *Aeronaut. Research Council R. & M.* **1944** (1943).

Tail flutter:

7.75   Broadbent, E. G., and W. T. Kirkby: Control Surface Flutter. *J. Roy. Aeronaut. Soc.* **56**, 355–381 (May 1952).

7.76   Jahn, H. A., G. H. L. Buxton, and I. T. Minhinnick: Fuselage Vertical Bending-Elevator Flutter on the Typhoon. *Aeronaut. Research Council R. & M.* **2121** (1944).

7.77   Scruton, C. (editor): Experiments on Tail Flutter. *Aeronaut. Research Council R. & M.* **2323** (monograph), 1–137 (1948).

7.78   Smilg, B.: The Response of Control Surfaces to Vibratory Excitation in Still Air. *U.S. Air Material Command Tech. Rept.* **4561** (1940).

7.79   Smilg, B.: A Statistical Survey of the Flutter of Airplane Control Surfaces. *U.S. Air Material Command Tech. Rept.* **4595** (1940).

7.80   Thomson, W. T.: Charts for Fuselage Torsion versus Control-Surface Flutter. *Trans. ASME* **68**, 51–56 (1946).

Tab flutter:

7.81   Buxton, G. H. L., and G. D. Sharpe: The Effect of Tab Mass-Balance on Flutter. I, Ternary Tailplane–Elevator-Tab Flutter. *Aeronaut. Research Council R. & M.* **2418** (1946).

7.82   Collar, A. R.: The Prevention of Flutter of Spring Tabs. *Aeronaut. Research Council R. & M.* **2034** (1943).

7.83 Collar, A. R., and G. D. Sharpe: A Criterion for the Prevention of Spring-Tab Flutter. *Aeronaut. Research Council R. & M.* **2637** (1946).

7.84 Naylor, G. A., and A. Pellew: Binary Aileron-Spring Tab Flutter. *Aeronaut. Research Council R. & M.* **2576** (1942).

7.85 Naylor, G. A.: Flutter of Control Surface Tabs. *Aeronaut. Research Council R. & M.* **2606** (1942).

7.86 Scruton, C., P. Ray, and D. V. Dunsdon: The Effect of Tab Mass-Balance on Flutter. II, Experiments on the Influence of Tab Mass-Balance on Flutter. *Aeronaut. Research Council R. & M.* **2418** (1946).

7.87 van de Vooren, A. I.: The Treatment of a Tab in Flutter Calculations, including a Complete Account of Aerodynamic Coefficients. *Natl. Luchtvaartlab. Amsterdam Rept.* **V1386** (1947).

7.88 van de Vooren, A. I.: Ternary Wing Bending–Aileron-Tab Flutter. *Natl. Luchtvaartlab. Amsterdam Rept.* **F86** (1951).

7.89 Voight, H., and F. Walter: Flutter Characteristics of a Wing Equipped with a Flettner Servo-tab. *U.S. Air Force Tech. Rept.* **6182** (1950). Translated from *Deut. Luftfahrt-Forsch. ForschBer.* **1204** (1940).

7.90 Voight, H., F. Walter, and W. Heger: Flutter of Control Systems with Geared Tabs. *U.S. Air Force Tech. Rept.* **6183** (1950). Translated from Instutut für Aerodynamik der deutschen Versuchsanstalt für Luftfahrt (Dec. 1940).

7.91 Wasserman, L. S., W. J. Mykytow, and I. Spielberg: Tab Flutter Theory and Applications. *U.S. Air Force Tech. Rept.* **5153** (Sept. 1944).

7.92 Wittmeyer, H.: Theoretical Investigations of Ternary Lifting Surface–Control Surface–Triming Tab Flutter and Derivation of a Flutter Criterion. *Aeronaut. Research Council R. & M.* **2671** (1952).

Servo-controlled aircraft and many degrees of freedom (see also § 6.12):

7.93 van de Vooren, A. I.: Theory and Practice of Flutter Calculations for Systems with Many Degrees of Freedom. *Natl. Luchtvaartlab. Amsterdam Rept.* **F100** (1951).

7.94 Winson, J.: The Flutter of Servo-controlled Aircraft. *J. Aeronaut. Sci.* **16**, 397–404 (1949).

Effect of concentrated masses:

7.95 Lambourne, N. C., and D. Weston: An Experimental Investigation of the Effect of Localized Masses on the Flutter and Resonances of a Model Wing. (Part I, Flutter Tests.) *Aeronaut. Research Council R. & M.* **2533** (1944).

7.96 Runyan, H. L., and J. L. Sewall: Experimental Investigation of the Effects of Concentrated Weights on Flutter Characteristics of a Straight Cantilever Wing. *NACA Tech. Note* **1594** (1948).

7.97 Runyan, H. L., and C. E. Watkins: Flutter of a Uniform Wing with an Arbitrarily Placed Mass according to a Differential Equation Analysis and a Comparison with Experiment. *NACA Tech. Rept.* **966** (1950). *NACA Tech. Note* **1848** (1949).

7.98 Serbin, H., and E. L. Costilow: Application of Response Function to Calculation of Flutter Characteristics of a Wing Carrying Concentrated Masses. *NACA Tech. Note* **2540** (1951).

7.99 Sewall, J. L., and D. S. Woolston: Preliminary Experimental Investigation of Effects of Aerodynamic Shape of Concentrated Weights on Flutter of a Straight Cantilever Wing. *NACA Research Memo. RM* **L9E17** (1949).

Effect of sweepback:

7.100   Babister, A. W.: Flutter and Divergence of Swept-Back and Swept-Forward Wings. Report No. 39, *College Aeronaut. Cranfield Rept.* **39** (June 1950); *Aeronaut. Research Council R. & M.* **2761** (1953).

7.101   Barmby, J. G., H. J. Cunningham, and I. E. Garrick: Study of Effects of Sweep on the Flutter of Cantilever Wings. *NACA Tech. Note* **2121** (1950); *NACA Rept.* **1014** (1951).

7.102   Nelson, H. C., and J. E. Tomassoni: Experimental Investigation of the Effects of Sweepback on the Flutter of a Uniform Cantilever Wing with a Variably Located Concentrated Mass. *NACA RM* **L9F24** (1949).

7.103   Radok, J. R.: Dynamic Aeroelasticity of Aircraft with Swept Wings. *College Aeronaut. Cranfield Rept.* **58** (1952).

7.104   Richards, E. J.: Practical Design Problems Arising from Sweepback. *Aeronaut. Conf. London,* 381–406 (1947). Royal Aeronautical Society, London.

Small-aspect ratio:

7.105   Halfman, R. L., and H. Ashley: Aeroelastic Properties of Slender Wings. *Proc. 1st U.S. Natl. Congr. Applied Mech.,* 907–916 (June 11–16, 1951). ASME, New York.

Effect of the compressibility of air:

7.106   Broadbent, E. G.: Flutter Problems of High-Speed Aircraft. *2d Intern. Aeronaut. Conf. N.Y.,* 556–581 (1949). Institute of Aeronautical Sciences.

7.107   Castile, G. E., and R. W. Herr: Some Effects of Density and Mach Number on the Flutter Speed of Two Uniform Wings. *NACA Tech. Note* **1989** (1949).

7.108   Erickson, A. L., and R. L. Mannes: Wing-Tunnel Investigation of Transonic Aileron Flutter. *NACA RM* **A9B28** (1949).

7.109   Frazer, R. A., and S. W. Skan: Influence of Compressibility on the Flexural-Torsional Flutter of Tapered Cantilever Wings. *Aeronaut. Research Council R. & M.* **2553** (1942).

7.110   Garrick, I. E.: Bending-Torsion Flutter Calculations Modified by Subsonic Compressibility Corrections. *NACA Rept.* **836** (1946); *NACA Tech. Note* **1034** (1946).

7.111   Ijff, J.: Influence of Compressibility on the Calculated Flexure-Torsion Flutter Speed of a Family of Rectangular Cantilever Wings. *Natl. Luchtvaartlab. Amsterdam Rept.* **F118** (1952).

7.112   Smilg, B.: The Prevention of Aileron Oscillations at Transonic Airspeeds. *U.S. Air Force Tech. Rept.* **5530** (1946).

7.113   Widmayer, E. Jr., W. T. Lauten Jr., and S. A. Clevensen: Experimental Investigation of the Effect of Aspect Ratio and Mach Number on the Flutter of Cantilever Wings. *NACA RM* **L50C15a** (1950).

7.114   Woolston, D. S., and G. E. Castile: Some Effects of Variations in Several Parameters Including Fluid Density on the Flutter Speed of Light Uniform Cantilever Wings. *NACA Tech. Note* **2558** (1951).

7.115   Woolston, D. S., and V. Huckel: A Calculation Study of Wing-Aileron Flutter in Two Degrees of Freedom for Two-Dimensional Supersonic Flow. *NACA Tech. Note* **3160** (1954).

Effect of small variation of parameters:

7.116   Serbin, H.: The Response of an Aerodynamic System under External Harmonic Force. *J. Applied Phys.* **22**, 1307–1315 (1951).

270	ENGINEERING FLUTTER ANALYSIS

7.117	van de Vooren, A. I.: A Method to Determine the Change in Flutter Speed Due to Small Changes in the Mechanical System. *Natl. Luchtvaartlab. Amsterdam Rept.* V1366 (1947).

7.118	van de Vooren, A. I.: The Change in Flutter Speed Due to Small Variations in Some Aileron Parameters. *Natl. Luchtvaartlab. Amsterdam Rept.* V1380 (1947).

Panel flutter:

7.119	Fung, Y. C.: The Static Stability of a Two-dimensional Curved Panel in a Supersonic Flow, with an Application to Panel Flutter. *J. Aeronaut. Sci.* 21, 556–565 (1954).

7.120	Goland, M., and Y. L. Luke: An Exact Solution for Two-Dimensional Linear Panel Flutter at Supersonic Speeds. *J. Aeronaut. Sci.* 21, 275–276 (Apr. 1954).

7.121	Hedgepeth, J. M., B. Budiansky, and R. W. Leonard: Analysis of Flutter in Compressible Flow of a Panel on Many Supports. *J. Aeronaut. Sci.* 21, 475–486 (1954).

7.122	Shen, S. F.: Flutter of a Two-Dimensional Simply Supported Uniform Panel in a Supersonic Stream. *MIT Aeronaut. Engg. Dept. Rept.* to Office of Naval Research, Contract N5 ori-07833 (1952).

Propeller and helicopter blades vibration and flutter.

7.123	Baker, J. E., and R. S. Paulnock: Experimental Investigation of Flutter of a Propeller with Clark Y Section Operating at Zero Forward Velocity at Positive and Negative Blade-Angle-Settings. *NACA Tech. Note* 1966 (1949).

7.124	Clifton, F., and L. H. G. Sterne: Strain Gauge Test of a Fluttering Propeller in a Wind Tunnel. *Aeronaut. Research Council R. & M.* 2072 (1943).

7.125	DeGroff, H.: Aerodynamic Forces on a Propeller in Non-Stationary Motion. Ph.D. thesis. California Institute of Technology (1949).

7.126	Ewing, H. G., J. Kettlewell, and D. R. Gaukroger: Comparative Flutter Tests on Two-, Three-, Four-, and Five-Blade Propellers. *Aeronaut. Research Council R. & M.* 2634 (1952).

7.127	Ficker, G.: The Flutter of Helicopter Blades. *U.S. Air Force Tech. Rept.* F-TR-1177-ND (1948).

7.128	Forshaw, J. R., H. B. Squire, and F. J. Bigg: Vibration of Propellers Due to Non-uniform Inflow. *Aeronaut. Research Council R. & M.* 2054, part II (1942).

7.129	Hohenemser, K.: Flattern von Drehflügeln im Standlauf. *Ing. Arch.* 10, 133–143 (1939).

7.130	Inglesby, J. V.: Flutter of Propeller Blades. *J. Roy. Aeronaut. Soc.* 50, 98–118 (1946).

7.131	Kettlewell, J., and H. Ewing: Strain Gauge Flutter Tests on a 4-Blade Propeller with Duralumin Blades. *Aeronaut. Research Council R. & M.* 2471 (1952).

7.132	Lilley, G. M.: An Investigation of the Flexure-Torsion Flutter Characteristics of Airfoils in Cascade. *College Aeronautics Cranfield Rept.* 60 (1952).

7.133	Mendelson, A.: Effect of Centrifugal Force on Flutter of Uniform Cantilever Beam at Subsonic Speeds with Application to Compressor and Turbine Blades. *NACA Tech. Note* 1893 (1949).

7.134	Minhinnick, I. T.: An Investigation of the Flutter Speed of an Airscrew Blade Taking into Account the Chordwise Flexibility and the Twist of the Blade. *Aeronaut. Research Council R. & M.* 2073 (1943).

7.135	Morris, J.: Airscrew Blade Vibration. *Aeronaut. Research Com. R. & M.* 1835 (1937).

7.136 Plunkett, R. T.: Free and Forced Vibrations of Rotating Blades. *J. Aeronaut. Sci.* **18**, 278–282 (1951).

7.137 Postlethwaite, F., B. C. Carter, W. G. A. Perring, and K. V. Diprose: Permissible Proximity of a Propeller to the Leading Edge of a Wing, as Decided by Propeller Blade Vibration. *Aeronaut. Research Council R. & M.* **2054**, part I (1942).

7.138 Rosenberg, R.: Aero-Elastic Instability in Unbalanced Lifting Rotor Blades. *J. Aeronaut. Sci.* **11**, 361–368 (1944).

7.139 Shannon, J. F.: Vibration Problems in Gas Turbines, Centrifugal and Axial Flow Compressors. *Aeronaut. Research Council R. & M.* **2226** (1945).

7.140 Sterne, L. H. G.: Spinning Tests on Fluttering Propellers. *Aeronaut. Research Council R. & M.* **2022** (1945).

7.141 Sterne, L. H. G., and R. H. Brown: The Elimination of Flutter from a Propeller. *Aeronaut. Research Council R. & M.* **2047** (1943).

7.142 Sterne, L. H. G., H. G. Ewing, and J. Kettlewell: Strain Gauge Investigation of Propeller Flutter. *Aeronaut. Research Council R. & M.* **2472** (1951).

7.143 Theodorsen, Th., and A. A. Regier: Effect of the Lift Coefficient on Propeller Flutter. *NACA Wartime Rept.* **L-161**, ACR L5F30 (1945).

7.144 Turner, M. J., and J. B. Duke: Propeller Flutter. *J. Aeronaut. Sci.* **16**, 323–336, 638–639 (1949).

Application of digital and analog computing machines:

7.145 Barrois, W., and J. Simon-Suisse: Mécanisation des Problèmes de Vibrations et de Flutter par le Calcul Matriciel. *Proc. 7th Intern. Congr. Applied Mech.* **4**, 63–80 (1948).

7.146 Baird, E. F., and H. J. Kelley: Formulation of the Flutter Problem for Solution on an Electronic Analog Computer. *J. Aeronaut. Sci.* **17**, 189–190 (1950).

7.147 Bell, W. D.: A Simplified Punch-Card Approach to the Solution of the Flutter Determinant. *J. Aeronaut. Sci.* **15**, 121–122 (1948).

7.148 Biot, M. A., and T. H. Wianko: Electric Network Model for Flexure Torsion Flutter. *GALCIT Flutter Rept.* **3** (1941). California Institute of Technology.

7.149 Biot, M. A., and T. H. Wianko: Theory of Electrical Flutter Predictor for Three Degrees of Freedom. *GALCIT Flutter Rept.* **8** (1943). California Institute of Technology.

7.150 Dennis, E. A., and D. G. Dill: Application of Simultaneous Equations Machines to Aircraft Structure and Flutter Problems. *J. Aeronaut. Sci.* **17**, 107–113 (1950).

7.151 Faure, G., J. Simon-Suisse, and Th. Rona: Deux Circuits analogiques pour l'inversion des matrices symétriques et la recherche de la vitesse critique de flutter. *Proc. 7th Intern. Congr. Applied Mech.* **4**, 81–95 (1948).

7.152 Landahl, M. T., and J. E. Stark: An Electrical Analogy for Solving the Oscillating-Surface Problem for Incompressible Nonviscid Flow. *Roy. Inst. Technol. Stockholm Rept. KTH AERO* **TN 34** (1953).

7.153 Leppert, E. L., Jr.: An Application of IBM Machines to the Solution of the Flutter Determinant. *J. Aeronaut. Sci.* **14**, 171–174 (Mar. 1947).

7.154 MacNeal, R. H., G. D. McCann, and C. H. Wilts: The Solution of Aeroelastic Problems by Means of Electrical Analogies. *J. Aeronaut. Sci.* **18**, 777–789 (1951).

7.155 Winson, J.: The Solution of Aeroelastic Problems by Electronic Analogue Computation. *J. Aeronaut. Sci.* **17**, 385–395 (1950).

Chapter 8

TRANSIENT LOADS, GUSTS

Transient loads occur in aeronautics either through airplane maneuver or through external excitations such as gusts and landing impacts. In general, the determination of the loading, as well as that of the response, requires the solution of the equations of motion of the aircraft.

In many dynamic-stress problems, of which the gust loads on aircraft structures is one, the time history of the external load assumes such a wide variety of shapes and magnitudes that any particular solution, derived with respect to a special load-time history, cannot characterize the whole situation. When an attempt is made to measure the external load-time history on an actual airplane flying through a turbulent atmosphere, the statistical nature of the gust problem is revealed. How to derive from the experimental data on atmospheric turbulences the statistical information that is useful in airplane design is an interesting problem. How to predict, theoretically, the airplane responses (acceleration, inertia load, stresses, etc.) with respect to such statistical information of atmospheric turbulences is of practical importance.

Some mathematical concepts useful in the dynamic-stress analysis will be discussed in § 8.1. The unit-step and unit-impulse functions, the indicial admittance, complex impedance, Duhamel integral, etc., are briefly explained. The response of an airplane to a gust of specified profile is treated in § 8.2. From § 8.3 on, the statistical aspects of the dynamic-stress problems are considered. In § 8.4, the concepts of the mathematical probability and distribution functions are explained. In § 8.5, the question of choosing proper statistical averages to be measured and calculated is considered.

As an illustrative example, the problem of gust loading is discussed. In § 8.6, the mean square value of the response is calculated on the basis of the power spectrum of the excitation. The interpretation and use of such results are discussed in § 8.7.

8.1 SOME MATHEMATICAL CONCEPTS

The Concept of Complex Impedance and Admittance. In § 1.8, it is shown that the steady-state solution of the equation

$$m\frac{d^2x}{dt^2} + \beta\frac{dx}{dt} + Kx = F_0 e^{i\omega t} \qquad (\beta > 0) \qquad (1)$$

may be written as

$$x(t) = \frac{F_0 e^{i\omega t}}{Z(i\omega)} \qquad (2)$$

where

$$Z(i\omega) = m(i\omega)^2 + \beta(i\omega) + K \qquad (3)$$

Equations 1 and 2 may be compared to the relations between the electromotive force $E(t)$ and the current $I(t)$ flowing in an electric circuit:

$$L\frac{dI}{dt} + RI + \frac{1}{C}\int_0^t I\, dt = E$$

$$I(t) = E(t)/Z \qquad (4)$$

where Z is the *impedance of the circuit* which depends on the frequency ω of the electromotive force, the inductance L, resistance R, and the capitance C of the circuit:

$$Z = i\omega L + R + \frac{1}{i\omega C} \qquad (5)$$

By the anology between Eqs. 2 and 4, $Z(i\omega)$ is called the *impedance* of the system represented by Eqs. 1. If Eq. 1 represents a mechanical system, $Z(i\omega)$ may be called the *mechanical impedance*. Its inverse, $1/Z(i\omega)$, is called the *admittance*.

The response of a linear system to an exciting force that varies harmonically is simply given by multiplying the exciting force by the admittance.

In engineering literature a different notation is often used by putting $s = i\omega$ and writing $Z(s)$ instead of $Z(i\omega)$. Note that $Z(s)$ can be obtained directly from the differential equation by a formal process of replacing the operator d/dt by the symbol s, and $\int_0^t dt$ by $1/s$.

The rules for calculating the impedance of a circuit are the same as those for combining the electrical resistance. Thus (Fig. 8.1), if Z_1 and Z_2 are in series, the resultant impedance is

$$Z = Z_1 + Z_2 \qquad (6)$$

On the other hand, if Z_1 and Z_2 are in parallel, the resultant Z is given by the equation

$$\frac{1}{Z} = \frac{1}{Z_1} + \frac{1}{Z_2} \qquad (7)$$

The Principle of Superposition. Since the differential equation 1 is linear, the principle of superposition holds. In particular, if the right-hand side of Eq. 1 is

$$F_1(t) + F_2(t) = F_{10}e^{i\omega t} + F_{20}e^{i2\omega t}$$

a particular solution will be

$$x(t) = \frac{F_1}{Z(i\omega)} + \frac{F_2}{Z(i2\omega)}$$

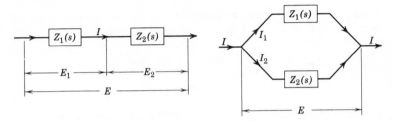

Two impedances in series   Two impedances in parallel

Fig. 8.1. Rules for combining impedances.

More generally, if the right-hand side is given by a Fourier series

$$F(t) = \sum_{n=-\infty}^{\infty} c_n e^{in\omega t} \tag{8}$$

then

$$x(t) = \sum_{n=-\infty}^{\infty} \frac{c_n}{Z(in\omega)} e^{in\omega t} \tag{9}$$

As a further generalization, if $F(t)$ is represented by a Fourier integral

$$F(t) = \frac{1}{\sqrt{2\pi}} \int_{-\infty}^{\infty} A(\omega)e^{i\omega t} \, d\omega \tag{10}$$

where

$$A(\omega) = \frac{1}{\sqrt{2\pi}} \int_{-\infty}^{\infty} F(t)e^{-i\omega t} \, dt$$

then a particular solution is

$$x(t) = \frac{1}{\sqrt{2\pi}} \int_{-\infty}^{\infty} \frac{A(\omega)}{Z(i\omega)} e^{i\omega t} \, d\omega \tag{11}$$

These processes can be justified if the right-hand sides of Eqs. 9 and 11 converge (see Chapters IX and X of Kármán and Biot[1.54]).

The Unit-Step and the Unit-Impulse Functions.   A *unit-step function* $\mathbf{1}(t)$ is a function defined as follows (Fig. 8.2)

$$\begin{aligned}
\mathbf{1}(t) &= 0 \quad \text{for} \quad t < 0 \\
&= \tfrac{1}{2} \quad \text{for} \quad t = 0 \quad\quad\quad (12) \\
&= 1 \quad \text{for} \quad t > 0
\end{aligned}$$

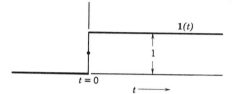

Fig. 8.2.   Unit step function.

If the point of jump is moved to the point $t = \tau$, the unit-step function is written as $\mathbf{1}(t - \tau)$ for which

$$\begin{aligned}
\mathbf{1}(t - \tau) &= 0 \quad \text{for} \quad t < \tau \\
&= \tfrac{1}{2} \quad \text{for} \quad t = \tau \quad\quad\quad (13) \\
&= 1 \quad \text{for} \quad t > \tau
\end{aligned}$$

A *unit-impulse function* $\delta(t)$ is a function that is zero for $t < -\varepsilon$ and $t > \varepsilon$, ε being any positive number, but tends to ∞ when $t = 0$, and

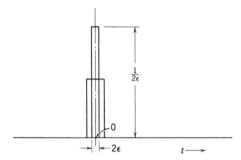

Fig. 8.3.   Unit impulse function.

the integral of $\delta(t)$ taken over the interval $-\varepsilon$ to ε is equal to 1 (Fig. 8.3). Thus, if $\delta(t)$ denotes the unit-impulse function, then

$$\begin{aligned}
\delta(t) &= 0 \quad \text{for} \quad t \neq 0 \\
\lim_{\varepsilon \to 0} \int_{-\varepsilon}^{\varepsilon} \delta(t)\, dt &= 1
\end{aligned} \quad\quad (14)$$

Whereas the unit-step function is well defined, the unit-impulse function, mathematically speaking, is not.   But, since its introduction by Dirac, the unit-impulse function (often called the Dirac δ-function) has become a powerful tool for physicists.*

Let us quote without proof (for a proof, see, for example, Kármán and Biot,[1.54] pp. 394–396) the following integral representations of the unit-step and unit-impulse functions

$$\mathbf{1}(t) = \frac{1}{2\pi} \lim_{\beta=0} \int_{-\infty}^{\infty} \frac{e^{i\omega t}}{\beta + i\omega} \, d\omega$$

$$= \frac{1}{2} + \frac{1}{2\pi} \int_{-\infty}^{\infty} \frac{\sin \omega t}{\omega} \, d\omega \tag{15}$$

$$\delta(t) = \lim_{\varepsilon \to 0} \frac{1}{2\pi\varepsilon} \int_{-\infty}^{\infty} \frac{\sin \omega\varepsilon}{\omega} \, e^{i\omega t} \, d\omega \tag{16}$$

Indicial Admittance. The response of a physical system to a unit-step function is called the *indicial admittance.* In some cases the indicial admittance can be found by elementary methods. In general, the method of Laplace transformation (Chapter 10) gives the solution readily.

Example 1. Consider the equations

$$\frac{d^2x}{dt^2} + \omega_0^2 x = \mathbf{1}(t)$$

$$x = \frac{dx}{dt} = 0 \quad \text{when} \quad t = 0 \tag{17}$$

The general solution of the differential equation is

$$x = \frac{1}{\omega_0^2} + C_1 \sin \omega_0 t + C_2 \cos \omega_0 t$$

Using the initial conditions, we find $C_1 = 0$, $C_2 = -1/\omega_0^2$. Hence, the indicial admittance is

$$A(t) = x = \frac{1}{\omega_0^2} (1 - \cos \omega_0 t) \, \mathbf{1}(t) \tag{18}$$

Example 2. Consider the same problem with the right-hand side replaced by a unit-impulse function.

$$\frac{d^2x}{dt^2} + \omega_0^2 x = \delta(t)$$

$$x = \frac{dx}{dt} = 0 \quad \text{for} \quad t \leqslant -\varepsilon \qquad (\varepsilon > 0) \tag{19}$$

---

* The δ function can be defined as a limit of some well-defined functions. See Chapter V of van de Pol and Bremmer.[8.2]

Integrating the differential equation from $-\varepsilon$ to ε, where ε is a small number, we obtain

$$\int_{-\varepsilon}^{\varepsilon} \frac{d^2x}{dt^2}\, dt + \omega_0{}^2 \int_{-\varepsilon}^{\varepsilon} x\, dt = 1 \tag{20}$$

The first integral in the above equation can be written as

$$\int_{-\varepsilon}^{\varepsilon} \frac{d}{dt}\left(\frac{dx}{dt}\right) dt = \int_{-\varepsilon}^{\varepsilon} d\left(\frac{dx}{dt}\right) = \frac{dx}{dt}\Big|_{-\varepsilon}^{\varepsilon} = \left(\frac{dx}{dt}\right)_{t=\varepsilon} \tag{21}$$

since $(dx/dt)_{t=-\varepsilon} = 0$. The second integral in Eq. 20 tends to zero when ε tends to zero, because, in the neighborhood of $t = 0$, $|x|$ is a finite number, say $< K$, and the integral is bounded by $2K\varepsilon$, which vanishes in the limit $\varepsilon \to 0$. Hence, in the limit, Eq. 20 becomes

$$\left(\frac{dx}{dt}\right)_{t=\varepsilon} = 1$$

The given system (Eqs. 19) is then equivalent to the following

$$\frac{d^2x}{dt^2} + \omega_0{}^2 x = 0, \qquad t > 0$$
$$x = 0, \qquad \frac{dx}{dt} = 1, \qquad t = (0+) \tag{22}$$

the solution of which is

$$h(t) = \frac{1}{\omega_0} \sin \omega_0 t\, \mathbf{1}(t) \tag{23}$$

Note that the results of Exs. 1 and 2 are related by the equation

$$h(t) = \frac{dA}{dt}$$

A more general result, valid also when $A(0) \neq 0$, is (see Kármán and Biot,[1.54] p. 402)

$$h(t) = A(0)\, \delta(t) + \frac{dA}{dt} \tag{24}$$

where $A(t)$ is the indicial admittance and $h(t)$ the response to a unit-impulse function.

Duhamel's Integral. When the governing equation is linear, the principle of superposition holds. Thus, the response to the sum of two step functions is the sum of the indicial admittances. In other words, the response to a function

$$c_1\,\mathbf{1}(t) + c_2\,\mathbf{1}(t - \tau)$$

is

$$c_1\,A(t) + c_2\,A(t - \tau)$$

where $A(t)$ is the indicial admittance to $\mathbf{1}(t)$.

Now any function $f(t)$, having a continuous derivative, can be represented in the integral form

$$f(t) = f(0) + \int_0^t \frac{df}{dt}(\tau)\,d\tau$$

$$= f(0)\,\mathbf{1}(t) + \int_0^t \frac{df}{dt}(\tau)\,\mathbf{1}(t - \tau)\,d\tau \qquad (25)$$

Since an integration is the limit of a summation, and since for each element $\frac{df}{dt}(\tau)\,\mathbf{1}(t - \tau)\,d\tau$ the response is

$$\frac{df}{dt}(\tau)\,A(t - \tau)\,d\tau$$

we obtain by the principle of superposition that the response to the function $f(t)$ is

$$x(t) = f(0)\,A(t) + \int_0^t \frac{df}{dt}(\tau)\,A(t - \tau)\,d\tau \qquad (26)$$

Integrating by parts, we obtain an equivalent form:

$$x(t) = f(t)\,A(0) + \int_0^t f(\tau)\,\frac{dA}{dt}(t - \tau)\,d\tau \qquad (27)$$

The integrals in Eqs. 25, 26, and 27 are known as *Duhamel integrals*.

By Eq. 24, the Duhamel integral 27 can be written as

$$x(t) = \int_0^t f(\tau)\,h(t - \tau)\,d\tau \qquad (28)$$

where $h(\tau)$ is the response to a unit-impulse function.

A graphical interpretation of these results is originated by von Kármán and Biot[1.54] and is given in Fig. 8.4 which seems to be self-explanatory.

Relation between the Admittance and the Indicial Admittance. Perhaps the harmonic function is the simplest periodic function, and the unit-step function is the simplest nonperiodic function. Nevertheless, any arbitrary function, under very mild mathematical restrictions on continuity and differentiability, can be resolved, either into simple-harmonic components in the form of a Fourier integral, or into unit-step functions by means of a Duhamel integral. Knowing either the impedance or the indicial admittance, we can derive the response to an arbitrary forcing function by a single integration. Hence, the problem of dynamic responses resolves ultimately into finding either the indicial admittance or the impedance.

In particular, the indicial admittance can be determined from the admittance by an integration, and vice versa. In fact, since the Fourier representation of the unit-step function is Eq. 15, and since the impedance

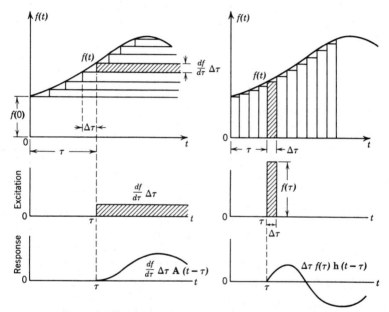

Fig. 8.4. Duhamel integral. Illustrations by von Kármán and Biot, Ref. 1.54. (Courtesy of McGraw-Hill Book Co.).

of a component $e^{i\omega t}/(\beta + i\omega)$ is $(\beta + i\omega)Z(i\omega)$, we can write the indicial admittance $A(t)$ as

$$A(t) = \frac{1}{2\pi} \lim_{\beta \to 0} \int_{-\infty}^{\infty} \frac{e^{i\omega t}}{(\beta + i\omega)Z(i\omega)} \, d\omega \qquad (29)$$

If the physical system is undamped, so that $Z(i\omega)$ vanishes at some real number ω, the integral in Eq. 29 becomes divergent. A convergent result can usually be obtained by introducing a "convergence factor," $\beta > 0$, into the impedance $Z(i\omega)$, and writing

$$A(t) = \frac{1}{2\pi} \lim_{\beta \to 0} \int_{-\infty}^{\infty} \frac{e^{i\omega t}}{(\beta + i\omega)Z(\beta + i\omega)} \, d\omega \qquad (30)$$

This is a Fourier integral.   If we put

$$i\omega = s$$

then

$$A(t) = \frac{1}{2\pi i} \int_{c-i\infty}^{c+i\infty} \frac{e^{st}}{s \, Z(s)} \, ds \tag{31}$$

where c is a positive number greater than the real part of all the roots of $Z(s) = 0$.   But Eq. 31 is the Bromwich's integral which represents the inversion of a Laplace transform (cf. Eq. 2 of § 10.1).   By the transform pair of Laplace transformation, we obtain

$$\frac{1}{s \, Z(s)} = \int_0^\infty e^{-st} A(t) \, dt \tag{32}$$

In other words, $1/s \, Z(s)$ is the Laplace transform of $A(t)$, and $A(t)$ is the inverse Laplace transform of $1/s \, Z(s)$.

8.2 RESPONSE OF AN AIRPLANE TO A GUST OF SPECIFIED PROFILE

One of the critical design conditions for airplane structures is the gust loading, which the airplane encounters when flying through a turbulent atmosphere.   It is customary to assume that the nonuniformity in the flow consists of small disturbances superimposed on a uniform steady flow.   Generally, only the component of the disturbing velocity normal to the flight path is considered.   Such normal disturbances are called gusts.

To study the response of an airplane to a gust, let us make the following assumptions

1. The airplane is rigid.

2. The disturbed motion is symmetrical with respect to the airplane's longitudinal plane of symmetry, but the pitching motion can be neglected.

3. The airplane is initially in horizontal flight at constant velocity U.

4. The gust is normal to the flight path, and is uniform in the spanwise direction.

5. The variation of the forward speed of the airplane can be neglected.

6. The quasi-steady lift coefficient may be used, and the chordwise distribution of the gust velocity may be regarded as constant at any instant and equal to the gust at the mid-chord point.

The disturbed motion of the airplane, consequently, has only the degree of freedom the vertical displacement z (measured at the airplane's center

of mass, positive downward). According to Newton's law, the equation
of motion is

$$m\ddot{z} = -L \tag{1}$$

where m is the total mass of the airplane, L is the total lift (positive
upward), and a dot indicates a differentiation with respect to time. To
derive an expression for the lift L, the gust profile must be specified.
Evidently, it is the gust distribution *relative* to the airplane that is of
significance. Hence, no generality is lost by regarding the gust speed,
$w(t)$, as a function of time. Then, according to assumption 6, the lift
can be written as

$$L = \frac{1}{2}\rho U^2 S \frac{dC_L}{d\alpha}\left(\frac{w}{U} + \frac{\dot{z}}{U}\right) \tag{2}$$

where ρ is the density of the air, and S is the wing area. Using Eq. 2 and
introducing a parameter λ, of physical dimension $[T^{-1}]$,

$$\lambda = \frac{\rho U S}{2m}\frac{dC_L}{d\alpha} \tag{3}$$

we can write Eq. 1 as

$$\ddot{z} = -\lambda(w + \dot{z}) \tag{4}$$

This equation is to be solved for the initial conditions

$$z = \dot{z} = 0 \quad \text{when} \quad t = 0 \tag{5}$$

An integrating factor of Eq. 4 is easily seen to be $e^{\lambda t}$. Equation 4 may
be written as

$$\frac{d}{dt}(\dot{z}e^{\lambda t}) = -\lambda w(t)\,e^{\lambda t}$$

Integrating, and using Eqs. 5, we obtain

$$\dot{z}(t) = -\lambda e^{-\lambda t}\int_0^t w(x)\,e^{\lambda x}\,dx$$

A second integration gives

$$z(t) = -\lambda \int_0^t e^{-\lambda\tau}\,d\tau \int_0^\tau w(x)\,e^{\lambda x}\,dx$$

Changing the order of integration, we have

$$z(t) = -\lambda \int_0^t w(x)\,e^{\lambda x}\,dx \int_x^t e^{-\lambda\tau}\,d\tau$$

i.e.,

$$z(t) = \int_0^t w(x)[e^{-\lambda(t-x)} - 1]\,dx \tag{6}$$

If $w(x)$ is a step function, the so-called *sharp-edged gust*, so that $w(x)$ is equal to a constant w_0 for $x > 0$ and vanishes for $x < 0$ (Fig. 8.5), then an integration of Eq. 6 leads to

$$z(t) = \frac{1}{\lambda} w_0 (1 - e^{-\lambda t}) - w_0 t \qquad (7)$$

and

$$\ddot{z}(t) = - \lambda w_0 \, e^{-\lambda t} \qquad (8)$$

The acceleration reaches the maximum when $t = 0$.

$$\ddot{z}_{\max} = - \lambda w_0 \qquad (9)$$

Dividing $\ddot{z}_{\max}$ by the gravitational acceleration g, we obtain the sharp-edged gust formula

$$\Delta n = \frac{\ddot{z}_{\max}}{g} = \lambda \frac{w_0}{g} = \frac{\rho U}{2} \frac{S w_0}{mg} \frac{dC_L}{d\alpha} \qquad (10)$$

where Δn denotes the increment of *load factor*. The product of Δn and the weight of the structures gives the acting inertia force.

Equation 7 is the indicial admittance of the displacement z for a sharp-edged gust. Equation 6 is the Duhamel integral for an arbitrary gust

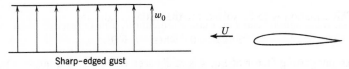

Sharp-edged gust

Fig. 8.5. Sharp-edged gust.

profile. According to § 8.1, the same problem can be as easily solved by the method of mechanical impedance, in which the response to a sinusoidal gust, $w_0 \sin \omega t$, is first obtained. The response to an arbitrary gust can then be obtained by a Fourier integral. It is easy to show that the results obtained by these two methods agree with each other.

The sharp-edged gust formula is derived under the six simplifying assumptions named above and the idealized gust profile of Fig. 8.5. In reality, none of these assumptions can be fulfilled. Nevertheless, the formula is convenient for use in airplane design. If the gust speed is based on an "effective" value which is derived by reducing the experimental acceleration data according to Eq. 10, the result can be used to predict the gust load factor on similar airplanes. The effective gust speed w_0, however, would have to be determined for each type of airplane, because the effects of the simplifying assumptions are different for different airplane size, geometry, flexibility, center-of-gravity location, dynamic stability characteristics, and flight Mach number.

Much work has been done in the direction of relaxing one or another of the assumptions made in deriving the sharp-edged gust formula. Using aerodynamics of an unsteady incompressible flow, Küssner[8.20] obtained in 1931 the response of a rigid airplane, restrained against pitch, to a gust with a finite velocity gradient. Küssner also extended his analysis to take into account the elasticity of the wing in bending, but assumed the deflection mode to be of the same form as the static deflection curve under a uniformly distributed load. He concluded that the stresses in an elastic wing may be considerably higher than that in a rigid wing. In addition, he showed that, for a gust of given intensity, the load factor reaches a maximum when the gust is inclined at 65° to 70° to the flight path, but the load factor due to a normal gust (gust velocity perpendicular to the flight path) differs from this maximum by less than 10 per cent. The response of a rigid airplane free to pitch is treated by Bryant and Jones in 1932[8.5] under the assumptions of quasi-steady lift, and in 1936[8.6] for a semirigid wing including the unsteady-flow characteristics as given by Wagner. Similar extensions were made by Williams and Hanson[8.36] in 1937, Sears and Sparks[8.32] in 1941, and Pierce[8.24] and Putnam[8.27] in 1947.

More extensive investigation on the effect of elastic deformation was made by Goland, Luke, and Kahn[8.10] in 1947. Jenkins and Pancu[8.16] in 1948, Bisplinghoff, Isakson, Pian, Flomenhoft, and O'Brien[8.3, 8.4] in 1949. In these studies the bending and torsion deflections of the wing are approximated by a number of deflection modes.

A comprehensive analysis of the effect of the gust gradient and the airplane pitching was made by Greidanus and van de Vooren[8.11] (1948). The allowance of the pitching degree of freedom introduces great complication into the analysis because of the phase lag in the downwash between the wing and the tail.

Further extensions were made by Bisplinghoff and his associates, Mazelsky, Diederich, Houbolt, and others, to account for aerodynamic forces in a flow of a compressible fluid.

Since transient problems can be best treated by the method of Laplace transformation, details of some of these extensions will be discussed in Chapter 10. In the remaining sections of the present chapter, let us turn to the statistical aspects of the dynamic-stress problem.

8.3   STATISTICAL ASPECTS OF DYNAMIC-STRESS PROBLEMS

In the preceding section the response of an airplane to a specific gust is discussed. In applying such an analysis to the practical design of an aircraft, we must know how to specify the gust profile.

Figure 8.6 shows a record of vertical gusts measured by a hot-wire anemometer carried on a stationary balloon (fastened to the ground by wires) 146 meters above the ground.* It is similar in nature to the anemometer records of wind-tunnel turbulences. From such a figure it is indeed impossible to say what kind of an isolated gust profile can characterize the real picture.

Similar difficulty arises in other dynamic-stress problems. In fact, one of the most difficult problems in structural design with regard to transient loads is the determination of the forcing function, or the selection, among

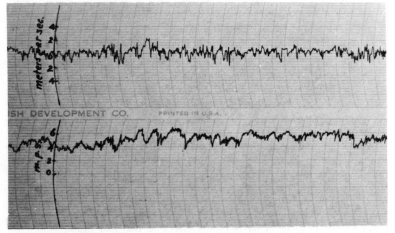

Fig. 8.6.   A gust record.   (Courtesy of Dr. P. MacCready
of the California Institute of Technology.)

the great variety of possible forcing functions, of those to be used as a basis for design. The exact form of the forcing function is always uncertain to some extent because of the large number of variables entering the problem. Thus, when a few important variables are considered in a calculation, other variables appear as disturbing influences. The random nature of the atmospheric turbulence as shown in the records of Fig. 8.6 are due to the influences of the viscosity, pressure, density, temperature, and humidity of the air and the initial velocity distributions, varying in such a complicated manner that a mechanistic prediction based on the hydrodynamic equations is practically impossible.

If the uncertainty of predicting the forcing function is recognized, the problem becomes statistical. One then tries to state the main features of the response, as well as those of the forcing function, in terms of

* This curve is given by P. MacCready.[8.52]

statistical averages and probability distributions. The statistical terminology will first be explained in the next three sections. The "gust" response of an airplane, to a continuous atmospheric turbulence such as the one recorded in Fig. 8.6, is then calculated.

8.4 THE MEANING OF PROBABILITY AND DISTRIBUTION FUNCTION

Consider a very simple experiment of tossing a coin and observing whether a "head" turns up. If we make n throws in which the "head" turns up v times, the ratio v/n may be called the *frequency ratio* or simply the *frequency* of the event "head" in the sequence formed by the n throws. It is a general experience that the frequency ratio shows a marked tendency to become more or less constant for large values of n. In the coin experiment, the frequency ratio of "head" approaches a value very close to $1/2$.

This stability of the frequency ratios seems to be an old experience for long series of repeated experiments performed under uniform conditions. It is thus reasonable to assume that, to any event E connected with a random experiment $\mathscr{E}$, we should be able to ascribe a number P such that, in a long series of repetitions of $\mathscr{E}$, the frequency of occurrence of the event E would be approximately equal to P. This number P is the *probability* of the event E with respect to the random experiment $\mathscr{E}$. Since the frequency ratio must satisfy the relation $0 \leqslant v/n \leqslant 1$, P must satisfy the inequality

$$0 \leqslant P \leqslant 1 \tag{1}$$

If an event E is an impossible event, i.e., it can *never* occur in the performance of an experiment $\mathscr{E}$, then its probability P is zero. On the other hand, if $P = 0$ for some event E, E is *not* necessarily an impossible event. But, if the experiment is performed one single time, it can be considered as practically certain that E will not occur.

Similarly, if E is a *certain* event, then $P = 1$. If $P = 1$, we cannot infer that E is certain, but we can say that, in the long run, E will occur in all but a very small percentage of cases.

The statistical nature of a random variable is characterized by its *distribution function*. To explain the meaning of the distribution function, let us consider a set of gust records similar to the one presented in Fig. 8.6. Let each record represent a definite interval of time. Suppose that we are interested in the maximum value of the gust speed in each record. This maximum value will be called "gust speed" for conciseness and will be denoted by y. The gust speed varies from one record to another. For a set of data consisting of n records, let v be the number in which the gust speed is less than or equal to a fixed number x. Then v/n is the frequency ratio for the statement $y \leqslant x$. If the total number of records n is increased without limit, then, according to the frequency interpretation of probability, the ratio v/n tends to a stationary value which represents the probability of the event "$y \leqslant x$." This probability, as a function of x, will be denoted by

$$P(y \leqslant x) = F(x) \tag{2}$$

The process can be repeated for other values of x until the whole range of x from $-\infty$ to ∞ is covered. The function $F(x)$ defined in this manner is called the *distribution function* of the random variable y.

Obviously $F(x)$ is a nondecreasing function, and

$$F(-\infty) = 0, \qquad 0 \leq F(x) \leq 1, \qquad F(+\infty) = 1 \tag{3}$$

If the derivative $F'(x) = f(x)$ exists, $f(x)$ is called the *probability density* or the *frequency function* of the distribution. Any frequency function $f(x)$ is nonnegative and has the integral 1 over $(-\infty, \infty)$. Since the difference $F(b) - F(a)$ represents the probability that the variable y assumes a value belonging to the interval $a < y \leqslant b$,

$$P(a < y \leqslant b) = F(b) - F(a) \tag{4}$$

In the limit it is seen that the probability that the variable y assumes a value belonging to the interval

$$x < y < x + \Delta x$$

is, for small Δx, asymptotically equal to $f(x)\,\Delta x$, which is written in the usual differential notation:

$$P(x < y < x + dx) = f(x)\,dx$$

In the following, we shall assume that the frequency function $f(x) = F'(x)$ exists and is continuous for all values of x. The distribution function is then

$$F(x) = P(y \leqslant x) = \int_{-\infty}^{x} f(t)\,dt \tag{5}$$

Moreover,

$$\int_{-\infty}^{\infty} f(t)\,dt = 1 \tag{6}$$

Moments. The integrals

$$\alpha_v = \int_{-\infty}^{\infty} x^v f(x)\,dx \qquad (v = 1, 2, 3, \cdots)$$

if absolutely convergent, are called the first, second, third, $\cdots$ moment of the distribution function according as $\nu = 1, 2, 3 \cdots$ respectively. The first moment, called the *mean*, is often denoted by the letter m

$$m = \int_{-\infty}^{\infty} x f(x) \, dx \tag{8}$$

The integrals

$$\mu_\nu = \int_{-\infty}^{\infty} (x - m)^\nu f(x) \, dx \tag{9}$$

are called the *central moments*.   Developing the factor $(x - m)^\nu$ according to the binomial theorem, we find

$$\begin{aligned}
\mu_0 &= 1 \\
\mu_1 &= 0 \\
\mu_2 &= \alpha_2 - m^2 \\
\mu_3 &= \alpha_3 - 3m\alpha_2 + 2m^3
\end{aligned} \tag{10}$$

$\cdots$

Measures of Location and Dispersion.   The *mean* m is a kind of measure of the "location" of the variable y.   If the frequency function is interpreted as the mass per unit length of a straight wire extending from $-\infty$ to $+\infty$, then the mean m is the abscissa of the center of gravity of the mass distribution.

The second central moment μ_2 gives an idea of how widely the values of the variable are spread on either side of the mean.   This is called the *variance* of the variable, and represents the centroidal moment of inertia of the mass distribution referred to above.   We shall always have $\mu_2 \geqslant 0$. In the mass-distribution analogy, the moment of inertia μ_2 vanishes only if the whole mass is concentrated in the single point $x = m$.   Generally, the smaller the variance, the greater the concentration, and vice versa.

In order to obtain a characteristic quantity which is of the same dimension as the random variable, *the standard deviation*, denoted by σ, is defined, as the nonnegative square root of μ_2

$$\sigma = \sqrt{\mu_2} = \sqrt{\alpha_2 - m^2} \tag{11}$$

Example, the Normal Distribution.   The *normal distribution function* is defined by the expression

$$F(x) = \frac{1}{\sqrt{2\pi}} \int_{-\infty}^{x} e^{-t^2/2} \, dt \tag{12}$$

The corresponding *normal frequency function* is

$$f(x) = \frac{1}{\sqrt{2\pi}} e^{-x^2/2} \tag{13}$$

Diagrams of these functions are given in Fig. 8.7. The mean value of the distribution is 0, and the standard deviation is 1.

A random variable ξ will be said to be *normally distributed with the parameters m and σ*, or briefly *normal (m, σ)* if the distribution function

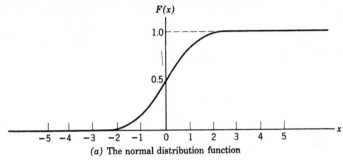

(a) The normal distribution function

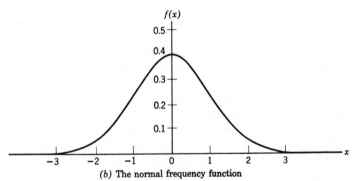

(b) The normal frequency function

Fig. 8.7. Normal distribution and normal frequency curves.

of ξ is $F\left(\dfrac{x-m}{\sigma}\right)$, where σ and m are constants ($\sigma > 0$). The frequency function is then

$$\frac{1}{\sigma} f\left(\frac{x-m}{\sigma}\right) = \frac{1}{\sigma\sqrt{2\pi}} e^{-(x-m)^2/2\sigma^2} \tag{14}$$

It is easy to verify that m is the mean, and σ is the standard deviation of the variable ξ.

Note that, in the normal distribution, the distribution function is completely characterized by the mean and the standard deviation.

8.5 STATISTICAL AVERAGES

Let $u(t)$ represent a random function of time, such as a velocity component at a point in a turbulent flow, the force acting in the landing-gear strut of an airplane during landing impact, the lift acting on an airfoil in passing through a gust, etc. By saying that $u(t)$ is a random function, we mean that at a given time t the value of u is not predictable from the data of the problem but takes random values which are distributed according to certain definite probability laws. We shall assume that the probability laws describing the randomness *are* determined by the data of the problem. Such a determination can sometimes be made through a suitable theoretical model, but in general it has to be obtained by experiments.

A set of observations forms an "ensemble" of events. In the example of gusts, each gust record, such as the one given in Fig. 8.6, is a member in an ensemble of such records. Imagine that a large number of observations be made simultaneously under similar conditions. Let the records be numbered and denoted by $u_1(t)$, $u_2(t)$, $\cdots$, $u_N(t)$. If the total number of records is N, the following averages may be formed

$$\widetilde{u(t)}^N = [u_1(t) + u_2(t) + \cdots + u_N(t)]/N$$
$$\widetilde{u^2(t)}^N = [u_1^2(t) + u_2^2(t) + \cdots + u_N^2(t)]/N \qquad (1)$$
$$\cdots \cdots$$

Assume that $\widetilde{u(t)}^N$ and $\widetilde{u^2(t)}^N$, etc., tend to definite limits, respectively, as $N \to \infty$; then these limiting values are *ensemble averages* of the random process $u(t)$.

In a similar manner, ensemble averages of the following nature can be formed:

$$\overline{u(t)u(t+\tau)} = \lim_{N \to \infty} \frac{1}{N} \sum_{i=1}^{N} u_i(t)u_i(t+\tau) \qquad (2)$$

$$\overline{u(t)u(t+\tau_1) \cdots u(t+\tau_m)} = \lim_{N \to \infty} \frac{1}{N} \sum_{i=1}^{N} u_i(t)u_i(t+\tau_1) \cdots u_i(t+\tau_m) \qquad (3)$$

These are ensemble averages of the *correlation functions* of the random function $u(t)$.

For a complete statistical description of a random process, ensemble averages of all orders are required. However, for practical dynamic-load problems in aeroelasticity, often the most important information is afforded by the mean value $\widewidehat{u(t)}$ and the mean "intensity" $(\widewidehat{u^2(t)})^{1/2}$. For these simplest kinds of average quantity, analysis can proceed in a simple manner. By definition, it is clear that $\widewidehat{u(t)}$ and $([u(t) - \widewidehat{u(t)}]^2)^{1/2}$ are the mean and the standard deviation of the random function $u(t)$ at any instant t.

If the ensemble averages of a function $u(t)$ are independent of the variable t, then $u(t)$ is said to be *stationary* in the ensemble sense.

A different kind of average is the well-known concept of *time average*. Thus, the mean and the mean square of a function of time $u(t)$ over a time interval $2T$, are

$$\overline{u(t)}^T = \frac{1}{2T} \int_{t-T}^{t+T} u(t)\, dt$$

$$\overline{u^2(t)}^T = \frac{1}{2T} \int_{t-T}^{t+T} u^2(t)\, dt$$

(4)

Similarly, higher-order averages $\overline{u^p(t)}^T$ ($p = 3, 4, \cdots$) can be formed, provided that the definition integrals converge. If the time averages tend to be independent of t and T when T is sufficiently large, then $u(t)$ is also said to be *stationary*, but in the sense of time-average. We shall write in this case the limiting value of $\overline{u^p(t)}^T$ as $T \to \infty$ as $\overline{u^p(t)}$.

The property of stationariness says, in effect, that all time instants are similar as far as the statistical properties of u are concerned. This suggests that the results of averaging over a large number of observations could be obtained equally well by averaging over a large time interval for *one* observation. In other words, for a stationary random process, one expects the time averages to be equal to the ensemble averages, and we do not have to distinguish the concepts of stationariness in time average or in ensemble average. The study of the exact conditions under which this equivalence of the time average over one observation and the ensemble average over many observations will indeed be true is called the ergodic theory. In aeroelastic problems which we are going to consider, this equivalence may be assumed.

It is natural to extend the above concepts to random functions of space. In the example of gusts, we may have to consider the velocity fluctuation **u**, itself a vector, as a function of space and time x, y, z, t. If records of the velocity fluctuation were taken over different regions of space and if the ensemble averages over space are independent of the spatial coordinates of the regions, then the velocity fluctuation is said to be *homogeneous*.

There are many flows, of interest to aeroelasticity, that are approximately homogeneous and stationary when the time and distance scales are properly chosen. Such is the case of atmospheric turbulence within a suitable expanse of time and space.

The Power Spectrum. Let us consider a stationary random process $u(t)$ which has a *mean value m equal to zero*[*] and define the mean square of $u(t)$ over an interval $2T$ as in Eq. 4. Let $u(t)$ be so "truncated" that it becomes zero outside the interval $(-T, T)$ (Fig. 8.8). Let the truncated function be written as $u_T(t)$.

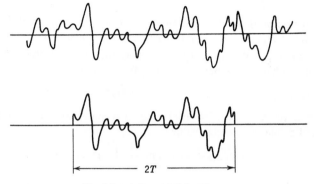

Fig. 8.8. A truncated function.

The Fourier integral of $u_T(t)$ exists provided that the absolute value of $u_T(t)$ is integrable and $u_T(t)$ has bounded variation,

$$u_T(t) = \frac{1}{\sqrt{2\pi}} \int_{-\infty}^{\infty} \mathscr{F}_T(\omega) e^{i\omega t} d\omega \tag{5}$$

where

$$\mathscr{F}_T(\omega) = \frac{1}{\sqrt{2\pi}} \int_{-T}^{T} u_T(t) e^{-i\omega t} dt \tag{6}$$

If $\mathscr{F}_T{}^*(\omega)$ denotes the complex conjugate, then $\mathscr{F}_T{}^*(-\omega) = \mathscr{F}_T(\omega)$ since $u_T(t)$ is real. It is well known (Parseval theorem) that

$$\int_{-\infty}^{\infty} u_T{}^2(t) dt = \int_{-T}^{T} u_T{}^2(t) dt = \int_{-\infty}^{\infty} |\mathscr{F}_T(\omega)|^2 d\omega \tag{7}$$

Hence

$$\overline{u^2(t)} = \lim_{T \to \infty} \frac{1}{2T} \int_{-\infty}^{\infty} |\mathscr{F}_T(\omega)|^2 d\omega = \lim_{T \to \infty} \int_{0}^{\infty} \frac{|\mathscr{F}_T(\omega)|^2}{T} d\omega \tag{8}$$

---

[*] If the mean value does not vanish, we may first subtract from $u(t)$ the mean value. In other words, we may consider only the *deviation* from the mean value.

For a stationary random process for which $\overline{u^2(t)}$ exists, the left-hand side of the above equation tends to a constant as $T \to \infty$; hence let

$$p(\omega) = \lim_{T \to \infty} \frac{|\mathscr{F}_T(\omega)|^2}{T} \tag{9}$$

then

$$\overline{u^2(t)} = \int_0^\infty p(\omega)\, d\omega \tag{10}$$

The function $p(\omega)$ is called the *power spectrum* or the *spectral density* of $u(t)$. When $u(t)$ is resolved into harmonic components by a Fourier analysis, as in Eq. 5, the element $p(\omega)\, d\omega$ gives the contribution to $\overline{u^2(t)}$ from components having frequencies ranging from ω to $\omega + d\omega$. The integral $\int_0^\omega p(\omega)\, d\omega$ represents the contribution to $\overline{u^2(t)}$ from frequencies less than ω.

Correlation Functions. Consider again a stationary random process $u(t)$, and define the average value

$$\psi(\tau) = \overline{u(t)\, u(t+\tau)} = \lim_{T \to \infty} \frac{1}{2T} \int_{-T}^{T} u(t)\, u(t+\tau)\, dt \tag{11}$$

$\psi(\tau)$ as a function of τ is a *correlation function* defined in the sense of time average. For a stationary random process it is the same as that defined in Eq. 2.

It follows from the definition of correlation function that

$$\psi(0) = \overline{u^2(t)} \tag{12}$$

Since the average values of a stationary random process are independent of the origin of t, it follows immediately that

$$\psi(\tau) = \psi(-\tau) \tag{13}$$

Furthermore, if $u(t)$ is a *continuous* function of t, then

$$\lim_{h \to 0} \overline{u(t)[u(t+\tau+h) - u(t+\tau)]} = 0$$

since the factor in [ ] tends to zero as $h \to 0$. This may be written as

$$\lim_{h \to 0} [\psi(\tau+h) - \psi(\tau)] = 0 \tag{14}$$

showing that $\psi(\tau)$ is continuous at all values of τ. According to the Schwarzian inequality,

$$\psi(\tau) \leqslant [\overline{u^2(t)} \cdot \overline{u^2(t+\tau)}]^{1/2} = [\psi(0) \cdot \psi(0)]^{1/2}$$

we have

$$\psi(\tau) \leqslant \psi(0) \tag{15}$$

In general, random processes having zero mean values satisfy the condition

$$\lim_{\tau \to \infty} \psi(\tau) = 0 \tag{16}$$

which means that the function $u(t)$ at two instants separated by a long time interval are uncorrelated with each other.

The correlation function $\psi(\tau)$ when plotted against τ is generally a bell-shaped curve. The interval of τ in which $\psi(\tau)$ differs significantly from zero is a measure of the "scale" of the random process. In the example of atmospheric turbulence, such a scale may be thought of as representing the mean size of eddies.

Relation between the Power Spectrum and the Correlation Function. The Fourier transform of $\psi(\tau)$ is

$$\phi(\omega) = \frac{1}{\sqrt{2\pi}} \int_{-\infty}^{\infty} \psi(\tau) e^{-i\omega\tau} d\tau \tag{17}$$

Since $\psi(\tau)$ is an even function of τ, $\phi(\omega)$ can be expressed as a real-valued integral

$$\phi(\omega) = \sqrt{\frac{2}{\pi}} \int_0^{\infty} \psi(\tau) \cos \omega\tau \, d\tau$$

because $e^{-i\omega t} = \cos \omega\tau - i \sin \omega\tau$ and the imaginary part of Eq. 17 vanishes. The inverse transform is

$$\psi(\tau) = \sqrt{\frac{2}{\pi}} \int_0^{\infty} \phi(\omega) \cos \omega\tau \, d\omega$$

When $\tau \to 0$, the left-hand side tends to $\overline{u^2(t)}$; hence

$$\overline{u^2(t)} = \sqrt{\frac{2}{\pi}} \int_0^{\infty} \phi(\omega) \, d\omega$$

A comparison with Eq. 10 shows that, aside from the numerical factor $\sqrt{2/\pi}$, $\phi(\omega)$ is identical with the power spectrum $p(\omega)$ defined before. A formal proof for the identity of $\sqrt{\frac{\pi}{2}}\phi(\omega)$ with $p(\omega)$ can be constructed.

Hence, we obtain the reciprocal relations between the power spectrum and the correlation function

$$\psi(\tau) = \int_0^{\infty} p(\omega) \cos \omega\tau \, d\omega \tag{18}$$

$$p(\omega) = \frac{2}{\pi} \int_0^{\infty} \psi(\tau) \cos \omega\tau \, d\tau \tag{19}$$

Equations 5 through 19 hold for any stationary function $u(t)$. Now consider an ensemble of functions $u_1(t)$, $u_2(t)$, $\cdots$. Each of these functions defines its $p(\omega)$ and $\psi(\tau)$, which can be averaged over the ensemble.

The resulting $\overline{p(\omega)}$ will be called the spectral density or the power spectrum of the random process. For a stationary random process it follows from Eqs. 18 and 19 that the correlation function and the power spectrum are each other's Fourier cosine transform.

Example. When

$$y(t) = A + B \sin (\omega_0 t + \alpha)$$

we have

$$\overline{y^2} = A^2 + \tfrac{1}{2}B^2$$

$$\psi(\tau) = \overline{y(t)y(t + \tau)} = A^2 + \tfrac{1}{2}B^2 \cos \omega_0 \tau$$

and, if we define a unit-impulse function $\delta(t)$ as in Eq. 14 of § 8.1, but impose further the condition $\delta(t) = \delta(-t)$ so that $\displaystyle\int_0^\infty \delta(t)dt = \tfrac{1}{2}$, then

$$p(\omega) = 2A^2\delta(\omega) + \tfrac{1}{2}B^2\delta(\omega - \omega_0)$$

This example shows that, if the mean value of a function is not zero, and if the function is periodic, the power spectrum will have singular peaks of the well-known Dirac δ-function type.

Example of Wind-Tunnel Turbulences. Consider the velocity fluctuations in the flow in a wind tunnel. Let the deviations from the respective mean values of the velocity components be written as u_1, u_2, u_3. These are functions of space and time $(x, y, z; t)$. Similar to the time correlation function $\psi(\tau)$, the spatial correlation functions such as

$$\overline{u_i(x, y, z; t)u_j(x + r, y, z; t)}$$
$$= \lim_{V \to \infty} \frac{1}{V} \int_V u_i(x, y, z; t)u_j(x + r, y, z; t) \, dx \, dy \, dz \quad (i, j = 1, 2, 3) \tag{20}$$

may be formed. If the field of flow is homogeneous, such a correlation function is well defined and is a function of r and t only. It is possible in wind-tunnel work to determine experimentally the general correlation function

$$\overline{u_i(x_1, y_1, z_1; t_1)u_j(x_2, y_2, z_2; t_2)} \qquad (i, j = 1, 2, 3) \tag{21}$$

which, for a stationary and homogeneous turbulence field, depends only on the relative position of the points (x_1, y_1, z_1) and (x_2, y_2, z_2), and the time interval $t_2 - t_1$, and not on the absolute location of the points or time.

Thus a large number of space and time correlation functions can be defined among various velocity components. It is here that the simplifying concept of an *isotropic* turbulence field enters. As was first shown by G. I. Taylor[8.54] and Th. von Kármán,[8.51] an isotropic turbulence field is one that can be specified by two "principal" velocity correlations, in a way similar to that two principal stresses at a point in an isotropic elastic solid define the state of stress at that point. The two principal velocity correlations are denoted by f and g and are defined pictorially in Fig. 8.9. $f(r)$ is the correlation function between the same velocity component measured at two points at a distance r apart, lying along a line in the direction of the velocity component. $g(r)$ is the correlation function between points at a distance r apart, lying along a direction normal to the direction of the velocity component. For example, let us write u, v, w in place of u_1, u_2, u_3, then

$$\overline{u(x, y, z; t)u(x + r, y, z; t)} = f(r)$$

$$\overline{u(x, y, z; t)u(x, y + r, z; t)} = g(r) \tag{22}$$

$$\overline{v(x, y, z; t)v(x + r, y, z; t)} = g(r)$$

If the field of fluctuation u, v, w is superimposed upon a mean flow of velocity U is the x direction, and if $\overline{u^2}/U^2$, $\overline{v^2}/U^2$, $\overline{w^2}/U^2$ are small quantities, it is possible to interchange time and space variables and consider the turbulence as simply being transported along with the velocity U in the x direction. Hence the time τ which enters into the time correlation function can be replaced by r/U, where r is chosen along the x axis. We may write

$$\overline{u(t)u(t + \tau)} = \psi_1(\tau) = f(\tau U)$$

$$\overline{v(t)v(t + \tau)} = \psi_2(\tau) = g(\tau U) \tag{23}$$

Experiments[8.1] have shown that in wind-tunnels the turbulences are nearly isotropic and the correlation functions $f(r)$ and $g(r)$ have the form

$$f(r) = f(0) \, e^{-r/L_1}$$

$$g(r) = g(0) \, e^{-r/L_2} \left(1 - \frac{r}{2L_2}\right) \tag{24}$$

The corresponding time correlation functions are

$$\psi_1(\tau) = \psi_1(0) \, e^{-\tau U/L_1}$$

$$\psi_2(\tau) = \psi_2(0) \, e^{-\tau U/L_2} \left(1 - \frac{U}{2L_2}\tau\right) \tag{25}$$

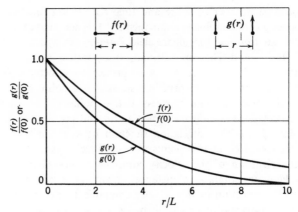

Fig. 8.9. Correlation functions f and g in isotropic
turbulences.

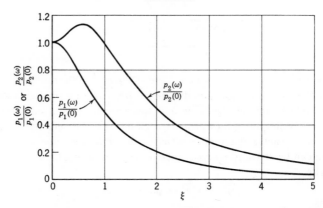

Fig. 8.10. Power spectra of atmospheric turbulences. $\xi = \omega L/U$, where ω
is the frequency, L is the scale of turbulence, U is the speed of flow.

and the power spectra are

$$p_1(\omega) = p_1(0)\,\frac{1}{1 + \xi_1^{\,2}}$$

$$p_2(\omega) = p_2(0)\,\frac{1 + 3\xi_2^{\,2}}{(1 + \xi_2^{\,2})^2} \tag{26}$$

where

$$\xi_1 = \frac{\omega L_1}{U}, \qquad \xi_2 = \frac{\omega L_2}{U}$$

The functions $f(r)$, $g(r)$, $p_1(\omega)$, and $p_2(\omega)$ are normalized and depicted in
Figs. 8.9 and 8.10.

The constants L_1 and L_2 in the above equations are quantities known as *scales of turbulence*. They are proportional to the areas under the normalized correlation curves

$$L_1 = \int_0^\infty \frac{f(r)}{f(0)}\, dr, \qquad L_2 = 2 \int_0^\infty \frac{g(r)}{g(0)}\, dr \qquad (27)$$

8.6   THE POWER SPECTRA OF THE EXCITATION AND THE RESPONSE

Suppose that the response y and the excitation f are connected by the linear differential equation

$$\frac{d^2y}{dt^2} + \beta \frac{dy}{dt} + \omega_0{}^2 y = f(t) \qquad (1)$$

According to Eqs. 10 and 11 of § 8.1, when the forcing function $f(t)$ is representable as a Fourier integral

$$f(t) = \frac{1}{\sqrt{2\pi}} \int_{-\infty}^{\infty} A(\omega)\, e^{i\omega t}\, d\omega \qquad (2)$$

the response is the sum of the complementary function and a particular solution given by

$$y(t) = \frac{1}{\sqrt{2\pi}} \int_{-\infty}^{\infty} \frac{A(\omega)}{Z(i\omega)}\, e^{i\omega t}\, d\omega \qquad (3)$$

provided that the integral converges. In Eq. 3, $Z(i\omega)$ is the impedance

$$Z(i\omega) = \omega_0{}^2 - \omega^2 + i\beta\omega \qquad (4)$$

If $f(t)$ represents a stationary fluctuation, it does not tend to zero as $t \to \infty$ and its Fourier transform does not exist. To overcome this difficulty the truncated function used in the last section can be used here again. Let $f(t)$ be truncated in such a manner that it is zero outside an interval $(-T, T)$. The Fourier transform of the truncated $f(t)$ can be defined provided that $f(t)$ is absolutely integrable in $(-T, T)$ and has bounded variations. Let

$$A(\omega) = \frac{1}{\sqrt{2\pi}} \int_{-T}^{T} f(t)\, e^{-i\omega t}\, dt \qquad (5)$$

When $A(\omega)$ is given by Eq. 5, the functions $f(t)$ and $y(t)$ given by Eqs. 2 and 3 will represent, respectively, the excitation and the response within the interval $(-T, T)$. From Eq. 3, we can derive the value of $\overline{y^2(t)}$ in the same interval. Comparing Eq. 3 with Eq. 5 of § 8.5, and following the reasoning of that section, we obtain

$$\overline{y^2(t)} = \int_0^\infty \frac{p(\omega)}{|Z(i\omega)|^2}\,d\omega \tag{6}$$

where $p(\omega)$ is the power spectrum of $f(t)$. Equation 6 shows that the power spectrum of the response is equal to that of the forcing function divided by the square of the absolute value of the impedance. The phase relationship between the excitation and the response, represented by the argument of the complex number $Z(i\omega)$, is completely obliterated in the power-spectra relations.

Equation 6 may be applied to general dynamic systems for which the impedance $Z(i\omega)$ can be defined. It holds generally for all dynamic systems described by differential equations, linear integral equations and linear integral–differential equations that occur in aeroelasticity.

8.7 THE RESPONSE OF A RIGID AIRPLANE TO ATMOSPHERIC TURBULENCES

As an illustration of the statistical approach, let us consider the motion of an airplane in response to atmospheric turbulences. For simplicity, let us use the assumptions listed in § 8.2 (except 6), that the airplane may be regarded as a rigid body, that the forward velocity U can be regarded as a constant, that the disturbed motion is symmetrical with respect to the longitudinal plane of symmetry, and that the pitching motion can be neglected and only the translational motion normal to the flight path is of significance. Moreover, we shall assume that only the velocity component of turbulence normal to the flight path need be considered. The effect of other velocity components will be neglected.

The characteristics of atmospheric turbulences may depend on the geographic and weather conditions. However, Clementson[8.47] has shown, that the correlation functions (and the power spectra) of atmospheric turbulences, in several different conditions (unstable air mass, water-land discontinuities, thunderstorms, and mountainous terrain), are, aside from a constant multiplier, remarkably similar to each other. They differ essentially only in intensity.

The following analysis will not refer to any particular weather condition but to isotropic turbulences as measured in wind tunnels.

We assume, therefore, that the mean value of the gust fluctuation is zero and that the gust has a power spectrum given by $p_2(\omega)$ of Eq. 26 of § 8.5, which can be written as

$$p(\omega) = \overline{w^2}\,\frac{l}{\pi U}\,\frac{1 + 3\xi^2}{(1 + \xi^2)^2}, \qquad \xi = \frac{\omega l}{U} \tag{1}$$

where $\overline{w^2}$ is the mean square intensity of the gust, l is the scale of turbulence, and U is the speed of the general flow, i.e., the speed of flight of the airplane. The scale of turbulence is defined by the area under the normalized correlation curve

$$R(r) = \frac{\overline{w(x, t)\, w(x + r, t)}}{\overline{w^2}} \tag{2}$$

so that

$$l = \int_0^\infty R(r)\, dr \tag{3}$$

Let us consider first an airfoil strip of unit span in a two-dimensional flow. Let the mean flow velocity be U which is parallel to the mean position of the airfoil chord. The airfoil is assumed thin, and the amplitude of its motion is assumed small in comparison with the chord. On the mean velocity of flow U is superposed a small turbulent fluctuation w, uniform across the span, and normal to the chord. The turbulent motion in the fluid may be regarded as transported with the mean flow, so that w is given by a stationary random function

$$w = w\left(t - \frac{x}{U}\right)$$

The lift that acts on the airfoil is partly due to the disturbance w and partly due to the motion of the airfoil itself. Within the framework of the linearized theory the lift induced by these two parts are superposable. The equation of motion of the airfoil (having the translational degree of freedom) can be written as

$$m\ddot{z} + L(z) = -L(w) \tag{4}$$

where $L(w)$ indicates the lift induced by $w(t)$, and $L(z)$ that due to the motion of the airfoil. Regarding $-L(w)$ as a known external force, Eq. 4 is the same as Eq. 10 of § 6.8, describing the disturbed motion of an airfoil. Thus it is seen that the problem of the response of an airfoil to a turbulent flow can be separated into two parts:* (1) the determination of the lift produced by the turbulent flow on an airfoil in a steady flow, and (2) the determination of the disturbed motion of the airfoil due to an exciting force, with the flow regarded as uniform and without turbulence.

---

* It must be observed that the possibility of such a separation into two independent problems depends on the assumption of small disturbances, so that the hydrodynamic equations may be linearized. Hence, the analysis is theoretically valid only for turbulences of small intensity.

According to the results in § 8.6, if the power spectrum of the turbulence $w(t)$ is $p_{gust}(\omega)$, that of the lift would be

$$p_{lift}(\omega) = \chi_a(\omega)\, p_{gust}(\omega) \tag{5}$$

where $\chi_a(\omega)$ denotes the square of the absolute value of the frequency response (admittance) of the lift to a sinusoidal gust. The power spectrum of the airfoil acceleration would be

$$p_{acc}(\omega) = \chi_s(\omega)\, \chi_a(\omega)\, p_{gust}(\omega) \tag{6}$$

where $\chi_s(\omega)$ denotes the square of the absolute value of the frequency response of the acceleration to a sinusoidal lift force. Hence, the problem is finally resolved to the determination of the quantities $\chi_s(\omega)$ and $\chi_a(\omega)$.

If we introduce the strip assumption that each section of an airfoil can be regarded as a two-dimensional airfoil, neglecting the effect of finite span, then the above argument can be applied to the whole airplane. We shall consider this problem in greater detail below.

Lift Due to a Sinusoidal Gust. Let the coordinate axes be fixed on the airfoil, with the origin $x = 0$ located at the mid-chord point. Let the vertical gust velocity be given by the expression

$$w(x, t) = w_0\, e^{i\omega(t - x/U)} \tag{7}$$

which shows a sinusoidal gust pattern moving past the airfoil with a speed U. In § 13.4, it will be shown that the lift induced by $w(x, t)$ on a two-dimensional airfoil of unit span is

$$L = 2\pi\rho b U w_0\, e^{i\omega t}\, \phi(k) \tag{8}$$

where ρ is the air density, b is the semichord length, k is the reduced frequency $\omega b/U$, and

$$|\phi(k)|^2 \doteq \frac{1}{1 + 2\pi k} \tag{9}$$

The resultant lift acts through the $^1/_4$-chord point from the leading edge. The factor $2\pi\rho b U\phi(k)$ represents the frequency response (admittance) of the lift to the gust.

Admittance of the Airplane. The equation of motion of a two-dimensional airfoil subject to a sinusoidal force has been derived in § 6.8. In a single degree of freedom of vertical translation, Eq. 10 of § 6.8 gives

$$m\,\frac{U^2}{b^2}\, z'' = -L(\tau) + P(\tau) \tag{10}$$

where m = the mass of the airfoil

τ = a dimensionless time = Ut/b

$z' = dz/d\tau, \qquad z'' = d^2z/d\tau^2$

The lift per unit span on the two-dimensional wing is given by an integral in Eq. 8 of § 6.8. Neglecting the lift on the tail surfaces, and using *strip assumption*, we can write the lift on the airplane as

$$L = 2\pi \frac{\rho U^2}{2} S \int_{-\infty}^{\tau} \Phi(\tau - \tau_0) \frac{z''(\tau_0)}{b} d\tau_0 + \pi \frac{\rho U^2}{2} S \frac{z''}{b}$$

where S = wing area, and $\Phi(\tau - \tau_0)$ is the Wagner's function.

In a simple-harmonic motion

$$z = z_0 e^{i\omega t} = z_0 e^{ik\tau}, \qquad P = P_0 e^{i\omega t} = P_0 e^{ik\tau} \tag{11}$$

the lift can be written as (compare Eqs. 7 and 9 of § 6.9)

$$L(\tau) = \pi \frac{\rho U^2}{2} S \left[2C(k) \frac{ik}{b} z_0 - k^2 \frac{z_0}{b} \right] e^{ik\tau} \tag{12}$$

Hence, the equation of motion is

$$\left\{ -m \frac{U^2}{b^2} k^2 + \frac{\pi \rho U^2}{2} \frac{S}{b} \left[2C(k)ik - k^2 \right] \right\} z_0 = P_0 \tag{13}$$

The quantity in { } is the complex impedance from $P(\tau)$ to $z(\tau)$. Its inverse, multiplied by $-\omega^2 = -(Uk/b)^2$, gives the admittance of the acceleration $\ddot{z}$. Introducing the airplane density ratio κ as a parameter,

$$\kappa = \frac{2m}{\pi \rho S b} \tag{14}$$

we obtain the admittance

$$\frac{\ddot{z}}{P} = \frac{2}{\pi \rho S b} \frac{k}{(1 + \kappa)k + 2G - 2iF} \tag{15}$$

The Gust Response. The intensity of the acceleration $\sqrt{\overline{\ddot{z}^2}}$ can now be computed. Assume the power spectrum of the gust given by Eq. 1. Neglecting the lift on the tail surfaces, and using strip theory, the lift on the wing due to a gust can be obtained by integrating Eq. 8 across the wing span. Then, in the notations of Eq. 6, we have

$$\chi_a = (\pi \rho S U)^2 |\phi(k)|^2 \tag{16}$$

$$\chi_s = \left| \frac{\ddot{z}}{P} \right|^2 = \left(\frac{2}{\pi \rho S b} \right)^2 \frac{k^2}{4(F^2 + G^2) + 4Gk(1 + \kappa) + k^2(1 + \kappa)^2} \tag{17}$$

The mean square of the acceleration is therefore

$$\overline{\ddot{z}^2} = \int_0^\infty \chi_a(k) \chi_s(k) p_{\text{gust}}(\omega) d\omega \quad \text{where} \quad k = \frac{\omega b}{U} \tag{18}$$

The exact integration of the above expression is difficult because of the complicated manner in which the Bessel functions are involved. To obtain an approximate solution, let us note that in the full range $(0, \infty)$ G^2 is much less than F^2. Hence we may neglect terms involving G^2 in Eqs. 17. Moreover, since

$$|F| < 1, \qquad |G| < 0.2$$

while in practice, κ varies from a number of order 40 for trainers and transports to 150 for high-speed fighters, a fair approximation of Eq. 17 is simply the quasi-steady result

$$\chi_s = \left(\frac{2}{\pi \rho S b}\right)^2 \frac{k^2}{4 + k^2(1 + \kappa)^2} \tag{19}$$

Using the approximations 9 and 19, and introducing the notations

$$s = b/L \tag{20}$$

$$\alpha = 2/s(1 + \kappa) \tag{21}$$

where s is the ratio of the wing semichord to the turbulence scale, we obtain

$$\overline{\ddot{z}^2} = \overline{w^2} \frac{4U^2}{\pi b^2(1 + \kappa)^2} \int_0^\infty \frac{\xi^2}{\alpha^2 + \xi^2} \frac{1}{1 + 2\pi s\xi} \frac{1 + 3\xi^2}{(1 + \xi^2)^2} d\xi \tag{22}$$

Let the integral in Eq. 22 be written as $I(\alpha, s)$ so that

$$\overline{\ddot{z}^2} = \overline{w^2} \frac{4U^2}{\pi b^2(1 + \kappa)^2} I(\alpha, s) \tag{23}$$

The variation of $I(\alpha, s)$ is shown in Fig. 8.11. Note that $\overline{\ddot{z}^2} \to 0$ both when $s \to 0$ and when $s \to \infty$. Hence, as the scale of turbulence becomes either negligibly small or infinitely large in comparison with the wing chord, the intensity of the acceleration experienced by the airplane will tend to zero. This can be expected, because, when the wing chord is very large in comparison with the turbulence scale, the "gusts" are smoothed out by canceling each other over the wing. On the other hand, when the chord length is very small in comparison with the scale of turbulence, the airfoil behaves quasi-stationarily. A rigid airplane having only the translational degree of freedom will experience no acceleration in a steady flight; hence, the limiting case $\overline{\ddot{z}^2} \to 0$ as $s \to 0$.

The importance of the critical speeds of aeroelastic stability on the dynamic-stress problem is evident from the airplane admittance to the sinusoidal lift. For example, when flutter speed is approached, $\chi_s(\omega)$ will become very great, thus causing very large dynamic stresses.

The mean square value of the acceleration $\ddot{z}$ is thus known as a function of the airplane speed, mass, and size, and the intensity and scale of the turbulence. From the assumption of stationariness of the gusts, the time averages are equal to the ensemble averages. Hence, $\overline{\ddot{z}^2}$ is also the standard deviation of an ensemble of gust responses. According to the interpretation given in § 8.4, it gives some idea of the dispersion of the induced acceleration caused by the gusts. The mean value being zero,

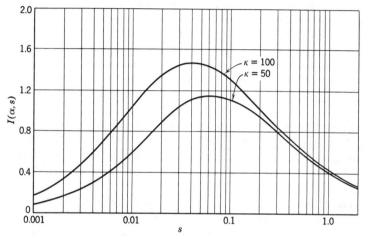

Fig. 8.11. The intensity factor $I(\alpha, s)$. κ is the airplane density ratio, s is the ratio of semichord to the scale of turbulence, and $\alpha = 2/s(1 + \kappa)$.

and the standard deviation being known, two of the most important parameters of the probability distribution of the gust response are determined.

In order to compare Eq. 23 with Eq. 10 of § 8.2, we may assume $\kappa \gg 1$ and reduce Eq. 23 into the form

$$\frac{\sqrt{\overline{\ddot{z}^2}}}{g} = \frac{2\pi\rho\sqrt{\overline{w^2}}\,US}{2mg}\sqrt{\frac{I(\alpha, s)}{\pi}}$$

whence, upon defining $(\overline{\ddot{z}^2})^{1/2}/g$ as the mean acceleration $\overline{\Delta n}$, and $(\overline{w^2})^{1/2}$ as a mean gust speed w_0, and replacing 2π by the lift curve slope $dC_L/d\alpha$, we obtain[8.44]

$$\overline{\Delta n} = \frac{\rho U}{2}\frac{Sw_0}{mg}\frac{dC_L}{d\alpha}\sqrt{\frac{I(\alpha, s)}{\pi}} \tag{24}$$

Comparing this with Eq. 10 of § 8.2, we see that they are identical but for the factor $\sqrt{I(\alpha, s)/\pi}$, which may be identified with the usual "gust alleviation factor." Williams[8.44] points out that, if we use the peak value of the

$I(\alpha, s)$ curve in (24), a proper account of the effect of airplane mass ratio on gust response is obtained. Moreover, the existence of a peak in the $I(\alpha, s)$ curve means that an airplane responds more readily to a scale of turbulence which is a constant multiple of wing chord, the constant being dependent upon the mass ratio. This explains a rather interesting experimental result that the so-called "gust gradient distance" is more closely related to the wing chord, rather than to the meteorological conditions.

If the probability-distribution function of the gust response is determined, the probability of encountering an acceleration of a specified magnitude can be found according to statistical methods. The probable number of times when the dynamic stresses in the structure exceed a specified stress level can be computed. Such information will be useful in designing an aircraft with respect to fatigue and service life. Engineering applications of this nature are discussed in Refs. 8.39–8.44, 8.63 and 8.86.

To obtain complete information about the probability distribution of dynamic stresses, higher-order correlation functions would have to be computed (see § 8.5). Such calculations are usually quite involved (see Mazelsky[8.83]). In practice, however, some idea about the probability distribution can often be obtained by experimental means, and a knowledge about the mean value and the root-mean-square deviation of the random variable is sufficient for engineering purpose.

The idea of statistical treatment is not new. The first application of statistical theory to dynamics goes back to Lord Rayleigh. It forms the basis of Taylor[8.54] and von Kármán's[8.51] theory of turbulence. It is also the foundation of the theory of Brownian motion, of noise in electric and acoustical systems, and of certain aspects of astrophysics. The mathematical theory of the stationary time series was developed by Wiener and Kintchine. The type of analysis used in this section was first made by Lin.[8.82] Its application to aeroelastic problems was first pointed out by Liepmann.[9.11]

The functions we are concerned with are functions of time and depend on chance. Such a function is known as a *stochastic process*. A stochastic process bears the same relation to a definite function that a random variable does to a definite number. Whereas an ensemble for a random variable may be regarded as the results of observation of a random experiment, an ensemble for a stochastic process should be regarded as a large number of experiments carried out under similar conditions, in which each experiment provides a function of time.

The application of the theory of stochastic process to dynamic problems raises many fundamental questions regarding the philosophy of structural design. Its successful application in the fields of engineering seismology, gust measurement,[8.41-8.43] fatigue-life studies, landing-impact loads,[8.70] and rough-water operations of seaplanes,[8.74-8.78] etc., may be noted. In many such applications it is not permissible to consider the random process as stationary. Examples of dynamic response analysis involving nonstationary stochastic process can be found in Ref. 8.70 in which the dynamic loads problem in aircraft structures during landing is treated.

BIBLIOGRAPHY

8.1 Dryden, H. L., G. B. Schubauer, W. C. Mock Jr., and H. K. Skramstad: Measurements of Intensity and Scale of Wind-Tunnel Turbulence and Their Relation to the Critical Reynolds Number of Spheres. *NACA Rept.* **581** (1937).

8.2 van de Pol, B., and H. Bremmer: *Operational Calculus Based on Two-Sided Laplace Integrals.* Cambridge Univ. Press, London (1950).

For aerodynamics of an airfoil encountering gusts, see Chapter 15.
For the analysis of aircraft response to gusts of specific profiles, see

8.3 Bisplinghoff, R. L., G. Isakson, and T. H. H. Pian: Methods in Transient Stress Analysis. *J. Aeronaut. Sci.* **17**, 259–270 (1950).

8.4 Bisplinghoff, R. L., G. Isakson, and T. F. O'Brien: Gust Loads on Rigid Airplanes with Pitching Neglected. *J. Aeronaut. Sci.* **18**, 33–42 (1951).

8.5 Bryant, L. W., and I. M. Jones: Acceleration of an Aeroplane upon Entering a Vertical Gust. *Aeronaut. Research Com. R. & M.* **1496** (1932).

8.6 Bryant, L. W., and I. M. Jones: Stressing of Aeroplane Wings Due to Symmetrical Gusts. *Aeronaut. Research Com. R. & M.* **1690** (1936).

8.7 Carta, F., T. Pian, and H. Lin: Lift and Moment Growths on a Swept, Tapered, Rigid Wing upon Entering a Gust. *U.S. Air Force Tech. Rept.* **6358**, part IX (1953). Wright Air Development Center.

8.8 Donely, P.: Summary of Information Relating to Gust Loads on Airplanes. *NACA Rept.* **997** (1950). Supersedes *NACA Tech. Note* **1976**.

8.9 Fisher, H. R.: Acceleration of Aeroplanes in Vertical Air Currents. I, Calculation of the Acceleration Experienced by an Aeroplane Flying through a Given Gust. *Aeronaut. Research Com. R. & M.* **1463** (1932).

8.10 Goland, M., Y. L. Luke, and E. A. Kahn: Prediction of Wing Loads Due to Gusts, including Aero-Elastic Effects. I, Formulation of the Method. *U.S. Air Force Tech. Rept.* **5706** (1947). II, Design Procedure, *U.S. Air Force Tech. Rept.* **5751** (1949). Wright Air Development Center.

8.11 Greidanus, J. H., and A. I. van de Vooren: Gust Load Coefficients for Wing and Tail Surfaces of an Aeroplane. *Natl. Luchtvaartlab. Amsterdam Rept.* **F 28** (1948).

8.12 Hoene, H.: Influence of Static Longitudinal Stability on the Behavior of Airplanes in Gust. *NACA Tech. Memo.* **1323** (1951).

8.13 Hooke, F. H.: The Response of a Simple Flexible Aeroplane to Gusts of Various Forms. *Australia Aeronaut. Research Lab. Rept.* **SM 170** (1951).

8.14 Houbolt, J. C.: A Recurrence Matrix Solution for the Dynamic Response of Elastic Aircraft. *J. Aeronaut. Sci.* **17**, 540–550 (1950). See also *NACA Rept.* **1010** (1951).

8.15 Houbolt, J. C., and E. E. Kordes: Gust-Response of an Airplane including Wing Bending Flexibility. *NACA Tech. Note* **2763** (1952).

8.16 Jenkins, E. S., and C. D. P. Pancu: Dynamic Loads on Airplane Structures. *SAE Quart. Trans.* **3**, 391–409 (1949).

8.17 Jewel, J. W., Jr., and P. J. Carpenter: A Preliminary Investigation of the Effects of Gusty Air on Helicopter-Blade Bending Moments. *NACA Tech. Note* **3074** (1954).

8.18 Jones, R. T.: Calculation of the Motion of an Airplane under the Influence of Irregular Disturbances. *J. Aeronaut. Sci.* **3**, 419–425 (1936).

8.19 Kordes, E. E., and J. C. Houlbolt: Evaluation of Gust Response Characteristics of Some Existing Aircraft with Wing Bending Flexibility Included. *NACA Tech. Note* **2897** (1953).

8.20 Küssner, H. G.: Stress Produced in Airplane Wings by Gusts. *NACA Tech. Memo.* **654**. Translated from *Z. Flugtech. u. Motorluftschif.* **22**, 579–596 (1931). Concerning an error, see Sears, *J. Aeronaut. Sci.* **5**, 379 (1939). See also *Luftfahrt-Forsch.* **13**, 410 (1936).

8.21 Mazelsky, B., and F. W. Diederich: A Method of Determining the Effect of Airplane Stability on the Gust Load Factor. *NACA Tech. Note* **2035** (1950).

8.22 Mazelsky, B.: Charts of Airplane Acceleration Ratio for Gusts of Arbitrary Shape. *NACA Tech. Note* **2036** (1950).

8.23 Minelli, C., and V. Gröbner: Indagini sulle vibrazioni dei velivoli (Vibrazioni alari prodotte da una raffica verticale). Investigations of Airplane Vibrations, (vibrations of a wing produced by a vertical gust). *Aerotecnica* **19**, 373–403 (1939).

8.24 Pierce, H. B.: Investigation of the Dynamic Response of Airplane Wings to Gusts. *NACA Tech. Note* **1320** (1947).

8.25 Pierce, H. B., and M. Trauring: Gust-Tunnel Tests to Determine Influence of Airfoil Section Characteristics on Gust-Load Factors. *NACA Tech. Note* **1632** (1948).

8.26 Pratt, K. G.: A Revised Formula for the Calculation of Gust Loads. *NACA Tech. Note* **2964** (1953).

8.27 Putnam, A. A.: An Improved Method for Calculating the Dynamic Response of Flexible Airplanes to Gusts. *NACA Tech. Note* **1321** (1947).

8.28 Radok, J. R. M., and L. F. Stiles: The Motion and Deformation of Aircraft in Uniform and Non-uniform Atmospheric Disturbances. *Australia Council Sci. & Ind. Research Div. Aeronaut. Rept.* **ACA-41** (1948).

8.29 Radok, J. R. M.: The Problem of Gust Loads on Aircraft. A survey of the theoretical treatment. *Australia Dept. Supply and Development, Div. Aeronaut. Rept.* **SM 133** (1949).

8.30 Reisert, T. D.: Gust-Tunnel Investigation of a Flexible-Wing Model with Semichord Line Swept Back 45°. *NACA Tech. Note* **1959** (1949). See also *NACA Tech. Note* **1794** for sweep angle of 30°.

8.31 Rhode, R. V.: Gust Loads on Airplanes. *SAE J.* **40**, 81–88 (1937).

8.32 Sears, W. R., and B. O. Sparks: On the Reaction of an Elastic Wing to Vertical Gusts. *J. Aeronaut. Sci.* **9**, 64–67 (1941).

8.33 Smilg, B., and Sibert: Dynamic Overstress in Airplane Wings. *U.S. Air Force Tech. Rept.* **4437** (1939).

8.34 Tye, W.: Gusts, *J. Roy. Aeronaut. Soc.* **51**, 721–732 (1947).

8.35 van de Vooren, A. I.: Loads on Wing and Tail Surfaces of an Aeroplane Due to a Sinusoidal Gust Wave. *Natl. Luchtvaartlab. Amsterdam Rept.* **F 33**.

8.36 Williams, D., and J. Hanson: Gust Loads on Tails and Wings. *Aeronaut. Research Com. R. & M.* **1823** (1937).

8.37 Williams, D., and R. P. N. Jones: Dynamic Loads in Aeroplanes under Given Impulsive Loads with Particular Reference to Landing and Gust Loads on a Large Flying Boat. *Aeronaut. Research Council R. & M.* **2221** (1945).

8.38 Wills, H. A.: The Life of Aircraft Structures. *2d Intern. Aeronaut. Conf. N.Y.*, 361–403 (1949). Institute of Aeronautical Sciences, New York.

For the statistical aspects of the aircraft gust load problem, see

8.39 Diederich, F. W.: The Response of an Airplane to Random Atmospheric Disturbances. Ph.D. thesis, California Institute of Technology (1954).

8.40 Fung, Y. C.: Statistical Aspects of Dynamic Loads. *J. Aeronaut. Sci.* **20**, 317–330 (May 1953).

8.41 Houbolt, J. C.: Correlation of Calculation and Flight Studies of the Effect of Wing Flexibility on Structural Responses to Gusts. *NACA Tech. Note* **3006** (1953).

8.42 Press, H., and B. Mazelsky: A Study of the Application of Power-Spectral Methods of Generalized Harmonic Analysis to Gust Loads on Airplanes. *NACA Tech. Note* **2853** (1953).

8.43 Press, H., and J. C. Houbolt: Some Applications of Generalized Harmonic Analysis to Gust Loads on Airplanes. *J. Aeronaut. Sci.* **22**, 17–26 (1954).

8.44 Williams, M. L.: An Expression for the Gust Alleviation Factor. *J. Aeronaut. Sci.* **20**, *Readers' Forum* 723 (1953).

For fundamental information concerning atmospheric turbulence, see

8.45 Bannon, J. K.: Severe Turbulence Encountered by Aircraft Near Jet Streams. *Meteorol. Mag.* **80**, 262–269 (1951).

8.46 Batchelor, G. K.: *The Theory of Homogeneous Turbulence.* Cambridge Univ. Press, London (1953).

8.47 Clementson, G. C.: An Investigation of the Power Spectral Density of Atmospheric Turbulence. D.Sc. thesis, MIT (1950).

8.48 Clement, W. F., and G. R. Chippendale: A Statistical Study of Atmospheric Turbulence by Flight Measurements. Master thesis, MIT (1951).

8.49 Dryden, H. L.: A Review of the Statistical Theory of Turbulence. *Quart. Applied Math.* **1**, 7–42 (1943).

8.50 Gringorten, I. I., and H. Press: A Meteorological Measure of Maximum Gust Velocities in Clouds. *NACA Tech. Note* **1569** (1948).

8.51 Kármán, Th. von: The Fundamentals of Statistical Theory of Turbulence. *J. Aeronaut. Sci.* **4**, 131–138 (1937).

8.52 MacCready, P. B., Jr.: Investigation of Atmospheric Turbulence. Ph.D. thesis, California Institute of Technology (1952).

8.53 Lange, K. O.: Measurements of Vertical Air Currents in the Atmosphere. *NACA Tech. Memo.* **648** (1931).

8.54 Taylor, G. I.: Statistical Theory of Turbulence. *Proc. Roy. Soc. London, A.* **151**, 421–444 (1935); also 455–464.

8.55 Winchester, J. H.: Jet Stream, High Altitude Winds. *Aviation Age* **17**, (Apr. 1952).

For airplane gust load data, and their method of analysis, see Ref. 8.8, and the following papers. Data analyses of the same nature as Ref. 8.57 are found also in the following *NACA* publications: H. C. Mikleboro, and C. C. Shufflebarger, *Tech. Notes* **2150** (1950); **2424** (1951); H. N. Murrow and C. B. Payne, *Tech. Note* **2951** (1953); A. M. Peiser and W. G. Walker, *Tech. Notes* **1141** (1946); **1142** (1946); H. Press, *Tech.*

Note **1645** (1948); Press and J. K. Thompson, *Tech. Note* **1917** (1949); Press and R. L. McDougal, *Tech. Note* **2663** (1952); R. Steiner and D. A. Persh, *RM* **L52A28** (1952); Steiner, *Tech. Note* **2833** (1952); J. K. Thompson, *RM* **L51H07** (1951); H. B. Tolefson, *Tech. Notes* **1233** (1947); **1628** (1948); Tolefson and C. A. Gurtler, *RM* **L50K29a** (1951); G. W. Walker, *Tech. Notes* **3041, 3051** (1953).

8.56 Binckley, E. T., and J. Funk: A Flight Investigation of the Effects of Compressibility on Applied Gust Loads. *NACA Tech. Note* **1937** (1949).

8.57 Coleman, T. L., and P. W. J. Schumacher: An Analysis of the Normal Acceleration and Airspeed Data from a Four-Engine Type of Transport Airplane in Commercial Operation on an Eastern United States Route from November 1947 to February 1950. *NACA Tech. Note* **2965** (1953). See also similar expositions in *NACA Tech. Note* **2176** (1950); *NACA RM* **L9E18a** (1950).

8.58 Freise, H.: Spitzenwerte und Häufigkeit von Böenbelastungen an Verkehrsflugzeugen. *Jahrb. deut. Luftfahrt-Forsch.* 289–302 (1938).

8.59 Funk, J., and E. T. Binckley: A Flight Investigation of the Effect of Center-of-Gravity Location on Gust Loads. *NACA Tech. Note* **2575** (1951).

8.60 Hislop, G. S., and D. M. Davies: An Investigation of High-Altitude Clear-Air Turbulence over Europe Using Mosquito Aircraft. *Aeronaut. Research Council R. & M.* **2737** (1953).

8.61 Kaul, H. W.: Statistische Ergebungen über Betriebsbeanspruchungen von Flugzeugflügeln. *Jahrb. deut. Luftfahrt-Forsch.* I 307 (1938). See also p. I-274.

8.62 Peiser, A. M., and M. Wilkerson: A Method of Analysis of V-G Records from Transport Operations. *NACA Rept.* **807** (1945). Supersedes *NACA ARR* **L5J04.**

8.63 Press, H.: The Application of the Statistical Theory of Extreme Values to Gust-Load Problems. *NACA Tech. Note* **1926** (1949); *NACA Rept.* **991** (1950).

8.64 Press, H.: An Approach to the Prediction of the Frequency Distribution of Gust Loads on Airplanes in Normal Operations. *NACA Tech. Note* **2660** (1952).

8.65 Taylor, J.: Measurement of Gust Loads in Aircraft. *J. Roy. Aeronaut. Soc.* **57**, 78–88 (1953).

For gust alleviation, see

8.66 Curry, N. S.: Gusts and Their Alleviation. *Interavia* **4**, 203–206 (April 1949).

8.67 Mickleboro, H. C.: Evaluation of a Fixed Spoiler as a Gust Alleviator. *NACA Tech. Note* **1753** (1948).

8.68 Phillips, W. H., and C. C. Kraft: Theoretical Study of Some Methods for Increasing the Smoothness of Flight through Rough Air. *NACA Tech. Note* **2416** (1951).

For aircraft landing impact analysis, see the following from which further references can be obtained. In general, methods used for gust response analysis are also applicable to the landing response analysis.

8.69 Biot, M. A., and R. L. Bisplinghoff: Dynamic Loads on Airplane Structures During Landing. *NACA Wartime Rept.* **W-92**, *ARR* **4H10** (1944).

8.70 Fung, Y. C.: A New Approach to the Analysis of Dynamic Stresses in Aircraft Structures during Landing. *Aercon Pasadena Report to U.S. Bureau of Aeronautics* (1954). *GALCIT Dynamic Loads Rept.* **54-1** (1954). California Institute of Technology.

8.71 McBrearty, J. F.: A Critical Study of Aircraft Landing Gears. *J. Aeronaut. Sci.* **15**, 263–280 (May 1948).

8.72 Pian, T. H. H., and H. I. Flomenhoft: Analytical and Experimental Studies on Dynamic Loads in Airplane Structures during Landing. *J. Aeronaut. Sci.* **17**, 765–774 (Dec. 1950).

8.73 Ramberg, W.: Transient Vibration in an Airplane Wing Obtained by Several Methods. *Natl. Bur. Standards J. Research*, **42**, 437–447 (May 1949). See also Ramberg and A. E. McPherson, **41**, 509–520 (Nov. 1948).

For the stochastic analysis of seaplane landing and planning over ocean waves, see

8.74 Denis, M. St., and W. J. Pierson, Jr.: On the Motions of Ships in Confused Seas. *Proc. Soc. Naval Architects and Marine Engrs.* (1954).

8.75 Locke, F. W. S. Jr.: A New Approach to Determining the Hydrodynamic Characteristics of Flying Boats during Flight Tests. *U.S. Navy Bur. Aeronaut. Research Div. Rept.* **1063**. Presented at 7th Intern. Congr. Applied Mech. London (Sept. 1948). See also *Rept.* **1184**.

For the statistical properties of ocean waves, see

8.76 Neumann, G.: Über die Komplexe Natur des Seeganges. *Deut. Hydrograph. Z.* **5**, 95–110; 252–277 (1952).

8.77 Neumann, G.: On Ocean Wave Spectra and a New Method of Forecasting Wind-Generated Sea. *Dept. Army Corps Engrs. Beach Erosion Board Tech. Memo.* **43** (Dec. 1953).

8.78 Seiwell, H. R.: The Principles of Time Series Analyses Applied to Ocean Wave Data. *Proc. Natl. Acad. Sci.* **35**, 518–528 (1949).

The statistical theory touched upon in this chapter is exposed in the following:

8.79 Cramér, H.: *Mathematical Methods of Statistics.* Princeton Univ. Press, New York (1946).

8.80 Gumbel, E. J.: Statistical Theory of Extreme Values and Some Practical Applications. *Natl. Bur. Standards Applied Math. Ser.* **33** (1954). Govt. Printing Office, Washington, D.C.

8.81 James, H. M., N. B. Nichols, and R. S. Phillips: *Theory of Servomechanisms.* McGraw-Hill, New York (1947).

8.82 Lin, C. C.: On the Motion of a Pendulum in a Turbulent Fluid. *Quart. Applied Math.* **1**, 43–48 (1943).

8.83 Mazelsky, B.: Extension of Power Spectral Methods of Generalized Harmonic Analysis to Determine Non-Gaussian Probability Functions of Random Input Disturbances and Output Responses of Linear Systems. *J. Aeronaut. Sci.* **21**, 145–153 (1954).

8.84 Rice, S. O.: Mathematical Analysis of Random Noise. *Bell System Tech. J.* **23**, 282–332 (1944), and **24**, 46–156 (1945).

8.85 Wiener, N.: *Extrapolation, Interpolation, and Smoothing of Stationary Time Series.* John Wiley & Sons, New York (1950).

8.86 Gumbel, E. J., and P. G. Carlson: Extreme Values in Aeronautics. *J. Aeronaut. Sci.* **21**, 389–398 (1954).

Chapter 9

BUFFETING AND STALL FLUTTER

9.1 BUFFETING PHENOMENON

Buffeting is an irregular motion of a structure or parts of a structure in a flow, excited by turbulences in the flow.

Historically, the term buffeting was originated in connection with an accident of a commercial airplane (type Junkers F13) at Meopham, England, on July 21, 1930. Four passengers and two pilots were killed. Eye witnesses to the accident could report only seeing the airplane enter a cloud, hearing a loud noise almost immediately, and seeing the fragments fall to the ground.[9.2] The unusual circumstances of the accident led scientific organizations in England and Germany to undertake detailed investigations of the possible causes. The British Aeronautical Research Committee conducted extensive laboratory investigations and concluded that the most probable cause of the accident was "buffeting" of the tail. In these wind-tunnel-model investigations it was established that at large angles of attack the tail vibrated intensely but irregularly, and the "mean" amplitude of the vibration increased with the increase of speed of flow. An investigation of the meteorological circumstances at Meopham indicated the presence of strong rising air currents. Hence, an explanation of the cause of the Meopham accident was offered as follows: The airplane, flying horizontally at high speed, suddenly entered a region of strong rising gusts; as a result there was a sharp increase in the angle of attack, with the formation of flow separation over the wing. The tail, situated in the wing wake, was subjected to intense forced vibrations caused by the turbulences in the separated flow, which brought about the accident. The term *buffeting* was used by the British investigators and was explained as irregular oscillations of the tail unit, in which the stabilizer bent rapidly up and down and the elevator moved in an erratic manner.

About the same time, German scientists Blenk, Hertel, and Thalau[9.5] conducted a series of laboratory and flight tests, using the same type of airplane as the one involved in the Meopham accident. The laboratory investigations showed that it was possible for the airplane to buffet, but in actual flight, except during a steep dive, buffeting of sufficient intensity to endanger the structures of the tail was not observed. For this reason,

Blenk concluded that the Meopham accident was probably caused by overstressing of the wing due to high gust or maneuvering load. Buffeting was investigated by means of motion pictures in flight. The following conclusions were reached:

1. The tail surfaces, at large angles of attack of the wing, entered a region of vortices springing from the intersection of the wing and fuselage,

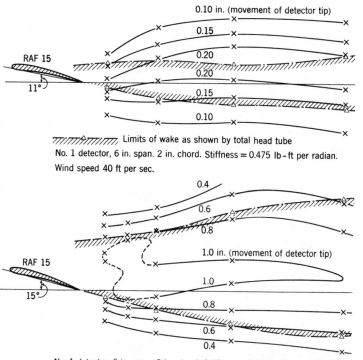

Fig. 9.1.   Duncan's buffeting-intensity contour.   From *Aeronaut. Research Council R. & M.* **1541.**   (Courtesy of the Aeronautical Research Council.)

and vibrations were observed.   The vortices arose on both sides of the fuselage, usually in an unsymmetrical manner.   The tail vibrations were irregular in amplitude and frequency; and, as a rule, large amplitudes were rare and continued only for a very short time.

2. The recorded amplitudes, plotted against the corresponding frequencies, showed clearly three resonances.   These resonance frequencies corresponded with the first three natural frequencies in the dynamic tests in the hangar.

3. The amplitudes increased slowly with the flight speed.

In 1933, Duncan and his associates published two papers on buffeting.[9.7, 9.8] In an attempt to separate numerous factors, Duncan investigated the case where the airfoil creating the disturbances was of infinite aspect ratio and where a "detector" (which played the part of a buffeting tail surface) was so arranged that its only possible movement was bending about a pair of flexural hinges, and, hence, it had only a single natural frequency. The oscillations of the detector were recorded optically. The relative position of the detector and the wing in front could be changed both vertically and horizontally. Hence, it was possible to explore a whole region behind the wing. The most interesting feature of these tests is illustrated in Fig. 9.1, in which contours of equal "buffeting intensity" are plotted. Here the "buffeting intensity" means the greatest total range of movement of the detector during a standard exposure of the camera. In the same figure is shown the width of the wake as obtained by a total-head tube.* Note that the buffeting-intensity contours do not coincide with the total-head wake.

Since these early researches show that tail buffeting is a result of flow separation, it is clear that tail buffeting can be prevented by preventing separation. This can be effected by a proper fillet† at the wing-fuselage junction, by boundary-layer control, and by limiting the operating conditions of the airplane. Tail buffeting can also be avoided by locating the airfoil outside the region where disturbances exist.

For modern high-speed airplanes, flow separation over the wing or the wing-fuselage junction at transonic speeds of flight causes a very serious

* Within a certain region the total head of the flow is smaller than that of the undisturbed stream. Although the transition to the normal value takes place gradually, the boundaries of this "total-head wake" are quite well defined. It is not clear how to interpret the total-head wake in a turbulent flow. However, the main indication of the experiment is that the buffeting intensity is related to the *turbulences* in the flow.

† A proper fillet, i.e., a smooth fairing at a wing-fuselage junction, is most effective in reducing flow separation caused by wing-fuselage interference. Von Kármán, in *Aerodynamics, Selected Topics in the Light of Their Historical Development*, p. 151 (Cornell Univ. Press, 1954), relates an interesting story about his reporting of the effectiveness of a fillet in preventing tail buffeting in a lecture in Paris in 1932. Some French engineers were apparently troubled by the same difficulty and immediately tried Kármán's fillets. Henceforth, the "fillet" was known in France as "karman," and expressions like "large karman" and "small karman" are used. Von Kármán himself attributes the development of fillet to the team work of himself, Clark Millikan, and Arthur Klein at the California Institute of Technology. See Kármám, Th. von.: Quelques problèmes actuels de l'aérodynamique. *J. techniques internationales de l'aéronaut*, 1–26 (Paris, 1933), Klein, A. L.: Effect of Fillets on Wing-Fuselage Interference. *Trans. A.S.M.E.* **56**, 1–7, AER–56–1 (1934).

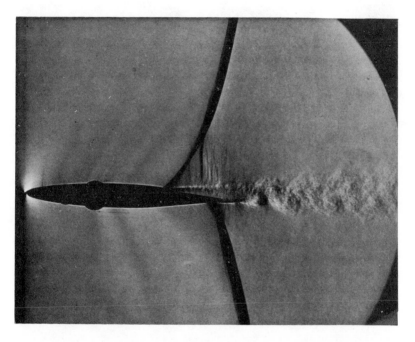

Schlieren photograph of the eddying wake following a shock-induced flow separation. The dark lines are shock waves. (Courtesy of National Physical Laboratory, England; photo by D. W. Holder)

Schlieren photograph of a 10 per cent scale model of the Nimbus
spacecraft with Snap-19 generator in the 50-inch Hypersonic Tunnel B
at Mach number 8 and Reynolds number of 0.42×10^6 per foot.
Photo by Optical Systems group. Von Karman Gas Dynamics
Facility, ARO, Inc. Arnold Engineering Development Center,
Arnold Air Force Station, Tennessee.

problem of tail buffeting. Tail buffeting of very high intensity has been observed. The general feature as demonstrated in Fig. 9.1 remains in a transonic flow, although the contour lines are different. The intensity of turbulence and the location of the highest intensity change with the wing thickness, camber, and angle of attack. Separation of the flow at the leading edge of a thin wing at moderate angles of attack causes the most severe tail buffeting. Control over such leading-edge separation, for instance, by proper camber near the leading edge, will be most helpful in reducing the intensity of tail buffeting at transonic speeds.

The general picture is therefore as follows: Whenever separation occurs in a flow, the turbulence level increases. If an airfoil is situated in a turbulent flow, it buffets. The term buffeting, however, will not be limited to airfoils. The oscillation of a smokestack in the wake of another smokestack is also an example of buffeting.

In § 9.2 the limits of the flight speed and the angle of attack beyond which tail buffeting may occur are discussed. In § 9.3 some remarks on the theories of tail buffeting are given. Buffeting of an airfoil caused by flow separation over the airfoil itself, in a flow which is otherwise free from turbulence, will be discussed in § 9.6.

9.2   TAIL-BUFFETING BOUNDARIES

Since tail buffeting is associated with flow separation over parts of the airplane ahead of the tail, it can be avoided by keeping the operation attitude of an airplane below the separation limits.

Figure 9.2 shows a plot of the lift coefficient versus the Mach number at various angles of attack of an NACA 2409–34 airfoil.[9.15] It is seen that, at each angle of attack, the lift coefficient increases with increasing Mach number until a maximum is reached; then it drops sharply with further increase in speed. This drop is associated either with a strong shock-wave formation or with flow separation. The airfoil is said to be shock-stalled when severe drop of C_L occurs.

At lower Mach number, Fig. 9.2 shows that a maximum C_L is reached at certain angle of attack. This angle (of the order of 12° for the NACA 2409–34 airfoil) is the stalling angle in the usual sense.

When the high-angle-of-attack stall and the shock-stall limits are plotted together, a figure showing the boundaries of the lift coefficients beyond which stall occurs is obtained (Fig. 9.3).

It is known that stall at high angle of attack is associated with flow separation. But a shock wave does not necessarily initiate complete breakoff of the flow from the airfoil. Moreover, high angle of attack and high speed of flow are not the only causes of separation. Improper fillets at the

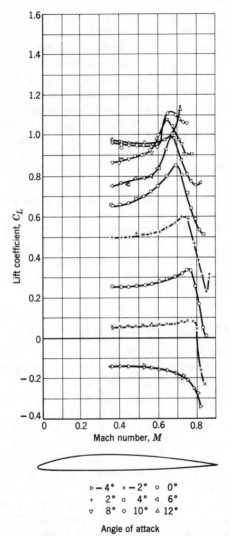

▷ −4°	× −2°	○ 0°
+ 2°	□ 4°	◁ 6°
▽ 8°	◇ 10°	△ 12°

Angle of attack

Fig. 9.2. C_L versus M curves for NACA 2409–34 airfoil.   From Stack and von Doenhoff, *NACA Rept.* **492.**   (Courtesy of the NACA.)

wing-fuselage junction, etc., cause separation readily. Thus for a given airplane, the boundaries indicating the onset of separation cannot be determined from the wing alone.

Nevertheless, for a given airplane, a boundary on a C_L vs. M chart can be determined which separates the region of possible tail buffeting from that of a smooth potential flow.　Such a *tail-buffeting boundary* (see, for example, Fig. 9.4) resembles in appearance the stall boundary of Fig. 9.3.　It is determined by flight testing.[9.16]

It should be noticed that two curves are shown in Fig. 9.4, which gives the buffeting boundaries for the horizontal tail of a fighter-type airplane that has a low-drag wing section.　One curve, labeled "abrupt pull-ups" was obtained by pulling up the airplane abruptly at various altitudes and Mach numbers, the degree of abruptness being limited by the inertia,

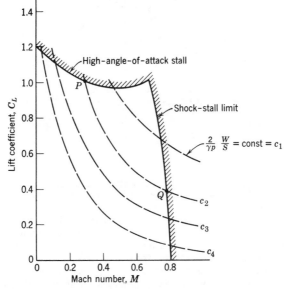

Fig. 9.3.　Stall boundary of the NACA 2409–34 airfoil (steady-state wind-tunnel tests).

control power, and the stability of the airplane.　The other curve was obtained by gradual stalls made in turns.　The difference between these two curves indicates the effect of the rate of change of angle of attack on the flow separation, which will be discussed further in Chapter 15.　For the test airplane from which Fig. 9.4 was obtained, buffeting of the horizontal tail in abrupt pull-ups occurred simultaneously with the attainment of maximum normal force at Mach numbers below about 0.64 (solid curve in figure).　Above this value of Mach number, tail buffeting in abrupt pull-ups occurred before the attainment of the maximum normal force (dotted curve).　In the range tested, altitude, and hence the Reynolds number, had no effect on the tail-buffeting boundary determined in abrupt pull-ups.

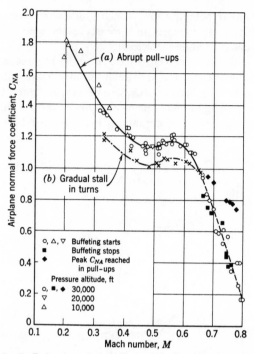

Fig. 9.4. Buffeting boundaries obtained in abrupt and gradual stalls for a test airplane. The airplane normal force coefficient C_{NA} is defined as the ratio $nW/(qS)$, where n is the load factor, i.e. the airplane normal acceleration measured at the center of gravity expressed as multiples of the gravitational acceleration. W is the airplane gross weight, q is the dynamic pressure, S is the wing area. The normal force (perpendicular to thrust line), rather than the lift (perpendicular to relative wind direction), is used because it is more convenient to be defined and measured in transient conditions. In steady flight $C_{NA} \doteq C_L$. Flight measurements by Stokke and Aiken, Ref. 9.16. (Courtesy of the NACA.)

Since in horizontal steady flight the lift L balances the weight W of the airplane, we have

$$W = L = \tfrac{1}{2}\rho U^2 S C_L \tag{1}$$

where S is the total wing area. Hence, in terms of the Mach number, we have

$$M^2 C_L = \frac{W}{S}\frac{2}{\rho a^2} = \frac{W}{S}\frac{2}{\gamma p} \tag{2}$$

where a denotes the speed of sound, p denotes the static pressure, and $\gamma = C_p/C_v$ denotes the ratio of specific heats at constant pressure and constant volume.

Thus the relation between the lift coefficient and the Mach number depends on the wing loading W/S and the static pressure p, which in turn depends on the altitude z. A curve showing the relation 2 appears as a dotted line in Fig. 9.3. When such curves are plotted on Fig. 9.4, the permissible range of airplane flight speed as limited by tail buffeting will be given by the intersections of the tail-buffeting boundaries with these curves. Since the atmospheric pressure p decreases with altitude, it is seen that the permissible range of flight speeds becomes narrower as the altitude increases. The segments of the $M^2 C_L = $ const lines between the tail buffeting boundary indicate the maneuverability of the airplane as far as tail buffeting is concerned. Clearly, it decreases with increasing altitude.

9.3 THEORIES OF TAIL BUFFETING

The similarity of the wake behind a stalled airfoil and that behind a circular cylinder suggests at once a "vortex-shedding" theory of tail buffeting. In Abdrashitov's analysis[9.1] the wing wake is regarded as a well developed vortex street in which the vortices are regularly spaced, and the buffeting motion of the tail is computed as a forced vibration. The random characteristics of tail buffeting are neglected.

A different idealization is made by Shih-chun Lo[9.12] who regards the wing wake as a vortex sheet—an "interface" without thickness, across which the flow undergoes a sudden change in velocity and density. Velocity potential is assumed to exist on both sides of the interface. It is shown that the vortex sheet reacts strongly with an oscillating airfoil located in the neighborhood of the interface, particularly for higher values of the reduced frequency. Large reduction of flutter speed, due to the presence of an interface, is indicated. It is impossible to verify experimentally Lo's mathematical theory, because such an idealized interface cannot be produced. However, wind-tunnel tests made by Dankworth and Walker,[9.6] using a "barrier" which blocks partly the flow in the working section, do not indicate any significant change in flutter speed of an airfoil when the distance between the airfoil and the "interface" (which is taken as the tangent to the barrier) exceeds 20 per cent of the chord length of the airfoil. When this distance is less than 20 per cent of the chord length, irregular motion of the airfoil is observed which is clearly due to turbulences in the jet mixing region.

Liepmann[9.11] points out that buffeting is the response of an elastic body to a turbulent flow, and hence is a stochastic process. A correct theory of buffeting, therefore, must account for the turbulent characteristics of the oncoming flow. If the power spectrum of the turbulences in the flow is known, the intensity of the buffeting motion can be calculated in a manner similar to the gust-response calculation of the previous chapter, provided that the intensity of turbulence is so low and the amplitude of the wing motion is so small that the wing is never stalled during buffeting. If stalling did occur, even only for a fraction of the time, the relation among the lift, the turbulent fluctuations, and the wing motion becomes nonlinear, and the calculation of the buffeting intensity is more difficult. We must point out that, turbulence being a stochastic process, large-velocity fluctuation may exist, though infrequently, even if the root mean square of the velocity fluctuation is small. Hence the chance of the wing's being stalled always exists.

One feature of the flow in the wake of a stalled wing seems undisputed. As with flow in the wake of a circular cylinder (see § 2.3), there may exist one or more frequencies at which the power spectrum has sharp peaks. These "predominant" frequencies are those at which the kinetic energy of the turbulent motion is more or less concentrated. Experimental values of the predominant frequency in the wake behind flat plates and airfoils in incompressible fluids are given by Fage and Johansen;[9.39] Tyler;[9.40] Blenk, Fuchs, and Liebers;[9.36] and Dunn and Finston.[9.38] All agree that, for angles of attack above about 30°, the reduced frequency, at the predominant peak of the power spectrum, is nearly a constant, independent of the angle of attack, provided that the projected length of the chord normal to the direction of flow is taken as the characteristic length. If the chord length is c and the angle of attack is α, the projected length is $c \sin \alpha$. The reduced frequency that remains constant is $nc \sin \alpha/U$, where n is the wave number (cycles per second) of the vortices, and U is the mean speed of flow. There are numerical discrepancies between various authors as shown in Table 9.1.

For smaller angles of attack, the variations of the values of the reduced frequency $nc \sin \alpha/U$ given by various authors become large, and seem strongly affected by the Reynolds number.* For angles of attack near and below the stalling angle, the reduced frequency $nc \sin \alpha/U$ decreases. The results of Dunn and Finston[9.38] are shown in Fig. 9.5. Tyler's result is similar, but the absolute values are smaller, whereas Blenk's result shows very little decrease in the reduced frequency $nc \sin \alpha/U$ for α down to 10°.

It may be noticed that the results as shown in Fig. 9.5 suggest that, for angles of attack α less than about 17°, the predominant vortex frequency is nearly proportional to α. Hence, a reduced frequency based on the chord length itself, nc/U, will remain nearly a constant. The average value of nc/U, for α between 8 to 17°, is approximately 0.54.

---

* See Chuan and Magnus.[9.21] The reduced frequency $nc \sin \alpha/U$ increases from 0.112 at $R = 9.3 \times 10^4$ to 0.178 at $R = 21 \times 10^4$, in the range of α between 18 and 29°.

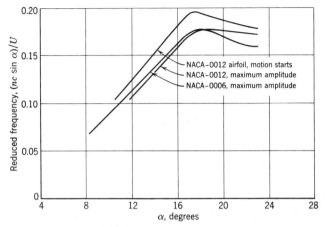

Fig. 9.5. The variation of the reduced frequency $(nc \sin \alpha)/U$ with angle of attack α. Measurements by Dunn and Finston, Ref. 9.38. Reynolds number 6×10^4 to 24×10^4.

Table 9.1   The Predominant Vortex Frequency in the Wake of Flat Plates and Airfoils

(a) For Flat Plate

Authors	$\dfrac{nc}{U} \sin \alpha$	Range of α, degrees	Reynolds No. (Based on Chord)
Fage, Johansen	0.148	30–90	$3–18 \times 10^4$
Tyler	0.158	30–90	150–4000
Blenk, Fuchs, Liebers	0.18	30–90	$0.6–2.4 \times 10^4$

(b) For Airfoils

Authors	Airfoil	$\dfrac{nc}{U} \sin \alpha$	Range of α, degrees	Reynolds No. (Based on Chord)
Tyler		0.150	30–90	150–4000
Blenk, Fuchs, Liebers	Göttingen 387, 409, 411	0.21	30–90	$0.6–2.4 \times 10^4$
Dunn, Finston	NACA–0012	0.17	20–45	$6–24 \times 10^4$

It may also be noted that the value of $nc \sin \alpha / U$ in Fig. 9.5, when multiplied by 2π, give values of reduced frequency of comparable magnitude to those shown in Fig. 2.4 for the wake of a circular cylinder.

9.4 STALL FLUTTER

The term flutter is applied categorically to oscillations of an elastic body in a flow that is steady in the absence of the body. If, during part or all of the time of oscillation the flow is separated, then the flutter phenomenon exhibits some characteristics different from those discussed in Chapters 5–7 and is called a stall flutter.

Stall flutter is a serious aeroelastic instability for rotating machineries such as propellers, turbine blades, and compressors, which sometimes have to operate at angles of attack close to the static stalling angle of the blades. Airplane wings and tails rarely suffer from stall flutter. However, the trend toward thin wing sections and large wing span has increasingly made stall flutter of wings a serious concern for the design of high-speed aircraft.

The phenomenon of stall flutter can be best illustrated by the spinning test of a propeller. Generally, when the speed of a propeller is gradually increased, the appearance of a peculiar noise normally associated with propeller-blade flutter can be quite definitely determined. As the speed is increased beyond a value at which flutter can first be detected aurally, considerable weaving of the propeller tips can be detected visually by an observer in the plane of the propeller disk. The blade motions can be recorded by attaching, to the blades, strain gages which measure the amplitude of strain as the propeller speed changes.

Figure 9.6 shows a typical result of the spinning test of a particular propeller obtained by Sterne.[7.140] The abscissa is the blade angle at 70 per cent radius of the propeller; the ordinate is the propeller speed. Above the solid curve is a region of flutter, below it, there is no flutter. Over the wide range of blade settings of this particular propeller, there are two very abrupt changes of flutter speed. These two abrupt changes divide the range of blade settings at which the propeller is tested into three regions—conveniently described as fine pitch, medium pitch, and coarse pitch. The characteristics in each region are as follows:

At very fine pitch settings, corresponding approximately to zero lift angle at the blade tips, there is no flutter in the speed range of the tests. In the fine-pitch range, the critical speed first decreases as the pitch angle is increased, and then tends to a constant value at larger angles of attack.

In the medium-pitch region, the critical flutter speed is constant, but is much lower than that in the fine- and coarse-pitch regions.

In the coarse-pitch region, the flutter speed is constant and is approximately equal to that at the coarser end of the fine-pitch region.*

Clearly, there are two distinct types of flutter. The blade settings in the medium-pitch range correspond to the stalling angles of the tip sections. In the medium-pitch range the blade stalls over part of the cycles of oscillation. In the coarse-pitch region the blade remains stalled throughout the cycles.† The flutter in the medium- and coarse-pitch regions is

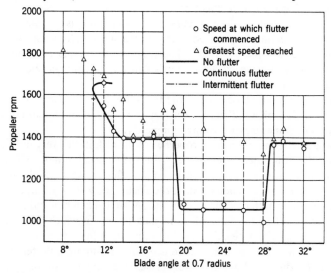

Fig. 9.6. Propeller flutter. Critical flutter speed against blade angle at 0.7 radius. Steady-state stalling angle of airfoil equal to 12 degrees, corresponding to a blade angle of 20 degrees at 0.7 radius of propeller. Measurements by Sterne, Ref. 7.140. (Courtesy of the Aeronautical Research Council.)

stall flutter. That in the fine-pitch region is classical flutter. The flutter motions of the blade in both regions are nearly sinusoidal.

In general, as the angle of attack is raised through the stall region, the following characteristics may be observed:

* This equality of the flutter speed in the fine-pitch and coarse-pitch regions must be regarded as a mere coincidence in this particular example.

† Stall flutter in the coarse-pitch range, i.e. at high angles of attack, resembles a forced vibration. In fact, if the speed of flow is gradually raised beyond the stall-flutter boundary at such a high angle of attack, the intensity of flutter increases rapidly at first, reaches a maximum, then decreases to practically flutter-free condition. The reason is that, at such high angles of attack, the predominant frequency in the wake (vortex street, cf. § 9.3) becomes clearly defined. The stall flutter of the wing is simply the reaction to the creation of the vortices in the wake. The phenomenon resembles the oscillation of smokestacks as analyzed in § 2.4.

1. The flutter speed drops severely.

2. The flutter frequency rises slowly toward the natural torsional oscillation frequency of the blade in still air.

3. The torsional motion predominates. Whereas in classical flutter the torsional strain and bending strain are of the same order of magnitude, in stall flutter the amplitude of the bending oscillation becomes negligibly small in comparison with that of the torsion, although the axis of rotation is in general not the elastic axis.[7.124]

4. Usually the flutter speed reaches a minimum and rises again as the blade becomes completely stalled. The flutter amplitude becomes very small in the coarse-pitch region. The range of the angle of attack in which the violent low-speed flutter persists increases as the elastic axis is moved backward along the chord.

5. There is a large phase shift at the transition from classical flutter to stall flutter. The phase difference between the bending and torsion drops by about 45°, sometimes vanishing completely.

6. Variation of the structural properties of the airfoils has very different effects on the stall flutter as compared with the classical flutter. Change of inertia-axis location has little effect on stall flutter, which could occur even when the inertia axis lies ahead of the elastic axis when it is impossible to obtain classical flutter. The ratio of the uncoupled bending and torsion frequencies in still air has little effect on stall flutter, for which the critical speed is often higher when the ratio is equal to 1 than at other values (whereas at low angles of attack there is usually a minimum flutter speed when the bending and torsion frequencies are equal). In one case, Stüder[9.26] found the critical speed to be the lowest when the wing is restrained from translatory motion and is allowed only freedom in pitch.

Stüder[9.26] was the first one who studied stall flutter experimentally in great detail. In a series of tests he examined the field of flow around an airfoil oscillating about the stalling angle. His result shows that, in a stroke of increasing amplitude, the separation is delayed to an angle of attack appreciably greater than that for a stationary airfoil. On the return movement, re-establishment of a smooth flow is also delayed. Stüder calls this an "aerodynamic hysteresis" and concludes that it is the basic cause for stall flutter. This observation is supported by the earlier works of Farren and others (see § 15.4).

Thus the influence of flow separation on stall flutter is revealed through the hysteresis effect. Since the aerodynamic characteristics is nonlinear, the principle of superposition does not hold. The aerodynamic forces corresponding to different modes of motion cannot be added. Hence, an extensive experimental evaluation of the aerodynamic forces is necessary (see surveys by Victory,[9.27] Halfman,[9.22] and Bratt[15.121]).

A sketch showing typical forms of the hysteresis loop corresponding to simple-harmonic oscillations in pitch is given in Fig. 9.7. At smaller angles of attack, the loop is elliptical, as predicted by the linearized theory. At angles of attack in the neighborhood of the static stalling angle, the loop is of the shape of a figure 8 (the left-hand side loop may become very small or disappear entirely, leaving a loop which indicates all positive work). At large angles of attack, the loop becomes oval again. The exact shape of the loops depend on the location of the axis of rotation,

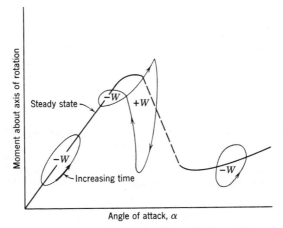

Fig. 9.7. Aerodynamic hysteresis. From Halfman, Johnson, and Hayley, Ref. 9.22. (Courtesy of the NACA.)

the reduced frequency, the amplitude of oscillation, the Reynolds number, and the airfoil shape.

The area within each loop represents the work done by the air on the airfoil in each cycle of oscillation. A positive sign in front of W in Fig. 9.7 indicates an energy gain by the airfoil; a negative sign, an energy loss. The oscillation is aerodynamically unstable with respect to pitch if the net gain of energy during each cycle is positive.

Since for a stalled airfoil the aerodynamic force produced by a simple-harmonic motion, though periodic, is not exactly simple harmonic, the work per cycle must be obtained by graphical integration of experimental results. In Fig. 9.8 is shown the work per cycle in pure pitching motion computed[9.22] from the data obtained by Bratt and his associates[9.27] (airfoil's steady-state stalling angle 12°, elastic-axis location mid-chord, Reynolds number 1.42×10^5). In this figure, α_i denotes the mean angle of attack, $\Delta\alpha$ denotes the difference between the mean angle of attack and the steady-state stalling angle. A positive value of $\Delta\alpha$ means that α_i is above the steady-state stalling angle.   The amplitude of pitching oscillation is approximately 6°.   Points above the $W = 0$ line correspond to aerodynamically unstable oscillations.   A curve corresponding to the linearized

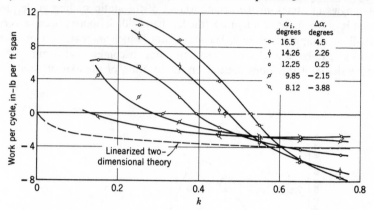

Fig. 9.8.   Work per cycle in pure pitch.   Elastic axis at mid-chord.   Measurements by Bratt et al., Ref. 9.27, for an airfoil whose static stalling angle is 12 degrees, at Reynolds number 1.42×10^5.   Figure by Halfman, Ref. 9.22. The work per cycle are computed for wind speed 95 mph, semichord 0.484 ft, and amplitude of oscillation 6.13 degrees.   (Courtesy of the NACA.)

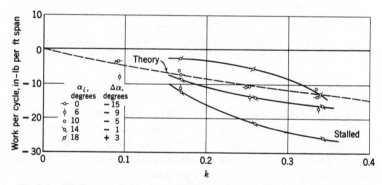

Fig. 9.9.   Work per cycle in pure translation.   Static stalling angle 12 degrees. Reynolds number 10^6.   Wind speed 95 mph.   Semichord 0.484 ft.   Vertical translation amplitude 0.9 inches.   From Halfman et al., Ref. 9.22.   (Courtesy of the NACA.)

two-dimensional theory is added for comparison.   When α_i is small, the experimental work per cycle agrees closely with the theoretical prediction (cf. Figs. 20–26 of Ref. 9.22).

Figure 9.9 shows the work per cycle in pure translation obtained by Halfman[9.22] for a wing designated as "blunt," whose steady-state stalling angle is 15°, at a Reynolds number near 10^6. It is seen that the work per cycle remains negative for the range of reduced frequencies tested. The oscillation in pure translation is therefore stable, a conclusion in agreement with the tests by von Kármán and Dunn[2.22] (§ 2.5).

Because of the nonlinear nature of the aerodynamic response, the British and American experimental results cannot be compared with each

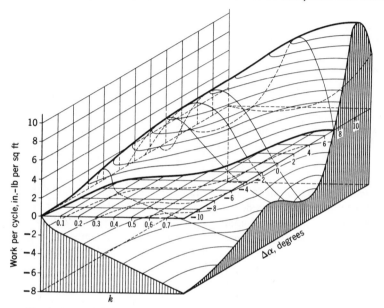

Fig. 9.10. Relief of typical variation of work per cycle in pure pitch with $\Delta\alpha$ and the reduced frequency k. Wind speed 95 mph. Elastic axis location 37 per cent chord aft leading edge. Semichord 0.484 ft. Amplitude of pitching oscillation 6.08 degrees. From Halfman, Ref. 9.22. (Courtesy of the NACA.)

other quantitatively, since the Reynolds number, the airfoil shape, and the reduced frequency range are different. But, using $\Delta\alpha$ as a basis,

$$\Delta\alpha = \alpha_i - \alpha_{stall}$$

where α_i is the mean angle of attack and α_{stall} is the steady-state stalling angle, Halfman[9.22] obtains an interesting qualitative comparison. From Halfman's results[9.22] in the lower k range and the British data[9.27] in the higher k range, Fig. 9.10 is constructed. This figure shows qualitatively the way in which work per cycle varies with k and α. First, for a low angle of attack, the curves of work per cycle against k remain negative, quite close to the theoretical curves. At given values of k the work per cycle gradually approaches zero as $\Delta\alpha$ increases. When $\Delta\alpha$ is zero, a positive work area appears at lower values of k. As $\Delta\alpha$ increases further, the maximum value of positive work increases, whereas the range of positive work is narrowed but continues to move to higher values of k.

Since stall flutter is predominantly torsional, it may be assumed, as a crude approximation, that the flutter frequency is the same as the natural vibration frequency of torsion in still air, and that stall flutter will occur when the aerodynamic work per cycle for pure pitch becomes positive and greater than the energy dissipated by the structure. From Fig. 9.10 it is seen that there is a region on the $(k, \Delta\alpha)$ plane in which the work per cycle is positive. For each $\Delta\alpha$, this region is bounded by two critical values of k. Let the upper critical value of k be denoted by k_{cr}. For a given airfoil, at a given angle of attack, the value of k gradually decreases as the speed of flow gradually increases ($k = \infty$ when $U = 0$, the frequency ω being that of the natural torsional vibration). When U reaches U_{cr}, where

$$U_{cr} = \frac{\omega b}{k_{cr}} \qquad (1)$$

a pitching oscillation of the airfoil causes no exchange of energy with the airstream, and gives a crude estimation of the critical stall flutter speed. More accurate methods of analysis are discussed in Refs. 9.22, 9.27.

Figure 9.11 shows a summary of the k_{cr} values obtained in various laboratories. The effects of location of the axis of rotation, airfoil shape, Reynolds number R, and amplitude of oscillation θ are revealed by the wide variation among the results.

9.5   PREVENTION OF THE STALL FLUTTER

According to Eq. 1 of § 9.4, the critical stall-flutter speed can be raised by increasing the frequency of free torsional oscillation of the blade, i.e., by increasing the torsional stiffness and reducing the mass moment of inertia.

Stall flutter can be delayed if the airfoil can be prevented from stalling. Thus, in designing propellers, care must be exercised to choose the proper airfoil section, and, if possible, the working angle of attack should be limited to below the stalling angle.

It is shown by Theodorsen and Regier[4.36] that the divergence speed of a propeller is an important parameter to consider in connection with the stall flutter (see § 4.8). If the rotational speed of a propeller is so high that the relative wind speed is close to the critical-divergence speed, the blade will be twisted excessively, possibly beyond the stalling angle, and thus causing stall flutter.   Theodorsen and Regier show that, for several models of wind-tunnel propellers tested in the high-angle-of-attack range (with initial settings below the stalling angle), flutter invariably occurs at a speed substantially below the classical flutter speed.   The angle of attack of the blade at which flutter occurs appears to be nearly constant and independent of the initial blade setting.   Apparently the blade simply

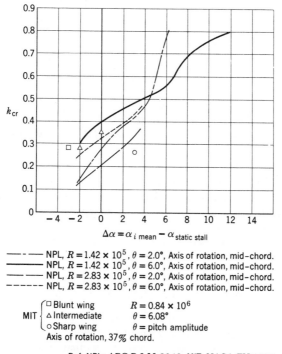

Fig. 9.11.   Critical reduced frequency.

twists to the stalling angle, and flutter starts.   Furthermore, Theodorsen and Regier show that the classical flutter speed and the divergence speed of a propeller are approximately the same because of the centrifugal force effect.   The problem of predicting propeller flutter is thus resolved primarily into the calculation of the speed at which the propeller will be twisted to stall.   This can be done by methods of Chapter 3.

As to the practical design measures to raise the critical stall-flutter speed, we may quote an interesting case reported by Sterne and Brown.[7.141]

They tested on a spinning tower a variable-pitch propeller fitted with compressed wood blades. The airfoil sections of the blade inboard of the 75 per cent radius were of conventional Clark shape, but outboard of the 75 per cent radius they had an undercamber. Serious flutter was encountered when the blades were set at angles larger than 20° at the 70 per cent radius. But, when the blades were set at 19° and less, there was no evidence of flutter. The difference in the flutter characteristics at the blade-angle settings of 20° and 19° was very pronounced. It appeared that the flutter was stall flutter. A pitch setting of 20° corresponded to a lift coefficient C_L of about 1.0 at the tip sections. The propeller was then modified by the removal of part of the leading edge of the sections outboard of 75 per cent radius, and reshaping the cross sections near the blade tips into the conventional Clark form. This modification had the effect of bringing the center of pressure further back, relative to the centroids of the inboard sections, so that the torsional deflection of the blades became smaller. The modified propeller was then retested. No flutter was encountered for blade settings up to 22° at the 70 per cent radius.

So far our attention is directed toward stall flutter at low speed of flow (incompressible fluid). The stall-flutter problem of high-speed aircraft is complicated not only by the effect of the compressibility of the air, but also by the geometrical factors often associated with high-speed wing designs: low aspect ratio, thin wing sections, sweep angle, large masses attached to the wings, etc. To determine the effects of these items on the stall-flutter characteristics of a wing is a challenging problem for future research.

9.6   BUFFETING FLUTTER OF A WING

A question arises naturally in connection with the study of tail buffeting: What happens to the wing that produces the turbulent wake? A separated flow over the wing creates aerodynamic forces that must be regarded as stochastic processes. The wing's elastic response is necessarily also stochastic and therefore may be properly called buffeting. However, as the unsteady aerodynamic forces are produced by the wing itself, in a flow that otherwise is steady, the motion of the wing may, according to the definition of § 9.4, be called flutter. The nature of the wing motion also borders between buffeting and stall flutter described in the preceding sections. Sometimes the motion is quite random, sometimes it is quite regular, and sometimes it is a mixture of the two: e.g., random in bending, while more or less a regular sinusoidal oscillation in torsion. In order to distinguish the last feature from those described in the preceding sections, we shall use a new term: *buffeting flutter*.

Buffeting flutter of a wing is a problem of grave concern in transonic flight. It frequently causes wing and aileron damages.

Regarding buffeting flutter as a mixture of buffeting and stall flutter, we anticipate a situation as follows. Consider a cantilever wing. The buffeting boundary of the wing will be similar to that of the tail, because of their common origin in flow separation. For a hypothetical wing this boundary is sketched in Fig. 9.12 (cf. Fig. 9.4). On the other hand, if the stall-flutter characteristics of the wing is similar to that shown in Fig. 9.6, and if we plot the stall-flutter boundary with angle of attack as

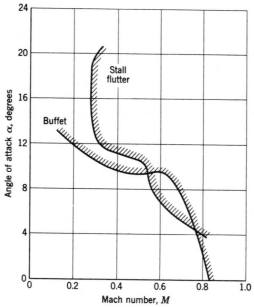

Fig. 9.12.   Buffeting flutter of a wing.

ordinate and Mach number as abscissa, the result of Fig. 9.12 may be obtained. The relative position of these two boundaries depends on the wing planform, airfoil section, wing-fuselage junction geometry, etc. The boundaries as shown in Fig. 9.12 intersect each other. The wing motion beyond the buffet boundary is random in amplitude and frequency, and in general is predominantly bending, with very small torsional motion. As the stall-flutter boundary is reached, torsional motion of more or less regular amplitude and frequency takes place, while random motion in bending may continue.

One may question the definitiveness of the terms buffeting, stall flutter, and buffeting flutter. Indeed, since fluctuations in the flow, whether caused by the wing itself or by other means, are the basic reason of all these phenomena, it is impossible to distinguish them with absolute clarity. Therefore they should be interpreted only as descriptive terms useful for engineering purpose: Buffeting characterizes the irregularity of the motion, stall flutter the more or less regular oscillations, and buffeting flutter a mixture of the two in different degrees of freedom.

As mentioned before, buffeting flutter is a serious problem in high-speed flight. Flight at high subsonic Mach numbers is often troubled by strong shock-wave formation and associated boundary-layer separation. A tendency of strong pressure pulsation over the wing exists, particularly in the region of the shock wave. This pressure pulsation becomes irregular when the angle of attack is sufficiently large. The intensity of this pressure pulsation can be mitigated by reducing the thickness of the airfoil and limiting the angle of attack.

As the basic cause of buffeting at high Mach number is the shock-wave formation and boundary-layer separation, the buffeting boundary should be related closely to the characteristics of the curve of lift coefficient versus Mach number. The "Mach number of divergence" of the wing (cf. § 4.3) may serve as a good estimate of the buffeting boundary at high speed. For the same reason, by reducing the wing thickness, by a proper control of the leading-edge camber which reduces the possibility of flow separation at the leading edge in moderate angles of attack, by sufficient sweepback, by using smaller aspect ratio, and by other auxiliary means, one can hope to design a wing that is practically free from buffeting throughout the transonic speed range. In supersonic flight, when the shock wave is attached to the leading edge of the wing, buffeting does not seem to be a serious problem.

An allied problem of buffeting of aircraft structures due to fluctuating loads induced by a jet is discussed by Miles.[9.42]

The phenomena of buffeting, stall flutter and buffeting flutter are profoundly interesting physical phenomena which still defy satisfactory mathematical analysis.

BIBLIOGRAPHY

Buffeting:

9.1 Abdrashitov, G.: Tail Buffeting. *NACA Tech. Memo.* **1041** (1943). Translated from *Central Aero-Hydrodynam. Inst. Moscow Rept.* **395** (1939).

9.2 Aerodynamics Staff of the N.P.L.: Technical Report by the Accident's Investigation Subcommittee on the Accident to the Aeroplane G-AAZK at Meopham, Kent (England), on 21 July 1930. *Aeronaut. Research Com. R. & M.* **1360** (1931).

9.3 Anderson, S. B.: Correlation of Pilot Opinion of Stall Warning with Flight Measurements of Various Factors Which Produce the Warning. *NACA Tech. Note* **1868** (1949).

9.4   Biechteler, C.: Tests for the Elimination of Tail Flutter. *NACA Tech. Memo.* **710** (1933). Translated from *Z. Flugtech.* **24**, 15–21 (Jan. 1933).

9.5   Blenk, H., H. Hertel, and K. Thalau: The German Investigation of the Accident at Meopham, Kent (England). *NACA Tech. Memo.* **669** (1932). Translated from *Z. Flugtech.* **23**, 73–86 (Feb. 1932).

9.6   Dankworth, E. G., and D. Walker: An Investigation of the Effects of a Sharp Velocity Gradient on the Flexure-Torsion Flutter Speed of an Airfoil. Aeronaut. Eng. thesis., California Institute of Technology (1951).

9.7   Duncan, W. J., D. L. Ellis, and C. Scruton: First Report on the General Investigation of Tail Buffeting. *Aeronaut. Research Com. R. & M.* **1457**, part I (1932).

9.8   Duncan, W. J., D. L. Ellis, and E. Smyth: Second Report on the General Investigation of Tail Buffeting. *Aeronaut. Research Com. R. & M.* **1541** (1933).

9.9   Frazer, R. A., W. J. Duncan, and V. M. Falkner: Experiments on the Buffeting of the Tail of a Model of a Low-Wing Monoplane. *Aeronaut. Research Com. R. & M.* **1457**, part II (1932).

9.10   Goethert, B.: Comments on Aileron Oscillations in the Shock-Wave Range. *U.S. Air Material Command, Tech. Rept.* **F-TR-2101-ND** (1947).

9.11   Liepmann, H. W.: On the Application of Statistical Concepts to the Buffeting Problem. *J. Aeronaut. Sci.* **19**, 793–800 (1952).

9.12   Lo, S. C.: Oscillating Airfoil in Parallel Streams Separated by an Interface. Ph.D. thesis, California Institute of Technology (1950).

9.13   Luskin, H., and E. Lapin: An Analytical Approach to the Fuel Sloshing and Buffeting Problems of Aircraft. *J. Aeronaut. Sci.* **19**, 217–228 (1952).

9.14   Muttray, H.: Investigation of the Effect of the Fuselage on the Wing of a Low-Wing Monoplane. *NACA Tech. Memo.* **517** (1929). Translated from *Luftfahrt-Forsch.*, 33–39 (June 11, 1928).

9.15   Stack, J., and A. E. von Doenhoff: Tests of 16 Related Airfoils at High Speeds. *NACA Rept.* **492** (1934).

9.16   Stokke, A. R., and W. S. Aiken, Jr.: Flight Measurements of Buffeting Tail Loads. *NACA Tech. Note* **1719** (1948).

9.17   van de Vooren, A. I., and H. Bergh: Spontaneous Oscillations of an Aerofoil Due to Instability of the Laminar Boundary Layer. *Natl. Luchtvaartlab. Amsterdam Rept.* **F 96** (1951).

9.18   White, J. A., and M. J. Hood: Wing-Fuselage Interference, Tail Buffeting, and Air Flow about the Tail of a Low-Wing Monoplane. *NACA Rept.* **482** (1934).

Stall flutter:

9.19   Baker, J. E.: The Effects of Various Parameters including Mach Number on Propeller-Blade Flutter, with Emphasis on Stall Flutter. *NACA RM* **L50 L12b** (1951).

9.20   Bollay, W., and C. D. Brown: Some Experimental Results on Wing Flutter. *J. Aeronaut. Sci.* **8**, 313–318 (1941).

9.21   Chuan, R. L., and R. J. Magnus: Vortex Shedding as Related to the Self-Excited Torsional Oscillations of an Airfoil. *NACA Tech. Note* **2429** (1951).

9.22   Halfman, R. L., H. C. Johnson, and S. M. Haley: Evaluation of High-Angle-of-Attack Aerodynamic Derivative Data and Stall-Flutter Prediction Techniques. *NACA Tech. Note* **2533** (1951).

9.23   Mendelson, A.: Aerodynamic Hysteresis as a Factor in Critical Flutter Speed of Compressor Blades at Stalling Conditions. *J. Aeronaut. Sci.* **16**, 645–652 (1949).

9.24   McCullough, G. B., and D. E. Gault: Examples of Three Representative Types of Airfoil-Section Stall at Low Speed. *NACA Tech. Note* **2502** (1951).

9.25   Sisto, F.: Stall-Flutter in Cascades. *J. Aeronaut. Sci.* **20**, 598–604 (1953).

9.26   Stüder, H. L.: Experimentelle Untersuchungen über Flügelschwingungen. *Mitt. Inst. Aerodynam. Eidgenössischen Tech. Hochschule Zurich*, Nr. **4**. Gebr. Leeman & Co. (1946). *Brit. Translation ARC* **2777**. A summary can be found in Victory's paper, Ref. 9.27.

9.27   Victory, M.: Flutter at High Incidence. *Aeronaut. Research Council R. & M.* **2048** (1943).

For flow separation, see

9.28   Cahen, G. L.: A Preliminary Gust-Tunnel Investigation of Leading-Edge Separation of Swept Wings. *NACA RM* **L52C20** (1952).

9.29   Howarth, L.: Concerning the Effect of Compressibility on Laminar Boundary Layers and their Separation. *Proc. Roy. Soc. London, A*, **194**, 16–42 (1948).

9.30   Jones, R. T.: Effect of Sweep Back on Boundary Layer and Separation. *NACA Rept.* **884** (1947).

9.31   Lindsey, W. F., B. N. Daley, and M. D. Humphreys: The Flow and Force Characteristics of Supersonic Airfoils at High Subsonic Speeds. *NACA Tech. Note* **1211** (1947).

9.32   McCormack, G. M., and W. L. Cook: A Study of Stall Phenomena on a 45° Swept-Forward Wing.. *NACA Tech. Note* **1797** (1949).

9.33   Moeckel, W.: Flow Separation Ahead of Blunt Bodies at Supersonic Speeds. *NACA Tech. Note* **2418** (1951).

9.34   Outman, V., and A. A. Lambert: Transonic Separation. *J. Aeronaut. Sci.* **15**, 671–674 (1948).

For detailed studies of the turbulence in the wake of a bluff body, see Refs. 2.25–2.28, and the following:

9.35   Batchelor, G. K.: Note on Free Turbulent Flows, with Special Reference to the Two-Dimensional Wake. *J. Aeronaut. Sci.* **17**, 441–445 (1950).

9.36   Blenk, H., D. Fuchs, and F. Liebers: Über Messungen von Wirbelfrequenzen. *Luftfahrt-Forsch.* **12**, 38–41 (1935).

9.37   Dryden, H. L.: A Review of the Statistical Theory of Turbulence. *Quart. Applied Math.* **1**, 7–42 (1943).

9.38   Dunn, L., and M. Finston: Self-Excited Oscillations of Airfoils. *GALCIT Rept.* **13** (1943). California Institute of Technology.

9.39   Fage, A., and F. C. Johansen: On the Flow of Air behind an Inclined Flat Plate of Infinite Span. *Proc. Roy. Soc. London, A*, **116**, 170–197 (1927).

9.40   Tyler, E.: Vortex Formation behind Obstacles of Various Sections. *Phil. Mag.* (7) **11**, 849–890 (1931).

For noise field and its effect on structures: see

9.41   Lighthill, M. J.: On Sound Generated Aerodynamically. I, General Theory. *Proc. Roy. Soc. London A.* **211**, 564–587 (1952). II, Turbulence as a Source of Sound, **222** (1954).

9.42   Miles, J. W.: On Structural Fatigue under Random Loading. *J. Aeronaut. Sci.* **21**, 753–762 (1954).

Chapter 10

APPLICATIONS OF LAPLACE TRANSFORMATION

The method of Laplace transformation can be applied to a number of aeroelastic problems. In this chapter the mathematical background will be outlined briefly, and its applications will be illustrated by examples in gust response and flutter.

10.1 LAPLACE TRANSFORMATION

The method of Laplace transformation is extremely powerful in treating certain response problems. Its mathematical foundation and refinements cannot be treated in this book because of the limited space. However, there exist many good books on this subject.[10.1–10.7] Hence, in the following, we shall only quote the rules of the method of Laplace transformation, and give some heuristic derivations to help understand them.

The *Laplace transform* of a function $F(t)$, which vanishes for $t < 0$, is defined by the integral

$$\mathscr{L}\{F(t)|s\} = \int_0^\infty e^{-st}F(t)\,dt \qquad (\mathrm{Rl}\ s > \alpha) \qquad (1)$$

The *inverse transform* is given by the equation

$$F(t) = \frac{1}{2\pi i}\int_{c-i\infty}^{c+i\infty} e^{st}\mathscr{L}\{F|s\}\,ds \qquad (c > \alpha) \qquad (2)$$

where α is a real number greater than the real parts of all the singular points of $\mathscr{L}\{F|s\}$. Whenever these integrals exist, $F(t)$ and $\mathscr{L}\{F|s\}$ are called the *original* and the *image*, respectively. If there is no confusion, the image $\mathscr{L}\{F|s\}$ will be written as $\mathscr{L}\{F\}$.

The conditions for the validity of the reciprocal relations 1 and 2 depend much on the sense of integration of the integrals. If the integrals are taken as Cauchy principal values in the Riemannian sense, a set of sufficient conditions which guarantees the validity of the Laplace transformation is

1. The integral 1 is absolutely convergent in the region $\mathrm{Rl}\ s > \alpha$.

2. $F(t)$ is of bounded variation in any finite interval, and $F(t) = 0$ for $t < 0$.

3. At a point of discontinuity t_0 the value of $F(t)$ is equal to the corresponding mean value (see Fig. 10.1):

$$F(t_0) = \lim_{\varepsilon \to 0} \frac{F(t_0 + \varepsilon) + F(t_0 - \varepsilon)}{2} \tag{3}$$

If these conditions are satisfied, the reciprocal relations between $F(t)$ and $\mathscr{L}\{F|s\}$ are essentiaily unique.

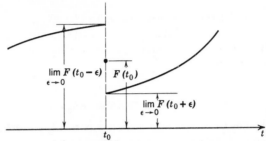

Fig. 10.1.   The mean value $F(t_0)$ at a point of discontinuity.

Example 1.   Let $F(t) = \mathbf{1}(t)$; i.e., $F(t) = 1$ when $t > 0$, $F(t) = \frac{1}{2}$ when $t = 0$, and $F(t) = 0$ when $t < 0$.   Then

$$\mathscr{L}\{\mathbf{1}(t)|s\} = \int_0^\infty e^{-st}\,dt = -\frac{1}{s}\,e^{-st}\,\Big|_0^\infty$$

Hence, when $\mathrm{Rl}\,s > 0$,

$$\mathscr{L}\{\mathbf{1}(t)\} = \frac{1}{s}$$

Example 2.   Let $F(t) = e^{kt}$ when $t > 0$.   Then

$$\mathscr{L}\{e^{kt}\} = \int_0^\infty e^{-st}\,e^{kt}\,dt = \frac{1}{s - k} \qquad (\mathrm{Rl}\,s > \mathrm{Rl}\,k)$$

Following van der Pol and Bremmer,[10.7] we shall regard the method of Laplace transformation as a language.   In actual applications we first transpose the problem under consideration, which is formulated in terms of the variable t, into a new problem in the variable s; i.e., the functional relations (such as differential and integral equations) in which the problem is first stated are transformed into their operational equivalent.   This is done by multiplying the given relation or relations by the function e^{-st} and then integrating from $t = 0$ to $t = \infty$.   This is a *transition from the t language to the s language*.   In many problems the new problem in the s language is considerably simpler than the original one in t, and therefore leads to an easier solution.   Once the problem is solved in terms of s, it only remains, as the second stage of the method, to translate the s solution back into the physical language of t.

For the purpose of such translation between the s and t languages, a "dictionary" is necessary.   An abridged one is given in Table 10.1.   A complete dictionary is given by Erdelyi.[10.8]

Table 10.1.   Laplace Transforms

$\mathscr{L}\{F|s\}$	$F(t)$
$\dfrac{1}{s}$	$\mathbf{1}(t)$
$\dfrac{1}{s^2}$	t
$\dfrac{1}{s^n}\qquad(n = 1, 2, \cdots)$	$\dfrac{t^{n-1}}{(n-1)!}$
$\dfrac{\Gamma(k)}{s^k}\qquad(k > 0)$	t^{k-1}
$\dfrac{\Gamma(k)}{(s-a)^k}\qquad(k > 0)$	$t^{k-1}e^{at}$
$\dfrac{1}{(s-a)(s-b)}\qquad(a \neq b)$	$\dfrac{1}{a-b}(e^{at} - e^{bt})$
$\dfrac{1}{s^2 + a^2}$	$\dfrac{1}{a}\sin at$
$\dfrac{s}{s^2 + a^2}$	$\cos at$
$\dfrac{1}{s^2 - a^2}$	$\dfrac{1}{a}\sinh at$
$\dfrac{s}{s^2 - a^2}$	$\cosh at$
$\dfrac{1}{(s^2 + a^2)^2}$	$\dfrac{1}{2a^3}(\sin at - at\cos at)$
$\sqrt{s - a} - \sqrt{s - b}$	$\dfrac{1}{2\sqrt{\pi t^3}}(e^{bt} - e^{at})$
$\dfrac{1}{\sqrt{s^2 + a^2}}$	$J_0(at)$
$\dfrac{1}{s}e^{-k/s}$	$J_0(2\sqrt{kt})$
$\dfrac{1}{s}e^{-k\sqrt{s}}\qquad(k \geqslant 0)$	$\operatorname{erfc}\left(\dfrac{k}{2\sqrt{t}}\right)$

The user of the language must know the "grammatical" rules by which the s and t languages are translated into each other. A few basic rules are quoted below. They are constructed on the basis of Eqs. 1 and 2. But, in applying the dictionary and grammar, these integrals need no longer be used explicitly.

Linearity. The Laplace transformation is *linear*; i.e., if A and B are constants, then

$$\mathscr{L}\{A\,F(t) + B\,G(t)\} = A\,\mathscr{L}\{F\} + B\,\mathscr{L}\{G\}$$

This follows from the definition of the transformation. Hence, a multiplication of the original by any constant A corresponds in the s language to a multiplication of the image by A, and the image of the sum of two originals is equal to the sum of the separate images.

Example 3.

$$\mathscr{L}\left\{\frac{1}{2}e^{kt} - \frac{1}{2}e^{-kt}\right\} = \frac{1}{2}\frac{1}{s-k} - \frac{1}{2}\frac{1}{s+k}$$

i.e.,

$$\mathscr{L}\{\sinh kt\} = \frac{k}{s^2 - k^2}$$

Similarity Rule. Let

$$\mathscr{L}\{F(t)\} = f(s) \qquad (\alpha < \text{Rl } s < \infty)$$

Then, if λ is a positive real constant,

$$\mathscr{L}\{F(\lambda t)\} = \frac{1}{\lambda}f\left(\frac{s}{\lambda}\right) \qquad (\lambda\alpha < \text{Rl } s < \infty)$$

Example 4.

$$\mathscr{L}\{\cos t\} = \frac{s}{s^2 + 1}; \quad \text{so} \quad \mathscr{L}\{\cos kt\} = \frac{s}{s^2 + k^2}$$

Shift Rule. If $f(s)$ is the Laplace transform of $F(t)$, then $e^{-\lambda s}f(s)$, $\lambda > 0$, is the transform of the function $F(t - \lambda)\,\mathbf{1}(t - \lambda)$, where $\mathbf{1}(t - \lambda)$ is the unit-step function.

This is seen from the following equation:

$$\int_0^\infty e^{-st}F(t - \lambda)\,\mathbf{1}(t - \lambda)\,dt = \int_\lambda^\infty e^{-st}F(t - \lambda)\,dt = e^{-\lambda s}\int_0^\infty e^{-st}F(t)\,dt$$

Note that

$$F(t - \lambda)\,\mathbf{1}(t - \lambda) = \begin{cases} 0 & \text{when } t < \lambda \\ F(t - \lambda) & \text{when } t > \lambda \end{cases}$$

The function $F(t - \lambda) \mathbf{1}(t - \lambda)$ is obtained by shifting the original function $F(t)$ to the right through a distance λ (Fig. 10.2).

Example 5. Since

$$\mathscr{L}\{\mathbf{1}(t)\} = \frac{1}{s}$$

we have

$$\mathscr{L}\{\mathbf{1}(t - \lambda)\} = e^{-\lambda s}/s$$

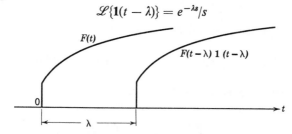

Fig. 10.2. The shift rule.

Fig. 10.3. A square wave.

and the Laplace transform of a square wave of band width $2a$ is (Fig. 10.3)

$$\mathscr{L}\{\mathbf{1}(t - \lambda) - \mathbf{1}(t - \lambda - 2a)\} = \frac{1}{s}(e^{-\lambda s} - e^{-\lambda s - 2as})$$

$$= \frac{2}{s}e^{-(\lambda + a)s}\sinh as$$

Example 6. In automatic control of airplane lateral stability, it is sometimes advantageous to gear the rudder angle δ_r to the yawing acceleration $\ddot{\psi}$, but lagging by a specified time interval, τ:

$$\delta_r(t) = K\ddot{\psi}(t - \tau)$$

Hence,

$$\mathscr{L}\{\delta_r(t)\} = Ke^{-\tau s}\mathscr{L}\{\ddot{\psi}(t)\}$$

The factor $e^{-\tau s}$ is often called a *lag operator*.

Attenuation Rule. If

$$\mathscr{L}\{F(t)\} = f(s) \qquad (\alpha < \mathrm{Rl}\ s < \infty)$$

then

$$\mathscr{L}\{e^{-\lambda t}F(t)\} = f(s + \lambda) \qquad [\alpha + \mathrm{Rl}\ \lambda < \mathrm{Rl}\ (s + \lambda) < \infty]$$

This follows directly from substitution into the definition integral (Eq. 1).

Convolution Rule. If $F_1(t)$ and $F_2(t)$ are piecewise continuous and

$$\mathscr{L}\{F_1(t)\} = f_1(s) \qquad (\alpha_1 < \text{Rl } s < \infty)$$

$$\mathscr{L}\{F_2(t)\} = f_2(s) \qquad (\alpha_2 < \text{Rl } s < \infty)$$

then

$$f_1(s) f_2(s) = \mathscr{L}\{F_1(t) * F_2(t)\} \qquad (\max (\alpha_1, \alpha_2) < \text{Rl } s < \infty)$$

where $F_1(t) * F_2(t)$ is the so-called *convolution integral*:

$$F_1(t) * F_2(t) = \int_0^t F_1(\tau) F_2(t - \tau) \, d\tau = \int_0^t F_1(t - \tau) F_2(\tau) \, d\tau$$

This is probably the most important of all rules.

A formal derivation of this rule is as follows:

$$f_1(s) f_2(s) = \int_0^\infty e^{-su} F_1(u) \, du \int_0^\infty e^{-sv} F_2(v) \, dv$$

$$= \int_0^\infty \int_0^\infty e^{-(u+v)s} F_1(u) F_2(v) \, du \, dv$$

Introducing a new variable $t = u + v$, and letting the variables (u, v) be transformed into (u, t), we have

$$f_1(s) f_2(s) = \int_0^\infty \int_{t=u}^\infty e^{-ts} F_1(u) F_2(t - u) \, du \, dt$$

$$= \int_0^\infty e^{-st} \, dt \int_0^t F_1(u) F_2(t - u) \, du$$

The role of F_1 and F_2 are obviously interchangeable. Hence, the last equation is equal to

$$f_1(s) f_2(s) = \int_0^\infty e^{-st} \, dt \int_0^t F_1(t - u) F_2(u) \, du$$

The last two equations state the convolution rule.

Differentiation Rule. Let

$$\mathscr{L}\{F(t)\} = f(s) \qquad (\alpha < \text{Rl } s < \infty)$$

If $F(t)$ has a derivative $F'(t)$ for $t > 0$, and if the Laplace transform of $F'(t)$ exists, then

$$\mathscr{L}\{F'(t)\} = s f(s) - F(0 +)$$

where $F(0 +)$ means the limit $\lim_{\varepsilon \to 0} F(0 + \varepsilon)$, $\varepsilon > 0$.

Hence, a differentiation of the original with respect to t corresponds to a multiplication of the image by a factor s.

A formal derivation of this rule is as follows.   By a formal integration by parts, we have

$$\mathscr{L}\{F'(t)\} = \int_0^\infty e^{-st} F'(t) \, dt$$

$$= e^{-st} F(t) \Big|_0^\infty + s \int_0^\infty e^{-st} F(t) \, dt$$

$$= -F(0+) + s \, \mathscr{L}\{F(t)\}$$

provided that the limit of $e^{-st}F(t)$ tends to zero as $t \to \infty$.

Let us remark that the continuity of the function $F(t)$ is important for the validity of the above formula.   If $F(t)$ is continuous except for an "ordinary" discontinuity at $t = t_0$, we would have

$$\mathscr{L}\{F'(t)\} = s \, f(s) - F(0+) - [F(t_0 + 0) - F(t_0 - 0)] \, e^{-t_0 s}$$

where the quantity in the brackets is the jump of $F(t)$ at $t = t_0$.

Repeated application of the above rule leads to the following:

Successive Differentiation Rule.   Let the function $F(t)$ be n times differentiable for $t > 0$, and let the Laplace transform of the nth derivative $F^{(n)}(t)$ exist for $\alpha < \text{Rl } s < \infty$.   Then $F(t)$ has a Laplace transform for all s, $\text{Rl } s > \alpha$, and

$$\mathscr{L}\{F^{(n)}(t)\} = s^n f(s) - s^{n-1} F(0+) - s^{n-2} F'(0+)$$

$$- s^{n-3} F''(0+) - \cdots - F^{(n-1)}(0+)$$

where $f(s) = \mathscr{L}\{F(t)\}$.

Integration Rule.   A specialization of the convolution rule to the case in which $f_1(s) = 1/s$ and therefore $F_1(t) = \mathbf{1}(t)$ leads to the following integration rule:

If

$$f(s) = \mathscr{L}\{F(t)\} \qquad (\alpha < \text{Rl } s < \infty)$$

then

$$\frac{1}{s} f(s) = \mathscr{L}\left\{\int_0^t F(\tau) \, d\tau\right\} \qquad (\max(\alpha, 0) < \text{Rl } s)$$

Multiplication by t^n.   If

$$f(s) = \mathscr{L}\{F(t)\} \qquad (\alpha < \text{Rl } s < \infty)$$

then

$$\frac{d^n}{ds^n} f(s) = \mathscr{L}\{(-t)^n F(t)\} \qquad (n = 1, 2, 3, \cdots)$$

$$(\alpha < \text{Rl } s < \infty)$$

10.2 ORDINARY LINEAR DIFFERENTIAL EQUATIONS WITH CONSTANT COEFFICIENTS

To solve the equation

$$\frac{d^n x}{dt^n} + a_1 \frac{d^{n-1}x}{dt^{n-1}} + \cdots + a_{n-1}\frac{dx}{dt} + a_n x = F(t), \qquad t > 0 \qquad (1)$$

with the initial values $x_0, x_1, \cdots, x_{n-1}$ for $x, dx/dt, \cdots, d^{n-1}x/dt^{n-1}$ when $t = 0$, where $a_1, a_2, \cdots$ are constants, we transform every term of the equation into its Laplace transform. Remembering the differentiation rule, we obtain the transformed equation:

$$(s^n + a_1 s^{n-1} + \cdots + a_{n-1}s + a_n)\,\mathscr{L}\{x\}$$
$$= \mathscr{L}\{F\} + (s^{n-1}x_0 + s^{n-2}x_1 + \cdots + sx_{n-2} + x_{n-1})$$
$$+ a_1(s^{n-2}x_0 + s^{n-3}x_1 + \cdots + sx_{n-3} + x_{n-2})$$
$$+ \cdots + a_{n-1}x_0 \qquad (2)$$

From this equation $\mathscr{L}\{x\}$ is solved. The function $x(t)$ is then obtained by finding the inverse of $\mathscr{L}\{x\}$.

Example 1.

$$\begin{cases} (D^2 + 3D + 2)x = 0, & t > 0 \\ x = x_0, & Dx = x_1 \quad \text{when} \quad t = 0 \end{cases}$$

where

$$Dx = \frac{dx}{dt}, \qquad D^2 x = \frac{d^2 x}{dt^2}, \quad \text{etc.}$$

The transformed equation is

$$(s^2 + 3s + 2)\,\mathscr{L}\{x\} = (sx_0 + x_1) + 3x_0$$

Thus

$$\mathscr{L}\{x\} = \frac{sx_0 + (x_1 + 3x_0)}{(s+1)(s+2)} = \frac{2x_0 + x_1}{s+1} - \frac{x_0 + x_1}{s+2}$$

Therefore,

$$x = (2x_0 + x_1)e^{-t} - (x_0 + x_1)e^{-2t}$$

Example 2.

$$\begin{cases} (D-1)(D-2)(D-3)x = 1, & t > 0 \\ x, Dx, \text{ and } D^2 x \text{ are zero} \quad \text{when} \quad t = 0 \end{cases}$$

The transformed equation is

$$(s-1)(s-2)(s-3)\,\mathscr{L}\{x\} = \frac{1}{s}$$

Thus,

$$\mathscr{L}\{x\} = \frac{1}{s(s-1)(s-2)(s-3)} = -\frac{1}{6s} + \frac{1}{2(s-1)} - \frac{1}{2(s-2)} + \frac{1}{6(s-3)}$$

Therefore,

$$x = -\tfrac{1}{6} + \tfrac{1}{2} e^t - \tfrac{1}{2} e^{2t} + \tfrac{1}{6} e^{3t}$$

Example 3.

$$\begin{cases} (3D+2)x + Dy = 1, & t > 0 \\ Dx + (4D+3)y = 0 \\ x_0 = y_0 = 0 \end{cases}$$

Transformed:

$$\begin{cases} (3s+2)\mathscr{L}\{x\} + s\mathscr{L}\{y\} = \dfrac{1}{s} \\ s\mathscr{L}\{x\} + (4s+3)\mathscr{L}\{y\} = 0 \end{cases}$$

Therefore,

$$\mathscr{L}\{x\} = \frac{4s+3}{s(s+1)(11s+6)} = \frac{1}{2s} - \frac{1}{5(s+1)} - \frac{33}{10(11s+6)}$$

Hence,

$$x = \tfrac{1}{2} - \tfrac{1}{5} e^{-t} - \tfrac{3}{10} e^{-6t/11}$$

Similarly,

$$\mathscr{L}\{y\} = -\frac{1}{(11s+6)(s+1)} = \frac{1}{5}\left(\frac{1}{s+1} - \frac{11}{11s+6} \right)$$

Thus,

$$y = \tfrac{1}{5}(e^{-t} - e^{-6t/11})$$

Heaviside's Expansion. When the equation to be solved is of the type

$$\phi(D)x = 1, \qquad t > 0 \tag{3}$$

with $x_0 = x_1 = \cdots = x_{n-1} = 0$, and

$$\phi(D) = D^n + a_1 D^{n-1} + \cdots + a_{n-1}D + a_n$$

we obtain

$$\mathscr{L}\{x\} = \frac{1}{s\,\phi(s)} \tag{4}$$

The solution is obtained by expanding $1/s\,\phi(s)$ into partial fractions. If the zeros of $\phi(s)$ are $\alpha_1, \alpha_2, \cdots, \alpha_n$, all different and none of them zero, we have

$$\mathscr{L}\{x\} = \frac{1}{s\,\phi(s)} = \frac{1}{s\,\phi(0)} + \sum_{r=1}^{n} \frac{1}{\alpha_r(s - \alpha_r)\,\phi'(\alpha_r)} \tag{5}$$

and thus

$$x(t) = \frac{1}{\phi(0)} + \sum_{r=1}^{n} \frac{1}{\alpha_r \, \phi'(\alpha_r)} \, e^{\alpha_r t} \tag{6}$$

where $\phi'(\alpha_r) = (d\phi/ds)_{s=\alpha_r}$.

The partial fraction expansion can be easily extended to the general case defined by Eq. 1. If $\mathscr{L}\{F\}$ is a polynomial in s, then, from Eq. 2, the Laplace transform of x can be written as a quotient of two polynomials

$$\mathscr{L}\{x(t)\} = \frac{M(s)}{N(s)} \tag{7}$$

Let the degree of $M(s)$ be lower than that of $N(s)$.* If $N(s)$ has n distinct roots $s_1, s_2, \cdots, s_n$, then $M(s)/N(s)$ can be expanded in the form

$$\frac{M(s)}{N(s)} = \sum_{k=1}^{n} \frac{B_k}{s - s_k} \tag{8}$$

where

$$B_k = \frac{M(s_k)}{N'(s_k)} \tag{9}$$

and $N'(s)$ is the first derivative of $N(s)$. Hence,

$$x(t) = \mathscr{L}^{-1}\left\{ \sum_{k=1}^{n} \frac{B_k}{s - s_k} \right\} = \sum_{k=1}^{n} B_k \, e^{s_k t} \tag{10}$$

The case of $N(s)$ with multiple roots can be treated in a similar way. For example, if s_1 is a double root, the expansion of $f(s)$ in partial fractions contains a term of the form

$$\frac{B_1 + B_2 s}{(s - s_1)^2} = \frac{B_1 + B_2 s_1 + B_2(s - s_1)}{(s - s_1)^2}$$

$$= \frac{B_1 + B_2 s_1}{(s - s_1)^2} + \frac{B_2}{s - s_1}$$

$$= \mathscr{L}\{(B_1 + B_2 s_1)t \, e^{s_1 t} + B_2 \, e^{s_1 t}\} \tag{11}$$

There are many problems in aeroelasticity that may be reduced to ordinary linear differential equations with constant coefficients. Familiar

---

* If the degree of $M(s)$ is equal or higher than that of $N(s)$, Eq. 7 can be written as $M'(s)/N(s) + P(s)$, where $P(s)$ is a polynomial and $M'(s)$ is of lower degree than $N(s)$.

examples appear in the classical theory of airplane dynamics. When the deformation of the main structure is described by a few generalized coordinates, ordinary linear differential equations are obtained. The principal assumptions responsible for such great simplification are (1) that the motion of the airplane consists of infinitesimal disturbances about a steady symmetrical flight, and (2) that the dependance of the aerodynamic forces on the rate of change of linear and angular velocities can be neglected except for the lag of the downwash between the tail and the wing. With the second assumption the unsteady aerodynamic action is expressed as a linear function of the linear and angular velocities, the coefficients being defined as "stability derivatives."

The extensive and intricate subject of airplane dynamics, however, will not be discussed here. (See Refs. 10.10–10.13.) In the following sections, illustrations of the application of Laplace transformation will be given in the gust-response and flutter problems.

10.3 RESPONSE TO GUSTS—RIGID AIRPLANE, PITCHING NEGLECTED

Let us assume, as in § 8.2, (1) that the airplane is initially in horizontal flight at a constant speed U, (2) that the gust is normal to the flight path of the airplane and has a velocity distribution w in the direction of flight and is uniform spanwise, (3) that the variation of the forward speed of the airplane as it traverses the gust can be neglected, and (4) that the gust intensity and the induced motion are so small that the linearized aerodynamic theory is valid. Whereas in § 8.2 the quasi-steady aerodynamic coefficients are used, we shall now show how the unsteady aerodynamic forces can be taken into account.

The equation of motion of the airplane having one degree of freedom is

$$m\ddot{z} = -L \tag{1}$$

where m is the total mass, z the downward displacement, and L the upward lift.

The lift consists of two parts; that arising from the motion of the airplane, and that induced by the gust.

The lift induced by the motion of a two-dimensional airfoil in an incompressible fluid has been derived in § 6.7. As in § 6.7, it is convenient to introduce the dimensionless time parameter τ:

$$\tau = \frac{2U}{c}t \tag{2}$$

where c denotes the chord length. A dot will indicate a differentiation with respect to the physical time t, whereas a prime will indicate a differentiation with respect to τ. Thus $\dot{z} = dz/dt$, $z' = dz/d\tau$, $\dot{z} = (2U/c)z'$. According to Eqs. 8 and 9 of § 6.7, we may write the lift induced by an airplane motion which starts at $\tau = 0$ as

$$L(\tau) = \frac{1}{2}\rho U^2 Sa \left[\frac{2}{c} z'(0)\Phi(\tau) + \frac{2}{c} \int_0^\tau \Phi(\tau - \sigma) z''(\sigma)\,d\sigma + \frac{1}{c} z''(\tau) \right] \quad (3)$$

In the above equation, ρ is the density of the fluid, S is the wing area, $a = dC_L/d\alpha$ is the steady-state lift curve slope (per radian), and $\Phi(\tau)$ is Wagner's function. Whereas the equations in § 6.7 are written for a two-dimensional flow of an incompressible fluid, Eq. 3 may be considered applicable to compressible fluids when $\Phi(\tau)$ is properly modified for Mach number. Strip assumption is assumed for the aerodynamic action, and the finite span effect will be corrected in an overall manner by correcting the lift curve slope a for aspect ratio as in a steady flow.

The lift induced by the gust can be expressed by a similar equation, based on a fundamental function $\Psi(t)$, which represents the ratio of the transient lift to the steady-state lift on an airfoil penetrating a sharp-edged gust normal to the flight path. Let the speed of the sharp-edged gust be w; then by definition the transient lift coefficient is

$$C_L(\tau) = a \frac{w}{U} \Psi(\tau) \quad (4)$$

where a is the steady-state lift-curve slope, and τ is the dimensionless time parameter as before. It is convenient to interpret τ here as the distance traveled by the airfoil, measured in semichords.

Tables 10.2 and 10.3 give the approximate expressions of $\Phi(\tau)$ and $\Psi(\tau)$ at Mach number zero, obtained by R. T. Jones[15.43] for elliptic wings. For $\Psi(\tau)$ it is assumed that the leading edge of the airfoil encounters the sharp-edged gust at the instant $\tau = 0$.

Table 10.2. Wagner's Function $\Phi(\tau)$ at $M = 0$

Æ	$\Phi(\tau)$	a
3	$1 - 0.283e^{-0.540\tau}$	1.2π
6	$1 - 0.361e^{-0.381\tau}$	1.5π
∞	$1 - 0.165e^{-0.045\tau} - 0.335e^{-0.300\tau}$	2π

Table 10.3.   Küssner's Function $\Psi'(\tau)$ at $M = 0$

$Æ$	$\Psi'(\tau)$	a
3	$1 - 0.679e^{-0.558\tau} - 0.227e^{-3.20\tau}$	1.2π
6	$1 - 0.448e^{-0.290\tau} - 0.272e^{-0.725\tau} - 0.193e^{-3.00\tau}$	1.5π
∞	$1 - 0.500e^{-0.130\tau} - 0.500e^{-\tau}$	2π

Figure 10.4 shows the results of a linearized theory for the indicial admittance of a two-dimensional airfoil entering a sharp-edged gust. In

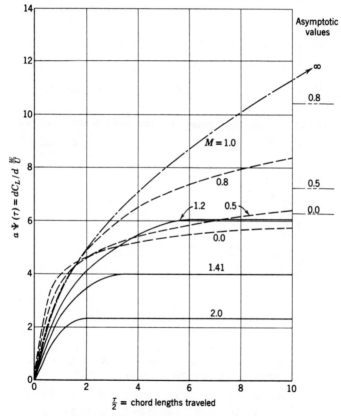

Fig. 10.4.   Indicial admittance of the lift of a restrained wing to unit sharp-edged gust.   Two-dimensional wing for several Mach numbers.   From Lomax, Ref. 15.84.   (Figure reproduced by courtesy of the NACA.)

order to indicate the effect of compressibility, the ratio $\partial C_L / \partial \left(\dfrac{w}{U} \right) = a \Psi'(\tau)$ is presented. Figure 10.5 shows the indicial admittance of a two-dimensional airfoil to a sudden sinking speed, i.e., $\partial C_L / \partial \left(\dfrac{\dot{z}}{U} \right) = a\, \Phi(\tau)$. The effect of finite-aspect ratio at Mach number 1.41 is shown in Fig. 10.6. These results are given by Lomax.[15.84]

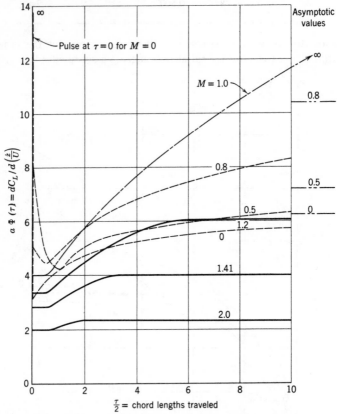

Fig. 10.5. Indicial admittance of the lift to a sudden sinking speed as a function of chord length traveled. Two-dimensional wing for several Mach numbers. (Figure reproduced by courtesy of the NACA.)

The lift induced by a variable gust can be written as a Duhamel integral. Since $\Psi'(0) = 0$, we obtain, according to Eq. 27 of § 8.1,

$$L(\tau) = \frac{1}{2} \rho U^2 S a \int_0^\tau \frac{w(\sigma)}{U} \Psi''(\tau - \sigma)\, d\sigma \tag{5}$$

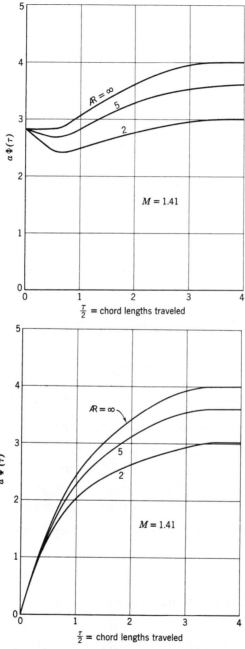

Fig. 10.6. The effect of finite-aspect ratio on $a\Phi(\tau)$ and $a\Psi'(\tau)$ at Mach number 1.41. Rectangular wings of aspect ratios 2, 5, and ∞. (Courtesy of the NACA.)

Again the strip assumption is used here and a is corrected for aspect ratio. The prime over Ψ'' indicates differentiation with respect to the variable τ. The equation of motion of the airplane is therefore

$$m \left(\frac{2U}{c}\right)^2 z'' = -\frac{1}{2}\rho U^2 Sa \left[\frac{2}{c} z'(0)\, \Phi(\tau) + \frac{2}{c}\int_0^\tau \Phi(\tau - \sigma)\, z''(\sigma)\, d\sigma \right.$$
$$\left. + \frac{1}{c} z''(\tau) + \int_0^\tau \frac{w(\sigma)}{U} \Psi''(\tau - \sigma)\, d\sigma\right] \quad (6)$$

Let the initial conditions be

$$z' = z = 0 \quad \text{when} \quad \tau = 0 \tag{7}$$

Then the Laplace transformation of Eq. 6 is

$$m \left(\frac{2U}{c}\right)^2 s^2 \mathscr{L}\{z\} = -\frac{1}{2}\rho U^2 Sa \left[\frac{2}{c} s^2 \mathscr{L}\{\Phi\}\, \mathscr{L}\{z\}\right.$$
$$\left. + \frac{1}{c} s^2 \mathscr{L}\{z\} + \frac{1}{U} s \mathscr{L}\{\Psi\}\, \mathscr{L}\{w\}\right] \tag{8}$$

i.e.,

$$\mathscr{L}\{z\} = -\frac{c}{4U} \frac{\mathscr{L}\{w\}\, \mathscr{L}\{\Psi\}}{s[\mu + \frac{1}{4} + \frac{1}{2}\mathscr{L}\{\Phi\}]} \tag{9}$$

where

$$\mu = \frac{2m}{\rho c Sa} \tag{10}$$

The motion $z(t)$ is then given by the inverse transform of Eq. 9. For a two-dimensional airfoil ($\mathcal{R} = \infty$) in an incompressible fluid, we have approximate expressions (see Tables 10.2 and 10.3):

$$\mathscr{L}\{\Psi\} = \frac{1}{s} - \frac{0.500}{s + 0.130} - \frac{0.500}{s + 1}$$
$$\mathscr{L}\{\Phi\} = \frac{1}{s} - \frac{0.165}{s + 0.0455} - \frac{0.335}{s + 0.300} \tag{11}$$

Consider a sharp-edged gust for which w is a step function with a jump w_0. Hence,

$$\mathscr{L}\{w\} = \frac{w_0}{s} \tag{12}$$

A substitution of Eqs. 11 and 12 into Eq. 9 gives

$$\mathscr{L}\{z''\} = s^2 \mathscr{L}\{z\}$$
$$= \frac{-0.1412 w_0 c}{(\mu + 0.25)U} \frac{s^3 + 0.5752 s^2 + 0.09292 s + 0.003107}{(s + 0.130)(s + 1)(s^3 + a_1 s^2 + a_2 s + a_3)} \tag{13}$$

where

$$a_1 = \frac{0.345\mu + 0.3363}{\mu + 0.25}, \qquad a_2 = \frac{0.0135\mu + 0.1436}{\mu + 0.25}$$

$$a_3 = \frac{0.006825}{\mu + 0.25} \tag{14}$$

$z''(\tau)$ can be found from Eqs. 13 according to Heaviside's expansion as demonstrated in § 10.2. From $z''(\tau)$ the acceleration $\ddot{z}$ in physical units is obtained:

$$\ddot{z} = U^2 \left(\frac{2}{c}\right)^2 z''(\tau) \tag{15}$$

The maximum acceleration calculated from Eq. 15 is smaller than that given by Eq. 9 of § 8.2, where quasi-steady aerodynamic coefficients are

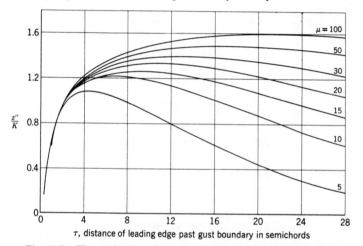

Fig. 10.7.   The acceleration of an airplane in entering a sharp-edged gust. K and μ are defined by Eqs. 16 and 10, respectively.

used. The ratio between the values of the maximum acceleration given by Eq. 15 and Eq. 9 of § 8.2 represents the effect of the lag in time for the circulation to grow after encountering the gust. This ratio can be obtained by dividing $\ddot{z}_{\max}$ by the quasi-steady value $-(U/\mu c)w_0$.

The calculated result of z'' in an incompressible fluid, expressed in units of K, where

$$K = \frac{0.1412}{\mu + 0.25} \frac{w_0 c}{U} \tag{16}$$

is shown in Fig. 10.7. The true value of z'' is the value shown in the figure times the value of K.   The acceleration ratio $\ddot{z}_{max}/$(maximum quasi-steady acceleration) is shown in Fig. 10.8.   The importance of the airplane density ratio μ on the gust-response characteristics of an airplane is evident.

For other gust profiles, we may either use Eq. 9 directly with proper $\mathscr{L}\{w\}$ to find the inverse transform, or use the response to a sharp-edged

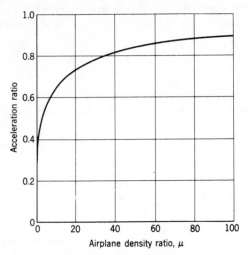

Fig. 10.8.   The acceleration ratio.

gust obtained above to generalize by a Duhamel integral.   The latter is a more practical procedure.

10.4   TRANSFER FUNCTIONS

It is convenient to introduce a terminology for the statement of physical problems in the s language.   Suppose that an input $Y_i(t)$ to a mechanical or electronic system is connected with the output $Y_0(t)$ by a linear differential operator:*

$$f(D)\,Y_0(t) = Y_i(t) \tag{1}$$

In general, the transformed equation can be written as

$$y_o(s) = F(s)\,y_i(s) + A(s) \tag{2}$$

where

$$y_o(s) = \mathscr{L}\{Y_o(t)|s\}, \quad y_i(s) = \mathscr{L}\{Y_i(t)|s\}$$

* An integral operator of the convolution type leads to the same result.

$F(s)$ is a function of s depending solely on the operator $f(D)$, and $A(s)$ is a linear function of the initial values of $Y_o(t)$ and its derivatives.

If $F(s)$ tends to a finite value when $s \to +0$, the limit

$$\lim_{s \to +0} F(s) = K \tag{3}$$

is defined as the *gain*, and the function

$$G(s) = \frac{1}{K} F(s) \tag{4}$$

is called the *transfer function* of the given operator.

To clarify the physical meaning of the gain and the transfer function, let us consider the following simple example:

$$\begin{cases} T\dfrac{dY_o}{dt} + Y_o = Y_i & (t > 0) \\ Y_o = 0 \quad \text{when} \quad t \to +0 \end{cases} \tag{5}$$

The transformed equation is

$$(Ts + 1)\, y_o = y_i$$

Hence

$$y_o = \frac{1}{Ts + 1}\, y_i \tag{6}$$

It follows that

$$F(s) = \frac{1}{Ts + 1}$$

$$\text{Gain} = K = 1 \tag{7}$$

$$\text{Transfer function} = G(s) = \frac{1}{Ts + 1}$$

In the special case $Y_i(t) = \mathbf{1}(t)$, so that $y_i(s) = 1/s$,

$$y_o(s) = \frac{1}{s(Ts + 1)} = \frac{1}{s} - \frac{T}{Ts + 1} \tag{8}$$

then

$$Y_o(t) = (1 - e^{-t/T})\, \mathbf{1}(t)$$

The input and output are shown in Fig. 10.9.   It is seen that the output lags behind the input, and that, at a time $t = T$, the output has reached 63 per cent of its asymptotic value for $t \to \infty$.   The limiting ratio

$$\lim_{t \to \infty} \frac{Y_o(t)}{Y_i(t)} = K = 1 \tag{9}$$

shows that the gain K represents the limiting value of the response to a unit step function as $t \to \infty$.*

In the special case $Y_i(t) = A\,e^{i\omega t}$ where A is a constant,

$$y_o(s) = \frac{1}{(Ts+1)(s-i\omega)}A$$

$$= \left(-\frac{T}{1+i\omega T}\frac{1}{Ts+1} + \frac{1}{1+i\omega T}\frac{1}{s-i\omega}\right)A$$

Hence,

$$Y_o(t) = \frac{A}{1+i\omega T}(-e^{-t/T} + e^{i\omega t}) \tag{10}$$

The first term tends to zero as t increases; it represents the transient

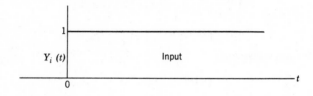

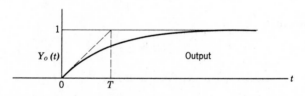

Fig. 10.9. The input and output for the differential system 5.

disturbance associated with the initial application of $Y_i(t)$. The second term represents a steady-state solution. The ratio

$$\frac{[Y_o(t)]_{\text{steady}}}{Y_i(t)} = \frac{1}{1+i\omega T} \tag{11}$$

is the same as $F(i\omega)$ obtained from $F(s)$ by replacing s with $i\omega$. Thus, for a sinusoidal input,

$$[Y_o(t)]_{\text{steady}} = F(i\omega)\,Y_i(t) \tag{12}$$

---

* This is a Tauberian theorem. A sufficient condition is that $F(s)$ be the Laplace transform of some function $F(t)$ for all Rl $s > 0$, $F(s) \to K$ as $s \to +0$, and $F(t)$ is of order $o(1/t)$ as $t \to \infty$. To apply this theorem to our example, we should consider the unit-step function $\mathbf{1}(t)$ as the limit of $e^{-\lambda t}\,\mathbf{1}(t)$ as $\lambda \to 0$.

It is clear then that $F(i\omega)$ represents the frequency response of the dynamic system, and that the gain K represents the limiting value of the response when $\omega \to 0$. The transfer function $G(i\omega)$ is the ratio of the frequency response to the response at zero frequency.

It is convenient to write the complex vector $G(i\omega)$ in terms of its magnitude M and its phase angle θ,

$$G(i\omega) = Me^{i\theta} \tag{13}$$

For the present example, Eq. 7 gives

$$M = \frac{1}{\sqrt{1 + \omega^2 T^2}}, \qquad \tan \theta = -\omega T \tag{14}$$

It is customary in electrical engineering to plot M and θ as functions of ω on a logarithmic paper (the Bode diagrams), and also to plot the locus

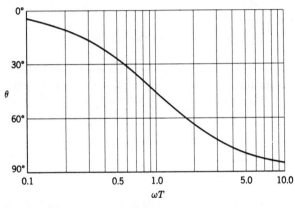

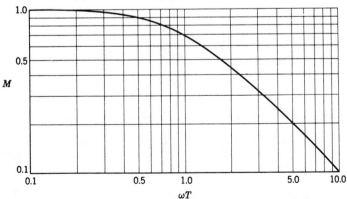

Fig. 10.10. The Bode diagrams for Eq. 14.

of $G(i\omega)$ as a vector on a complex plane, or the locus of $1/G(i\omega) = (1/M)e^{-i\theta}$ (the Nyquist diagrams).   For the example defined by Eqs. 5, the Bode and Nyquist diagrams are shown in Figs. 10.10 and 10.11.

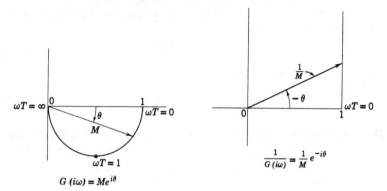

Fig. 10.11.   The Nyquist diagrams for Eq. 14.

Evidently $F(i\omega)$ is exactly the mechanical admittance defined in § 8.1.

Let us consider the combination of several mechanical systems into a circuit.   For a series arrangement of n elements as shown in Fig. 10.12, let the rth element have a gain K_r and a transfer function $G_r(i\omega) = M_r e^{i\theta_r}$.

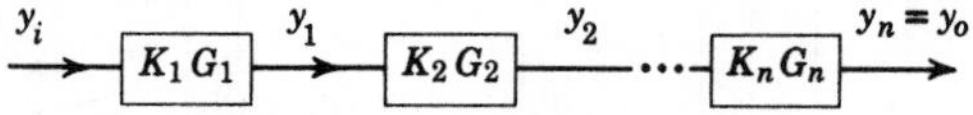

Fig. 10.12.   Combined elements.

This series array is equivalent to a single block having a transfer function $G(i\omega)$ and gain K:

$$K = K_1 K_2 \cdot \cdot \cdot K_n$$

$$G(i\omega) = (M_1 e^{i\theta_1})(M_2 e^{i\theta_2}) \cdot \cdot \cdot (M_n e^{i\theta_n}) \qquad (15)$$

$$= (M_1 M_2 \cdot \cdot \cdot M_n)e^{i(\theta_1 + \theta_2 + \cdots + \theta_n)}$$

The over-all transfer function $G(i\omega) = Me^{i\theta}$ can thus be determined:

$$\log_{10} M = \log_{10} M_1 + \log_{10} M_2 + \cdot \cdot \cdot + \log_{10} M_n$$

$$\theta = \theta_1 + \theta_2 + \cdot \cdot \cdot + \theta_n \qquad (16)$$

As another example a simple feedback servo is shown in Fig. 10.13.   The input y_i passes through a block with gain K_1 and transfer function $G_1(s)$

and produces the output y_o which is fed back through an "adder" to the input, so that the resultant input becomes $y_i - y_o$. In this case

$$y_o = K_1 G_1 (y_i - y_o)$$

we have

$$y_o = \frac{K_1 G_1}{1 + K_1 G_1} y_i \qquad (17)$$

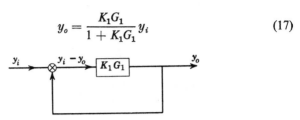

Fig. 10.13. A simple feedback system.

A somewhat more general case is shown in Fig. 10.14, for which

$$y_o = \frac{K_1 G_1}{1 + K_1 G_1 K_2 G_2} y_i \qquad (18)$$

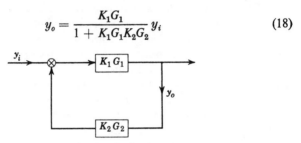

Fig. 10.14. Another feedback system.

Thus the resultant gain and transfer functions are, respectively,

$$K = K_1/(1 + K_1 K_2)$$

$$G = \frac{1 + K_1 K_2}{K_1} \frac{K_1 G_1}{1 + K_1 K_2 G_1 G_2} \qquad (19)$$

10.5 CRITERIA OF STABILITY

If the transfer function $G(s)$ between the input $y_i(s)$ and the output $y_o(s)$ is a rational function, it will be completely characterized by its zeros and poles; for such a function can be written as

$$G(s) = A \frac{(s - a_1)^{\mu_1}(s - a_2)^{\mu_2} \cdots (s - a_m)^{\mu_m}}{(s - b_1)^{\lambda_1}(s - b_2)^{\lambda_2} \cdots (s - b_n)^{\lambda_n}} \qquad (1)$$

where A is a constant, $a_1, a_2, \cdots, a_m$ are the zeros, and $b_1, b_2, \cdots, b_n$

are the poles; the positive integers μ_1, μ_2, $\cdots$; λ_1, λ_2, $\cdots$ are the orders of the zeros and poles, respectively. We assume $\Sigma\mu_i < \Sigma\lambda_i$.

The location of the poles is important in determining the stability of the solution $Y_0(t)$ which can be derived according to Heaviside's expansion theorem. Expanding $G(s)/s$ into partial fractions, we have terms of the form

$$\frac{A_1}{(s - b_i)^{\lambda_i}} + \frac{A_2}{(s - b_i)^{\lambda_i - 1}} + \cdots + \frac{A_{\lambda_i}}{s - b_i} \qquad (2)$$

associated with each pole b_i. Therefore, when the input $Y_i(t)$ is a unit-step function, $y_i(s) = 1/s$, the inverse transform involves the term

$$K[A_1 t^{\lambda i-1}/(\lambda_i - 1)! + A_2 t^{\lambda i-2}/(\lambda_i - 2)! + \cdots + A_{\lambda i}]e^{b_i t} \qquad (3)$$

Obviously the solution (3) is "stable" (remains finite as $t \to \infty$) if the real part of b_i is negative, and is "unstable" (becomes unbounded) if the real part of b_i is positive. Thus a necessary condition for stability is that the transfer function should have no poles in the right half-plane $\mathrm{Rl}\ s > 0$. This conclusion holds also for transcendental transfer functions: If $y_0(s)$ has no singularity on the right half-plane $\mathrm{Rl}\ s > 0$, then the inverse $Y_0(t)$ remains finite as $t \to \infty$. If $y_0(s)$ has a pole on the right half-plane, then $Y_0(t)$ becomes infinitely large as $t \to \infty$. (See footnote on p. 360.)

Hence, if the transfer function $G(s)$ is a rational function, the problem of stability is to find whether the real parts of all the roots of the denominator are negative.

The conditions for a polynomial with real coefficients

$$P(s) = p_0 s^n + p_1 s^{n-1} + \cdots + p_{n-1} s + p_n \qquad (4)$$

to have only pseudo-negative roots (i.e., the real parts of all the roots are negative) are the well-known Routh-Hurwitz conditions. They are given in Appendix 2. This algebraic problem has been generalized to include polynomials with complex coefficients by Sherman, DiPaola, and Frissell.[11.20, 11.21] An application of these results solves the stability problem completely when $G(s)$ is a rational function.

A different method is introduced by Bode and Nyquist (see Ref. 10.10) and is based on Cauchy's theorem in the theory of functions of a complex variable. It is applicable to transcendental transfer functions.

Let S be a region on the (x, y) plane bounded by a simple closed curve C. A complex-valued function $f(x, y) = \phi(x, y) + i\,\psi(x, y)$ defined in S may be regarded as a function of a complex variable $z = x + iy$ and written as $f(z)$. When z traces the curve C, the locus of the complex number $f(z)$, when plotted on a complex plane, gives a curve $\bar{C}$ (Fig. 10.15). Now there is a theorem in the theory of functions of a complex variable which states that if $f(z)$ is analytic in S and continuous on C, and does not vanish on C, then *the excess of the number of zeros over the number of poles of $f(z)$ within C is* $(1/2\pi)$ *times the increase in* arg $f(z)$ *as z goes once around C in the positive direction.** This result is sometimes called *the principle of the argument*, and is easily derived from the theorem of residue, considering the residue of the function $f'(z)/f(z)$ in C. A zero of order r is counted as r simple zeros, and a pole of order s is counted as s simple poles. The positive direction of C is so defined that, if an observer

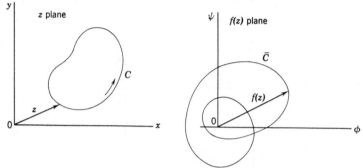

Fig. 10.15.   Locus of $f(z)$ as z traces a closed curve C.

moves along the curve C in the positive direction, the region S enclosed by C appears to his left-hand side.

This principle can be applied as follows. If we wish to find the number of zeros of $f(z)$ in a contour C, we determine (a) the change in the argument of $f(z)$ as z goes around C in the positive direction, and (b) the number of poles of $f(z)$ in C. Then

$$\text{No. of zeros in } C = \frac{\text{change of argument of } f(z)}{2\pi} + \text{no. of poles in } C$$

The number of poles can usually be determined by inspection. In particular, if $f(z)$ is a polynomial, it has no pole in any contour C.

As a particular application, if the number of zeros of $f(z)$ in the right half-plane, Rl $z > 0$, is to be found, we may consider a contour C con-

---

* The argument of z, written as arg z, is the angle between the radius vector $z = x + iy$ and the x axis; i.e., arg $z =$ arc tan y/x. Thus arg $f(z) =$ arc tan ψ/ϕ. The argument of $f(z)$ is increased by 2π if the end point of the radius vector $f(z)$ encircles the origin once in the positive direction; it is decreased by 2π if the radius vector encircles the origin once in the negative direction. Thus in Fig. 10.15, if $f(z)$ traces $\bar{C}$ in the positive direction, arg $f(z)$ is increased by 4π as z goes once around C in the positive direction.

sisting of a large semicircle of radius R in the right half-plane (Fig. 10.16). In the limits when R tends to infinity, the entire right half-plane will be enclosed in C. Observe the locus of $f(z)$ as z goes round C once. If it encircles the origin of the $f(z)$ plane n times in the counterclockwise direction, $f(z)$ changes its argument by $2n\pi$. If the number of poles in the right half-plane is known to be p, then the number of zeros in the right half-plane is $n + p$.

The contour C must not pass through any pole or zero of $f(z)$. If there are zeros and poles on the desired contour C, the difficulty can be

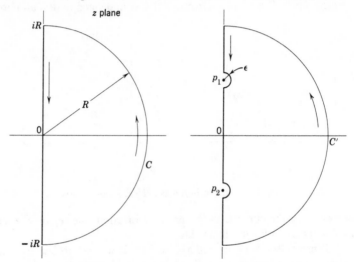

z plane

Fig. 10.16.   A semicircular   Fig. 10.17.   A deformed semicircular
contour.   contour.

avoided by making a deformation of the contour in the neighborhood of these points. Thus, as shown in Fig. 10.17, if p_1, p_2 are poles on the imaginary axis and C is chosen as semicircle, we may take a deformed contour C' which circumvents the poles by small semicircular arcs of radius ε and let $\varepsilon \to 0$. If p_1 is a pole of order n, we may write, in the neighborhood of p_1,

$$f(z) = \frac{1}{(z - p_1)^n} g(z)$$

where $g(z)$ is finite in the neighborhood of p_1. Let $z - p_1 = \varepsilon e^{i\theta}$, ·then

$$f(z) = \frac{1}{\varepsilon^n} [g(p_1) + O(\varepsilon)] e^{-in\theta}$$

Thus, when z goes around the small arc from $\theta = \pi/2$ to $\theta = -\pi/2$, the change in phase of $f(z)$ is $n\pi$.

When the contour C is symmetrical with respect to the real axis, it is useful to remember that, when $f(z)$ is a polynomial or a rational function with real coefficients, or is a combination of elementary functions with real coefficients,

$$f(\bar{z}) = \bar{f}(z)$$

where $\bar{z}$ and $\bar{f}$ denote the complex conjugate of z and f; i.e., $z = x + iy$, $\bar{z} = x - iy$, etc. Thus, if $f(z)$ corresponding to the upper half of the contour C is known, that corresponding to the lower half can be obtained by a mirror reflection in the real axis.

10.6 FLUTTER

The method of Laplace transformation can be applied to study the stability of oscillating airfoils. Let us consider the motion of a two-dimensional flat-plate airfoil of unit span having two degrees of freedom h and α as considered in § 6.8. The equations of motion are given by Eqs. 8, 9 and 10 of § 6.8. The Laplace transformation of these equations can be obtained easily. Assuming that

$$h = \dot{h} = \alpha = \dot{\alpha} = 0 \quad \text{when} \quad t = 0 \tag{1}$$

we obtain

$$A(s)\mathcal{L}\{h/b\} + B(s)\mathcal{L}\{\alpha\} = \mathcal{L}\{P\}/s^2$$
$$D(s)\mathcal{L}\{h/b\} + E(s)\mathcal{L}\{\alpha\} = \mathcal{L}\{Q\}/s^2 \tag{2}$$

where, with the symbols μ, x_a, r_a as defined on p. 216,

$$A(s) = -\mu - \mu\frac{k_h^2}{s^2} - 1 - \frac{2C(-is)}{s}$$

$$B(s) = -\mu x_\alpha + a_h - \frac{1}{s} - \frac{2C(-is)}{s^2}\left[1 + \left(\frac{1}{2} - a_h\right)s\right]$$

$$D(s) = -\mu x_\alpha + a_h + \left(\frac{1}{2} + a_h\right)2\frac{C(-is)}{s} \tag{3}$$

$$E(s) = -\mu r_\alpha^2\left(1 + \frac{k_\alpha^2}{s^2}\right) - a_h^2 - \left(\frac{1}{2} - a_h\right)\frac{1}{s}$$
$$-\frac{1}{8} + \left(\frac{1}{2} + a_h\right)2\frac{C(-is)}{s^2}\left[1 + \left(\frac{1}{2} - a_h\right)s\right]$$

The function $C(-is)/s$ denotes the Laplace transform of the Wagner's function $\Phi(\tau)$:

$$\int_0^\infty \Phi(\tau)e^{-s\tau}d\tau = \frac{C(-is)}{s} \tag{4}$$

A comparison of this equation with Eq. 6 of § 6.9 shows that $C(-is)$ can be obtained from the Theodorsen's function $C(k)$ by replacing k by $-is$. From the form of Theodorsen's function given in Eqs. 8, 8a, 8b of § 6.9, it is clear that $C(k)$ as a function of the complex variable k is analytic in the entire upper half-plane, including the real axis. Correspondingly $C(-is)$ is analytic for the right half-plane Rl $s \geqslant 0$. In other words, if we put $ik = s$, then the expressions of lift and moment given in Eqs. 9 of § 6.9 hold for a divergent oscillation $e^{s\tau}$, (Rl $s \geqslant 0$). The validity of this generalization has been discussed by W. P. Jones[15.80] and Van de Vooren in his discussion of Ref. 15.26.

The solution of Eq. 2 may be written as*

$$\mathscr{L}\{h/b\} = \frac{\Delta_{11}(s)}{s^2\Delta(s)}\mathscr{L}\{P\} + \frac{\Delta_{12}(s)}{s^2\Delta(s)}\mathscr{L}\{Q\}$$

$$\mathscr{L}\{\alpha\} = \frac{\Delta_{21}(s)}{s^2\Delta(s)}\mathscr{L}\{P\} + \frac{\Delta_{22}(s)}{s^2\Delta(s)}\mathscr{L}\{Q\} \tag{5}$$

where

$$\Delta_{11} = E, \quad \Delta_{12} = -B, \quad \Delta_{21} = -D, \quad \Delta_{22} = A \tag{6}$$

$$\Delta(s) = \begin{vmatrix} A(s) & B(s) \\ D(s) & E(s) \end{vmatrix} \tag{7}$$

The functions $\Delta(s)$, $\Delta_{11}(s)$, $\cdots$, $\Delta_{22}(s)$ have no poles in the right half-plane, Rl $s > 0$. Hence, in the right half-plane, the poles of $\mathscr{L}\{h/b\}$, etc., if any, must arise from the zeros of $\Delta(s)$, provided that $\mathscr{L}\{P\}$ and $\mathscr{L}\{Q\}$ has no poles on the right half-plane. If $\Delta(s)$ has a root with a positive real part, the motion h and α will eventually become infinitely large. If $\Delta(s)$ has no root with a positive real part, the motion will be convergent. Hence, the stability question is reduced to an investigation of the zeros of $\Delta(s)$ in the right half-plane.†

Dugundji[10.9] uses the Nyquist diagram for this purpose. It is necessary to examine the number of times the curve of $\Delta(s)$ encircles the origin as s traces an infinite semicircle on the right half-plane (Fig. 10.16) in the counterclockwise direction. Along the imaginary axis, $s = ik$, $\Delta(s)$

* The functions $\Delta_{11}(s)/\Delta(s)$, $\Delta_{12}(s)/\Delta(s)$, etc., are proportional to the transfer functions. From the general interpretation of the transfer functions it is clear that, if s is replaced by ik, then $\Delta_{11}(ik)/\Delta(ik)$, etc., represent the admittance of the dynamical system to

reduces into $\Delta(ik)$, which is exactly the flutter determinant Δ as given in Eq. 18, § 6.9. Along the semicircle, $s = Re^{i\theta}$, $\left(-\dfrac{\pi}{2} \leqslant \theta \leqslant \dfrac{\pi}{2} \right)$

$$\lim_{R \to \infty} C(-is) \to \tfrac{1}{2} \tag{8}$$

which can be easily seen from Eqs. 8a, 8b of § 6.9 or deduced from the asymptotic expansion of the Hankel functions in Eq. 8, § 6.9. Therefore $\Delta(s)$ tends to a constant on the semicircle as $R \to \infty$:

$$\lim_{R \to \infty} \Delta(s) \to \begin{vmatrix} -(\mu + 1) & -\mu x_\alpha + a \\ -\mu x_\alpha + a & -\mu r_\alpha{}^2 - a^2 - \tfrac{1}{8} \end{vmatrix} \tag{9}$$

At the origin, $s = 0$, Eqs. 3, 7 show that $\Delta(s)$ has a pole of order four. The contour of mapping must be deformed by describing a small semicircle from εi to $-\varepsilon i$ on the right half-plane. The corresponding change in the phase angle of $\Delta(s)$ is 4π.

The mapping of $\Delta(s)$ can now be made without difficulty. When $s = ik$, the real and imaginary part of $\Delta(ik)$ can be computed from Eq. 19, § 6.9. Note that $\Delta(ik)$ need be calculated for positive values of k only. Let $\bar{C}(k)$ denote the complex conjugate of $C(k)$, then we have

$$C(-k) = \bar{C}(k) \tag{10}$$

i.e.

$$C(k) = F + iG, \qquad C(-k) = F - iG$$

---

harmonic excitations. Therefore the explicit form of the functions $\Delta(s)$, $\Delta_{11}(s)$, etc., can be obtained from Eqs. 11 of § 6.9, by replacing ik by s.

† This conclusion is based on an important theorem by A. Erdelyi (*Lecture Notes on Laplace Transformation*, 1947, California Institute of Technology). Let $S_{-\Delta}$ denote a region that consists of all points satisfying the condition $\left| \arg (s - s_0) \right| < (\pi/2) + \Delta$, where $0 < \Delta \leqslant \pi/2$. Then the theorem states that under the assumptions:

(1) The function $f(s)$ is analytic, and regular in $S_{-\Delta}$ (but not necessarily at s_0 or at ∞).

(2) $f(s) \to 0$, uniformly in $S_{-\Delta}$, as $s \to \infty$.

(3) $f(s)$ has the following asymptotic representation in the sense of Poincare:

$$f(s) \sim \sum_{n=0}^{N} c_n(s - s_0)^{\lambda_n}, \qquad (-1 < \lambda_0 < \lambda_1 < \cdots < \lambda_N)$$

uniformly in $S_{-\Delta}$, as $s \to s_0$.

One concludes that the inverse Laplace transform of $f(s)$ has the asymptotic representation

$$F(t) \sim e^{s_0 t} \sum_{n=0}^{N} \frac{c_n}{\Gamma(-\lambda_n)} \, t^{-\lambda_n - 1} \qquad \text{as } t \to \infty.$$

This is obvious from the definition of $C(k)$ given in Eq. 6 of § 6.9, by replacing k by $-k$, i by $-i$. It can also be seen from Eqs. 8, § 6.9. So

$$\Delta(-ik) = \bar{\Delta}(ik)$$

The mapping of the negative imaginary axis is just the mirror image in the real axis of that of the positive imaginary axis.

Example.   Consider a wing with the following physical parameters:

$$\mu = 10, \qquad a_h = -0.2, \qquad x_\alpha = 0.1, \qquad r_\alpha{}^2 = \tfrac{1}{4}$$

$$\omega_h/\omega_\alpha = \tfrac{1}{5}, \qquad b = 6/2\pi = 0.9549 \text{ ft}, \qquad \omega_\alpha = 12 \text{ cycles per second}$$

The map of the semicircle on the $\Delta(s)$ plane is sketched in Fig. 10.18. It is clearly seen that, for the cases $U = 80$ and 120 ft per sec, the curve

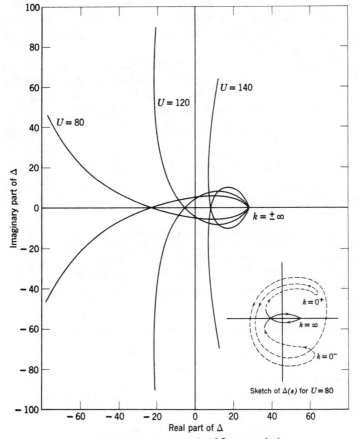

Fig. 10.18.   An example of flutter analysis.

encircles the origin once in the positive direction and once in the negative direction. The net encirclement is zero. Hence, the motion is stable. In case $U = 140$ ft per sec, the curve encircles the origin twice in the positive direction. Hence, $\Delta(s)$ has two zeros with positive real part. The wing is unstable at $U = 140$. An examination based on the method of § 6.9 shows that the critical flutter speed is 129 ft per sec.

The difference between the present analysis and that in § 6.9 is that, whereas in § 6.9 the equation $\Delta = 0$ is to be solved for a pair of roots k_{cr} and U_{cr}, in the present analysis the values of the flutter determinant Δ are examined as k varies from $-\infty$ to ∞.

The present method has a definite advantage when the flutter speed is so high that the effect of compressibility becomes important. As the contours are calculated for each specified speed U, the Mach number is known and the appropriate aerodynamic coefficients can be obtained directly from existing tables. Hence, the result of each calculation is a positive statement about the stability of the system, and the iterative correction for the Mach number effect is unnecessary.

BIBLIOGRAPHY

Books on Laplace transformation:

10.1   Carslaw, H. S., and J. C. Jaeger: *Operational Methods in Applied Mathematics.* Oxford Univ. Press, London (1941).

10.2   Courant, R., and D. Hilbert: *Methoden der mathematischen Physik*, Vol. 2. Springer, Berlin (1937).

10.3   Churchill, R. V.: *Modern Operational Mathematics in Engineering.* McGraw-Hill, New York (1944).

10.4   Doetsch, G.: *Theorie und Anwendung der Laplace-Transformation.* Springer, Berlin (1937).

   Handbuch der Laplace-Transformation, Band I, *Theorie der Laplace Transformation.* Verlag Birkhäuser, Basel, Switzerland (1950).

10.5   McLachlan, N. W.: *Complex Variable and Operational Calculus with Technical Applications.* Cambridge Univ. Press, London (1942).

10.6   Titchmarsh, E. C.: *Theory of Fourier Integrals.* Oxford Univ. Press, London (1937).

10.7   van der Pol, B., and H. Bremmer: *Operational Calculus Based on the Two-Sided Laplace Integral.* Cambridge Univ. Press, London (1950).

Table of Laplace transforms:

10.8   Erdélyi, A.: *Tables of Integral Transforms*, Vol. 1. McGraw-Hill, New York (1954).

Application to flutter:

10.9   Dugundji, J.: A Nyquist Approach to Flutter. *J. Aeronaut. Sci.* **19**, Reader's Forum, 422 (1952).

364   APPLICATIONS OF LAPLACE TRANSFORMATION

Principles of stability and control of an airplane:

10.10 Bcllay, W.: Aerodynamic Stability and Automatic Control. *J. Aeronaut. Sci.* **18**, 569–623 (1951).
10.11 Duncan, W. J.: *The Principles of the Control and Stability of Aircraft.* Cambridge Univ. Press, London (1952).
10.12 Jones, B. M.: *The Dynamics of an Airplane*, Vol. V of *Aerodynamic Theory*, edited by Durand. Julius Springer, Berlin (1934).
10.13 Perkins, C. D., and R. E. Hage: *Airplane Performance, Stability, and Control.* John Wiley & Sons, New York (1949).

Chapter 11

GENERAL FORMULATION OF AEROELASTIC PROBLEMS

In the previous chapters several types of aeroelastic problems are treated. A large number of other problems can be formulated, the practical importance of which depends on the type of the structure and the conditions of the flow. In order to clarify the basic features of any problem, and to determine the degree of approximation of an analysis, it is desirable to have the formulation as general as possible at the beginning, and then to specialize in particular cases by introducing further assumptions. In this way the relationship between various approaches to the problem can be understood, and, if an improvement on a known theory is desired, one can locate the restrictive assumptions and decide which one should be relaxed.

In § 11.1 the feedback nature of aeroelastic systems and their representation by *functional diagrams* are considered. The functional operators concept is then introduced in § 11.2 to reduce the pictorial representations of functional diagrams into algebraic equations. The operators entering aeroelastic analysis are examined in greater detail in §§ 11.3 to 11.5.

Knowing the form of the operators involved, we will have no difficulty in writing down the governing equations according to the functional diagrams and the operational equations. The accuracy of the formulation of each problem is clearly indicated by the assumptions used in deriving the operators.

The mathematical characteristics of aeroelastic problems are briefly discussed in § 11.6.

11.1 FUNCTIONAL DIAGRAMS

In analyzing an aeroelastic system, it is convenient to distinguish the various functions that each element performs. Sometimes many elements together perform a single function, such as an electronic amplifier, which converts an input signal into another signal, but is composed of many resistors, capacitors, transformers, and vacuum tubes. Sometimes a single element performs several functions, such as an airplane wing, which produces lift force, elastic deformation, and inertia force. Whether several elements should be grouped together to be considered as a single unit or a single element should be analyzed into several *functional* units depends on the particular problem under consideration. But in all cases it is helpful to represent the entire system in a pictorial form so that the interaction among various elements can be clearly seen. Such pictorial representations are called *functional diagrams*. They are also called *block diagrams* if no single physical element is analyzed into several functional units.

The meaning of functional diagrams can be best illustrated by examples.

Lift of an Elastic Wing. If a change of angle of attack α (measured at the wing root) of a cantilever wing is regarded as an input, and the total

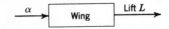

Fig. 11.1. Functional diagram of a wing.

lift force L acting on the wing as an output, the function of the wing may be represented symbolically as in Fig. 11.1. The relation between L and α, however, depends on the elastic deformation of the wing. It is convenient to consider the wing as composed of two elements: the wing as a lift-producing mechanism, and the wing as an elastic structure. With respect to aerodynamics, each configuration of the wing may be considered as rigid. With respect to the elastic deformation, the action of

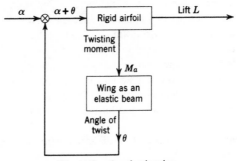

Fig. 11.2. An elastic wing.

aerodynamic forces is the same as any other system of exterior forces. The partial problems in aerodynamics and elasticity may be solved in the classical manner, but these two solutions must be properly combined to account for the behavior of an elastic wing. An elastic wing may be represented functionally as in Fig. 11.2. The airfoil, at an angle of attack α and having a deflection surface θ, produces an aerodynamic pressure distribution p. The elastic wing, in response to the pressure distribution, produces a deflection surface θ. Let us denote the wing surface corresponding to the initial angle of attack α also by the symbol α; then the geometrical configuration may be simply written as $\alpha + \theta$, with respect to which the aerodynamic force is computed. Thus the rigid-airfoil–elastic-structure system forms a loop. It is a feedback system.

Wing Divergence. In the preceding example let α and θ be so measured that $L = 0$ when $\alpha = \theta = 0$. Then, in general, $\theta = 0$ is the only solution

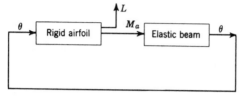

Fig. 11.3. Critical divergence condition.

when $\alpha = 0$. Let us now ask whether there exists a nontrivial solution $\theta \neq 0$ when $\alpha = 0$ (Fig. 11.3). If such a solution exists, the wing is said to be critically divergent. In the critical divergent condition the amplitude of θ is indeterminate. Since the system represented in Fig. 11.3 can be superposed on that in Fig. 11.2, it is clear that, at the critical-divergent condition, the lift problem has no *unique* solution. Conversely, when the

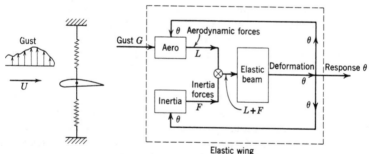

Fig. 11.4. Gust response of an elastic wing.

wing is nondivergent, the unique solution of Fig. 11.3 is $\theta \equiv 0$, and the corresponding solution of Fig. 11.2 exists.

Gust Loading on an Airfoil. In dynamic problems the wing performs three distinct functions: It produces (1) aerodynamic force, (2) inertia force, (3) elastic deformation. It is convenient to represent the wing's triple functions as three independent operators. In an example of the gust loading on a wing which is restrained against pitching, the functional diagram may be shown in Fig. 11.4.

Flutter is a companion problem. Instead of finding the response of the wing to a gust, we ask whether the homogeneous system obtained by setting $G = 0$ in Fig. 11.4, has a nontrivial solution.

Again, as in the corresponding problems of lift and divergence, here the alternative theorem is: Either the homogeneous system has a nontrivial solution (flutter occurs), or the gust-response problem has a solution.

11.2 THE FUNCTIONAL OPERATORS AND THEIR ALGEBRA

The functional diagrams can be represented algebraically by operators. Let a functional block that converts an "input" θ_i into an "output" θ_o (Fig. 11.5) be written symbolically as

$$\theta_o = \mathbf{O} \, \theta_i \tag{1}$$

The symbol $\mathbf{O}$ is called an *operator*, and the process of finding θ_o from θ_i is called an *operation* on θ_i by the operator $\mathbf{O}$. It is to be noted that θ_o

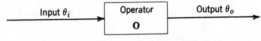

Fig. 11.5.   A functional block.

and θ_i may represent quantities of different physical dimensions, for example, θ_i an angle, and θ_o a force. The physical dimensions of an operator are determined by the dimensions of θ_o and θ_i.

If it is possible to solve the inverse problem of finding the input from a known output, then we say that an inverse of the operator exists, and write

$$\theta_o = \mathbf{O} \, \theta_i, \qquad \theta_i = \mathbf{O}^{-1} \theta_o \tag{2}$$

If there exists a one-to-one correspondence between θ_i and θ_o for all allowable θ_i, the operator is said to be *regular*.* Otherwise it is *singular*. A regular operator has a unique inverse. In the examples named above, the operator that transforms the angle of attack into lift is regular in ranges of flow speed that excludes the critical-divergence speeds, and is singular at the critical-divergence speeds. Similarly, the operator that transforms the gust into the wing response is regular when the speed of flow is not a flutter speed, and is singular at flutter speeds. In aero-elasticity, we generally consider regular operators that become singular only at certain special values of a set of parameters (e.g., the dynamic

---

* Clearly it is necessary to specify the regions in which the quantities θ_i and θ_o are considered. However, the explicit statements of the regions of interest in our examples below are generally omitted in favor of conciseness. For example, if the linearized airfoil theory is used, the angle of attack, the downwash, and the lift force, etc., must be small. But the range of applicability of the linearized theory will not be mentioned every time. It is hoped that this lack of rigor will not cause confusion.

pressure, the reduced frequency). The set of parameters at which an operator becomes singular are *eigenvalues* of that operator. The determinations of the response from a regular operator and the eigenvalues of a singular operator are the two main problems in aeroelasticity.

An operator is said to be *linear* if it has the following property:

$$\mathbf{O}(x_1 + x_2) = \mathbf{O}x_1 + \mathbf{O}x_2 \tag{3}$$

where x_1 and x_2 both belong to the region over which the operator $\mathbf{O}$ is defined. It is said to be *nonlinear* if this relation does not hold over the entire region of definition. For mathematical simplicity, only linear operators will be considered hereafter.

Modern mathematical theory of functional transformations can be applied to the functional operators to render the mathematical treatment exact. But for the present purpose a heuristic account will be sufficient.

The operators occurring in aeroelasticity may be algebraic, differential, integral, or integral–differential.* They will be discussed in greater detail later. First let us consider some of their algebraic properties.

Consider again the steady-state lift-distribution problem. Let us designate the operators by subscripts as defined by the following equations:

$$p = \mathbf{O}_1(\alpha + \theta), \qquad \theta = \mathbf{O}_2 p \tag{4}$$

where p is the aerodynamic pressure distribution, α is the wing surface without elastic deformation, and θ is the elastic deflection corresponding to p. Obviously we may consider θ as being generated by two successive operations $\mathbf{O}_1$ and $\mathbf{O}_2$ and write the entire process in a single equation:

$$\theta = \mathbf{O}_2\mathbf{O}_1(\alpha + \theta) \tag{5}$$

The successive operations $\mathbf{O}_2\mathbf{O}_1$ may be regarded as a single operation generated by $\mathbf{O}_2$ and $\mathbf{O}_1$ in the specified order. It is called a *composition product* of the original operators. Note that it is in general *noncommutative*; i.e., $\mathbf{O}_2\mathbf{O}_1 \neq \mathbf{O}_1\mathbf{O}_2$. Strict attention must be given to the order in which a composition product is formed.

Let us assume that $\mathbf{O}_1$ and $\mathbf{O}_2$ are both linear. Equation 4 can then be written as

$$p = \mathbf{O}_1\alpha + \mathbf{O}_1\theta$$

and

$$\theta = \mathbf{O}_2 p = \mathbf{O}_2(\mathbf{O}_1\alpha + \mathbf{O}_1\theta) = \mathbf{O}_2\mathbf{O}_1\alpha + \mathbf{O}_2\mathbf{O}_1\theta$$

---

* An element of a functional diagram may represent a single number, a continuous function, a matrix of numbers, or a matrix of continuous functions of space and time. In particular, the interpretation of the variables as matrices is very important. By such an interpretation the operational equations are made very concise.

Comparing this with Eq. 5, we find

$$O_2O_1(\alpha + \theta) = O_2O_1\alpha + O_2O_1\theta \tag{6}$$

Therefore *the composition product of two linear operators is linear.*

The lift distribution on the elastic wing as represented in Fig. 11.2 can then be characterized, under the linearity assumption, by the equation

$$\theta = O_2O_1\alpha + O_2O_1\theta$$

or

$$\theta - O_2O_1\theta = O_2O_1\alpha$$

Let us write I as an *identical operator*, which transforms any quantity into itself:

$$I\theta = \theta \tag{7}$$

The relation between θ and α can then be written as

$$(I - O_2O_1)\theta = O_2O_1\alpha \tag{8}$$

A formal solution of this equation is

$$\theta = (I - O_2O_1)^{-1} O_2O_1\alpha \tag{9}$$

where $(I - O_2O_1)^{-1}$ is the inverse operator (assumed unique) of $I - O_2O_1$. In order to interpret the meaning of the inverse operator $(I - O_2O_1)^{-1}$, let us develop the expression formally into a power series in O_2O_1 by means of the binomial theorem:

$$(I - O_2O_1)^{-1} = I + O_2O_1 + (O_2O_1)^2 + (O_2O_1)^3 + \cdots \tag{10}$$

where

$$(O_2O_1)^2 = (O_2O_1)(O_2O_1)$$
$$(O_2O_1)^{k+1} = (O_2O_1)(O_2O_1)^k \qquad (k = 2, 3, \cdots) \tag{11}$$

Since O_2O_1 transforms an angle into an angle, so do all the successive powers of O_2O_1. Applying Eq. 10 to Eq. 9, we may write

$$\theta = \sum_{k=1}^{\infty} (O_2O_1)^k \alpha \tag{12}$$

If the process can be justified, the solution θ can be obtained by summing the infinite series.

The summation of the infinite series 12 actually amounts to a process of successive approximations. The first term gives the elastic deformation of the wing corresponding to the lift acting on a rigid wing. The second term gives the increment of the elastic deformation due to the change of lift corresponding to the elastic deformation first computed. The third term gives the second increment of elastic deformation due to the lift corresponding to the first correction of the elastic deformation, and so on. Hence, in this particular case, the process 10 is justifiable for sufficiently small dynamic pressure of the flow.   The series 12 is known to converge (Chapter 3) whenever the dynamic pressure q is less than $|q_{\text{div}}|$.

Example 1.   *The Lift Acting on a Two-Dimensional Airfoil* (*Fig.* 11.2). Consider the two-dimensional case of § 3.1.   Here

$$M_a = Lec = qc^2ea(\alpha + \theta)$$
$$M_e = K\theta = M_a$$

Hence,

$$\mathbf{O}_1 = qc^2ea$$
$$\mathbf{O}_2 = 1/K$$

Equations 12 and 9 gives at once

$$\theta = \sum_{n=1}^{\infty} \left(\frac{qc^2ea}{K}\right)^n \alpha = \frac{qc^2ea}{K(1 - qc^2ea/K)} \alpha$$

The lift per unit span is given by

$$L = qca(\alpha + \theta) = \frac{qca\alpha}{1 - qc^2ea/K}$$

in agreement with § 3.1.

Example 2.   *Divergence* (*Fig.* 11.3).   Here the functional relation is represented by

$$\mathbf{O}_2\mathbf{O}_1\theta = \theta$$

In the two-dimensional case (Ex. 1), we obtain at once the critical condition

$$\frac{qc^2ea}{K} - 1 = 0$$

which yields

$$q_{\text{div}} = \frac{K}{c^2ea}$$

Example 3.   *Gust Loading* (*Fig.* 11.4).   Let the gust be represented by G and the other symbols denote quantities as shown in Fig. 11.4.   Let the operators be defined as follows:

$$L = \mathbf{O}_{LG}G + \mathbf{O}_{L\theta}\theta$$
$$F = \mathbf{O}_I\theta$$
$$\theta = \mathbf{O}_E(L + F)$$

From the loops shown in Fig. 11.4, we obtain

$$\mathbf{O}_E[(\mathbf{O}_{LG}G + \mathbf{O}_{L\theta}\theta) + \mathbf{O}_I\theta] = \theta$$

Hence,

$$\mathbf{O}_E\,\mathbf{O}_{LG}G = (\mathbf{I} - \mathbf{O}_E\mathbf{O}_{L\theta} - \mathbf{O}_E\mathbf{O}_I)\,\theta$$

or

$$\theta = (\mathbf{I} - \mathbf{O}_E\mathbf{O}_{L\theta} - \mathbf{O}_E\mathbf{O}_I)^{-1}\,\mathbf{O}_E\mathbf{O}_{LG}G$$

Example 4. *Flutter.* If $G = 0$, the last equation of Ex. 3 becomes a homogeneous one:

$$(\mathbf{I} - \mathbf{O}_E\mathbf{O}_{L\theta} - \mathbf{O}_E\mathbf{O}_I)\,\theta = 0 \tag{13}$$

which gives the critical-flutter condition. (Divergence may be considered as flutter of zero frequency, and hence is included in the above equation.) Since $\mathbf{O}_E$ depends on the rigidity of the wing, while $\mathbf{O}_{L\theta}$ depend on the dynamic pressure and the reduced frequency, the problem of flutter is to determine the eigenvalues of the rigidity, dynamic pressure, or reduced frequency at which Eq. 13 has a nontrivial solution.

11.3 NATURE OF THE OPERATORS IN AEROELASTICITY

As an aeroelastic system may be composed of a large number of electrical, hydraulic, mechanical, as well as aerodynamic elements, it is evident that the operators involved are much varied. From the point of view of analysis, it is convenient to classify the operators into two kinds: (1) those relating quantities that are essentially independent of the space coordinates, such as an electric voltage across two terminals, and (2) those relating quantities that are functions of space, such as the elastic displacements of an airplane. In operators of the second kind, space integrals are generally involved. Control-system operators, mechanical, electric,

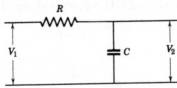

Fig. 11.6.   A simple R–C circuit.

or hydraulic, are generally of the first kind, while aerodynamic, structural, and inertia operators are generally of the second kind.

For example, a simple resistor and capacitor network as shown in Fig. 11.6 is governed by the following equation:

$$\left(1 + RC\frac{d}{dt}\right) V_2 = V_1 \tag{1}$$

The interested variables V_1 and V_2 are essentially independent of the space coordinates. More complicated networks can be built up by such elementary ones. The resulting relations between the input and output can be expressed as ordinary linear differential equations. The *analysis* of such a system presents little difficulty in principle. The *design* of a satisfactory system to perform a specified function, (the so-called *synthesis* problem) of course is more difficult.*

On the other hand, the inertia force, aerodynamic force, and elastic deformation, being functions of both space and time, must be governed either by partial differential equations or by integral equations.

In the following sections, the structural and aerodynamic operators occurring in aeroelasticity will be considered. The inertia operator describes inertia forces. In airplane dynamics, it is convenient to use a system of reference coordinates attached to the airplane, and thus moving with respect to an "inertial" frame of reference. The expression of the inertia operator referring to moving axes can be quite complicated, but it has been treated exhaustively in books of theoretical mechanics.† The case of small disturbances from a steady symmetric motion is of particular importance. The inertia operator can be linearized under the assumption that the square and higher-order products of the small disturbances (linear and angular velocities, as well as the elastic displacements) are negligible in comparison with the disturbances themselves.

11.4   THE STRUCTURAL OPERATOR

The structural operator gives the elastic deformation of a structure under a system of external forces. If the structure is perfectly elastic, so that Hooke's law applies, the structural operator can be concisely expressed in terms of influence functions.

A real material deviates from Hooke's law and shows *anelastic* or *inelastic* behavior to some extent. As an important case, let us assume that the fundamental relation between stress and strain is linear, but depends not only on the instantaneous values of stress and strain, but also on the rate of change of stress and strain.

Under the linearity assumption, no permanent set remains after the removal of all stresses. To derive an expression relating the force and

---

* There are numerous books and papers on automatic-control systems. See, for example, a book by H. M. James, N. B. Nichols, R. S. Phillips,[6.81] and a paper by Bollay,[10.10] from which further references can be found.

† For example, A. G. Webster, *The Dynamics of Particles and of Rigid, Elastic, and Fluid Bodies*, §§ 76–79, 91, and 102, reprinted by G. E. Stechert & Co., New York (1942). For the linearization of the inertia operator, see Refs. 10.10–10.13.

deformation, let us consider a simple bar which is initially free from stress. At an instant of time t_0 let a tensile force f be suddenly applied and then maintained for $t > t_0$. The corresponding elongation of the bar, (the indicial admittance) can be expressed in the following form:

$$\begin{aligned} \varepsilon(t - t_0) &= [K + \phi(t - t_0)]f \quad \text{for} \quad t \geqslant t_0 \\ &= 0 \qquad\qquad\qquad \text{for} \quad t < t_0 \end{aligned} \quad (1)$$

where K is a constant and $\phi(t - t_0)$ is a function of time. This is represented in Fig. 11.7. The function $\phi(t - t_0)$ is generally so chosen that $\lim_{t \to \infty} \phi(t - t_0) = 0$. Then K represents the limiting value of the elongation ε under unit tension after the transient effect is damped out. The function

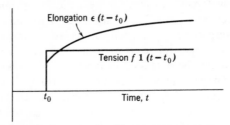

Fig. 11.7. Indicial response of a rod subject to tension.

$\phi(t - t_0)$ is sometimes called a "deformation function." Note that, if Hooke's law holds, $\phi(t - t_0) \equiv 0$.

Since the stress-strain relation is linear, the principle of superposition is applicable. The response of the bar to an arbitrary loading $f(t)$ can be obtained by the Duhamel integral

$$\varepsilon(t) = K f(t) + \int_{-\infty}^{t} \phi(t - t_0) \frac{df(t_0)}{dt_0} \, dt_0 \quad (2)$$

This result may be generalized to a three-dimensional body to show that the effect of anelasticity on the elastic deformation may be expressed by a tensor of generalized deformation function.

Anelasticity is revealed by the damping characteristics of an engineering structure (see Zener[11.5]). If $f(t)$ represent a cyclicly varying force, Eq. 2 will indicate a hysteresis loop the area of which represents the dissipation of energy, and is a measure of the *internal friction* of the material. For a harmonic motion, a linear material has an elliptic hysteresis loop.

The internal friction may become nonlinear and depend on the life history of a material if the stress level is sufficiently high. Experiments show that the energy dissipated per cycle when a body is subjected to cyclic stresses varies with the number of cycles it has been subjected. This phenomenon was observed in 1865 by Lord Kelvin[11.2] and was called the "fatigue of elasticity." There seems to exist a mechanism by which a metal can remember its experience, can show fatique by overworking and recovery by resting. For example, a mild steel at room temperature subjected to a cyclic stress whose extreme values lie below 80 per cent of the endurance limit shows little change of internal friction, whereas the same steel subjected to repeated stresses between this value and the endurance limit will cause an increase of internal friction as much as 25 times that of the initial value in the annealed condition (see Lazan[11.3]). Cyclic stresses above the endurance limit have more pronounced effects.

For built-up structures, energy dissipation may result from play in the riveted joints, from relative motion in the cracks of welding, or from the hinges of movable parts. The damping force may behave either as a viscous fluid or as dry friction, and generally depends on the amplitude of the motion. There are indications that, for built-up beams, the nonlinear component of the load-deflection curve contains mainly a second-power term, and the energy loss per cycle varies approximately as the third power of the amplitude of vibration.[11.1, 11.4]

Since aeroelasticity concerns built-up structures, and yet the definitive laws of damping are unknown, more careful experiments would have to be performed in order to isolate the effect of the various parameters involved. However, for metal airplanes of conventional design, the internal friction is small. It may be assumed that the effect of internal friction is revealed essentially through the energy it dissipates, rather than the exact form of the stress-deformation relationship. Under this assumption we may replace the real material by a linear solid with the stipulation that the corresponding energy dissipated per cycle be the same.

In airplane structures, the elastic buckling of the sheet metal is always an important factor. As the region of buckling develops when the stress level (or the load factor) increases, the load-deflection relationship becomes a function of the load. In some oscillation problems, the amplitude of oscillation can be regarded as infinitesimal, and a correct solution can be obtained by evaluating the effective influence function at the steady-state load factor. In response problems the nonlinearity cannot always be overlooked.

11.5 AERODYNAMIC OPERATORS

An aerodynamic operator relates the aerodynamic force acting on an airfoil and the motion of the airfoil. In aeroelasticity, the drag force and the skin friction being generally neglected, the most important aerodynamic operator is concerned with the aerodynamic lift and moment.

The complexity of the aerodynamic operator is the major difficulty in aeroelasticity. The explicit form of the aerodynamic operator, giving pressure distribution as a function of arbitrary wing deflection, is yet unknown even for such an idealized surface as an infinitely thin rectangular plate of finite span. It is here that a number of simplifying assumptions must be introduced. First, linearization of the hydrodynamic equations is imperative; hence, the theory will be applicable only to thin flat lifting surfaces. Second, it is necessary to tabulate the results numerically; hence, the form of the wing-deflection surface must be specified.

It is therefore evident why the method of generalized coordinates and the method of iteration are particularly suitable for aeroelastic problems. In each step of the iteration, as well as for each degree of freedom in generalized coordinates, the elastic deformation is known. Hence, the aerodynamic problem can be solved beforehand and the results tabulated.

It is also evident why the strip assumption on aerodynamic-force distribution across the span of a finite wing is so often made. When the strip assumption is used, it is necessary to tabulate only the two-dimensional flow cases. Without the strip assumption, it would be necessary to tabulate the results for every particular planform and every mode of motion of the wing.

11.6 MATHEMATICAL CHARACTERISTICS OF AEROELASTIC PROBLEMS

A step that is always involved in an aeroelastic analysis is the determination of elastic deformation from exterior forces. This can be expressed in operational form by an equation:

$$\xi = O_{stru} \{ O_{aero}\, \xi + O_{iner}\, \xi + F_o \} \tag{1}$$

where ξ is a vector describing the elastic deformation, O_{stru} is a structural operator giving ξ for any system of exterior forces. O_{aero} and O_{iner} are, respectively, the aerodynamic and inertia operators which give the exterior forces caused by a given elastic deformation, and F_o is the acting exterior force that is independent of the elastic deformation. The operators O_{stru}, O_{aero} can be expressed as integrals with proper kernels.

When ξ represents the true elastic displacement in the body, the structural operator (or rather the kernel of the integral representing the structural operator) is *symmetric* according to Maxwell's reciprocal relation, which states that the elastic displacement ξ at a point A caused by a force F acting at a point B is equal to the displacement (in the direction of F) at B caused by a force (acting in the direction of ξ) at A. But the kernel of the aerodynamic operator is in general *unsymmetric*, because the aerodynamic force at A caused by a displacement at B is, in general, different from that at B caused by a "corresponding" displacement at A. Thus Eq. 1, as an integral equation, has an unsymmetrical kernel.

The unsymmetry of the kernel of the basic integral equation 1 may also arise from the structural operator when ξ represents the displacement in generalized coordinates. For example, the deformation of a slender wing may be described by a deflection w perpendicular to the plane of the wing and a rotation θ about a reference axis. Then ξ may be considered as having two components w and θ. Correspondingly the exterior forces are generalized forces: a lift corresponding to w and a moment corresponding to θ. In this case the structural operator, which consists of integrals of the products of the exterior forces with proper influence functions, becomes unsymmetric in general, because the deflection at a point A due to a couple acting at a point B is in general unequal to the deflection at B due to a couple at A. In other words, the influence function connecting w with the moment about the elastic axis is unsymmetric.

An integral equation with a real symmetric kernel possesses many nice properties that are lost when the symmetry of the kernel is lost. The main mathematical difficulty in aeroelasticity lies in the unsymmetry of the kernels of the governing integral equations.

Symmetric cases are exceptions rather than the rule in aeroelasticity. It is interesting to consider the conditions under which the operators become symmetric. For concreteness let us consider a slender wing. The structural operator becomes symmetric when bending and torsion are "separated," i.e., when a torsional moment about the elastic axis induces no deflection of that axis and vice-versa. This occurs for an unswept cantilever wing with a straight elastic axis. The aerodynamic operator is simplified when the strip assumption is introduced, according to which the aerodynamic forces and moments at any section of the wing simply depend on the local ξ and wing chord. The inertia operator is symmetric if linearized. Thus the kernel of Eq. 1, as an integral equation, may become symmetric when bending and torsion of the wing are elastically uncoupled and when the strip assumption for aerodynamic forces is introduced. The problems of torsional divergence and aileron reversal of a normal wing as given in § 3.2 and § 4.2 are examples of this case. The divergence and reversal of a swept wing (§ 4.5) are examples of the unsymmetric case. In the flutter problem, the aerodynamic forces are complex functions of the reduced frequency; the kernel is no longer real valued.

Most problems in aeroelasticity can also be formulated as boundary-value problems in differential equations, which are connected with the integral equations by proper Green's functions (i.e., the influence functions). It is well known that the Green's functions are symmetric if the boundary-value problem defined by the differential equation is "self-adjoint" (see Collatz).[11.6] The unsymmetry of the aeroelastic operators is associated with the *non-self-adjointness* of the boundary-value problem.

The stability problems in aeroelasticity are eigenvalue problems. Under certain conditions, among which the Hermitian self-adjointness or the Hermitian symmetry of the kernel is the most important, it can be shown that eigenvalues always exist and are real valued, that the eigenfunctions form a "complete" set of functions, that the iteration procedure for the calculation of the eigenvalues and eigenfunctions is valid, and that the bounds to the eigenvalues can be estimated. The same is not all true with regard to non-self-adjoint problems. The eigenvalues are in general complex and may not always exist. The completeness of the eigenfunctions is questionable, and, consequently, ordinary proofs of the convergence of the iteration procedure requires re-examination. A simple estimation of the bounds of the eigenvalues is yet unknown.

It is beyond the scope of this book to discuss the mathematical problems connected with the non-self-adjoint equations in aeroelasticity. There exists an extensive mathematical literature, but simple and decisive theorems useful for practical calculations are few. It may be pointed out, however, that recent studies initiated by the flutter research have already yielded many significant results. Of greatest importance is Wielandt's proof[11.24] that the classical iteration procedure can be used to find the eigenvalues (if they exist at all) and eigenfunctions of non-self-adjoint equations. Of practical methods of calculation, Lanczos's "minimized iterations" method[11.13, 11.14] is powerful and labor-saving, particularly when several eigenvalues and eigenfunctions are desired. Wielandt's "iterative transformation" procedure[11.24-11.26] is applicable to flutter and similar eigenvalue problems. (A partial but much more readable account of Wielandt's method is given by Gossard[11.9].) By a simple extension, Wielandt also gives a "broken (*gebrochene*) iteration" procedure[11.24] which can be used to correct a given approximation for any arbitrary higher eigenvalue and the corresponding eigenfunctions without the knowledge of the preceding eigenvalues. Both Wielandt's and Lanczos's methods are applicable to algebraic (matrix), differential, or integral operators, and are of importance in studying the fundamental questions in aeroelasticity and in checking approximate solutions. On the other hand, solution of nonhomogeneous equations on the basis of expanding an arbitrary function in series of biorthogonal functions, the concept of "adjoint energy function," and a variational principle which leads to a procedure of the Rayleigh-Ritz type have been introduced by Flax.[7.1] A method of calculating the eigenvalues of complex matrices is given by Wielandt,[11.24] who gives also a simple algorithm to determine whether some of the eigenvalues have a positive imaginary part. A different form of the last mentioned generalization of Routh's rules for discriminating the pseudo-negative roots of a polynomial with complex coefficients is given independently by Sherman, DiPaola, and Frissell.[11.20, 11.21]

BIBLIOGRAPHY

References related to structural operators:

11.1 Fearnow, D. O.: Investigation of the Structural Damping of a Full-Scale Airplane Wing. *NACA RM* **L51A04** (1951). *NACA Tech. Note* **2594** (1952).

11.2 Lord Kelvin: *Mathematical and Physical Papers*, Vol. III, p. 22 et seq. London (1890).

11.3 Lazan, B. J.: A Study with New Equipment of the Effects of Fatigue Stress on the Damping Capacity and Elasticity of Mild Steels. *Trans. Am. Soc. Metals* **42**, 499–558 (1950).

11.4 Pian, T. H. H., and F. C. Hollowell, Jr.: Structural Damping in a Simple Built-up Beam. *Proc. 1st Natl. Congr. Applied Mech.* Chicago (1951).

11.5 Zener, C.: *Elasticity and Anelasticity of Metals*. Chicago Univ. Press (1948).

The following references are related to the mathematical problems raised in aeroelasticity. See also Refs. 6.5, 7.1, and 7.3.

11.6 Collatz, L.: *Eigenwertaufgaben, mit technischen Anwendungen*. Akademische Verlag, Leipzig (1949).

11.7 Davis, H. T.: The Present Status of Integral Equations. *Indiana Univ. Studies* **70** (June 1926).

11.8 Duncan, W. J.: The Principles of the Galerkin Method. *Aeronaut. Research Com. R. & M.* **1848** (1938) see also *R. & M.* **1798** (1937).

11.9 Gossard, M. L.: An Iterative Transformation Procedure for Numerical Solution of Flutter and Similar Characteristic-Value Problems. *NACA Tech. Note* **2346** (1951).

11.10 Hamilton, H. J.: Roots of Equations by Functional Iteration. *Duke Math. J.* **13**, 113–121 (1946).

11.11 Hilbert, D.: *Gesammelt Abhandlungen*, Vol. 3, p. 107, et seq. Berlin (1935).

11.12 Ince, E. L.: *Ordinary Differential Equations*. Longmans, Green & Co., London (1926). Reprinted by Dover Publications, New York (1945).

11.13 Lanczos, C.: An Iteration Method for the Solution of the Eigenvalue Problem of Linear Differential and Integral Operators. *Natl. Bur. Standards J. Research* **45**, 255–282 (1950).

11.14 Lanczos, C.: Solution of Systems of Linear Equations by Minimized Iterations. *Natl. Bur. Standards J. Research* **49**, 33–53 (1952).

11.15 Minorsky, N.: Self-Excited Oscillations in Systems Possessing Retarded Actions. *Proc. 7th Intern. Congr. Applied Mech.* **4**, 43–51 (1948).

11.16 Paige, L. J., and O. Taussky: Simultaneous Linear Equations and the Determination of Eigenvalues. *Natl. Bur. Standards Applied Math. Ser.* **29** (Aug. 1953). U.S. Govt. Printing Office, Washington, D.C.

380 GENERAL FORMULATION OF AEROELASTIC PROBLEMS

11.17 Pell-Weeler, A. J.: Applications of Biorthogonal Systems of Functions to the Theory of Integral Equations. *Trans. Am. Math. Soc.* **12**, 165–180 (1911).

11.18 Pell-Weeler, A. J.: Linear Equations with Unsymmetric Systems of Coefficients. *Trans. Am. Math. Soc.* **20**, 23–39 (1919).

11.19 Pell-Weeler, A. J.: Biorthogonal Systems of Functions. *Trans. Am. Math. Soc.* **12**, 135–164 (1911).

11.20 Sherman, S., J. DiPaola, and H. F. Frissell: The Simplification of Flutter Calculations by Use of an Extended Form of the Routh-Hurwitz Determinant. *J. Aeronaut. Sci.* **12**, 385–392 (1945).

11.21 Sherman, S.: Generalized Routh-Hurwitz Discriminant. An extension of the theorems of Sturm, Routh, and Hurwitz on the roots of polynomial equations. *Phil. Mag.* (7) **37**, 537–551 (1946).

11.22 Sponder, E.: On the Representation of the Stability Region in Oscillation Problems with the Aid of the Hurwitz Determinants. *NACA Tech. Memo.* **1348** (1952). Translated from *Schweiz. Arch.* (Mar. 1950).

11.23 Wayland, H.: Expansion of Determinantal Equations into Polynomial Form. *Quart. Applied Math.* **2**, 277 (1945).

11.24 Wielandt, H.: Contributions to the Mathematical Treatment of Complex Eigenvalue Problems: I, Computation of the Eigenvalues of Complex Matrices. *Deut. Luftfahrt-Forsch. Forsch Ber.* **1806/1**. *AAF Translation* **F-TS-1510-RE** (1948). Wright Air Development Center. II, The Method of Iteration for Non-Self-Adjoint Linear Eigenvalue Problems. *Aerodynam. Vers.-Anst. Goettingen AVA B* **43/J/21**. *Math. Z.* **50** (1944). *Brit. Translation* **MAP-VG-155**. III, Method of Iteration in Flutter Computation. *Deut. Luftfahrt-Forsch. UM* **3138**. *AVA B* **44/J/21**. *AAF Translation* **F-TS-904-RE**. *Brit. Translation* **MAP-VG-124**. IV, Determination of a Lower and Upper Bound for Eigenvalues. *AVA B* **44/J/20**. V, Determination of Higher Eigenvalues by Broken Iteration. *Deut. Luftfahrt-Forsch. UM* **3177**. *Brit. Translation* **MAP-VG-125**. VI, Proof of Convergence for the Iteration Method. *AVA B* **44/J/38** (1944), *UM* **3169**.

11.25 Wielandt, H.: Reflections of Schwarz as to Solve Singular Integral Equation of the First Kind. *Aerodynam. Vers.-Anst. Goettingen AVA B* **44/J/22**.

11.26 Wielandt, H.: Solution of General Linear Eigenvalue Problems. *Aerodynam. Vers.-Anst. Goettingen AVA B* **44/J/19**, (**ATI-18804**, *AFF Translation* **F-TS-976-RE**).

11.27 Williams, J.: The Distribution of the Roots of a Complex Polynomial Equation. *Aeronaut. Research Council R. & M.* **2238** (1946).

Chapter 12

FUNDAMENTALS OF NONSTATIONARY AIRFOIL THEORY

Four distinctive ranges of speed, classified according to the magnitude of the free-stream Mach number M, are considered in aerodynamics. These are:

1. The "incompressible" speed range, in which $M^2 \ll 1$ and the fluid may be considered incompressible.
2. Subsonic-speed range, $M < 1$, and below the Mach number of "divergence" (cf. § 4.3).
3. Transonic-speed range, $M \sim 1$.
4. Supersonic-speed range, $M > 1$.

This classification is based on a theoretical point of view, particularly on the method of analysis. There is no essential difference in the flow patterns between ranges 1 and 2. But, in range 1, which is a limiting case of 2, considerable mathematical simplifications are possible.

Since airfoil characteristics vary with the speed ranges, the most efficient configuration of an airplane is likely to be different for different design speeds. Thus sweptback wings become favorable for airplanes designed for high subsonic and transonic speeds, and thin delta wings and wings of small-aspect ratios are favored in transonic and supersonic designs. Such differences in the geometrical configurations and the corresponding differences in the structural constructions have important effects on the method of aeroelastic analysis.

In this and the next three chapters the aerodynamics of an oscillating airfoil is studied. We shall consider harmonic oscillations of small amplitudes only, so that in most cases the principle of superposition is applicable. In case the aerodynamic equations may be linearized, the aerodynamic response to an arbitrary motion can be obtained from the response to harmonic oscillations by an integration.

Of the four speed ranges mentioned above, the aerodynamics of an incompressible flow has been exhaustively developed; that of the subsonic and supersonic flows is also developed to certain extent. But, in the transonic-speed range, much theoretical and experimental work remains to be done before a reliable analysis can be made.

12.1 FUNDAMENTAL EQUATIONS OF AERODYNAMICS

In the following the tensor notation will be used.   The components of a vector or a tensor are referred to a system of rectangular Cartesian coordinates (x_1, x_2, x_3) which will be written as (x, y, z) if convenient. The components of a velocity vector will be denoted by (u_1, u_2, u_3) or by (u, v, w).   Similarly the components of other quantities will be represented either by a subscript or by a self-explanatory triplet of letters, whichever be the more convenient in special instances.   The Roman indices always range through 1, 2, 3, unless otherwise stated.   The summation convention (p. 5) will be used: Any index repeated twice in the same term indicates a summation over the total range of that index.

In the tensor notation the fundamental equations of aerodynamics of a nonviscous fluid are

1. The Eulerian equation of motion (law of conservation of momentum)

$$\rho \left(F_i - \frac{Du_i}{Dt} \right) = \frac{\partial p}{\partial x_i} \tag{1}$$

2. The equation of continuity (law of conservation of mass)

$$\frac{\partial \rho}{\partial t} + \frac{\partial(\rho u_i)}{\partial x_i} = 0 \tag{2}$$

where ρ is the fluid density, p the pressure, F_i the force per unit volume acting on the fluid (such as gravitation), u_i the velocity components, and t the time. Du_i/Dt denotes the acceleration of a particle of the fluid.   To express Du_i/Dt in terms of the space and time derivatives of the velocity field, note that, if the position of an element of the fluid is described by $x_i(t)$, the velocity of the fluid element is

$$u_i = \frac{dx_i}{dt} = u_i(x_1, x_2, x_3; t)$$

which is a function of space and time.   By the usual rule of differentiation we obtain

$$\frac{Du_i}{Dt} = \frac{\partial u_i}{\partial t} + \frac{\partial u_i}{\partial x_j}\frac{\partial x_j}{\partial t} = \frac{\partial u_i}{\partial t} + u_j \frac{\partial u_i}{\partial x_j} \tag{3}$$

The acceleration of a fluid particle is written as Du_i/Dt to distinguish it from the partial derivative $\partial u_i/\partial t$.   Using Eq. 3, the equation of motion can be written as

$$\frac{\partial u_i}{\partial t} + u_j \frac{\partial u_i}{\partial x_j} = F_i - \frac{1}{\rho}\frac{\partial p}{\partial x_i} \tag{4}$$

The derivation of these equations can be found in any book on theoretical aerodynamics (e.g., Ref. 1.46). In the airfoil theory considered below, the body force F_i can be omitted, since the only significant body force, the gravitation, introduces only a field of hydrostatic pressure which does not concern us. The assumption that the fluid is nonviscous will be made throughout the following discussion, not because the effect of viscosity is unimportant, but because we shall consider only the flow over a thin airfoil at a small angle of attack without separation, in which case the boundary-layer theory shows that the fluid outside the boundary layer may be regarded as nonviscous and the effect of viscosity can be stated in a phenomenological rule that the velocity must remain finite and tangent to the airfoil at the sharp trailing edge. This assumption was put forward by M. Wilhelm Kutta (1867–1944) and Nikolai E. Joukowski (1847–1921) independently, and is called the *Kutta-Joukowski condition*.

For a compressible fluid, Eqs. 1 and 2 do not suffice in defining uniquely the flow. It is necessary to know also the thermal and caloric states of the fluid and the heat transfer. For an example of the ideal gas, the thermal equation of state is $p/\rho = RT$, and the caloric equation of state is given by the relationship between the internal energy and the temperature. These, in addition to an equation expressing the balance of heat and mechanical energy (the first law of thermodynamics), define a flow uniquely for proper boundary conditions.

The analysis can be greatly simplified if it is possible to assume that the fluid is *piezotropic*, for which the density ρ is a unique function of pressure. For a piezotropic fluid the potential energy of the fluid can be defined by pressure alone and an integration of the equation of motion along a stream line defines the energy balance completely. Then Eqs. 1 and 2 are sufficient to determine the flow. Fortunately, this is the case in *thin airfoil theory*, which deals with airfoils of infinitesimal thickness at small angle of attack performing motions of infinitesimal amplitude. The disturbances caused by the airfoil in a flow is thus infinitesimal, and shock waves, if any, will be of infinitesimal strength. No external heat source will be considered. Under these circumstances it can be shown that the change of entropy in the entire field of flow is an infinitesimal quantity of higher order of smallness. Thus the flow may be correctly regarded as isentropic, and the relation

$$p/\rho^\gamma = \text{const} \tag{5}$$

holds for the entire field, γ being the ratio of the specific heats c_p/c_v. c_p is the specific heat of the gas at constant pressure, and c_v is that at constant volume.

12.2 VORTICITY AND CIRCULATION

The *circulation* $I(\mathscr{C})$ in any closed circuit $\mathscr{C}$ is defined by the line integral

$$I(\mathscr{C}) = \int_{\mathscr{C}} \mathbf{u} \cdot d\mathbf{l} \qquad (1)$$

where $\mathscr{C}$ is any closed curve in the fluid, and the integrand is the scalar product of the velocity vector $\mathbf{u}$ and the vector $d\mathbf{l}$, which is tangent to the

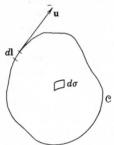

Fig. 12.1.   Circulation.   Notations.

curve $\mathscr{C}$ and of length dl (Fig. 12.1).   Clearly, the circulation is a function of both the velocity field and the chosen curve $\mathscr{C}$.

Using Stokes's theorem, the line integral can be transformed into the surface integral

$$I(\mathscr{C}) = \int_{S} (\mathbf{\nabla} \times \mathbf{u})_n \, d\sigma \qquad (2)$$

where S is any surface in the fluid bounded by the curve $\mathscr{C}$, provided that there is no discontinuity in the velocity field.   The symbol $\mathbf{\nabla}$ means the vectorial operator $\left(\dfrac{\partial}{\partial x_1}, \dfrac{\partial}{\partial x_2}, \dfrac{\partial}{\partial x_3} \right)$.   The vector product $\mathbf{\nabla} \times \mathbf{u}$, sometimes written as **curl u**, has the following three components in the direction of x_1, x_2, x_3:

$$\frac{\partial u_3}{\partial x_2} - \frac{\partial u_2}{\partial x_3}, \qquad \frac{\partial u_1}{\partial x_3} - \frac{\partial u_3}{\partial x_1}, \qquad \frac{\partial u_2}{\partial x_1} - \frac{\partial u_1}{\partial x_2}$$

If the direction cosines of the outer normal of the surface element $d\sigma$ are denoted as n_1, n_2, n_3; then the normal component of the vector $\mathbf{\nabla} \times \mathbf{u}$ is

$$(\mathbf{\nabla} \times \mathbf{u})_n = \mathbf{n} \cdot (\mathbf{\nabla} \times \mathbf{u})$$

$$= n_1 \left(\frac{\partial u_3}{\partial x_2} - \frac{\partial u_2}{\partial x_3} \right) + n_2 \left(\frac{\partial u_1}{\partial x_3} - \frac{u_3}{\partial x_1} \right) + n_3 \left(\frac{\partial u_2}{\partial x_1} - \frac{\partial u_1}{\partial x_2} \right)$$

We define the *vorticity* at a point in a velocity field as the quantity

$$\chi = \nabla \times \mathbf{u} = \mathbf{curl\ u} \tag{3}$$

The law of change of circulation with time, when it is taken around a *fluid line*, i.e., a curve $\mathscr{C}$ formed by definite fluid particles, is given by the *theorem of Lord Kelvin*:

Theorem: If the fluid is nonviscous and the body force is conservative, then

$$\frac{DI}{Dt} = -\int_{\mathscr{C}} \frac{dp}{\rho} \tag{4}$$

If, in addition to the above conditions, the fluid is *piezotropic*, then the last integral vanishes because $\mathscr{C}$ is a closed curve, and we have the *Helmholtz theorem* that

$$\frac{DI}{Dt} = 0 \tag{5}$$

The flow is said to be *irrotational* if the vorticity is zero throughout the region under consideration.

In the thin-airfoil theory, the conditions of the Helmholtz theorem are satisfied. Hence, the circulation I about any fluid line never changes with

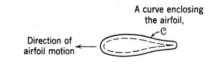

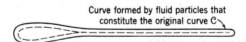

Fig. 12.2. Fluid line $\mathscr{C}$ enclosing an airfoil and its wake.

time. Since the motion of the fluid is caused by the motion of the airfoil and since at the beginning the fluid is at rest and $I = 0$, it follows that I vanishes at all times. Note, however, that the volume occupied by the airfoil is exclusive of the fluid. A fluid line $\mathscr{C}$ enclosing the boundary of the airfoil becomes elongated when the airfoil moves forward as shown in Fig. 12.2. According to the Helmholtz theorem, the circulation about $\mathscr{C}$ is zero, so that the total vorticity inside $\mathscr{C}$ vanishes, but one cannot conclude that the vorticity actually vanishes everywhere inside $\mathscr{C}$. Hence, in the region occupied by the airfoil, and in the wake behind the airfoil, vorticity may exist. However, the Helmholtz theorem applies to the region outside the airfoil and its wake, and the vanishing of circulation about every possible fluid line clearly shows that the flow is irrotational outside the airfoil and its wake.

12.3 VELOCITY POTENTIAL OF AN IRROTATIONAL FIELD

If the velocity vector can be derived from the gradient of a scalar function Φ:

$$\mathbf{u} = \boldsymbol{\nabla}\Phi \qquad \left(\text{i.e., } u_i = \frac{\partial \Phi}{\partial x_i}\right) \tag{1}$$

then the flow field is *irrotational*,

$$\boldsymbol{\nabla} \times \mathbf{u} = 0 \qquad (\text{i.e., } \mathbf{curl}\ \mathbf{u} = 0) \tag{2}$$

since $\boldsymbol{\nabla} \times \boldsymbol{\nabla}\Phi \equiv 0$ is an identity for any scalar function Φ. The function Φ is called the *velocity potential* of the flow field.

If the fluid is incompressible, the equation of continuity

$$\frac{\partial u_i}{\partial x_i} = 0 \tag{3}$$

may be written, in an irrotational flow, as

$$\nabla^2\Phi = 0 \qquad \left(\text{i.e., } \frac{\partial^2\Phi}{\partial x^2} + \frac{\partial^2\Phi}{\partial y^2} + \frac{\partial^2\Phi}{\partial z^2} = 0\right) \tag{4}$$

which is the equation governing Φ for an incompressible fluid.

According to § 12.2, in thin airfoil theory the flow outside the airfoil and its wake is irrotational and the velocity potential exists.

12.4 COMPLEX POTENTIAL IN TWO-DIMENSIONAL IRROTATIONAL FLOW OF AN INCOMPRESSIBLE FLUID

In a two-dimensional flow in the x, y plane, the velocity has two components u, v in the x and y direction, respectively. If the fluid is incompressible, the equation of continuity

$$\frac{\partial u}{\partial x} + \frac{\partial v}{\partial y} = 0 \tag{1}$$

can be satisfied identically if u, v are derived from an arbitrary function Ψ so that

$$u = \frac{\partial \Psi}{\partial y}, \qquad v = -\frac{\partial \Psi}{\partial x} \tag{2}$$

The function Ψ is called a *stream function*. The reason for this name follows the definition of a streamline, which is a curve the tangent of which is everywhere parallel to the local velocity vector. Thus a stream-line is defined by the equation.

$$\frac{dy}{dx} = \frac{v}{u} \quad \text{or} \quad u\,dy - v\,dx = 0 \tag{3}$$

or, from Eqs. 2,

$$\frac{\partial \Psi}{\partial y}\,dy + \frac{\partial \Psi}{\partial x}\,dx = d\Psi = 0 \tag{4}$$

Thus, along a streamline, $\Psi(x, y)$ is a constant. Conversely, every constant defines a streamline

$$\Psi(x, y) = \text{const} \tag{5}$$

If, in addition to being incompressible, the flow is irrotational, then

$$\frac{\partial u}{\partial y} - \frac{\partial v}{\partial x} = 0 \tag{6}$$

Substituting Eqs. 2, we see that

$$\frac{\partial^2 \Psi}{\partial x^2} + \frac{\partial^2 \Psi}{\partial y^2} = 0 \tag{7}$$

Thus, for an irrotational incompressible flow, both the velocity potential and the stream function satisfy the Laplace equation.

By definitions of the velocity potential Φ and the stream function Ψ, we have

$$u = \frac{\partial \Psi}{\partial y}, \qquad v = -\frac{\partial \Psi}{\partial x}$$

$$u = \frac{\partial \Phi}{\partial x}, \qquad v = \frac{\partial \Phi}{\partial y} \tag{8}$$

Hence,

$$\frac{\partial \Psi}{\partial y} = \frac{\partial \Phi}{\partial x}, \qquad \frac{\partial \Psi}{\partial x} = -\frac{\partial \Phi}{\partial y} \tag{9}$$

which shows that the streamlines ($\Psi = \text{const}$) and the potential lines ($\Phi = \text{const}$) are orthogonal to each other.

Equations 9 are the Cauchy-Riemann differential equations of a function of a complex variable. If $W(z) = \Phi + j\Psi$ is an analytic function of a complex variable $z = x + jy$, then Eqs. 9 are satisfied. Conversely, if Eqs. 9 are satisfied, and if the partial derivatives are continuous in a region, then there exists a function $W(z) = \Phi + j\Psi$ which is analytic in that region. In other words, the velocity potential and the stream function are the real and imaginary parts of an analytic function. Furthermore, the real or imaginary part of any arbitrary analytic function can be taken as a velocity potential or a stream function. Φ and Ψ are said to be *conjugate* to each other, so that still another way of stating the above fact is that, to every solution of the Laplace equation,

$$\nabla^2\Phi = 0$$

there corresponds a conjugate function $\Psi(x, y)$ satisfying the same equation, and that the system of curves $\Phi = $ const and $\Psi = $ const are orthogonal.

In some occasions, Φ may be associated with other physical quantities, such as an acceleration potential or pressure. The corresponding conjugate function Ψ may not have an apparent physical meaning, but it serves as a useful medium in the analysis.

The derivatives of the *complex potential* $W(z) = \Phi + j\Psi$ has a simple interpretation. Since

$$\frac{dW}{dz} = \frac{\partial\Phi}{\partial x} + j\frac{\partial\Psi}{\partial x} = u_x - ju_y \tag{10}$$

when Φ is interpreted as a velocity potential, dW/dz is the *image in the x axis of the velocity vector* $u_x + ju_y$. Hence, dW/dz is often called a *complex velocity function*.

12.5 THE ACCELERATION POTENTIAL OF A PIEZOTROPIC FLUID

Euler's equations of motion for a piezotropic fluid

$$\frac{Du_i}{Dt} = -\frac{1}{\rho}\frac{\partial p}{\partial x_i} = -\frac{\partial}{\partial x_i}\int_{p_0}^{p}\frac{dp}{\rho} \qquad (i = 1, 2, 3) \tag{1}$$

indicate that the acceleration vector of a fluid particle,

$$a_i = \frac{Du_i}{Dt} \qquad (i = 1, 2, 3) \tag{2}$$

is a gradient of a scalar quantity. Hence a_i can be derived from a scalar function $\phi(x_1, x_2, x_3, t)$ so that

$$a_i = \frac{\partial\phi}{\partial x_i} \qquad (i = 1, 2, 3) \tag{3}$$

The function ϕ is then called an *acceleration potential* of the flow field.

Substituting Eqs. 3 into Eq. 1 and integrating with respect to space, we obtain

$$\phi + \int_{p_0}^{p} \frac{dp}{\rho} = \text{const} \qquad (4)$$

The integration constant may be a function of time. But, if there is no change in the conditions at infinity, it is a pure constant.

For a piezotropic fluid, the integral $\int_{p_0}^{p} \frac{dp}{\rho}$ is a continuous function of p which is continuous everywhere except at a shock wave. In the thin-airfoil theory, shock waves of finite intensity do not exist; hence, in view of Eq. 4, the acceleration potential is continuous everywhere. This is in sharp contrast with the velocity potential, which does not exist in the wake. When the wake is idealized into a surface without thickness, the velocity potential is discontinuous across that surface, but the acceleration potential is continuous. This advantageous fact is very important in the application of the acceleration potential to the airfoil theory.

Further simplifications are obtained by linearization. Let us consider small disturbances in an otherwise uniform, rectilinear, steady flow. Let the deviation of each velocity component u_i ($i = 1, 2, 3$) from the uniform, steady value be so small that u_i^2 is negligible in comparison with $|u_i|$ itself. The corresponding changes in pressure and density of the fluid, according to the equations of motion, are also infinitesimals of the first order. Hence writing

$$\rho = \rho_0 + \rho', \qquad p = p_0 + p'$$

where ρ_0, p_0 are the initial density and pressure and ρ', p' are the small perturbations, then $\rho'^2 \ll |\rho'|$, $p'^2 \ll |p'|$. We have, from Eq. 4, taking ϕ at p_0 to be zero,

$$\phi + \int_{p_0}^{p_0+p'} \frac{dp}{\rho_0(1 + \rho'/\rho_0)} = \phi + \int_{0}^{p'} \frac{dp'}{\rho_0}\left[1 - \frac{\rho'}{\rho_0} + \left(\frac{\rho'}{\rho_0}\right)^2 + \cdots\right] = 0$$

or

$$\phi = -\frac{p'}{\rho_0} \qquad (5)$$

when small quantities of the second and higher orders are neglected. Hence, the acceleration potential ϕ is directly proportional to the pressure disturbance in the fluid. This offers a simple physical interpretation of the acceleration potential in a linearized theory.

The linearized field equations of the acceleration potential can be derived as follows.

Let us take a reference coordinate system which has *no relative motion with respect to the fluid at infinity*, and consider small perturbations of the velocity field so that the squares and higher powers of the velocity components or their space and time derivatives are negligible in comparison with their first power. Let p_0 and ρ_0 be the pressure and density of the undisturbed fluid, p', ρ' the small perturbations, and let u_i be the velocity components; then the Eulerian equations of motion and the equation of continuity are *linearized* into

$$\frac{\partial u_i}{\partial t} = -\frac{1}{\rho_0}\frac{\partial p'}{\partial x_i} \qquad (i = 1, 2, 3) \tag{6}$$

and

$$\rho_0 \frac{\partial u_i}{\partial x_i} + \frac{\partial \rho'}{\partial t} = 0 \tag{7}$$

respectively. Differentiating Eq. 6 with respect to x_i, and Eq. 7 with respect to t, and eliminating the sum $\dfrac{\partial^2 u_i}{\partial t\,\partial x_i}$, we obtain

$$\frac{\partial^2 p'}{\partial x_i\,\partial x_i} - \frac{\partial^2 \rho'}{\partial t^2} = 0 \tag{8}$$

For a piezotropic fluid, if we write

$$\frac{dp}{d\rho} = a^2 \tag{9}$$

then Eq. 8 can be written as

$$\frac{\partial^2 p'}{\partial x_i\,\partial x_i} - \frac{1}{a^2}\frac{\partial^2 p'}{\partial t^2} = 0 \tag{10}$$

which is the *fundamental wave equation in acoustics* for a fluid stationary at infinity.

When small perturbations are considered,

$$a^2 = \frac{dp}{d\rho} = \left(\frac{dp}{d\rho}\right)_{\rho'=0} + \rho'\left(\frac{d^2 p}{d\rho^2}\right)_{\rho'=0} + \cdots$$
$$= a_0{}^2 + O(\rho')$$

If we neglect the second- and higher-order terms, Eq. 10 becomes

$$\frac{\partial^2 p'}{\partial x_i\,\partial x_i} - \frac{1}{a_0{}^2}\frac{\partial^2 p'}{\partial t^2} = 0 \tag{11}$$

The quantity a_0 is the speed of propagation of a sound wave in the undisturbed fluid.   Generally a_0 is written as

$$a_0 = \sqrt{\left(\frac{\partial p}{\partial \rho}\right)_S} \tag{12}$$

to signify that the derivative $\partial p/\partial \rho$ is taken at the condition of constant entropy.   That a_0 is the velocity of propagation of a sound wave can be seen from the following example.   Consider a spherical disturbance.   By spherical symmetry, Eq. 11 can be transformed in spherical polar coordinates to

$$\frac{\partial^2 \phi}{\partial r^2} + \frac{2}{r} \frac{\partial \phi}{\partial r} - \frac{1}{a_0^2} \frac{\partial^2 \phi}{\partial t^2} = 0 \tag{13}$$

A general solution of this equation is

$$\phi = \frac{1}{r} f(r - a_0 t) + \frac{1}{r} g(r + a_0 t) \tag{14}$$

where f and g are arbitrary functions.   Clearly the first term represents a wave radiating out from the origin, and the second term one converging toward the origin, both propagating with a speed a_0.

Since the acceleration potential is proportional to the perturbation pressure p', the acceleration potential ϕ is governed by the same equation

$$\frac{\partial^2 \phi}{\partial x_i \partial x_i} - \frac{1}{a_0^2} \frac{\partial^2 \phi}{\partial t^2} = 0 \tag{15}$$

Differentiating Eq. 15 with respect to x_j, we see that the components of acceleration $a_j = \partial \phi / \partial x_j$ satisfy the same equation.   Further, since $\rho' = (1/a_0^2)p' +$ higher-order infinitesimals, we see that ρ' is governed by the same equation.   Finally, since $a_i = \partial u_i / \partial t$ in the linearized theory, the wave equation for the components of acceleration can be integrated to show that the velocity components u_i as well as the velocity potential Φ are also governed by the same equation.

If the fluid is incompressible, $\rho = $ const; then a is infinity, and Eq. 15 becomes

$$\frac{\partial^2 \phi}{\partial x_i \partial x_i} = 0 \tag{16}$$

For an incompressible fluid in a two-dimensional flow, the governing equation

$$\frac{\partial^2 \phi}{\partial x^2} + \frac{\partial^2 \phi}{\partial y^2} = 0 \tag{17}$$

shows that there exists a function ψ conjugate to ϕ so that $w = \phi + j\psi$ is an analytic function of $z = x + jy$. $w(z)$ may be called a *complex acceleration potential*. Similar to the complex velocity function of § 12.4, the complex *acceleration function*

$$\frac{dw}{dz} = a_x - j\,a_y \tag{18}$$

is the reflection of the acceleration vector in the x axis

12.6 BOUNDARY CONDITIONS IN AIRFOIL PROBLEMS

An indisputable boundary condition is that the fluid must not penetrate the solid body so that the flow must be *tangent* to the solid surface at all times.* The other boundary condition, which in the case of a fluid of infinite extent refers to the conditions of flow at infinitely large distance from the airfoil, requires a careful consideration.

In an incompressible fluid, the influence of any disturbance is instantly transmitted in all directions to infinity. Consider the motion of a fluid generated by the motion of an airfoil which begins at a time t_0. During a finite time interval $t - t_0$, the airfoil moves about and sweeps out a region of space, every point of which is occupied by the airfoil at one time or another. Let this region be denoted by $\mathscr{A}$. Consider now a spherical surface with radius R_0 so large that the entire region $\mathscr{A}$ is enclosed in the sphere. Introduce the spherical polar coordinates, R, θ, ψ, where θ and ψ are the polar and azimuth angles, respectively. Let the resultant velocity at a point (R, θ, ψ) be $V(R, \theta, \psi)$. The total kinetic energy of the fluid in the region outside the sphere R_0 is

$$\text{K.E.} = \frac{1}{2} \int_{R_0}^{\infty} \int_0^{\pi} \int_0^{2\pi} \rho V^2 R^2 \sin\theta \, dR \, d\theta \, d\psi$$

This must be a finite quantity because it is only part of the energy imparted to the fluid by the motion of the airfoil. Therefore the improper integral must be convergent, and it is seen that the velocity V must decrease to

* The condition of tangency does not suffice to determine uniquely the flow around the airfoil. In addition, the Kutta-Joukowsky condition (§ 12.1) at the trailing edge must be satisfied. This condition determines the circulation around the airfoil.

The leading edge of an actual airfoil is rounded, and no special edge condition is needed. In a linearized theory, the airfoil being regarded as infinitely thin, the singularity at the leading edge may be regarded as the limiting form of such a rounded edge. In the small perturbation theory it suffices to require that the total integrated force be finite, and that the infinity in the pressure distribution at the leading edge be of a proper order, in analogy to the steady-flow theory.

zero as R increases indefinitely at a rate faster than $1/R^{3/2}$. Thus the condition at infinity for an incompressible fluid is that the velocity disturbance decreases to zero, or that the velocity potential tends to a constant. The relation between the velocity and acceleration then shows that the acceleration of the fluid particles decreases to zero at infinity, while the acceleration potential and pressure tend to constant values at infinity.

If the airfoil is moving in a compressible fluid, the disturbance in the fluid due to the motion of the airfoil is propagated with the speed of sound. If the speed of motion of the airfoil (relative to a coordinate system which is at rest with respect to the fluid at infinity) is less than the speed of sound, the disturbance will be felt in all directions. After a sufficiently long period of time, the region of the fluid influenced by the motion of the airfoil will be much larger than the region $\mathscr{A}$ swept out by the airfoil, and the argument of the last paragraph can be used again to conclude that the velocity disturbances decrease to zero at infinity.

If the motion of the airfoil is a small oscillation about a rectilinear translation, which has taken place for an indefinitely long time, the region $\mathscr{A}$ named above consists of the volume occupied by the airfoil and a wake which extends indefinitely behind the airfoil. The conclusions reached above may be so stated that the velocity disturbances must vanish at least as fast as $1/R^{3/2}$ as $R \to \infty$, where R is the shortest distance between the point in question and the airfoil and its wake.

Consider finally the case in which the airfoil is moving at a supersonic speed. The speed of motion being higher than the speed of propagation of sound, the disturbances cannot be felt in front of the envelope of Mach waves generated by the leading edge of the airfoil. Hence, there is no disturbance upstream of the airfoil. On the other hand, at supersonic speeds, energy can be propagated to infinity in the form of shock waves. In a two-dimensional supersonic flow, the region influenced by the shock waves may be limited in extent, and the strength of disturbance does not necessarily diminish toward zero when the distance from the airfoil increases indefinitely. However, we may impose the condition that there are no sources of disturbance in the fluid other than the airfoil under consideration. This, together with the condition of no disturbance in front of the Mach-wave envelope from the leading edge, suffices to determine a unique solution in the supersonic case.

In the following discussions, we shall restrict ourselves to the linearized theory, which is valid only if the disturbance caused by the airfoil is infinitesimal. This implies that the wing is infinitely thin and of infinitesimal camber and has an infinitesimal angle of attack. Such a wing is said to be *planar*.

Let the free-stream direction be parallel to the mean position of the planar wing, which lies on the (x_1, x_2) plane. Then the distance from the wing surface to the (x_1, x_2) plane is an infinitesimal quantity of the first order. On the assumption that the quantities ρ', p', u_i, a_i, etc., are continuous functions of space and time, it is easy to see (by expansion into power series) that the *difference* of any of these quantities on the surface of the airfoil from that on the (x_1, x_2) plane is an infinitesimal of higher order than that quantity itself. Therefore, in the linearized theory it is permissible to apply the boundary conditions of a planar system in a region on the (x_1, x_2) plane that represents the projection of the airfoil on that plane, instead of on the actual airfoil surface.

12.7 METHODS OF SOLVING THIN-AIRFOIL PROBLEMS

The thin-airfoil problems can be treated by the usual methods of mathematical physics. The three commonly used methods are:

1. Superposition of proper singularities. Examples are the superposition of sources and sinks to obtain a symmetric body or of the source-sink doublet layer to obtain a lifting surface.

2. By a transformation which reduces a given boundary-value problem to an easier problem, or to a problem the solution of which is known. For two-dimensional incompressible flow, the conformal transformation is a powerful method, because the theory of functions of a complex variable furnishes the necessary tool to carry out the operations. The well-known Lorentz transformation for the wave equation may be regarded as an example of conformal transformation in a four-dimensional space whose metric is defined as

$$ds^2 = dx^2 + dy^2 + dz^2 - a^2 \, dt^2$$

3. Operational methods, such as Laplace transformation or the Fourier transformation.

Each method has its advantages and difficulties. Most of the airfoil problems can be (and have been) solved by all three methods. In the following presentation, however, we shall choose only the shortest ones.

Chapter 13

OSCILLATING AIRFOILS IN TWO-DIMENSIONAL INCOMPRESSIBLE FLOW

13.1 THE PROBLEM SPECIFIED

The problem of an oscillating thin airfoil in a two-dimensional incompressible flow will be considered in this chapter. The mean motion of the airfoil is a rectilinear translation of speed U with respect to the fluid at infinity. To this mean motion a simple-harmonic oscillation of infinitesimal amplitude is superposed. Since an arbitrary motion of an airfoil can be analyzed into harmonic components by means of a Fourier analysis, and, conversely, by a synthesis of the simple-harmonic components, any general motion can be established, the analysis of harmonic oscillations actually forms the basis of a general airfoil theory in unsteady motion (of small amplitude).

The airfoil to be considered is a planar system. The treatment will be limited to a linearized theory.

Within the framework of a linearized theory, solutions may be superposed to generate another solution. The solution of an oscillating airfoil with finite but small thickness and camber at a given mean angle of attack can be obtained by a superposition of an unsteady solution for an oscillating airfoil of zero thickness and zero camber at zero mean angle of attack, and a steady-state solution for an airfoil of the given thickness and camber at the given mean angle of attack. The steady-state solution can be found in many textbooks on aerodynamics. Therefore, in discussing the aerodynamics of an oscillating airfoil, it is sufficient to consider an airfoil of zero thickness and zero camber at zero mean angle of attack.

The two-dimensional nonstationary airfoil theory was first formulated by Birnbaum and Wagner. A short historical review of the earlier works of Küssner, Glauert, and Theodorsen is given in § 5.8. More recent developments were made by Cicala, Küssner, Schwarz, von Kármán, Sears, Dietze, W. P. Jones, Biot, and others. See bibliography.

13.2 GENERAL EQUATIONS

Let a system of rectangular coordinates (x, y) be taken, with the x axis parallel to the flow at infinity, and with the origin at a fixed point which is the mean position of the mid-chord point of the airfoil. The mean chord of the airfoil lies along the x axis. Let the scale be so chosen that the semichord length is equal to unity. Therefore the projection of the leading edge of the airfoil is at $x = -1$, and that of the trailing edge is at $x = +1$. Let the airfoil be described by the equation

$$y = Y(x, t) \qquad (-1 \leqslant x \leqslant 1) \tag{1}$$

(See Fig. 13.1.) It is assumed that Y and $\partial Y/\partial x$ are so small that the skeleton airfoil does not differ appreciably from a horizontal line. The disturbances caused by the motion of the airfoil are therefore small.

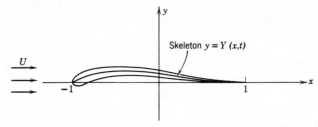

Fig. 13.1.   Airfoil profile and coordinate system.

The velocity (u, v) of the fluid consists of a uniform mean velocity $(U, 0)$ in the direction of the positive x axis, and a small perturbation (u', v'):

$$
\begin{aligned}
u &= U + u' \qquad (|u'| \ll U) \\
v &= v' \qquad\quad (|v'| \ll U)
\end{aligned}
\tag{2}
$$

The acceleration components a'_x and a'_y are linearized into

$$
\begin{aligned}
a'_x &= \frac{\partial u'}{\partial t} + (U + u')\frac{\partial u'}{\partial x} + v'\frac{\partial u'}{\partial y} = \frac{\partial u'}{\partial t} + U\frac{\partial u'}{\partial x} \\
a'_y &= \frac{\partial v'}{\partial t} + (U + u')\frac{\partial v'}{\partial x} + v'\frac{\partial v'}{\partial y} = \frac{\partial v'}{\partial t} + U\frac{\partial v'}{\partial x}
\end{aligned}
\tag{3}
$$

when small quantities of the second order are neglected. Since the fluid is assumed to be incompressible, $\rho = $ const; the *acceleration potential* ϕ' is proportional to the change of pressure p'

$$\rho\phi' = -p' \tag{4}$$

and is governed by the equation

$$\frac{\partial^2 \phi'}{\partial x^2} + \frac{\partial^2 \phi'}{\partial y^2} = 0 \tag{5}$$

As shown in § 12.5, there exists a pair of conjugate harmonic functions ϕ' and ψ', so that

$$w = \phi' + j\psi' = f(x + jy) \tag{6}$$

is an analytic function of a complex variable $z = x + jy$, and the systems of curves $\phi'(x, y, t) = \text{const}$ and $\psi'(x, y, t) = \text{const}$ are orthogonal. w is the *complex acceleration potential*.

The boundary condition at infinite distance from the airfoil is that u', $v' \to 0$, as shown in § 12.6. That on the airfoil is the tangency of the flow to the solid surface. To express the latter condition mathematically, we may assume that during oscillation every point of the airfoil moves in the vertical direction only. The velocity vector of a point on the airfoil is therefore $\mathbf{v}_1 = (0, \partial Y/\partial t)$. The velocity vector of the fluid is $\mathbf{v}_2 = (U + u'_a, v'_a)$ where the subscript a indicates that the corresponding quantities are evaluated on the airfoil. If we resolve these velocity vectors into directions tangent and normal to the airfoil, then the aforesaid boundary condition is that the normal components of $\mathbf{v}_1$ and $\mathbf{v}_2$ must be equal. Now $\mathbf{n} = (-\partial Y/\partial x, 1)$ is a vector in the direction of the normal; hence, we must have $\mathbf{v}_1 \cdot \mathbf{n} = \mathbf{v}_2 \cdot \mathbf{n}$, or, in terms of the components of these vectors,

$$\frac{\partial Y}{\partial t} = -(U + u'_a)\frac{\partial Y}{\partial x} + v'_a$$

Neglecting small quantities of the second order, we obtain the boundary condition

$$v'_a = \frac{DY}{Dt} = \frac{\partial Y}{\partial t} + U\frac{\partial Y}{\partial x} \tag{7}$$

Similarly, the normal components of the acceleration vectors of the fluid and the airfoil must agree. This leads to the boundary condition

$$a'_y = \frac{D^2 Y}{Dt^2} = \frac{\partial^2 Y}{\partial t^2} + 2U\frac{\partial^2 Y}{\partial x\,\partial t} + U^2\frac{\partial^2 Y}{\partial x^2} \tag{8}$$

An acceleration potential ϕ' satisfying the boundary condition 8 does not necessarily satisfy the velocity boundary condition 7, which actually amounts to an additional restriction on ϕ'. To express Eq. 7 in terms of ϕ', we have, from Eqs. 3,

$$\frac{\partial \phi'}{\partial y} = a'_y = \frac{\partial v'}{\partial t} + U\frac{\partial v'}{\partial x} \tag{9}$$

For harmonic oscillations we may use the complex representation (§ 1.8) and write

$$\phi' = \phi(x, y)e^{i\omega t}, \qquad u' = u(x, y)e^{i\omega t}, \qquad v' = v(x, y)e^{i\omega t} \tag{10}$$

where $\phi(x, y)$, $u(x, y)$, $v(x, y)$ are functions of space variables alone. Then Eq. 9 can be written as

$$\frac{\partial}{\partial y} \phi(x, y) = i\omega v + U \frac{\partial v}{\partial x} \tag{11}$$

Solving for v, and using the condition $v = 0$ when $x = -\infty$, we obtain

$$v(x, y) = \frac{1}{U} e^{-i\omega(x/U)} \int_{-\infty}^{x} \frac{\partial \phi(\xi, y)}{\partial y} e^{i\omega(\xi/U)} d\xi \tag{12}$$

This must be equal to the right-hand side of Eq. 7 on the airfoil.*

In addition to the boundary conditions 7 and 8, the flow must also satisfy the Kutta-Joukowski condition that the velocity be finite at the trailing edge of the airfoil. An equivalent form of this condition is that the pressure (and hence ϕ) be continuous at the trailing edge of the airfoil.

13.3   THE FORM OF THE ACCELERATION POTENTIAL

A two-dimensional airfoil is conveniently treated by conformal mapping. The transformation between the complex numbers z and ζ

$$z = \frac{1}{2} \left(\zeta + \frac{1}{\zeta} \right) \tag{1}$$

maps a circle of unit radius on the ζ plane into a straight-line segment from $x = -1$ to $x = 1$ on the real axis of the z plane (Fig. 13.2). A point with polar coordinates r ($= 1$) and θ on the circle corresponds to the point $x = \cos\theta$, $y = 0$ on the line segment. The space outside the circle maps into the whole z plane, and that inside the circle is mapped into a second sheet of Riemann surface.

According to the linearized theory of a planar system (§ 12.6), the boundary conditions on the airfoil may be applied to its projection on the x axis instead of to the airfoil itself. Hence, the conditions 7 and 8 of § 13.2 should be satisfied on the segment $(-1, 1)$ of the x axis. On the corresponding ζ plane, these conditions are to be satisfied on the unit circle.

The complex acceleration function is transformed as

$$\frac{dw}{dz} = \frac{dw}{d\zeta} \frac{d\zeta}{dz}, \qquad \frac{dw}{d\zeta} = \frac{dw}{dz} \frac{dz}{d\zeta} \tag{2}$$

---

* The symbol i comes from the complex representation of harmonic oscillation. It is entirely different from the symbol j in Eq. 6. The functions $\phi(x, y)$, $u(x, y)$ do not involve j, but they may involve i. The symbol j is used to separate the potential and stream functions, but i is used to denote a change of phase angle in time, $ij \neq -1$.

The function $\left|\dfrac{dz}{d\zeta}\right|$ is the scale factor of the transformation. Physically it is the ratio of the length of an acceleration vector on the z plane to that of the transformed vector on the ζ plane. Now on the airfoil (the unit circle on the ζ plane),

$$\left|\frac{dz}{d\zeta}\right|_{\zeta=e^{j\theta}} = \frac{1}{2}\left|1 - \frac{1}{\zeta^2}\right|_{\zeta=e^{j\theta}} = |\sin\theta| \tag{3}$$

Therefore, if the vertical component of the acceleration on the airfoil in

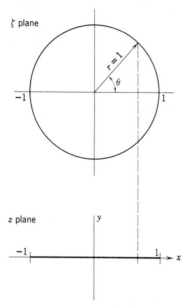

Fig. 13.2. Conformal mapping of a line segment into a circle.

z plane is $a'_y(x, 0, t)$, the component of acceleration normal to the unit circle in the ζ plane must be

$$a'_n(r = 1, \theta = \cos^{-1} x, t) = a'_y(x, 0, t)\sin\theta \tag{4}$$

This correspondence must be remembered in imposing boundary conditions on the ζ plane.

The condition at infinity on the z plane is that the acceleration potential ϕ tends to a constant when $z \to \infty$. Since the limit of $dz/d\zeta$ as z tends to infinity is finite $(= 1/2)$, the condition at infinity on the ζ plane is

$$\lim_{|\zeta|\to\infty} \text{Rl}(w) = \text{const} \tag{5}$$

Our problem is then to find a complex acceleration potential which satisfies Eqs. 4 and 5 above, Eq. 7 of § 13.2, and the Kutta-Joukowski condition.

According to Eq. 5, the function w can be posed in the form of a Laurent series

$$w(\zeta) = \frac{A_1}{\zeta} + \frac{A_2}{\zeta^2} + \cdots \tag{6}$$

where the constant term has been put to zero, because it contributes nothing to the acceleration field. The coefficients A_1, A_2, $\cdots$ can be determined according to the boundary condition 4 on the unit circle $\zeta = e^{j\theta}$. The Kutta-Joukowski condition is satisfied if series 6 converges at the point $\zeta = 1$ corresponding to the trailing edge.

It remains to satisfy the velocity boundary condition 7 of § 13.2, the velocity being given by Eq. 12 of that section. It will generally be found, however, that this boundary condition cannot be satisfied by the function $w(\zeta)$ in the form of Eq. 6 with the A's chosen according to Eq. 4. To satisfy the additional condition it is expedient to add to series 6 a term that contributes nothing to the normal acceleration on the unit circle, but yet does contribute to the normal velocity on the airfoil. Such a term can be readily found. It can be shown that a source-sink doublet, with an axis tangent to a circle, will have that circle as one of its streamlines. Hence a doublet (in acceleration potential), with axis tangent to a unit circle, produces zero acceleration normal to that circle. If such a doublet lies at the leading edge of the airfoil, its complex potential is

$$w(\zeta) = \frac{jA_0}{\zeta + 1} \tag{7}$$

where A_0 is an arbitrary real constant. Such a term makes the leading edge of the airfoil a singular point, where the pressure tends to infinity like $R^{-1/2}$, R being the distance from the leading edge. A singularity at the leading edge of this nature is found in the thin-airfoil theory in the subsonic steady-state case. Hence, its presence in the unsteady subsonic case is to be expected. With Eq. 7, we have

$$w(\zeta) = \frac{jA_0}{\zeta + 1} + \frac{A_1}{\zeta} + \frac{A_2}{\zeta^2} + \cdots \tag{8}$$

No other singular point is permitted on the airfoil. Hence the series 8 must converge everywhere on the unit circle, except at the leading edge, where it diverges logarithmically. It will be shown that by assuming the form 8 all the boundary conditions can be satisfied. That the solution so obtained is the unique solution can be proved.

13.4 AIRFOIL PERFORMING VERTICAL-TRANSLATION AND ROTATIONAL OSCILLATIONS*

Let us consider first the vertical-translation oscillations. Using the complex representation of harmonic oscillations, we may describe the airfoil surface by the equation

$$y = y_0 b e^{i\omega t} \tag{1}$$

where y_0 is a real number representing the ratio of the amplitude of the vertical motion to the semichord b of the airfoil which is taken as 1 in the following analysis. y_0 is therefore dimensionless. It is convenient for the following derivation to express the circular frequency ω in terms of the nondimensional reduced frequency k. Using the semichord $b = 1$ as the characteristic length, we define

$$k = \frac{\omega b}{U} = \frac{\omega}{U}$$

Hence, Eq. 1 may be written as

$$y = y_0 \, e^{iUkt} \tag{1a}$$

On the airfoil, the boundary values of the vertical velocity and acceleration of the flow are, according to Eqs. 7 and 8 of § 13.2,

$$v' = \frac{Dy}{Dt} = iUky_0 \, e^{iUkt}$$

$$a'_v = \frac{D^2 y}{Dt^2} = -U^2 k^2 y_0 \, e^{iUkt} \tag{2}$$

In all the equations that follow, the factor e^{iUkt} occurs in every term; hence, it will be omitted in writing.

The conformal mapping method described in § 13.3 may be used. In the transformed plane (i.e. the ζ plane), the normal component of the acceleration on the airfoil (the unit circle) is, according to Eq. 4 of § 13.3,

$$a'_n(|\zeta| = 1) = -U^2 k^2 y_0 \sin \theta \tag{3}$$

An inspection of series 8 of § 13.3 suggests that, for the present problem, the proper form of the complex acceleration potential $w(\zeta)$ is

$$w(\zeta) = \frac{jA}{\zeta + 1} + \frac{jB}{\zeta} \tag{4}$$

* The following method is given by M. A. Biot.[13.1]

This will be verified below. According to Eq. 4, the real and imaginary parts of $w(\zeta)$, i.e., the acceleration potential and its conjugate function are

$$\phi = \phi_1 + \phi_2, \qquad \psi = \psi_1 + \psi_2 \tag{5}$$

where

$$\phi_1 = A\,\frac{\sin\theta_1}{r_1}, \qquad \psi_1 = A\,\frac{\cos\theta_1}{r_1}$$

$$\phi_2 = B\,\frac{\sin\theta}{r}, \qquad \psi_2 = B\,\frac{\cos\theta}{r}$$

the meaning of r, θ, r_1, θ_1 being shown in Fig. 13.3. On the airfoil, i.e., on the unit circle in ζ plane, we have

$$r = 1, \qquad r_1 = 2\cos\theta_1$$

The corresponding normal components of acceleration are therefore

$$\left(\frac{\partial\phi_1}{\partial r}\right)_{r_1 = 2\cos\theta_1} = 0, \qquad \left(\frac{\partial\phi_2}{\partial r}\right)_{r=1} = -B\left(\frac{\sin\theta}{r^2}\right)_{r=1} = -B\sin\theta \tag{6}$$

The second relation of 6 is evident. The first of Eqs. 6 can be verified by direct substitution. It can also be recognized through the fact that ψ_1 is

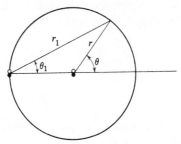

Fig. 13.3. Definitions of r, θ, and r_1, θ_1.

a constant on the unit circle (a streamline of the doublet), and that $\phi_1 = \text{const}$ and $\psi_1 = \text{const}$ curves are orthogonal. The constant potential lines are therefore normal to the unit circle, and the normal derivative of ϕ_1 vanishes on the circle.

Comparing Eqs. 6 and 3, we see that the normal acceleration on the airfoil is satisfied by taking

$$B = U^2 k^2 y_0 \tag{7}$$

The constant A is left undetermined, giving us freedom to satisfy the kinematic condition on the velocity component.

Now the y component of the velocity is related to the y component of acceleration by Eq. 12 of § 13.2:

$$v = \frac{e^{-ikx}}{U} \int_{-\infty}^{x} a_y(x) e^{ikx} \, dx = -\frac{e^{-ikx}}{U} \int_{-\infty}^{x} \left(\frac{\partial \psi_1}{\partial x} + \frac{\partial \psi_2}{\partial x} \right) e^{ikx} \, dx \qquad (8)$$

Hence, at a point $x = -1 + \xi$ on the airfoil ($0 \leqslant \xi \leqslant 2$), the boundary condition on velocity is, from Eqs. 2 and 8,

$$iUky_0 = -\frac{e^{-ik(-1+\xi)}}{U} \int_{-\infty}^{-1+\xi} e^{ikx} \left(\frac{\partial \psi_1}{\partial x} + \frac{\partial \psi_2}{\partial x} \right) dx \qquad (9)$$

This equation must be satisfied for all values of ξ. Differentiating Eq. 9 with respect to ξ and using Eqs. 2, we can readily show that the derivatives on both sides of the equation vanish identically. Hence, Eq. 9 is a true identity in the variable ξ. This means that, if Eq. 9 is satisfied at one point, it is satisfied for the entire airfoil. In particular, we can take ξ to be a small number tending to zero. To evaluate the integral in Eq. 9, since the term $\partial \psi_1/\partial x$ introduces a divergent singularity at the leading edge, we shall first integrate the terms containing $\partial \psi_1/\partial x$ by parts to obtain a convergent integral. The limit for $\xi \to 0$ in Eq. 9 then becomes

$$iUky_0 = -e^{ik} \left[e^{ikx} \frac{\psi_1}{U} \right]_{x=-\infty}^{x=-1} + \frac{e^{ik}}{U} \int_{-\infty}^{-1} e^{ikx} \left(ik\psi_1 - \frac{\partial \psi_2}{\partial x} \right) dx \qquad (10)$$

ψ_1 and ψ_2 in this expression must be evaluated on the physical plane, (i.e., the z plane). They must be transformed from their values on the ζ plane, given by Eqs. 5, back to the z plane. Now, on the circle $|\zeta| = 1$, ψ_1 is a constant

$$(\psi_1)_{|\zeta|=1} = \frac{A}{2}$$

while $\psi_1 \to 0$ when $r_1 = |x + 1| \to \infty$. The first term on the right-hand side of Eq. 10 is equal to $-A/2U$. To evaluate the remaining term in Eq. 10, note that the conformal transformation 1 of § 13.3,

$$z = \frac{1}{2} \left(\zeta + \frac{1}{\zeta} \right)$$

has the inverse transformation

$$\zeta = z + \sqrt{z^2 - 1}$$

where $\sqrt{z^2 - 1}$ is taken as the principal branch that assumes real positive values when z is real and > 1. On this branch, $\sqrt{z^2 - 1}$ takes negative real values when z is real and < -1. Hence, on the negative real axis, where $r = -\zeta$, $r_1 = -\zeta - 1$ and $z = x$, $(x < -1)$ we have

$$r = -x + \sqrt{x^2 - 1}$$

$$r_1 = -x - 1 + \sqrt{x^2 - 1}$$

Note further that $\theta = \theta_1 = \pi$ in the same range on the negative real axis. Substituting these values of r, r_1, θ, θ_1 into Eq. 5, we see that, when $x < -1$,

$$\psi_1 = \frac{-A}{-x - 1 + \sqrt{x^2 - 1}}$$

$$\psi_2 = \frac{-B}{-x + \sqrt{x^2 - 1}} = B(x + \sqrt{x^2 - 1}) \tag{11}$$

Hence,

$$\int_{-\infty}^{-1} e^{ikx}\, \psi_1\, dx = \frac{A}{2} \int_1^\infty e^{-ikx} \left(1 - \sqrt{\frac{x+1}{x-1}}\right) dx \tag{12}$$

The integral in Eq. 12 can be expressed in terms of the modified Bessel functions of the second kind of order zero, K_0, and that of order one, K_1. By definition:

$$K_0(iz) = \int_1^\infty \frac{e^{-iz\xi}}{\sqrt{\xi^2 - 1}}\, d\xi \qquad (-\pi \leqslant \arg z \leqslant 0) \tag{13}$$

$$K_1(iz) = -\frac{dK_0(iz)}{d(iz)} \tag{14}$$

Differentiating Eq. 13 with respect to iz, we have

$$K_1(iz) = \int_1^\infty \frac{\xi e^{-iz\xi}}{\sqrt{\xi^2 - 1}}\, d\xi = \int_1^\infty \left(\sqrt{\frac{\xi+1}{\xi-1}} - \frac{1}{\sqrt{\xi^2 - 1}}\right) e^{-iz\xi}\, d\xi \tag{15}$$

Comparing Eqs. 12, 13, and 15, we see that, if k is a complex number whose argument satisfies the condition $(-\pi < \arg k < 0)$:

$$\int_{-\infty}^{-1} e^{ikx}\, \psi_1\, dx = -\frac{A}{2} \left[K_1(ik) + K_0(ik) - \frac{e^{-ik}}{ik}\right] \tag{16}$$

Similarly,

$$\int_{-\infty}^{-1} e^{ikx} \frac{\partial \psi_2}{\partial x}\, dx = B \int_{-\infty}^{-1} e^{ikx} \left(1 + \frac{x}{\sqrt{x^2 - 1}}\right) dx$$

$$= -U^2 k^2 y_0 \left[K_1(ik) - \frac{e^{-ik}}{ik}\right] \tag{17}$$

But the final expressions in Eqs. 16 and 17 are analytic functions of k, regular over the entire Argend plane if that plane is cut along the negative real axis. Hence, by the principle of analytic continuation, Eqs. 16 and 17 are valid also when $\arg k = 0$. Combining Eqs. 10, 16, and 17, we obtain

$$A = -2iU^2 k y_0 \, C(k) \qquad (18)$$

where

$$C(k) = \frac{K_1(ik)}{K_1(ik) + K_0(ik)} = F(k) + i\, G(k) \qquad (19)$$

The function $C(k)$ is often referred to as *Theodorsen's function*. Its numerical value is given in Table 6.2.

All boundary conditions concerning the velocity and acceleration are now satisfied. Since the function $w(\zeta)$, given by Eq. 4, is continuous at the trailing edge, the Kutta condition is also satisfied. Thus the solution is completed. From Eq. 5 the acceleration potential is

$$\phi = -2iU^2 k y_0 \, C(k)\, \frac{\sin \theta_1}{r_1} + U^2 k^2 y_0 \, \frac{\sin \theta}{r} \qquad (20)$$

The pressure distribution on the airfoil is obtained from the relation $p = -\rho\phi$ by putting $r = 1$, $r_1 = 2\cos\theta_1$, in the above equation. Since ϕ is antisymmetric in y, the pressures acting on the upper and lower side of the airfoil are of opposite sign and the lift distribution l (positive upward) is equal to $-2p$.

$$l = 2\rho U^2 y_0 \left[-ik\, C(k) \tan\frac{\theta}{2} + k^2 \sin\theta \right] \qquad (21)$$

The complex amplitude of the total lift can be obtained by an integration*

$$L = \int_{-1}^{1} l \, dx = \int_{0}^{\pi} l \sin\theta \, d\theta = \pi\rho U^2 y_0 k^2 \left[1 - \frac{2i}{k} C(k) \right] \qquad (22)$$

The moment about the mid-chord point is (positive in the nose-up sense)

$$M_{1/2} = -\int_{-1}^{1} lx \, dx = -\int_{0}^{\pi} l \cos\theta \sin\theta \, d\theta$$

$$= -\pi\rho U^2 i y_0 k \, C(k) \qquad (23)$$

A comparison between Eq. 22 and Eq. 23 shows that part of the lift that is proportional to $C(k)$ has a resultant acting at the $1/4$-chord point. This part of the lift can be identified as that caused by the bound vorticity over the airfoil. The other part of the lift has a resultant that acts through the mid-chord point. This latter term arises from a noncirculatory

* Lift $= Le^{i\omega t}$.

origin, and is equal to the product of the apparent mass and the vertical acceleration. The apparent mass is independent of the flight speed. For a flat plate the mass of the fluid enclosed in a circumscribing cylinder having the airfoil chord as a diameter is the theoretical apparent mass associated with the vertical motion.

The rotational oscillations can be solved in a similar manner. Let the skeleton airfoil, which executes rotational oscillation with a small amplitude about the origin (the mid-chord point), be represented by the equation

$$y = - \alpha_0 x e^{i\omega t} = - \alpha_0 x e^{iUkt} \tag{24}$$

The boundary values of the vertical velocity and acceleration of the flow on the airfoil are, accordingly,

$$v' = - \alpha_0 U e^{ikUt}(ikx + 1) \tag{25}$$

$$a'_y = \alpha_0 U^2 k e^{ikUt}(kx - 2i) \tag{26}$$

When the airfoil is transformed into a unit circle on the ζ plane, the complex acceleration potential in this case assumes the form

$$w(\zeta) = \frac{jA}{\zeta + 1} + \frac{jB}{\zeta} + \frac{jC}{\zeta^2} \tag{27}$$

The constants B and C are easily seen to be

$$B = 2ikU^2\alpha_0, \qquad C = - \tfrac{1}{4}k^2U^2\alpha_0 \tag{28}$$

The constant A must be determined from the boundary condition of velocity as before. The result is

$$A = 2U^2\alpha_0 \frac{K_1(ik) - \dfrac{ik}{2} K_0(ik)}{K_1(ik) + K_0(ik)} = 2U^2\alpha_0 \left[\left(1 + \frac{ik}{2}\right) C(k) - \frac{ik}{2} \right] \tag{29}$$

The complex acceleration potential being determined, the pressure distribution over the airfoil can be obtained from $p = - \rho\phi$, and the total lift and moment about the mid-chord are given by a simple integration:

$$L = \pi\rho U^2 k \left[i + \left(i + \frac{2}{k}\right) C(k) \right] \alpha_0 e^{iUkt} \tag{30}$$

$$M_{1/2} = \frac{1}{2} \pi\rho U^2 k \left[- i + \frac{k}{4} + \left(i + \frac{2}{k}\right) C(k) \right] \alpha_0 e^{iUkt} \tag{31}$$

Writing

$$\alpha = \alpha_0 e^{iUkt}, \qquad \dot\alpha = \frac{d\alpha}{dt}, \qquad \text{etc.} \tag{32}$$

we obtain

$$L = \pi \rho U \dot{\alpha} + 2\pi \rho U^2 \left(1 + \frac{ik}{2}\right) C(k)\, \alpha \qquad (33)$$

$$M_{1/2} = -\frac{\pi \rho U}{2} \dot{\alpha} - \frac{\pi \rho}{8} \ddot{\alpha} + \pi \rho U^2 \left(1 + \frac{ik}{2}\right) C(k)\, \alpha \qquad (34)$$

Comparing the expressions L and $M_{1/2}$, and remembering that the wing semichord is taken as 1 in the analysis, we see that the term $\pi \rho U \dot{\alpha}$ represents a lift that acts at the $3/4$-chord point, the term proportional to $C(k)$ represents a lift that acts at the $1/4$-chord point, and the term $(\pi \rho / 8)\ddot{\alpha}$ is a pure couple. It can be shown that the term proportional to $C(k)$ represents the lift due to circulation. The other two terms are of noncirculatory origin.

The lift due to circulation $2\pi \rho U^2 \left(1 + \dfrac{ik}{2}\right) C(k)\, \alpha$ may be compared with the corresponding term $-2i\pi\rho U^2\, C(k)k y_0 e^{iUkt}$ due to translation (Eq. 22). The "upwash" *at the* $3/4$*-chord point* due to translation is

$$w_y = iUk y_0 e^{iUkt} \qquad (35)$$

That due to rotation is

$$w_\alpha = -U\alpha_0 e^{iUkt} - \tfrac{1}{2} iUk\alpha_0 e^{iUkt} \qquad (36)$$

It is seen that, in both translation and rotation cases, the lift due to circulation can be written as

$$L_1 = -2\pi\rho U\, C(k) w \qquad (37)$$

where w stands for either w_y or w_α. Thus the upwash at the $3/4$-chord point has a unique significance. For this reason the $3/4$-chord point is called the *rear aerodynamic center*.

13.5 KÜSSNER-SCHWARZ' GENERAL SOLUTION

By following the method of the previous sections, the aerodynamic force acting on a skeleton airfoil oscillating with arbitrary mode can be computed. Let the airfoil be located from $x = -1$ to $x = +1$ as in § 13.2, and let the harmonic oscillation of the points on the airfoil be described by

$$y = f(x)e^{i\omega t} \qquad (1)$$

The "upwash" v on the airfoil is therefore given by

$$v = U\frac{\partial y}{\partial x} + \frac{\partial y}{\partial t} = U\left(\frac{\partial y}{\partial x} + iky\right) \qquad (2)$$

It is convenient to introduce a new variable $x = \cos \theta^*$ and consider v as a function of θ and t. We can express v by a Fourier series:

$$v(\theta, t) = - U e^{i\omega t} \left(P_0 + 2 \sum_{1}^{\infty} P_n \cos n\theta \right) \tag{3}$$

where

$$U e^{i\omega t} P_0 = - \frac{1}{\pi} \int_0^{\pi} v(\theta, t) \, d\theta$$

$$U e^{i\omega t} P_n = - \frac{1}{\pi} \int_0^{\pi} v(\theta, t) \cos n\theta \, d\theta \tag{4}$$

With the upwash given in this form, the acceleration potential ϕ and the pressure distribution p can be determined as in the previous sections. The lift distribution can be written as

$$l(\theta, t) = \rho U^2 e^{i\omega t} \left(2a_0 \tan \frac{\theta}{2} + 4 \sum_{1}^{\infty} a_n \sin n\theta \right) \tag{5}$$

Küssner and Schwarz[13.23] obtained the following relations:

$$a_0 = C(k)(P_0 + P_1) - P_1$$

$$a_n = \frac{ik}{2n} P_{n-1} + P_n - \frac{ik}{2n} P_{n+1} \qquad (n \geqq 1) \tag{6}$$

where $C(k)$ is the Theodorsen's function defined by Eq. 19 of § 13.4. (Küssner writes $(1 + T)/2$ for $C(k)$.)

Using Eqs. 6, Küssner and Schwarz derive the following general solution which is independent of the Fourier coefficients:†

$$l(\Theta, t) = \rho U \frac{2}{\pi} \int_0^{\pi} v(\theta, t) \left[\{ C(k)(1 + \cos \theta) - \cos \theta \} \tan \frac{\Theta}{2} \right.$$

$$\left. + ik \Lambda(\cos \Theta, \cos \theta) \sin \theta + \frac{\sin \Theta}{\cos \theta - \cos \Theta} \right] d\theta \tag{7}$$

where

$$\Lambda(\cos \Theta, \cos \theta) = \log \left| \frac{\sin \dfrac{\Theta + \theta}{2}}{\sin \dfrac{\Theta - \theta}{2}} \right| = \frac{1}{2} \log \frac{1 - \cos (\Theta + \theta)}{1 - \cos (\Theta - \theta)} \tag{8}$$

---

* $\theta = \pi$ at the leading edge, $\theta = 0$ at the trailing edge.
† Schwarz gives a second derivation of this equation in Ref. 13.29.

The integral in Eq. 7 is defined by its Cauchy principal value. The total lift is then given by

$$L = \int_{-1}^{1} l \cdot dx = \int_{0}^{\pi} l(\Theta, t) \sin \Theta \, d\Theta \qquad (9)$$

and the moment about a point $x_0 = \cos \chi_0$ is

$$M_{x_0} = \int_{-1}^{1} l \cdot (x - x_0) \, dx = \int_{0}^{\pi} l(\Theta, t)(\cos \Theta - \cos \chi_0) \sin \Theta \, d\Theta \qquad (10)$$

The result of integration gives the following expressions for the total lift and the stalling moment about the mid-point corresponding to the upwash given by Eq. 3:

$$L = 2\pi\rho U^2 e^{i\omega t} \left[(P_0 + P_1) \, C(k) + (P_0 - P_2) \frac{ik}{2} \right] \qquad (11)$$

$$M_{1/2} = \pi\rho U^2 e^{i\omega t} \left\{ P_0 \, C(k) - P_1[1 - C(k)] - (P_1 - P_3) \frac{ik}{4} - P_2 \right\} \qquad (12)$$

It is interesting to notice that the resultant lift depends on only the first three coefficients in Eq. 3, and the resultant moment on the first four. This is, in fact, connected with the general expression of the complex potential as given by Eq. 8 of § 13.3. It may be verified that the total lift depends only on the imaginary part of the coefficients A_0 and A_1, and the total moment depends only on the imaginary part of the coefficients A_0 and A_2.

Example. Lift Force Due to a Sinusoidal Gust. Let us consider the lift acting on an airfoil, flying at a uniform speed and entering a sinusoidal gust. If the coordinate axes are fixed on the airfoil, the vertical gust may be represented by a velocity distribution

$$w(x, t) = We^{i\omega(t - x/U)} \qquad (13)$$

which expresses the fact that a sinusoidal gust pattern, with amplitude W (a constant), moves past the airfoil with the speed of flight U. If the wave length of the gust is l, the frequency ω with which the waves pass any point of the airfoil is

$$\omega = 2\pi U/l \qquad (14)$$

If the gust velocity $w(x, t)$ is considered positive upward, the relative velocity at any point on the airfoil to the fluid (measured positive upward) is

$$v(x, t) = - We^{i\omega t}e^{-i(\omega/U)\cos\theta} = - We^{i\omega t}e^{-ik\cos\theta} \qquad (15)$$

where the transformation $x = \cos\theta$ has been made for points on the airfoil.   Equation 15 may be put into the general form of Eq. 3 by the identity*

$$e^{iz \cos \theta} = J_0(z) + 2 \sum_{n=1}^{\infty} i^n J_n(z) \cos n\theta \tag{16}$$

where $J_n(z)$'s are Bessel functions of the first kind.   Putting $z = -k$ and noting that $J_n(-z) = (-)^n J_n(z)$, it is seen that, according to Eqs. 15 and 16,

$$v(x, t) = - W e^{i\omega t} \left[J_0(k) + 2 \sum_{n=1}^{\infty} (-i)^n J_n(k) \cos n\theta \right] \tag{17}$$

By comparing Eqs. 3 and 17, the expressions for the lift and moment can immediately be written according to Eqs. 11 and 12.   If the chord length of the airfoil is c instead of 2, the lift and moment per unit span are

$$L = \pi \rho c U W e^{i\omega t} \left\{ [J_0(k) - i J_1(k)] C(k) + [J_0(k) + J_2(k)] \frac{ik}{2} \right\} \tag{18}$$

$$M_{1/2} = \left(\frac{\pi}{4} \right) \rho c^2 U W e^{i\omega t} \left\{ J_0(k) C(k) + i J_1(k)[1 - C(k)] \right.$$
$$\left. + [i J_1(k) + i J_3(k)] \frac{ik}{4} + J_2(k) \right\} \tag{19}$$

These expressions can be simplified by means of the recurrence formula for Bessel functions,

$$2n J_n(z)/z = J_{n-1}(z) + J_{n+1}(z) \tag{20}$$

from which the following results are obtained:

$$[J_0(k) + J_2(k)] \frac{ik}{2} = i J_1(k)$$

$$[i J_1(k) + i J_3(k)] \frac{ik}{4} = - J_2(k) \tag{21}$$

Therefore Eqs. 18 and 19 are reduced to

$$L = \pi \rho c U W e^{i\omega t} \phi(k)$$

and

$$M_{1/2} = L \cdot \frac{c}{4} \tag{22}$$

---

* Cf. Gray, Mathews, MacRobert, *Bessel Functions*, p. 32, Eq. 6, Macmillan Co., London (1931), or H. B. Dwight, *Tables of Integrals*, Eqs. 818, Macmillan Co.

where

$$\phi(k) = [J_0(k) - i J_1(k)]C(k) + i J_1(k) \tag{23}$$

Clearly, the resultant lift acts through the $1/4$-chord point from the leading edge. The factor $\pi \rho c U \phi(k)$ represents the frequency response (admittance) of the lift to the gust. Writing

$$C(k) = F + iG \tag{24}$$

where F, G are real functions of the reduced frequency k, we have

$$|\phi(k)|^2 = (J_0^2 + J_1^2)(F^2 + G^2) + J_1^2 + 2J_0J_1G - 2J_1^2F \tag{25}$$

The function $\phi(k)$ is plotted as a vector diagram in Fig. 13.4. The function $|\phi(k)|^2$ is plotted in Fig. 13.5. An approximate expression,

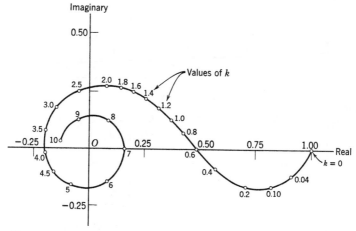

Fig. 13.4. Vector diagram showing the real and imaginary parts of Sears' $\phi(k)$ function as a function of the reduced frequency k. A radius vector drawn from the origin O to any point on the curve gives the value of the function $\phi(k)$. From Sears, Ref. 13.42. (Courtesy of the Institute of the Aeronautical Sciences.)

which agrees with the power-series expansion of Eq. 25 up to the first power in k and with the first term of the asymptotic expansion of Eq. 25 is

$$|\phi(k)|^2 = \frac{a + k}{a + (\pi a + 1)k + 2\pi k^2} \qquad (a = 0.1811) \tag{26}$$

Expression 26 approximates Eq. 25 closely over the whole range of k from 0 to ∞. A simpler approximation is

$$|\phi(k)|^2 = \frac{1}{1 + 2\pi k} \tag{27}$$

A comparison of Eqs. 25, 26, and 27 is given in Fig. 13.5. It is seen that Eq. 27 agrees fairly well with the exact expression 25, except for small values of k.

The solution 22 is due to Sears.[13.42] The approximation 27 is due to Liepmann.[9.11]

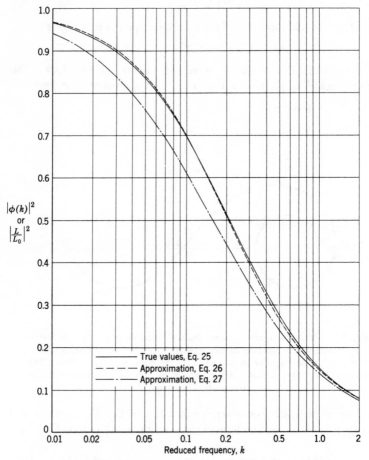

Fig. 13.5. Comparison of approximate expressions of $|\phi(k)|^2$.

13.6 TABULATION OF RESULTS—INCOMPRESSIBLE FLOW

Formulas and numerical tables of the aerodynamic coefficients for an airfoil with flap and tab, whose hinge lines do not necessarily coincide with their respective leading edges, have been published by many authors. The most important references are:

Dietze:[13.11, 13.12] *Luftfahrt-Forsch.* **16**, 84–96 (1939); **18**, 135–141 (1941).
W. P. Jones:[13.16, 13.17] *Aeronaut. Research Council R. & M.* **1948** (1941); *R. & M.* **1958** (1942).
Küssner and Schwarz:[13.23] *Luftfahrt-Forsch.* **17**, 337–354 (1940). Translated as *NACA Tech. Memo.* **991**.
Theodorsen[13.32] and Garrick:[13.33] *NACA Rept.* **496** (1934); *Rept.* **736** (1942).

The most comprehensive numerical tables are published by Küssner and Schwarz in the reference named above. The relations between the special functions tabulated by various authors are listed in a paper by W. P. Jones.[13.17]

For flutter calculations, Smilg and Wasserman's tables of L_h, L_α, etc., which are derived from Küssner and Schwarz's results, are widely used. These tables are contained in the following references:

AAF Tech. Rept. 4798, U.S. Air Force, by Smilg and Wasserman,[6.20] (1942).
Introduction to the Study of Aircraft Vibration and Flutter, book by Scanlan and Rosenbaum.[6.19]

While both sealed and unsealed gaps between the main wing and the flap and the control tab are considered in Küssner and Schwarz's paper, only unsealed gaps are considered in Smilg and Wasserman's. Furthermore, the tab hinge line is assumed to be located at the tab leading edge in

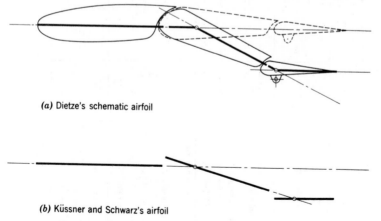

(a) Dietze's schematic airfoil

(b) Küssner and Schwarz's airfoil

Fig. 13.6.   Comparison of Dietze's aerodynamically balanced flap with Küssner's.

Table 13.1. Comparison of Notations

Reference	Duncan, Collar, Ref. 13.34	Theodorsen, Ref. 13.32	Küssner, Schwarz, Ref. 13.23	Dietze, Ref. 13.11, 13.12	Frazer, Ref. 7.31	Cicala, Ref. 13.7, 13.8	Lyon, Ref. 15.3	Kassner, Fingado, Ref. 6.15	Jones, Ref. 13.17	Scanlan, Rosenbaum, Smilg, Wasserman, Ref. 6.19, 6.20	This Book
Chord	c	$2b$	$2l$	$l(=t_P l)$	c	$L(=2D \text{ or } l)$	$c(=2D)$	t	c	$2b$	$c=2b$
Free-stream velocity	W	v	v	v	V	V	V	v	V	v	U
Circular frequency	p	$\dfrac{kv}{b}$	ν	ω	$\omega(=2\pi f)$	Ω	ν	ν	$\bar{\omega}(=2\pi f)$	ω	ω
Reduced frequency	$\dfrac{\lambda}{2}$	k	$-i\omega$	ω_r	$\dfrac{\lambda}{2}$	ω	ω	$\dfrac{1}{2V}$	$\dfrac{\omega}{2}$	k	k
Translational motion at reference point	z	h	$Al e^{i\nu t}$	δ	z	$\eta L \text{ or } \eta$	hl	y_A	z	h	h
Rotational motion of airfoil	θ	α	$B e^{i\nu t}$	β	θ	α	θ	ϕ	θ	α	α
Lift vector (circulation function)	$1-\dfrac{H}{2}-\dfrac{i\lambda}{2}G$	C	$\dfrac{1+T}{2}$	$\dfrac{1+T}{2}$	$-iC$	$1-\lambda$	$1-\lambda$	$\bar{P}$	C	C	C
Real part of $C(k)$	$1-\dfrac{H}{2}$	F	$\dfrac{1+T'}{2}$	$\dfrac{1+T'}{2}$	A	$1-\lambda'$	$1-\lambda'$	A	A	F	F
Imaginary part of $C(k)$	$-\dfrac{\lambda G}{2}$	G	$\dfrac{T''}{2}$	$\dfrac{T''}{2}$	$-B$	$-\lambda''$	λ''	$-B$	$-B$	G	G
Lift (+ upward)	$-Z$	$-P$	K	$-P$	$-Z$	P	P		$-Z$	$-L'$	L
Pitching moment (+ nose up)	M	M_α	$-M_0$	M	M	$-M$	M		M	M'	Mx_0 *

* x_0 is the point about which the moment is taken.

414

AAF Tech. Rept. 4798. The tables in *AAF Tech. Rept.* 4798 for tab oscillations and tables for aileron with $e < 0$, where eb is the coordinate of the aileron leading edge aft the mid-chord point, are not reproduced in Scanlan and Rosenbaum's book.

Dietze's results for the aerodynamic overhang of the control surfaces differ from those of other authors by the assumption that the overhang part of the flap is bent and faired with the main wing, and that of the tab is faired with the flap. See the comparison in Fig. 13.6.

The notations used by several authors are listed in Table 13.1.

BIBLIOGRAPHY

13.1 Biot, M. A.: Some Simplified Methods in Airfoil Theory. *J. Aeronaut. Sci.* **9**, 186–190 (1942).

13.2 Birnbaum, W.: Die tragende Wirbelfläche als Hilfsmittel zur Behandlung des ebenen Problems der Tragflügeltheorie. *Z. angew. Math. u. Mech.* **3**, 290–297 (1923).

13.3 Birnbaum, W.: Das ebene Problem des schlagenden Flügels. *Z. angew. Math. u. Mech.* **4**, 277–292 (1924).

13.4 Borbely, S. von: Mathematischer Beitrag zur Theorie der Flügelschwingungen. *Z. angew. Math. u. Mech.* **16**, 1–4 (1936).

13.5 Borkmann, and F. Dietze: Tafeln zum Luftkraftgesetz des harmonisch schwingenden Flügels. *ZWB ForschBer.* **1417** (1941).

13.6 Brower, W. B., and R. H. Lassen: Additional Values of $C(k)$. *J. Aeronaut. Sci.* **20**, 148–150 (1953).

13.7 Cicala, P.: Le azioni aerodinamiche sui profili di ala oscillanti in presenza di corrente uniforme. *Mem. R. Accad. Sci. Torino II* **68**, 73. Also **70**, 356–371 (1935).

13.8 Cicala, P.: Le azioni aerodinamiche sul profilo oscillante. *Aerotecnica* **16**, 652–655 (1936).

13.9 Cicala, P.: Le oscilazioni flesso-torsionali di un'ala in corrente uniforme. *Aerotecnica* **16**, 735 (1936).

13.10 Dietze, F.: Zur Berechnung der Auftriebskraft am schwingenden Ruder. *Luftfahrt-Forsch.* **14**, 361–362 (1937).

13.11 Dietze, F.: Die Luftkräfte der harmonisch schwingenden in sich verformbaren Platte (Ebenes Problem). *Luftfahrt-Forsch.* **16**, 84–96 (1939).

13.12 Dietze, F.: Zum Luftkraftgesetz der harmonisch schwingenden, knickbaren Platte. (Flügel mit Ruder und Hilfsruder.) *Luftfahrt-Forsch.* **18**, 135–141 (1941).

13.13 Ellenberger, G.: Bestimmung der Luftkräfte auf einen ebenen Tragflügel mit Querruder. *Z. angew. Math. u. Mech.* **16**, 199–226 (1936).

13.14 Glauert, H.: The Force and Moment of an Oscillating Aerofoil. *Aeronaut. Research Com. R. & M.* **1242** (1929).

13.15 Jaeckel, K.: Über die Bestimmung der Zirkulationsverteilung für den zweidimensionalen Tragflügel bei beliebigen periodischen Bewegungen. *Luftfahrt-Forsch.* **16**, 135–138 (1939).

13.16 Jones, W. P.: Aerodynamic Forces on an Oscillating Aerofoil-Aileron-Tab Combination. *Aeronaut. Research Com. R. & M.* **1948** (1941).

416   OSCILLATING AIRFOILS IN INCOMPRESSIBLE FLOW

13.17   Jones, W. P.: Summary of Formulae and Notations Used in Two-Dimensional Derivative Theory. *Aeronaut. Research Com. R. & M.* **1958** (1942).

13.18   Jordan, P.: Zum luftkraftgesetz der geschlossenen Stufe nach Küssner und Schwarz. *ZWB ForschBer.* **1537** (1941).

13.19   Jordan, P.: Vereinfachte Integration der Luftkräfte bei Flatteruntersuchungen. *ZWB ForschBer.* **1538** (1941).

13.20   Jordan, P.: Tafeln zum Luftkraftgesetz des harmonisch schwingenden Flügels mit aerodynamisch ausgeglichenen Rudern. *ZWB ForschBer.* **1539** (1942).

13.21   Kármán, Th. von, and W. R. Sears: Airfoil Theory for Non-uniform Motion. *J. Aeronaut. Sci.* **5**, 379–390 (1938).

13.22   Küssner, H. G.: Zusammenfassender Bericht über den instationären Auftrieb von Flügeln. *Luftfahrt-Forsch.* **13**, 410–424 (1936).

13.23   Küssner, H. G., and L. Schwarz: The Oscillating Wing with Aerodynamically Balanced Elevator. *NACA Tech. Memo.* **991** (1941). Translated from *Luftfahrt-Forsch.* **17**, 337–354 (1940).

13.24   Lapin, E., R. Crookshanks, and H. F. Hunter: Downwash behind a Two-Dimensional Wing Oscillating in Plunging Motion. *J. Aeronaut. Sci.* **19**, 447–450 (1952).

13.25   Luke, Y., and M. A. Dengler: Tables of the Theodorsen Circulation Function for Generalized Motion. *J. Aeronaut. Sci.* **18**, 478–483 (1951).

13.26   Morris, R. M.: The Two-Dimensional Hydrodynamical Theory of Moving Airfoils. Part I, *Proc. Roy. Soc. A* **161**, 406–419 (1937); Part II, **164**, 346–368 (1938); Part III, **172**, 213–230 (1939); Part IV, **188**, 439–463 (1947).

13.27   Postel, E. E., and E. L. Leppert, Jr.: Theoretical Pressure Distributions for a Thin Airfoil Oscillating in Incompressible Flow. *J. Aeronaut. Sci.* **15**, 8, 486–492 (1948).

13.28   Schmieden, C.: Die Strömung um einen ebenen Tragflügel mit Querruder. *Z. angew. Math. u. Mech.* **16**, 193–198 (1936).

13.29   Schwarz, L.: Berechnung der Druckverteilung einer harmonisch sich verformenden Tragfläche in ebener Strömung. *Luftfahrt-Forsch.* **17**, 379–386 (Dec. 1940).

13.30   Schwarz, L.: Untersuchung einiger mit den Zylinderfunktionen nullter Ordnung Verwandten Funktionen. *Luftfahrt-Forsch.* **20**, 341–372 (Feb. 1944).

13.31   Söhngen, H.: Auftrieb und Moment der geknickten Platt mit Spalt. *Luftfahrt-Forsch.* **17**, 17–22 (1940).

13.32   Theodorsen, Th.: General Theory of Aerodynamic Instability and the Mechanism of Flutter. *NACA Rept.* **496** (1934).

13.33   Theodorsen, Th., and I. E. Garrick: Nonstationary Flow about a Wing-Aileron-Tab Combination Including Aerodynamic Balance. *NACA Rept.* **736** (1942).

For resistance derivatives in British usage, see

13.34   Duncan, W. J., and A. R. Collar: Resistance Derivatives of Flutter Theory. Part I, *Aeronaut. Research Com. R. & M.* **1500** (1932). Part II, Results for Supersonic Speeds, *Aeronaut. Research Council R. & M.* **2139** (1944).

13.35   Duncan, W. J.: Some Notes on Aerodynamic Derivatives. *Aeronaut. Research Council R. & M.* **2115** (1945).

13.36   Temple, G.: The Representation of Aerodynamic Derivatives. *Aeronaut. Research Council R. & M.* **2114** (1945).

13.37   Williams, J.: Some Approximate Representations for the Airload Coefficients of Two-Dimensional Vortex Sheet Theory. *Aeronaut. Research Council R. & M.* **1968** (1944).

For semi-experimental methods designed to achieve better agreement between theory and experiment for an airfoil having finite thickness and finite angle of attack, see

13.38   Jones, W. P.: Aerofoil Oscillations at High Mean Incidences. *Aeronaut. Research Council R. & M.* **2654** (1953).
13.39   Jordan, P. F.: Note on Semi-experimental Methods for the Determination of Aerodynamic Derivatives for an Oscillating Wing-Aileron System. *Aeronaut. Research Council R. & M.* **2706** (1952).
13.40   Schwarz, L.: On a Half-Experimental Method for the Determination of Unsteady Pressure Distributions. *Tech. Ber.* **11**, no. 5, 133–137 (1944).

For the reaction of an airfoil to sinusoidal gust, see

13.41   Kemp, N. H.: On the Lift and Circulation of Airfoils in Some Unsteady Flow Problems. *J. Aeronaut. Sci.* **19**, 713–714 (1952).
13.42   Sears, W. R.: Some Aspects of Non-stationary Airfoil Theory and Its Practical Application. *J. Aeronaut. Sci.* **8**, 104–108 (1941).

For flow pulsation in the direction of flight, see

13.43   Greenberg, J. M.: Airfoil in Sinusoidal Motion in Pulsating Stream. *NACA Tech. Note* **1326** (1947).
13.44   Greenberg, J. M.: Some Considerations on an Airfoil in an Oscillating Stream. *NACA Tech. Note* **1372** (1947).
13.45   Isaacs, R.: Airfoil Theory for Flows of Variable Velocity. *J. Aeronaut. Sci.* **12**, 113–118 (1945).

The virtual mass of Joukowski airfoils and the forces acting on airfoil in a flow containing vortex elements are treated in

13.46   Sedov, L. I.: On the Theory of the Unsteady Motion of an Airfoil. *NACA Tech. Memo.* **1156**. Translated from Russian Central Aero-Hydrodynamic Institute Report.

Chapter 14

OSCILLATING AIRFOILS IN TWO-DIMENSIONAL COMPRESSIBLE FLOW

An oscillating airfoil in a two-dimensional compressible flow will be treated in this chapter. *All the hypotheses made in § 13.1 regarding the linearization are again assumed here.*

The linearized equation for the acceleration potential ϕ, referred to a frame of reference at rest relative to the fluid at infinity is given by Eq. 15 of § 12.5. If the coordinate system is moving with a speed U in the negative x-axis direction relative to the fluid at infinity, the field equation of small disturbances can be obtained by transforming that equation according to the *Galilean transformation*

$$x = x' - Ut', \qquad y = y', \qquad t = t' \tag{1}$$

The following field equation is obtained in this new coordinate system for a two-dimensional flow in the (x,y) plane:

$$\frac{\partial^2 \phi}{\partial t^2} + 2U \frac{\partial^2 \phi}{\partial x\, \partial t} + U^2 \frac{\partial^2 \phi}{\partial x^2} = a^2 \left(\frac{\partial^2 \phi}{\partial x^2} + \frac{\partial^2 \phi}{\partial y^2} \right) \tag{2}$$

where the primes are omitted. For an observer fixed on the moving coordinates, the fluid at infinity has a velocity U in the positive x-axis direction. a is the speed of sound in the undisturbed flow.

As in the last chapter, the principle of superposition is valid, and it suffices to treat oscillating airfoils of zero camber, zero thickness, and at zero mean angle of attack.

In a subsonic flow, $(U < a)$, an elementary solution of Eq. 2 was obtained by Possio who derived an integral equation governing ϕ and obtained some numerical results by a method of collocation in 1937. These calculations were repeated and extended by Frazer (1941) and Frazer and Skan (1942). The kernel of Possio's equation was tabulated by Schwarz (1943). Approximate solutions were proposed by Schade (1944), Dietze (1942–44), and Fettis (1952). An exact solution of Possio's equation in closed form is yet unknown.

A different approach to the subsonic oscillating-airfoil theory was pursued independently by Reissner and Sherman (1944), Biot (1946),

Timman (1946), Haskind (1947), and Küssner (1953). The boundary-value problem was directly attacked by the introduction of (confocal) elliptic coordinates. An explicit solution of the problem can then be obtained in terms of Mathieu functions. A great deal of mathematical work is required, however, to bring the solution into a form suitable for numerical calculations.

In contrast to the subsonic case, the linearized supersonic case is extremely simple. This is so because of the simple physical condition that (in the two-dimensional case) the flows above and below the airfoil are independent of each other, and that the flow over the airfoil is independent of the conditions in the wake. The corresponding differential equation, of the hyperbolic type, can be solved by a number of methods. The first solution was given by Possio in 1937 by a method of superposition of sources and sinks. Von Borbely solved the problem by the method of Laplace transformation (1942), and Temple and Jahn solved it by Riemann's method (1945). Extensive numerical results were obtained by Schwarz, Temple and Jahn, Garrick and Rubinow, and others (see bibliography).

In §§ 14.1 through 14.4, Possio's integral equation for the subsonic flow will be derived in a manner used by Biot, Karp, Shu, and Weil.[14.1] Approximate methods of solution of Possio's equation will be outlined in § 14.5.

The supersonic case is discussed in § 14.6, where, following Stewartson,[14.40] the Laplace transformation method is used. In § 14.7, the results known so far are tabulated for a quick reference.

The lift of an oscillating wing can also be computed from the indicial responses to suddenly started motions. It turns out that, if the main-stream Mach number is close to one, it is simpler to calculate the indicial responses. Heaslet, Lomax, and Spreiter[14.24] have shown (1949) how the lift due to a harmonic vertical-translation oscillation of a flat plate in a flow with main-stream Mach number equal to one can be calculated according to the linearized theory.

14.1 GENERAL EQUATIONS AND ELEMENTARY SOLUTIONS

In a subsonic flow, the main-stream Mach number $M = U/a$ is less than 1, and the quantity

$$\beta = \sqrt{1 - M^2} \tag{1}$$

is real and positive. Equation 2 of the last section can be written as

$$\frac{1}{a^2}\frac{\partial^2 \phi}{\partial t^2} + \frac{2M}{a}\frac{\partial^2 \phi}{\partial x\, \partial t} - \beta^2 \frac{\partial^2 \phi}{\partial x^2} = \frac{\partial^2 \phi}{\partial y^2} \tag{2}$$

which can be simplified by the following transformation:

$$\tau = t + \frac{M}{\beta^2 a} x = t + \alpha x \tag{3}$$

$$\xi = \frac{x}{\beta^2 a}, \qquad \eta = \frac{y}{\beta a}, \tag{4}$$

where

$$\alpha = M/(\beta^2 a)$$

Equation 2 then becomes the classical wave equation

$$\phi_{\tau\tau} = \phi_{\xi\xi} + \phi_{\eta\eta} \tag{5}$$

Introducing polar coordinates

$$\xi = r' \cos \theta, \qquad \eta = r' \sin \theta \tag{6}$$

so that

$$r' = \sqrt{\xi^2 + \eta^2} = \frac{1}{\beta^2 a} \sqrt{x^2 + \beta^2 y^2}$$

$$\theta = \tan^{-1}\left(\frac{\eta}{\xi}\right) = \tan^{-1}\left(\frac{\beta y}{x}\right) \tag{7}$$

we have

$$\phi_{\tau\tau} = \phi_{r'r'} + \frac{1}{r'}\phi_{r'} + \frac{1}{r'^2}\phi_{\theta\theta} \tag{8}$$

Assuming a solution of the form

$$\phi(r', \theta, \tau) = e^{i\omega\tau} e^{in\theta} R(r') \tag{9}$$

we see that the equation governing $R(r')$ is

$$\frac{d^2 R}{dr'^2} + \frac{1}{r'}\frac{dR}{dr'} + \left(\omega^2 - \frac{n^2}{r'^2}\right) R = 0 \tag{10}$$

which is the Bessel's differential equation of order n.   Taking the Hankel functions of the first and second kind of order n, $H_n^{(1)}(\omega r')$ and $H_n^{(2)}(\omega r')$, as the fundamental system of solutions of Eq. 10, we obtain a general solution of Eq. 8:

$$\phi(r', \theta, t) = A_n \begin{Bmatrix} H_n^{(1)}(\omega r') \\ H_n^{(2)}(\omega r') \end{Bmatrix} \begin{Bmatrix} \sin n\theta \\ \cos n\theta \end{Bmatrix} e^{i\omega\tau} \tag{11}$$

The brace { } means that either of the two functions enclosed in it may be taken.

The Hankel functions $H_n^{(1)}(\omega r')$ and $H_n^{(2)}(\omega r')$ (also called Bessel functions of the third kind) are tabulated in several treatises.*   They are

---

* For example, see Jahnke and Emde, *Tables of Functions*.   Dover Publications.

complex valued for real arguments. We shall not discuss the general properties of the Hankel functions here.* But the following salient properties are needed for the following discussions:

1. The conjugate complex of $H_n^{(1)}(re^{i\theta})$ is $H_n^{(2)}(re^{-i\theta})$ [and not $H_n^{(1)}(re^{-i\theta})$]. If $H_n^{(1)}(z)$ is known, $H_n^{(2)}(z)$ can be obtained from the formula

$$H_n^{(2)}(z) = 2J_n(z) - H_n^{(1)}(z) \tag{12}$$

where $J_n(z)$ is the Bessel function of the first kind of order n.

2. For very small arguments ($|z| \ll 1$),

$$H_0^{(2)}(z) \sim -\frac{2i}{\pi} \log \frac{cz}{2} \qquad (c = e^\gamma \doteq e^{0.5772} \doteq 1.781)$$

$$H_1^{(2)}(z) \sim \frac{2i}{\pi z} \tag{13}$$

$$H_n^{(2)}(z) \sim \frac{i}{\pi}(n-1)! \left(\frac{2}{z}\right)^n \qquad (n,\ \text{integer},\ \neq 0)$$

The $\sim$ sign means that, in the Laurent's series expansion of the Hankel functions, the sum of all the other terms are of order smaller than the first term.

3. For very large argument, the dominant terms of the asymptotic representation of the Hankel functions are ($|z| \gg 1$):

$$H_n^{(1)}(z) \sim \left(\frac{2}{\pi z}\right)^{1/2} \exp\left\{i\left(z - \frac{1}{2}n\pi - \frac{1}{4}\pi\right)\right\}, \qquad -\pi < \arg z < 2\pi$$

$$H_n^{(2)}(z) \sim \left(\frac{2}{\pi z}\right)^{1/2} \exp\left\{-i\left(z - \frac{1}{2}n\pi - \frac{1}{4}\pi\right)\right\}, \qquad -2\pi < \arg z < \pi \tag{14}$$

We can now examine the physical meaning of the solution 11. Consider the case $n = 0$. The solution is

$$\phi(r', \theta, t) = A_0 \begin{Bmatrix} H_0^{(1)}(\omega r') \\ H_0^{(2)}(\omega r') \end{Bmatrix} e^{i\omega(t + \alpha x)} \tag{15}$$

According to Eqs. 12 and 13, it is seen that the absolute values of $H_0^{(1)}(r')$ and $H_0^{(2)}(r')$ become infinitely large when $r' \to 0$. Hence, both terms of Eq. 15 behave like a source at the origin. To see the properties of the waves corresponding to these two elementary solutions, we examine the solution at large values of $\omega r'$. The asymptotic representation for the

---

* An exhaustive treatment is given by Watson: *Bessel Functions*, Cambridge Univ. Press. An elegant introduction sufficiently developed for physical applications can be found in Chapter IV of Sommerfeld's book: *Partial Differential Equations*, Academic Press (1949).

functions $H_0^{(1)}(\omega r')$ and $H_0^{(2)}(\omega r')$ are given by Eqs. 14 by setting $n = 0$. Hence, asymptotically, at large distance from the source, the acceleration potentials are

$$\phi_1 = A_0 H_0^{(1)}(\omega r')e^{i\omega(t+\alpha x)} \sim A_0 \sqrt{\frac{2}{\pi\omega r'}}\, e^{-(\pi/4)i + i\omega\alpha x}\, e^{i\omega(t+r')}$$

$$\phi_2 = A_0 H_0^{(2)}(\omega r')e^{i\omega(t+\alpha x)} \sim A_0 \sqrt{\frac{2}{\pi\omega r'}}\, e^{(\pi/4)i + i\omega\alpha x}\, e^{i\omega(t-r')}$$

(16)

The factor $t + r'$ remains constant when t increases if the distance r' decreases with suitable speed. Hence, ϕ_1 represents an incoming wave, converging toward the source at the origin. Similarly the factor $t - r'$ in ϕ_2 indicates that ϕ_2 represents a wave radiating from the source. If the source at the origin is regarded as the only cause of disturbance in the flow field, we must impose the condition that waves can only radiate out from the source, and, hence, we must discard $H_0^{(1)}(\omega r')$ as a physical solution.

The case $n = 1$ can be similarly examined. The solution that represents waves radiating out from the singularity is given by

$$\phi(r', \theta, t) = A_1 H_1^{(2)}(\omega r') \begin{Bmatrix} \sin\theta \\ \cos\theta \end{Bmatrix} e^{i\omega(t+\alpha x)}$$

(17)

The $\cos\theta$ term on the right-hand side is antisymmetrical with respect to the y axis, while the other, $\sin\theta$, is antisymmetrical with respect to the x axis. Since $H_1^{(2)}(\omega r')$ tends to infinity when $r' \to 0$, so the solution

$$\phi(r', \theta, t) = B \sin\theta\, H_1^{(2)}(\omega r')e^{i\omega(t+\alpha x)}$$

(18)

represents a "doublet" at the origin, with axis perpendicular to the x axis, while the other solution involving $\cos\theta$ represents a doublet at the origin with axis perpendicular to the y axis.

If the singularity (a source, a doublet, etc.) is situated at the point (x_0, y_0) instead of at the origin, it is necessary to replace the x and y occurring in the above solutions by $(x - x_0)$ and $(y - y_0)$, respectively. In particular, r' should be replaced by

$$r' = \frac{1}{\beta^2 a} \sqrt{(x - x_0)^2 + \beta^2(y - y_0)^2}$$

(19)

14.2 LIFT AND THE STRENGTH OF DOUBLET IN ACCELERATION POTENTIAL

In the linearized thin-airfoil theory, the discontinuity of the pressure field across the airfoil can be represented by a layer of doublets, the strength of which varies with time. In order to find the relation between the lift distribution and the oscillation mode, the lift force and the induced downwash velocity corresponding to a given distribution of doublets must be known. These relations will be derived in the present and the next sections.

First let us remark that, in the airfoil theory, the doublets are distributed over a surface (a line segment along the x axis in the two-dimensional theory). Consider, then, a line distribution of "doublets" along the x axis from $x = -1$ to $x = +1$. Let $B(x_0)$ be the strength per unit length of the doublet distribution at a point $(x = x_0, y = 0)$ on the airfoil. The total acceleration potential is then

$$\phi(x, y, t) = e^{i\omega t} \int_{-1}^{1} B(x_0) \sin \theta\, H_1^{(2)}(\omega r') e^{i\omega\alpha(x - x_0)}\, dx_0 \tag{1}$$

where

$$r' = \frac{1}{\beta^2 a} \sqrt{(x - x_0)^2 + \beta^2 y^2} \tag{2}$$

$$\theta = \tan^{-1} \left(\frac{\beta y}{x - x_0} \right) \tag{3}$$

$$\alpha = \frac{M}{\beta^2 a} \tag{4}$$

Let us evaluate the local lift distribution acting on a doublet element of length 2ε situated at $x = \xi$, ε being a small number compared with 1. Let $L(x, t)$ be the lift per unit length acting on the doublet layer at $(x, 0)$. Then the lift force on the element concerned is $2\varepsilon L(\xi, t)$. Referring to Fig. 14.1,

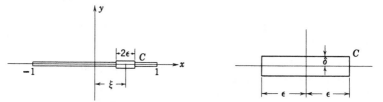

Fig. 14.1. The region of integration.

let us take a curve C which is a narrow rectangle of height 2δ and length 2ε. According to Newton's law, the rate of change of momentum of the fluid enclosed in C is equal to the external force acting on it. The latter consists of the pressure force acting on the boundaries of C and the lift force exerted by the doublet layer. Now, when the strength $B(x_0)$ is so chosen as to make the resulting flow field correspond to a physical problem, the acceleration potential $\phi(x, y, t)$ must satisfy the kinematic boundary conditions on velocity and acceleration. In particular, the vertical acceleration $\partial \phi/\partial y$ must be finite over the airfoil. Therefore, the rate of change of momentum of the fluid enclosed in C is a finite number times the area enclosed in C. As the height of the rectangular boundary of C shrinks to zero, the area $4\varepsilon\delta$ enclosed in C also tends to zero. On the other hand, the lift and the pressure force are proportional to ε. Hence, if $\delta \ll \varepsilon$, the momentum change of the fluid enclosed in C will be negligible in comparison with the pressure force acting on it. This being the case, we obtain

$$2\varepsilon L(\xi, t) = \int_C p\,dx = -\rho_0 \int_C \phi(x, y, t)\,dx \tag{5}$$

i.e., since $\phi(x, -\delta, t) = -\phi(x, \delta, t)$,

$$2\varepsilon L(\xi, t) = 2\rho_0 \lim_{\delta \to 0} \int_{\xi-\varepsilon}^{\xi+\varepsilon} \phi(x, \delta, t)\,dx \tag{6}$$

The limit $\delta \to 0$ can be taken inside of the integral. Now

$$\lim_{\delta \to 0} \phi(x, \delta, t) = \lim_{\delta \to 0} e^{iwt} \int_{-1}^{1} B(x_0) \sin \theta \, H_1^{(2)}(\omega r') e^{i\omega\alpha(x-x_0)}\,dx_0$$

$$= \lim_{\delta \to 0} e^{iwt}\left\{ \int_{-1}^{x-\varepsilon} + \int_{x-\varepsilon}^{x+\varepsilon} + \int_{x+\varepsilon}^{1} \right\} \tag{7}$$

When $|x - x_0| > \varepsilon > 0$, $H_1^{(2)}(\omega r')$ is finite, but $\sin \theta$ tends to zero as $\delta \to 0$. Hence, the first and the last integrals in the brackets of Eq. 7 vanish in the limit as long as $\delta \ll \varepsilon$. Applying the mean-value theorem to the middle integral in the brackets of Eq. 7, we obtain

$$\lim_{\delta \to 0} \phi(x, \delta, t) = \lim_{\delta \to 0} e^{iwt} B(x + \lambda\varepsilon) e^{i\omega\alpha\lambda\varepsilon} \int_{x-\varepsilon}^{x+\varepsilon} \sin \theta \, H_1^{(2)}(\omega r')\,dx_0 \tag{8}$$

where λ is some number between -1 and 1. To evaluate the last integral, notice that

$$H_1^{(2)}(z) = \frac{2i}{\pi z} - \frac{i}{\pi} z \log z + \text{(power series in } z) \tag{9}$$

and

$$\sin \theta = \frac{\beta y}{\sqrt{(x - x_0)^2 + \beta^2 y^2}} \tag{10}$$

Hence, when ε is small in comparison with 1,

$$\lim_{\delta \to 0} \phi(x, \delta, t) = e^{i\omega(t+\alpha\lambda\varepsilon)} B(x + \lambda\varepsilon) \lim_{\delta \to 0} \int_{x-\varepsilon}^{x+\varepsilon} \frac{2i}{\pi} \frac{\beta^3 a}{\omega} \frac{\delta}{[(x - x_0)^2 + \beta^2\delta^2]}\,dx_0$$

$$= \frac{2i}{\pi} \frac{\beta^3 a}{\omega} e^{i\omega(t+\alpha\lambda\varepsilon)} B(x + \lambda\varepsilon) \lim_{\delta \to 0} \frac{2}{\beta} \tan^{-1} \frac{\varepsilon}{\beta\delta}$$

$$= \frac{2i\beta^2 a}{\omega} e^{i\omega(t+\alpha\lambda\varepsilon)} B(x + \lambda\varepsilon) \tag{11}$$

Substituting this result into Eq. 6, we obtain

$$2\varepsilon L(\xi, t) = 2\rho_0 \int_{\xi-\varepsilon}^{\xi+\varepsilon} \frac{2i\beta^2 a}{\omega} e^{i\omega(t+\alpha\lambda\varepsilon)} B(x + \lambda\varepsilon)\, dx \qquad (-1 < \lambda < 1)$$

Since $B(x + \lambda\varepsilon)$ is a continuous function, we obtain, in passing to the limit $\varepsilon \to 0$

$$L(\xi, t) = \frac{4i\rho_0 \beta^2 a}{\omega} e^{i\omega t} B(\xi) \tag{12}$$

This formula relates the local strength of a continuously distributed doublet layer to the local lift force per unit length. It is seen that they are directly proportional.

The horizontal force acting on the distributed doublets with vertical axes is zero.

14.3 DOWNWASH CORRESPONDING TO A DOUBLET IN ACCELERATION POTENTIAL

Let us consider the velocity field corresponding to the acceleration potential of a vertical doublet. Without loss of generality we shall put the doublet at the origin.

For an oscillating field with a time factor $e^{i\omega t}$, the vertical component of velocity v is related to the vertical component of acceleration a_y through the equation (§ 13.2, Eq. 12)

$$v = \frac{1}{U} e^{-i\omega(x/U)} \int_{-\infty}^{x} a_y\, e^{i\omega(x/U)}\, dx \tag{1}$$

Replacing a_y by $\partial\phi/\partial y$, where ϕ is given by Eq. 18, § 14.1, we have

$$v(x, y, t) = \frac{1}{U} e^{-i\omega(x/U)} \int_{-\infty}^{x} B e^{i\omega(t+\kappa x)} \frac{\partial}{\partial y} [\sin\theta\, H_1^{(2)}(\omega r')]\, dx \tag{2}$$

where we have introduced a new notation:

$$\kappa = \alpha + \frac{1}{U} = \frac{1}{U\beta^2} \tag{3}$$

Now, according to the definitions of r' and θ,

$$\frac{\partial\theta}{\partial y} = \frac{\cos\theta}{\beta a r'}, \qquad \frac{\partial\theta}{\partial x} = -\frac{\sin\theta}{\beta^2 a r'}$$
$$\frac{\partial r'}{\partial y} = \frac{\sin\theta}{\beta a}, \qquad \frac{\partial r'}{\partial x} = \frac{\cos\theta}{\beta^2 a} \tag{4}$$

and, by putting $n = 0$ into the recurrence formula

$$\frac{n}{z} H_n(z) - H'_n(z) = H_{n+1}(z) \tag{5}$$

we obtain

$$\frac{\partial}{\partial y} H_0^{(2)}(\omega r') = \frac{dH_0^{(2)}(\omega r')}{d(\omega r')} \, \omega \, \frac{\partial r'}{\partial y} = - H_1^{(2)}(\omega r') \, \omega \, \frac{\sin \theta}{\beta a}$$

Hence,

$$v(x, y, t) = - \frac{1}{U} \frac{B\beta a}{\omega} e^{i\omega(t - x/U)} \int_{-\infty}^{x} e^{i\omega\kappa x} \frac{\partial^2}{\partial y^2} [H_0^{(2)}(\omega r')] \, dx \tag{6}$$

This integral is divergent when $y = 0$ and $x \geqslant 0$. One can replace it, however, by an integral that does not diverge. If we remember that

$$\phi = e^{i\omega\tau} H_0^{(2)}(\omega r')$$

is a solution of the fundamental equation

$$\phi_{\tau\tau} = \phi_{\xi\xi} + \phi_{\eta\eta}$$

it would become clear, by a change of appropriate variables, that

$$\frac{\partial^2}{\partial y^2} [H_0^{(2)}(\omega r')] = - \frac{\omega^2}{\beta^2 a^2} H_0^{(2)}(\omega r') - \beta^2 \frac{\partial^2}{\partial x^2} [H_0^{(2)}(\omega r')] \tag{7}$$

Substituting this into Eq. 6 and putting

$$A = \frac{1}{U} B \frac{\beta a}{\omega} e^{i\omega(t - x/U)} = \frac{1}{U} \frac{L}{4i\rho_0 \beta} e^{-i\omega(x/U)} \tag{8}$$

where B is expressed in terms of L by means of Eq. 12, § 14.2, we obtain

$$v(x, y, t) = A \int_{-\infty}^{x} e^{i\omega\kappa x} \left[\frac{\omega^2}{\beta^2 a^2} H_0^{(2)}(\omega r') + \beta^2 \frac{\partial^2}{\partial x^2} H_0^{(2)}(\omega r') \right] dx = I_1 + I_2 \tag{9}$$

Integrating the second member by parts twice, we obtain

$$I_2 = A\beta^2 \left\{ \omega \left[- H_1^{(2)}(\omega r') \frac{\cos \theta}{\beta^2 a} - i\kappa H_0^{(2)}(\omega r') \right] e^{i\omega\kappa x} \right.$$
$$\left. - \omega^2 \kappa^2 \int_{-\infty}^{x} e^{i\omega\kappa x} H_0^{(2)}(\omega r') \, dx \right\} \tag{10}$$

because $H_0^{(2)}$ and $H_1^{(2)}$ vanish at infinity.

Hence, Eq. 9 becomes

$$v(x, y, t) = A\beta^2\omega e^{i\omega\kappa x}\left\{-H_1^{(2)}(\omega r')\frac{\cos\theta}{\beta^2 a} - i\kappa H_0^{(2)}(\omega r')\right\}$$

$$+ A\left(\frac{1}{\beta^2 a^2} - \beta^2\kappa^2\right)\omega^2\int_{-\infty}^x e^{i\omega\kappa x}H_0^{(2)}(\omega r')\,dx \qquad (11)$$

which is valid for all values of y.   Now, when $y = 0$,

$$r' = \frac{1}{\beta^2 a}|x| \qquad (12)$$

$$\cos\theta = \pm 1 \quad \text{according as} \quad x \gtrless 0$$

Hence, after some simplification and introducing another notation

$$W = \frac{\omega}{\beta^2 a}x = \frac{Mk}{\beta^2}x \qquad (13)$$

where k is the reduced frequency, we obtain

$$v(x, 0, t) = -\frac{iA\omega e^{i\omega\kappa x}}{U}\left\{-iM H_1^{(2)}(|W|)\frac{W}{|W|} + H_0^{(2)}(|W|)\right\}$$

$$-\frac{A\omega^2}{U^2}\int_{-\infty}^x e^{i\omega\kappa x}H_0^{(2)}\left(\omega\frac{|x|}{\beta^2 a}\right)dx \qquad (14)$$

But it is shown in Appendix 4, that

$$\int_{-\infty}^0 e^{i\omega\kappa x}H_0^{(2)}\left(\omega\frac{|x|}{\beta^2 a}\right)dx = \frac{2U\beta}{\pi\omega}\log\frac{1 + \sqrt{1 - M^2}}{M} \qquad (15)$$

Hence, on substituting A from Eq. 8, and writing

$$v(x, 0, t) = \frac{\omega L}{\rho_0 U^2}K(M, x) \qquad (16)$$

we obtain

$$K(M, x) = \frac{1}{4\beta}e^{iMW}\left\{iM\frac{W}{|W|}H_1^{(2)}(|W|) - H_0^{(2)}(|W|)\right\}$$

$$+ \frac{i\beta}{4}e^{-i\omega(x/U)}\left\{\frac{2}{\pi\beta}\log\frac{1 + \beta}{M} + \int_0^{W/M}e^{iu}H_0^{(2)}(M|u|)\,du\right\} \qquad (17)$$

This is the velocity downwash field corresponding to a *doublet* at the origin.   If the doublet is situated at $(\xi, 0)$, we have to replace x in the above formulas by $x - \xi$.

14.4 POSSIO'S INTEGRAL EQUATION

The results of the preceding sections can be combined to give an integral equation expressing the boundary-value problem of an oscillating airfoil.   The total vertical component of velocity due to a lift distribution $L(\xi, t)$ is, according to Eq. 16 of § 14.3, given by the Cauchy integral:

$$v(x, 0, t) = \frac{\omega}{\rho_0 U^2} \int_{-1}^{1} L(\xi, t)\, K(M, x - \xi)\, d\xi \tag{1}$$

where

$$v(x, 0, t) = \bar{v}(x)e^{i\omega t}, \qquad L(\xi, 0, t) = L(\xi)e^{i\omega t}$$

The function $v(x, 0, t)$ is given by the boundary condition (Eq. 7 of § 13.2) for a given airfoil in a given oscillation mode.   The problem is then to determine $L(\xi, t)$ from Eq. 1 under the side condition that $L \to 0$ at the trailing edge $\xi = 1$ (the Kutta-Joukowski condition).*   Equation 1 is called the *Possio's integral equation*.†   The *kernel* $K(M, x - \xi)$ depends on the Mach number M, the reduced frequency $k = l\omega/U$, and the distance $x - \xi$.   It has a singular point at $x = \xi$.

Possio's equation is a special case of a general type of singular integral equation studied earlier by Carleman.‡   It is a Cauchy-type integral equation of the first kind.   With the imposition of the Kutta condition at the trailing edge, a unique solution is known to exist.   Several methods of reducing this equation into an ordinary (nonsingular) integral equation are known.   But so far no solution of Possio's equation has been obtained in closed form.   Numerical or graphical methods must be employed.

---

* The pressure perturbation p is continuous outside the airfoil, is antisymmetric in y, and hence vanishes behind the trailing edge when $y = 0$.   Therefore, the lift, which is the difference of pressure across the airfoil, vanishes at the trailing edge. It does not vanish at the leading edge because the leading edge is a singular point for L.

† A shorter and more general derivation is given by Küssner[15.12] on the basis of Lorentz transformation of the wave equation.   Küssner's integral equation holds for three-dimensional flow in both supersonic and subsonic cases.   Specialization into Possio's integral equation in the subsonic case, into Birnbaum's equation in the incompressible case, and into Prandtl's equation in the steady-state case are demonstrated. But the form of Küssner's integral equation is simple only if it is stated in terms of divergent integrals.   In practice, an involved limiting process is required to transform such divergent integrals into those for which Cauchy's principal value can be formed.

‡ See M. Muskhlishvili: *Singular Integral Equations*, in Russian, translated by Radok and Woolnough, Ministry of Supply, Australia.   Published by P. Noordhoff, Groningen, Holland (1953).

For this purpose the singularities of $K(M, z)$ may be isolated in the following manner according to Schwarz:[14.16]

$$K(M, z) = \frac{F(M)}{z} + i\,G(M) \log |z| + K_1(M, z) \qquad (2)$$

where

$$z = kx \qquad (k = \text{reduced frequency})$$

$$F(M) = -\frac{1}{2\pi}\sqrt{1 - M^2}, \qquad G(M) = \frac{1}{2\pi\sqrt{1 - M^2}} \qquad (3)$$

In particular, at zero Mach number,

$$K(0, z) = -\frac{1}{2\pi z} + \frac{i}{2\pi}\,e^{-iz}\left[Ci(z) + i\left\{ Si(z) + \frac{\pi}{2} \right\} \right] \qquad (4)$$

where $Ci(z)$ and $Si(z)$ are the cosine and sine integrals, respectively.* $K_1(M, z)$, defined by Eq. 2, is a continuous function of z. Numerical tables of the functions K and K_1 are published by Possio, Dietze, Schwarz, and Schade.†

A different separation is given by Dietze:[14.2]

$$K(M, z) = \Delta K_1(M, z) + \Delta K_2(M, z) + K(0, z) \qquad (5)$$

where

$$\Delta K_1(M, z) = \frac{k_{10}}{z} + k_{11} + k_{12} \log |z| + z(k_{13} + k_{14} \log |z|) \qquad (6)$$

$$\Delta K_2(M, z) \doteq \sum_{n=2}^{9} k_{2n} z^n \qquad (7)$$

The real and imaginary parts of the coefficients k_{2n} are tabulated in Dietze's paper.[14.2]

14.5 APPROXIMATE SOLUTIONS OF POSSIO'S EQUATION

It is a general feature of Carleman's integral equation that the solution admits a singularity. For an airfoil, such a singularity is located at the leading edge where the acceleration potential should tend to infinity like $\sqrt{(1 - x)/(1 + x)}$ when $x \to -1$. This particular form of the singularity can be derived from Carleman's general theory, or from analogy with the incompressible-flow case. The intensity of the singularity can

* Jahnke and Emde, *Tables of Functions*, p. 3. Dover Publications.
† See § 14.7 and Bibliography. Summaries of numerical tables are given in Refs. 14.1 and 14.8.

be determined by the Kutta-Joukowski condition that the lift be zero at the trailing edge.

When we write $L(x, t) = L(x)e^{i\omega t}$, the function $L(x)$ consists of a term $const \cdot \sqrt{(1 - x)/(1 + x)}$ and a nonsingular part. Possio[14.12] and Frazer[14.6] write the nonsingular part as a series involving a number of constant coefficients and determine these constants by the collocation method. Schade[14.15] writes the nonsingular part of $L(x)$ as a series in Legendre polynomials. The nonsingular part of the kernel $K_1(M, z)$ as well as the upwash distribution $\bar{v}(x, 0)$ are also expressed as series in Legendre polynomials. The undetermined coefficients are then obtained on the basis of the orthogonality of Legendre polynomials in the interval $-1 \leqslant x \leqslant 1$.

The best-known method is probably Dietz's iteration procedure,[14.2] which will be outlined below. Let us introduce the notation of the *composition product*

$$\frac{\omega}{\rho_0 U^2} \int_{-1}^{1} K(M, x - \xi) L(\xi) \, d\xi = K \cdot L \tag{1}$$

$$\frac{\omega}{\rho_0 U^2} \int_{-1}^{1} K(0, x - \xi) L(\xi) \, d\xi = K_0 \cdot L \tag{2}$$

Then Possio's equation can be written as

$$K \cdot L = \bar{v} \tag{3}$$

where $\bar{v}(x)$ is the given vertical velocity component on the airfoil. If the fluid is incompressible, the corresponding lift distribution is given by L_0, and Eq. 3 is reduced to

$$K_0 \cdot L_0 = \bar{v} \tag{4}$$

The solution of this equation is known explicitly. (See § 13.5.) Let us write $L = L_0 + \Delta L_0$ which defines ΔL_0; we have

$$K_0 \cdot L_0 = \bar{v} = K \cdot L = K \cdot L_0 + K \cdot \Delta L_0 \tag{5}$$

or

$$K \cdot \Delta L_0 = \bar{v}_1 \tag{6}$$

if

$$\bar{v}_1 \equiv (K_0 - K) \cdot L_0 \tag{7}$$

The function $\bar{v}_1$ being known, Eq. 6 is completely analogous to Eq. 3. Hence, the same procedure can be applied. Let

$$\Delta L_0 = L_1 + \Delta L_1 \tag{8}$$

where L_1 is the solution of the incompressible-fluid problem

$$\bar{v}_1 = K_0 \cdot L_1$$

and ΔL_1 is governed by the equation

$$K \cdot \Delta L_1 = \bar{v}_2 \equiv (K_0 - K) \cdot L_1$$

Continuing this process, we obtain the approximate solution

$$L = L_0 + L_1 + L_2 + \cdots + L_n + \Delta L_n \tag{9}$$

where L_m is the solution of the incompressible-fluid problem

$$K_0 \cdot L_m = \bar{v}_m \qquad (m \leqslant n) \tag{10}$$

while

$$\bar{v}_m \equiv (K_0 - K) \cdot L_{m-1} \tag{11}$$

$\bar{v}_0 = \bar{v}$ being specified, and

$$K \cdot \Delta L_n = (K_0 - K)L_n$$

The convergence of this procedure has not been proved, but is indicated by Dietz's numerical examples. The rapidity of convergence is found to deteriorate for increasing M and k.

Recently, Fettis[14.4] introduced a method that avoids the iterative procedure and gives a relatively simple solution on the basis of an approximate kernel, in which the nonsingular part of Possio's kernel is replaced by a polynomial.

14.6 OSCILLATING AIRFOILS IN A TWO-DIMENSIONAL SUPERSONIC FLOW

The problem of an oscillating airfoil in a two-dimensional supersonic flow can be solved in several different ways. In this section the Laplace-transformation method, in a form due to Stewartson,[14.40] will be used. The theory will be limited to the linearized case, so that the airfoil must be infinitesimally thin, and executing harmonic oscillations of small amplitudes. The principle of superposition holds. It is sufficient to consider airfoils of zero thickness and zero camber, with stationary mean position. The fluid moves over it with an undisturbed velocity U at infinity. The x axis is taken in the direction of the free stream, and the origin of coordinates is taken at the leading edge of the airfoil. In the first-order theory the wing may be assumed to lie in the plane $y = 0$. The coordinate z, in the spanwise direction, does not appear in the problem. The flow is assumed to be irrotational, with a velocity potential $Ux + \Phi$, and deviations in velocity components, pressure, and density are so small that squares and products of these deviations may be neglected in comparison with the first-order terms.

The equation of the velocity potential, referred to a frame of reference at rest relative to the fluid at infinity, is (§ 12.5):

$$\frac{\partial^2 \Phi}{\partial x^2} + \frac{\partial^2 \Phi}{\partial y^2} - \frac{1}{a^2} \frac{\partial^2 \Phi}{\partial t^2} = 0 \tag{1}$$

where a is the velocity of sound, which, in our order of approximation, is a constant. Transforming to axes moving with speed U in the negative x direction, so that x is replaced by $x + Ut$, Eq. 1 becomes (Eq. 2, p. 418):

$$a^2 \left(\frac{\partial^2 \Phi}{\partial x^2} + \frac{\partial^2 \Phi}{\partial y^2} \right) = \frac{\partial^2 \Phi}{\partial t^2} + 2U \frac{\partial^2 \Phi}{\partial x\, \partial t} + U^2 \frac{\partial^2 \Phi}{\partial x^2} \tag{2}$$

The increment of pressure at any point due to the disturbance is given by the Eulerian equations of motion

$$\frac{\partial u_i}{\partial t} + u_j \frac{\partial u_i}{\partial x_j} = -\frac{1}{\rho} \frac{\partial}{\partial x_i} p$$

Substituting into this equation

$$u_1 = U + \frac{\partial \Phi}{\partial x}, \qquad u_2 = \frac{\partial \Phi}{\partial y}$$

and linearizing the result, one obtains

$$p = -\rho_0 \left(\frac{\partial \Phi}{\partial t} + U \frac{\partial \Phi}{\partial x} \right) \tag{3}$$

ρ_0 being the density of the fluid at infinity. Hence a determination of Φ on the airfoil is sufficient to determine the pressure acting on it.

When the wing executes a simple-harmonic motion, the time t enters as an exponential factor $e^{i\omega t}$. Let

$$\Phi = \Psi(x, y)e^{i\omega t} \tag{4}$$

Then Eq. 2 becomes

$$a^2 \left(\frac{\partial^2 \Psi}{\partial x^2} + \frac{\partial^2 \Psi}{\partial y^2} \right) = U^2 \frac{\partial^2 \Psi}{\partial x^2} + 2i\omega U \frac{\partial \Psi}{\partial x} + (i\omega)^2 \Psi \tag{5}$$

or

$$\frac{\partial^2 \Psi}{\partial y^2} = \bar{\beta}^2 \frac{\partial^2 \Psi}{\partial x^2} + 2i \frac{\omega M}{a} \frac{\partial \Psi}{\partial x} + \left(\frac{i\omega}{a} \right)^2 \Psi \tag{6}$$

where M is the Mach number U/a and

$$\bar{\beta}^2 = M^2 - 1 \tag{7}$$

The boundary conditions must be formulated according to the following considerations:

1. The disturbances created at the leading edge propagate along a wedge which is called the *Mach wedge*. In front of the Mach wedge the disturbances cannot be felt, and the flow is uniform relative to the wing. Hence, one may put $\Phi = 0$ for $x \leqslant 0$.

2. Inside the Mach wedge, the velocity of flow normal to the airfoil must conform to the actual motion of the airfoil. If the equation of the airfoil surface is specified by

$$y = Y(x, t) \tag{8}$$

then the normal velocity of the flow on the airfoil must satisfy the following equation (§ 13.2):

$$\frac{\partial \Phi}{\partial y} = \frac{\partial Y}{\partial t} + U \frac{\partial Y}{\partial x} \tag{9}$$

In a simple-harmonic motion,

$$Y(x, t) = Z(x)e^{i\omega t} \tag{10}$$

we have, on the airfoil (part of the plane $y = 0$),

$$\frac{\partial \Psi}{\partial y} = i\omega Z + U \frac{\partial Z}{\partial x} \tag{11}$$

On the rest of the plane $y = 0$, the pressure must be continuous.

Equation 6 with the boundary condition 11 may be solved by Laplace transformation. Define

$$\bar{\psi} = \mathscr{L}\{\Psi\} = \int_0^\infty e^{-sx} \Psi(x) \, dx \tag{12}$$

Then $\bar{\psi}$ satisfies

$$\frac{\partial^2 \bar{\psi}}{\partial y^2} = \left[\bar{\beta}^2 s^2 + 2Ms \frac{i\omega}{a} + \left(\frac{i\omega}{a} \right)^2 \right] \bar{\psi} \tag{13}$$

since Ψ and $\partial \Psi / \partial x$ vanish when $x = 0$. The boundary condition becomes

$$\frac{\partial \bar{\psi}}{\partial y} = g(s) \tag{14}$$

where $g(s)$ is the Laplace transform of the right-hand side of Eq. 11. Thus we require a solution of Eq. 13 so that $\bar{\psi} \to 0$ as $|y| \to \infty$ and such that Eq. 14 is satisfied. Let

$$\mu^2 = \bar{\beta}^2 s^2 + 2Ms \frac{i\omega}{a} + \left(\frac{i\omega}{a} \right)^2 \tag{15}$$

The general solution of Eq. 13, if we take μ to be the branch on the right half plane, i.e., with $\mathscr{R}l\mu > 0$, is

$$\bar{\psi} = Ae^{-\mu y} + Be^{\mu y}$$

The constants A and B are determined by the boundary conditions. It is necessary to distinguish the solution on the upper and lower half-space. For the upper half-space ($y \geqslant 0$), the condition $\bar{\psi} \to 0$ as $y \to \infty$ requires $B = 0$; and the condition 14 requires $A = -g(s)/\mu$. Similarly the solution for $y < 0$ can be determined. Hence,

$$\bar{\psi} = -\frac{g(s)}{\mu} e^{-\mu y} \quad \text{for} \quad y > 0 \tag{16}$$

$$\bar{\psi} = \frac{g(s)}{\mu} e^{\mu y} \qquad \text{for} \quad y < 0 \tag{17}$$

and

$$(\bar{\psi})_{y=0} = -\frac{g(s)}{\mu} \operatorname{sgn} y \tag{18}$$

where the symbol sgn y indicates a sign to be taken as positive on the positive side ($y > 0$) of $y = 0$ plane, and as negative on the negative side.

It is now necessary to find the inverse transform of Eq. 18. The inverse transform of $g(s)$ is $(\partial \Psi / \partial y)_{y=0}$ and is given by Eq. 11. From Table 10.1, we find

$$\mathscr{L}\{J_0(ax)\} = \frac{1}{\sqrt{s^2 + a^2}} \tag{19}$$

Now

$$\frac{1}{\mu} = \frac{1}{\sqrt{\bar{\beta}^2 s^2 + 2Ms \dfrac{i\omega}{a} + \left(\dfrac{i\omega}{a}\right)^2}} = \frac{1}{\bar{\beta}\sqrt{\left(s + \dfrac{i\omega}{a}\dfrac{M}{\bar{\beta}^2}\right)^2 + \left(\dfrac{\omega}{a\bar{\beta}^2}\right)^2}}$$

Hence, according to the attenuation rule,

$$\mathscr{L}^{-1}\left\{\frac{1}{\mu}\right\} = \frac{1}{\bar{\beta}} \exp\left[-\frac{i\omega}{a}\frac{M}{\bar{\beta}^2} x\right] J_0\left(\frac{\omega}{a\bar{\beta}^2} x\right) \tag{20}$$

An application of the convolution rule leads to the final formula

$$(\Psi)_{y=0} = -\frac{\operatorname{sgn} y}{\bar{\beta}} \int_0^x \left(\frac{\partial \Psi}{\partial y}\right)_{\substack{x=\xi \\ y=0}} \exp\left[-\frac{iM\omega(x-\xi)}{a\bar{\beta}^2}\right] J_0\left(\frac{\omega(x-\xi)}{a\bar{\beta}^2}\right) d\xi \tag{21}$$

The pressure change on the airfoil is, according to Eq. 3,

$$p(x) = -\rho_0 e^{i\omega t}\left[i\omega\Psi(x, 0) + U\frac{\partial \Psi}{\partial x}(x, 0)\right] \tag{22}$$

Since the pressure changes on the upper and lower surfaces of the airfoil are equal and opposite, the lift distribution is simply

$$l(x) = 2\rho_0 e^{i\omega t} \left[i\omega \Psi(x, +0) + U \frac{\partial \Psi}{\partial x}(x, +0) \right] \tag{23}$$

Example. Consider the vertical-translational oscillation. Let the equation of the airfoil be

$$Y(x, t) = -h = -h_0 e^{i\omega t} \tag{24}$$

where h_0 is a constant. Then, on the airfoil

$$\left(\frac{\partial \Psi}{\partial y} \right)_{y=0} = -i\omega h_0 \tag{25}$$

Hence,

$$\Psi(x, 0) = i\omega h_0 \frac{\operatorname{sgn} y}{\bar\beta} \int_0^x \exp\left[-\frac{iM\omega(x - \xi)}{a\bar\beta^2} \right] J_0 \left[\frac{\omega(x - \xi)}{a\bar\beta^2} \right] d\xi \tag{26}$$

Introducing the nondimensional parameters

$$x' = \frac{x}{2b} \quad u = \frac{x - \xi}{2b} \quad k = \frac{\omega b}{U}$$

$$\Omega = \frac{2kM^2}{M^2 - 1} \tag{27}$$

where $2b$ is the chord length of the airfoil, one obtains

$$\Psi(x', 0) = \frac{i\omega h_0}{\bar\beta} 2b \operatorname{sgn} y \int_0^{x'} e^{-i\Omega u} J_0 \left(\frac{\Omega}{M} u \right) du \tag{28}$$

Since

$$\frac{\partial \Psi}{\partial x} = \frac{1}{2b} \frac{\partial \Psi}{\partial x'} = \frac{i\omega h_0}{\bar\beta} e^{-i\Omega x'} J_0 \left(\frac{\Omega}{M} x' \right) \operatorname{sgn} y$$

the lift distribution over the airfoil is

$$l(x') = 2\rho_0 e^{i\omega t} \left[-\frac{\omega^2 h_0 2b}{\bar\beta} \int_0^{x'} e^{-i\Omega u} J_0 \left(\frac{\Omega}{M} u \right) du \right.$$

$$\left. + \frac{i\omega h_0 U}{\bar\beta} e^{-i\Omega x'} J_0 \left(\frac{\Omega}{M} x' \right) \right] \tag{29}$$

Writing

$$\dot h = i\omega h_0 e^{i\omega t}, \qquad \ddot h = (i\omega)^2 h_0 e^{i\omega t} \tag{30}$$

we have

$$l(x') = \frac{2\rho_0}{\bar\beta} \left[2b\ddot h \int_0^{x'} e^{-i\Omega u} J_0 \left(\frac{\Omega}{M} u \right) du + U\dot h e^{-i\Omega x'} J_0 \left(\frac{\Omega}{M} x' \right) \right] \tag{31}$$

Note that the integral is a function of the parameters M and Ω, or, alternatively, M and k.

The total lift on the airfoils is

$$L = 2b \int_0^1 l(x')\,dx' \tag{32}$$

The moment (positive nose-up) on the airfoil about the leading edge is

$$M_{\text{L.E.}} = -4b^2 \int_0^1 x'\,l(x')\,dx' \tag{33}$$

The integration of Eqs. 32 and 33 has been discussed by von Borbély,[14.29] Schwarz,[14.39] Garrick and Rubinow.[14.32]

14.7 TABULATION OF RESULTS—COMPRESSIBLE FLOW

In the case of supersonic flow, many authors give numerical tables of aerodynamic coefficients for oscillating airfoils. For flutter-calculation purpose the following reference, prepared by E. C. Kennedy, is the most comprehensive:

1. *Handbook of Supersonic Aerodynamics.*[14.44] M: 1.1 (0.1) 2.0 (0.2) 4.0 (0.5) 5.0 (1.0) 12. Ω: 0.01 (0.01) 0.04 (0.02) 0.10 (0.05) 0.40 (0.10) 1.00 (0.20) 3.0 (0.50) 5.0, 7.5, 10, 15, 20.   (8 fig.)

The coefficients C_{Lh}, $C_{L\alpha}$, C_{Mh}, $C_{M\alpha}$ listed in this *Handbook* are, respectively, the L_h, L_α, M_h, M_α defined in Chapter 6. The independent entry in the *Handbook* is the frequency parameter Ω, which is related to the reduced frequency k and Mach number M by the relation

$$\Omega = \frac{2M^2}{M^2 - 1}k$$

In the following reference, compiled by Y. L. Luke on the basis of tables published by Garrick and Rubinow and Jordan, the independent entries are the reduced velocity $U/b\omega = 1/k$, and the coefficients L_h, L_α, M_h, M_α:

2. *Tables of Coefficients for Compressible Flutter Calculations.*[14.8] M: $\frac{10}{9}$, $\frac{5}{4}$, $\frac{10}{7}$, $\frac{5}{3}$, 2, $\frac{5}{2}$, $\frac{10}{3}$, 5. $1/k$: approx. 0.1 (irreg.) 100.   (5 dec.)

The scope of these tables is indicated above in the ranges of M and Ω.* Note that in the supersonic case it suffices to tabulate the four fundamental aerodynamic coefficients named above. The lift and moment coefficients

---

* The notation is as follows: If a table is given for a range of the argument from 0.10 to 0.50 at intervals of 0.05; this is indicated by writing 0.10 (0.05) 0.50. The notation (5 fig.), (6 dec.), etc., implies that most of the figures in the table in question are given to the apparent accuracy of 5 figures or 6 decimals, respectively.

Table 14.1. Tabulated Subsonic Coefficients

Reference	Luke, Ref. 14.8	Timman, et al., Ref. 14.19–14.20	Fettis, Ref. 14.3
L_h	(L_h)	$\dfrac{1}{k^2}(k_a)$	(L_h)
L_α	(L_α)	$\dfrac{1}{k^2}(k_b)$	(L_α)
M_h	(M_h)	$\dfrac{1}{k^2}(m_a)+\dfrac{1}{2k^2}(k_a)$	(M_h)
M_α	(M_α)	$\dfrac{1}{k^2}(m_b)+\dfrac{1}{2k^2}(k_b)$	(M_α)
			$(L_\beta), (M_\beta), (L_z), (M_z)$
Flap coefficients	$(L_\beta), (M_\beta)$ $(T_h), (T_\alpha), (T_\beta)$ $(P_h), (P_\alpha), (P_\beta)$	$L_\beta = \dfrac{1}{k^2}(k_c)$ $M_\beta = \dfrac{1}{k^2}(m_c)+\dfrac{1}{2k^2}(k_c)$ $T_h = \dfrac{1}{k^2}(n_a)$ $T_\alpha = \dfrac{1}{k^3}(n_b)$ $T_\beta = \dfrac{1}{k^3}(n_c)$	$(T_h), (T_\alpha), (T_\beta), (T_z)$ $(P_h), (P_\alpha), (P_\beta), (P_z)$
Scope of tables*	For $M = 0.7$, all coefficients, at $e_\beta = 0.16,\ 0.34,\ 0.52,\ 0.70,\ 1.0$. $k = 0\ (0.02)\ 0.1\ (0.1)\ 0.7$. For $M = 0.5,\ 0.6$, coefficients $L_h,\ M_h,\ L_\alpha,\ M_\alpha,\ M_\beta$, $L_\beta,\ T_h,\ T_\alpha,\ T_\beta$ at $e_\beta = 0.16,\ 0.70$. Coefficients $P_h,\ P_\alpha,\ P_\beta$ at $e_\beta = 0.70$. Same k range.	$M = 0.35,\ 0.5\ (0.1)\ 0.8$ $k =$ approximately 0.1 (irreg.) 3 for $L_h,\ L_\alpha,\ M_h,\ M_\alpha$. But $k > 1.0$ are omitted for flap coefficients. (5 fig. for $L_h,\ L_\alpha,\ M_h,\ M_\alpha$) (5 dec. for flap coefficients). $e_\beta = 0.4,\ 0.6,\ 0.8$.	$M = 0.7$ $k = 0.04\ (0.04)\ 0.52$ $e_\beta = -0.2\ (0.1)\ 0.9$
Remarks		Corrected and extended tables in Ref. 14.48.	

* $e_\beta =$ location of flap leading edge measured from the wing mid-chord point, as a fraction of the wing semichord ($+$ aft, see Fig. 6.13)

Table 14.2. Comparison of Notations

References	Possio, Ref. 14.12	Possio, Ref. 14.38	von Borbély, Ref. 14.29	Küssner, Ref. 15.12	Schwarz, Ref. 14.16	Dietze, Ref. 14.2	Temple Jahn, Ref. 14.42
Chord	l	1	t	$2l$	2	l	c
Free-stream velocity	V_0	V_1	U_0	v	v	v	V
Density, undisturbed fluid	ρ	ρ_1	ρ_0	ρ	ρ	ρ_∞	ρ_0
Circular frequency	Ω	Ω	σ	ν	ν	ω	$2\pi f$
Reduced frequency	ω	ω	—	$\bar{\omega}$	ω_r	ω_r	$\dfrac{\lambda}{2}$
Frequency parameter $\left(\dfrac{2M^2 k}{M^2-1}\right)$	$\dfrac{V_f + iV_g}{a\omega}$	—	ω	—	—	—	y
Kernel of Possio's equation	—	—	—	—	$\mathscr{K}$	$\mathscr{K}$	—
Mach number	λ	λ	Mach angle ϑ	β	λ	M	M
Downward displacement at reference point	$-\eta$	a	$-Y$	—	—	δ	z'_0
Pitching angle (+ nose up)	α	b	E	—	—	—	α'
Lift on wing (+ upward)	P	P	P	—	—	—	L'
Pitching moment (+ nose up)	$-M$	$-M$	$-M$	—	—	—	M'

Table 14.2 (*continued*)

References	Garrick Rubinow, Ref. 14.32	Karp, et al., Ref. 14.1	Luke, Ref. 14.8	Turner, Ref. 14.21	Timman, Ref. 14.17	Fettis, Ref. 14.4	Supersonic Handbook Ref. 14.44	This Book
Chord	$2b$	$2l$	$2b$	l	$2l$	$2b$	$2b$	$2b$
Free-stream velocity	v	U	v	V	v	v	V	U
Density, undisturbed fluid	ρ	ρ_0	ρ	ρ	ρ_0	ρ	ρ	ρ
Circular frequency	ω	ω	ω	ω	ν	ν	ω	ω
Reduced frequency	k	ω_r	k	ω_r	ω	ω	k	k
Frequency parameter $\left(\dfrac{2M^2k}{M^2-1}\right)$	$\bar{\omega}$	λ	—	—	—	μ	Ω	Ω
Kernel of Possio's equation	—	K	K	K	—	K	—	K
Mach number	M	M	M	M	β	λ	M	M
Downward displacement at reference point	h_0	$-Y$	h	q_s	A	h	h'	h
Pitching angle ($+$ nose up)	α_0	E	α	q_D	B	α	α	α
Lift on wing ($+$ upward)	$-P$	L_0	$-L$	ΔP_s	$-K$	$-L$	$-L$	L
Pitching moment ($+$ nose up)	M_α	M_0	M	ΔM_D	M	M	M	M_{x_0}*

* x_0 is the point about which the moment is taken. Moment about the mid-chord is written as $M_{l/2}$, etc.

439

involving the motion of a flap and a tab can be expressed in terms of these four fundamental coefficients. Explicit formulas expressing these relations involving a flap are given in Ref. 14.44. A complete list of formulas including both the flap and the tab can be found in Ref. 14.1.

For subsonic flow, the published data are meager. The principal sources are the papers by Possio, Frazer, Frazer and Skan, Dietze, Schade, Schwarz, Turner and Rabinowitz, Timman, Van de Vooren and Greidanus, and Fettis (see bibliography). The numerical results of the first eight authors named above have been compiled and converted into the L_h, M_h, $\cdots$ coefficients by Luke in Ref. 14.8. A summary of the published tables is given in Table 14.1. In Table 14.2, the notations used in some of the most important references, for both the supersonic and the subsonic cases, are listed.

In Table 14.1, the symbols L_h, M_h, etc., are defined in § 6.10. These coefficients are referred to the $^1/_4$-chord axis for both the rotation α and the moment M_α. In Timman, van de Vooren and Greidanus's papers, the rotation and moment are referred to the mid-chord axis; hence, a transformation is needed when comparison of the data is to be made. This is indicated in Table 14.1. Under each column the factors in parentheses are the tabulated quantity expressed in the author's notation. The adjacent coefficients are the factors necessary to convert to the corresponding L_h, L_α, M_h, or M_α, which is listed at the left on the same horizontal line. Thus, under Timman and opposite L_h, we find the entry $\dfrac{1}{k^2}(k_a)$.

This indicates that (k_a) is given by Timman and that $L_h = (k_a)/k^2$. With the exception of the symbols for the quantities actually tabulated in the references, the notation of this book is used throughout. The notations of the original authors may be found in Table 14.2.

BIBLIOGRAPHY

Oscillating wing, subsonic flow. References to wartime German contributions by Borkemann, Eichler, etc., can be found in Ref. 14.1 or 7.3.

14.1 Biot, M. A., S. N. Karp, S. S. Shu, and H. Weil: Aerodynamics of the Oscillating Airfoil in Compressible Flow. A review prepared by Brown Univ. Applied Math. Dept. for Air Material Command. I, Theory, *U.S. Air Force Tech. Rept. F-TR* **1167-ND**. II, Graphical and Numerical Data, *U.S. Air Force Tech. Rept. F-TR* **1195-ND** (GDAM A-9-M III/II).

14.2 Dietze, F.: Die Luftkrafte des harmonisch schwingenden Flügels im kompressibaren Medium bei Unterschallgeschwindigkeit (Ebene Probleme). I, Method of Computation, *ZWB ForschBer.* **1733** (1943), *Air Force Translation* **F-TS 506-Re.** (1946). II, Tables and Curves, *ZWB ForschBer.* **1733/2** (1944). *Air Force Translation* **F-TS-948-Re** (1947).

14.3 Fettis, H. E.: Tables of Lift and Moment Coefficients for an Oscillating Wing-Aileron Combination in Two-Dimensional Subsonic Flow. *U.S. Air Force Tech. Rept.* **6688** (1951). *First Supplement* (1953). *Supplement 1* (1954) contains corrections and data from Ref. 14.47. Wright Air Development Center.

14.4 Fettis, H. E.: An Approximate Method for the Calculation of Non-Stationary Air Forces at Subsonic Speeds. *Proc. 1st U.S. Natl. Congr. Applied Mech.*, 715–722 (June 11–16, 1951). See also *WADC Tech. Rept.*, 52–56 (1952). Wright Air Development Center.

14.5 Fettis, H. E. Reciprocal Relations in the Theory of Unsteady Flow over Thin Airfoil Sections. *Proc. 2d Midwestern Conf. Fluid Mech.*, 145–154 (Mar. 1952). Ohio State Univ., Columbus, Ohio.

14.6 Frazer, R. A., and S. W. Skan: Possio's Derivative Theory for an Infinite Aerofoil Moving at Subsonic Speeds. *Aeronaut. Research Council R. & M.* **2553** (1942).

14.7 Haskind, M. D.: Oscillations of a Wing in a Subsonic Gas Flow. *Prikl. Mat. i Mekh. Moskow* **XI**, 1, 129–146 (1947). In Russian. *Air Material Command and Brown Univ. Translation* **A9-T-22**.

14.8 Luke, Y. L.: Tables of Coefficients for Compressible Flutter Calculations. *Air Force Tech. Rept.* **6200** (1950). Wright Air Development Center.

14.9 Miles, J. W.: Quasi-stationary Airfoil Theory in Compressible Flow. *J. Aeronaut. Sci.* **16**, Reader's Forum, 509 (1949).

14.10 Miles, J. W.: On Virtual Mass and Transient Motion in Subsonic Compressible Flow. *Quart. J. Mech. Applied Math.* **4**, 388–400 (1951).

14.11 Miles, J. W.: On the Kirchhoff Solution for an Oscillating Wing in Subsonic Compressible Flow. *J. Aeronaut. Sci.* **19**, Reader's Forum, 785 (1952).

14.12 Possio, C.: L'azione aerodinamica sul profilo oscillante in un fluido compressible a velocita iposonora. *Aerotecnica*, **XVIII**, fasc. 4, 441–458 (Apr. 1938). *Brit. Air Ministry Translation* **830**.

14.13 Radok, J. R. M.: An Approximate Theory of the Oscillating Wing in a Compressible Subsonic Flow for Low Frequencies. *Natl. Luchtvaartlab. Amsterdam Rept.* F 97 (1951).

14.14 Reissner, E.: On the Application of Mathieu Functions in the Theory of Subsonic Compressible Flow Past Oscillating Airfoils. *NACA Tech. Note* **2363** (1951). Tables by Reissner's method, see Ref. 14.47.

14.15 Schade, Th.: Numerische Lösung des Possioschen Integralgleichung der schwingenden Tragfläche in ebener Unterschallströmung. *ZWB, UM* **3209**, **3210**, and **3211** (1944). Curtiss-Wright Corp. *Translation* N-CGD-621, 622, and 623 (1946). *Brit. Translation* **MAP-VG-197**.

14.16 Schwarz, L.: Zahlentafeln zur Luftkraftberechnung der schwingenden Tragfläche im kompressibler ebener Unterschallströmung. *ForschBer.* **1838** (1943).

14.17 Timman, R., and A. I. van de Vooren: Theory of the Oscillating Wing with Aerodynamically Balanced Control Surface in a Two-Dimensional, Subsonic, Compressible Flow. *Natl. Luchtvaartlab. Amsterdam Rept.* F 54 (1949).

14.18 Timman, R.: Approximate Theory of the Oscillating Wing in Compressible Subsonic Flow for High Frequencies. *Natl. Luchtvaartlab. Amsterdam Rept.* F 99 (1951).

14.19 Timman, R., A. I. van de Vooren, and J. H. Greidanus: Aerodynamic Coefficients of an Oscillating Airfoil in Two-Dimensional Subsonic Flow. *J. Aeronaut. Sci.* **18**, 797–802 (1951). See comments by H. E. Fettis, *J. Aeronaut. Sci.* **19**, 353 (1952). See extended and corrected tables, Ref. 14.48.

442   OSCILLATING AIRFOILS IN COMPRESSIBLE FLOW

14.20   Timman, R.: Linearized Theory of the Oscillating Airfoil in Compressible
        Subsonic Flow. *J. Aeronaut. Sci.* **21**, 230–236 (1954).
14.21   Turner, M. J., and S. Rabinowitz: Aerodynamic Coefficients for an Oscillating
        Airfoil with Hinged Flap, with Tables for a Mach Number of 0.7. *NACA
        Tech. Note* **2213** (1950).
14.22   van Spiegel, E.: An Approximate Theory of the Oscillating Wing with Control
        Surface in a Compressible Subsonic Flow for Low Frequencies. *Natl.
        Luchtvaartlab. Amsterdam Rept.* **F 114** (1952).

Oscillating wing, transonic flow:

14.23   Biot, M. A.: Transonic Drag of an Oscillating Body. *Quart. Applied Math.* **7**,
        101–105 (1949).
14.24   Heaslet, M. A., H. Lomax, and J. R. Spreiter: Linearized Compressible-Flow
        Theory for Sonic Flight Speeds. *NACA Rept.* **956** (1950). Supersedes
        NACA Tech. Note **1824**.
14.25   Lin, C. C., E. Reissner, and H. S. Tsien: A Two-Dimensional Nonsteady
        Motion of a Slender Body in a Compressible Fluid. *J. Math. & Phys.* **27**,
        220–231 (1948).
14.26   Nelson, H. C., and J. H. Berman: Calculations on the Forces and Moments
        for an Oscillating Wing-Aileron Combination in Two-Dimensional Potential
        Flow at Sonic Speed. *NACA Rept.* **1128** (1953). Supersedes *NACA Tech.
        Note* **2590**.
14.27   Rott, N.: Oscillating Airfoils at Mach Number 1. *J. Aeronaut. Sci.* **16**,
        Reader's Forum, 380 (1949). See also Ref. 5.44.
14.28   Rott, N.: Transient Phenomena at Sonic Speed. *J. Aeronaut. Sci.* **16**, Reader's
        Forum, 439 (1949).

Oscillating wing, supersonic flow: (see also Refs. 14.1, 14.8, and 13.34). References
to wartime contributions of H. Hönl, P. Jordan, and L. Schwarz can be found in
Ref. 14.1.

14.29   Borbély, S. von: Über die Luftkräfte, die auf einen harmonisch schwingenden
        zweidimensionalen Flügel bei Überschallgeschwindigkeit wirken. *Z. angew.
        Math. u. Mech.* **22**, 190–205 (1942). *ZWB Forsch. Ber.* **1071**. *RTP Trans-
        lation* **2019** (1942).
14.30   Collar, A. R.: Theoretical Forces and Moments on Thin Aerofoil with Hinged
        Flap at Supersonic Speeds. *Aeronaut. Research Council R. & M.* **2004** (1943).
14.31   Fuller, F. E.: Computation of Possio's Integral for Linearized Supersonic
        Flow. *J. Aeronaut. Sci.* **19**, 9, 640–642 (1952).
14.32   Garrick, I. E., and S. I. Rubinow: Flutter and Oscillating Air-Force Calcula-
        tions for an Airfoil in a Two-Dimensional Supersonic Flow. *NACA Rept.*
        846. Supersedes *NACA Tech. Note* **1158** (1946).
14.33   Huckel, V., and B. Durling: Tables of Wing-Aileron Coefficients of Oscillating
        Air Forces for Two-Dimensional Supersonic Flow. *NACA Tech. Note* **2055**
        (1950).
14.34   Jones, W. P.: Negative Torsional Damping at Supersonic Speeds. *Aeronaut.
        Research Council R. & M.* **2194** (1946).
14.35   Jones, W. P., and S. W. Skan: Aerodynamic Forces on Biconvex Aerofoils
        Oscillating in a Supersonic Airstream and Calculation of Forces for Aerofoil
        with Flap. *Aeronaut. Research Council R. & M.* **2749** (1953).
14.36   Lighthill, M. J.: Oscillating Airfoils at High Mach Number. *J. Aeronaut.
        Sci.* **20**, 402–406 (June 1953).

14.37   Miles, J. W.: The Aerodynamic Forces on an Oscillating Airfoil at Supersonic
        Speeds. *J. Aeronaut. Sci.* **14**, 351–358 (1947). Errata, **15**, 343 (1948). For
        swept wing, **15**, 343–346 (1948). Oscillating flap, **15**, 565–568 (1948).
        Errata, **16**, 442–443 (1949).
14.38   Possio, C.: L'azione Aerodinamica sul Profilo Oscillante alle Velocita Ultra-
        sonore. *Acta. Pont. Acad. Sci.* **1**, 11, 93–105 (1937).
14.39   Schwarz, L.: Ebene Instationäre Theorie des Tragfläche bei Überschallgesch-
        windigkeit. *Göttingen AVA B* **43/J/17** (July 1943). *Air Force Translation*
        F-TS-934-RE.
14.40   Stewartson, K.: On the Linearized Potential Theory of Unsteady Supersonic
        Motion. *Quart. J. Mech. Applied Math.* **3**, 182–199 (1950).
14.41   Taussky, O. (Todd): A Boundary Value Problem for a Hyperbolic Differential
        Equation Arising in the Theory of the Non-Uniform Supersonic Motion of an
        Aerofoil. *Studies and essays presented to R. Courant on his 60th birthday*,
        421–436. Interscience Publishers, New York (1948).
14.42   Temple, G., and H. A. Jahn: Flutter at Supersonic Speeds. I, Derivative
        Coefficients for a Thin Airfoil at Zero Incidence. *Aeronaut. Research
        Council R. & M.* **2140** (1945).
14.43   Todd, O.: On Some Boundary Value Problems in the Theory of the Non-
        uniform Supersonic Motion of an Aerofoil. *Aeronaut. Research Council
        R. & M.* **2141** (1945).
14.44   U.S. Government Printing Office: *Handbook of Supersonic Aerodynamics.*
        Vol. 4, *Aeroelastic Phenomena. NAVORD Rept.* **1488.**
14.45   Van Dyke, M. D.: Supersonic Flow Past Oscillating Airfoils including Non-
        linear Thickness Effects. *NACA Tech. Note* **2982** (1953).
14.46   Weber, R.: Tables of Unsteady Aerodynamic Coefficients for Two-Dimensional
        Supersonic Flow. *Off. Nat. Étud. Rech. Aéro. Rep. Publ.* **41** (1950).

Additional References:

14.47   Blanch, Gertrude: Tables of Life and Moment Coefficients for Oscillating
        Airfoils in Subsonic Compressible Flow. *Nat. Bureau of Standards Rept.*
        2260 (1953).
14.48   Nationaal Luchtvaartlaboratorium, Amsterdam: Tables of Aerodynamic
        Coefficients for an Oscillating Wing-Flap System in a Subsonic Compressible
        Flow. *Rept. F* **151** (1954).

Chapter 15

UNSTEADY MOTIONS IN GENERAL.   EXPERIMENTS

The aerodynamics of unsteady motions of airfoils with arbitrary time history is described in § 15.1 as a simple generalization, involving one integration, of the harmonic-oscillation case.   In particular, the indicial admittance (response to a unit-step function) is derived from the oscillating case.   This does not imply that there are no shorter methods of deriving the indicial admittance, but rather emphasizes the reciprocal relation between the admittance and the indicial admittance.   The effects of finite span is briefly mentioned in § 15.2.

The ultimate test of a theory lies in experiment.   Some general considerations in experiments are given in § 15.3.   Some of the established experimental results are outlined in § 15.4.   Owing to the limitation of space, the experiments cannot be described in detail.   Moreover, since the numerical uncertainty of the experimental results available at present is such that a scheme of empirical correction of the theoretical results is not yet generally acceptable, the aim of our discussion will be to point out the conditions under which the linearized theory is inapplicable and to describe some of the phenomena yet unexplained by theory.

15.1   UNSTEADY MOTIONS OF A TWO-DIMENSIONAL AIRFOIL

The aerodynamic response of an airfoil performing an arbitrary unsteady motion about a mean uniform rectilinear translation can be calculated from that of an oscillating airfoil by means of a Fourier analysis. In § 8.1 it is shown that, if the response y to a forcing function $F = F_0 e^{i\omega t}$ is

$$y(t) = \frac{F(t)}{Z(i\omega)} \tag{1}$$

then the response to a periodic function $F(t)$

$$F(t) = \sum_{n=-\infty}^{\infty} C_n e^{in\omega t} \tag{2}$$

is

$$y(t) = \sum_{n=-\infty}^{\infty} \frac{C_n}{Z(in\omega)} e^{in\omega t} \tag{3}$$

and that to a nonperiodic function $F(t)$

$$F(t) = \frac{1}{\sqrt{2\pi}} \int_{-\infty}^{\infty} \mathscr{G}(\omega)e^{i\omega t}\, d\omega \tag{4}$$

is

$$y(t) = \frac{1}{\sqrt{2\pi}} \int_{-\infty}^{\infty} \frac{\mathscr{G}(\omega)}{Z(i\omega)}\, e^{i\omega t}\, d\omega \tag{5}$$

where

$$\mathscr{G}(\omega) = \frac{1}{\sqrt{2\pi}} \int_{-\infty}^{\infty} F(t)\, e^{-i\omega t}\, dt \tag{6}$$

These results can be applied directly to the airfoil problem if the linearized theory is accepted. The airfoil displacements (change of angle of attack, velocity of translation, aileron angle, etc.) may be considered as the forcing function and the induced lift, moment, or pressure distribution as the response. The admittance $1/Z(i\omega)$ is given by the theory of harmonically oscillating airfoils, as applied to $y\,(x, t) = f\,(x)\, e^{i\omega t}$

The procedure can be expressed also in Laplace transformation. In Eqs. 5 and 6, put

$$i\omega = s \tag{7}$$

and assume $F(t) = 0$ for $t < 0$; then

$$y(t) = \frac{1}{2\pi i} \int_{-\infty i}^{\infty i} \frac{\sqrt{2\pi}\, \mathscr{G}(-is)}{Z(s)}\, e^{st}\, ds \tag{8}$$

$$\sqrt{2\pi}\, \mathscr{G}(-is) = \int_0^{\infty} F(t)\, e^{-st}\, dt \tag{9}$$

Hence, formally, $\sqrt{2\pi}\mathscr{G}(-is)$ is the Laplace transformation of $F(t)$, and Eq. 8 shows that $y(t)$ is the inverse Laplace transformation of $\mathscr{L}\{F\}/Z(s)$, i.e.:

$$y(t) = \mathscr{L}^{-1}\left\{ \frac{\mathscr{L}\{F\}}{Z(s)} \right\}$$

or

$$\mathscr{L}\{y(t)\} = \mathscr{L}\{F(t)\}/Z(s) \tag{10}$$

These formal steps can be mathematically justified for suitable classes of forcing functions $F(t)$. The Laplace transformation of the response is equal to the Laplace transformation of the forcing function multiplied by $1/Z(s)$, which is obtained by replacing $i\omega$ by s in the admittance to harmonic oscillation.

When $F(t)$ is a unit-step function,

$$\mathscr{L}\{\mathbf{1}(t)\} = \frac{1}{s} \tag{11}$$

the response, called indicial admittance and denoted by $A(t)$, is given by

$$\mathscr{L}\{A(t)\} = \frac{1}{s\,Z(s)} \tag{12a}$$

or

$$A(t) = \mathscr{L}^{-1}\left\{\frac{1}{s\,Z(s)}\right\} \tag{12b}$$

in agreement with Eq. 32 of § 8.1.

Note that the downwash distribution over the airfoil must be the same in the unsteady motion as in the harmonic-oscillation case in order that the above formulas be applicable.

There are other methods of deriving the indicial admittance. For special problems special methods may be devised that are much shorter than the Fourier-transformation method mentioned above. Recently, great advance has been made in the calculation of indicial admittance with respect to a sudden motion of the airfoil or to a sharp-edged gust, for both the two-dimensional case and the finite-aspect ratio case. The success is remarkable, particularly at high subsonic Mach numbers and at $M = 1$ (linearized theory). On the other hand, the second- and higher-order theories are investigated in the supersonic case, and significant corrections to the first-order linearized theory (which is presented in the last chapter) are revealed for Mach numbers near and below $\sqrt{2}$. See bibliography, and in particular, Lomax,[15.84] and van Dyke.[14.45]

Example. Wagner's Problem. Consider an airfoil moving recti-linearly in a fluid with a relative speed U which is so small that the fluid may be considered incompressible. Let the angle of attack be suddenly increased by an amount α. Owing to this change the relative velocity of the fluid will have a component normal to the airfoil. This normal velocity is uniformly distributed along the chord and is a unit-step function of time:

$$v(x, t) = -\,U\alpha\,\mathbf{1}(t) \qquad (-1 \leqslant x \leqslant 1) \tag{13}$$

The lift and moment induced by this upwash can be found by the Laplace transformation. Note that the same problem arises if an airfoil suddenly starts to move in a stationary fluid with a constant velocity U and angle of attack α.

In § 13.4 it is shown that, if

$$v(x, t) = v_0\,e^{i\omega t} = i\omega y_0\,e^{i\omega t}$$

then the total lift force acting on the airfoil is

$$L = \pi \rho U^2 y_0 k^2 \left[1 - \frac{2i}{k} C(k) \right] e^{ik\tau}$$

$$= - i\pi \rho b U v_0 k \left[1 - \frac{2i}{k} C(k) \right] e^{ik\tau} \tag{14}$$

where

$$\tau = \frac{Ut}{b}, \qquad k = \frac{\omega b}{U}, \qquad b = 1 \tag{15}$$

and the stalling moment about the mid-chord point is

$$M_{1/2} = - \pi \rho U^2 i y_0 k \, C(k) \, e^{ik\tau} \tag{16}$$

$C(k)$ is Theodorsen's function.

Replacing ik by s, we obtain, for the lift force,

$$\frac{1}{Z(s)} = - \pi \rho b U \left[1 + \frac{2}{s} C(-is) \right] \qquad \text{(Rl } s > 0) \tag{17}$$

The indicial admittance of the lift force is therefore

$$L(\tau) = v_0 \mathscr{L}^{-1} \left\{ - \pi \rho b U \left[1 + \frac{2}{s} C(-is) \right] \right\}$$

$$= - \pi \rho b U v_0 \, \delta(\tau) - 2\pi \rho b U v_0 \mathscr{L}^{-1} \left\{ \frac{C(-is)}{s} \right\} \tag{18}$$

The first term gives an impulse function. The total impulse is obtained by an integration over an infinitesimal time interval. If we remember the scale factor b/U (Eq. 15) in changing from τ to t, the magnitude of the total impulse is seen to be $\pi \rho b^2 v_0$. This is the total impulse that is needed to move abruptly a mass $\pi \rho b^2$ to a velocity v_0. The quantity $\pi \rho b^2$ is the apparent mass of the fluid associated with the vertical motion (cf. § 6.7, Eq. 9).

The second term gives the lift due to circulation. Define a special function $\Phi(\tau)$:

$$\Phi(\tau) = \mathscr{L}^{-1} \left\{ \frac{C(-is)}{s} \right\} \qquad \text{(Rl } s > 0) \tag{19}$$

Then the circulatory lift can be written as

$$L_1(\tau) = - 2\pi \rho b U v_0 \, \Phi(\tau) \tag{20}$$

The function $\Phi(\tau)$ is the Wagner's function defined in § 6.7, and is shown graphically in Fig. 6.6.

15.2   THE EFFECT OF FINITE SPAN

The problem of estimating the spanwise distribution of lift and moment on an unswept lifting surface of finite span executing simple-harmonic oscillations in an incompressible fluid has been studied by many authors, among them the most notable being Cicala, Lyon, W. P. Jones, Skan, Sears, R. T. Jones, Küssner, Biot, Boehnlein, Wasserman, Reissner, Zartarian, Hsu, Ashley, Dengler, Goland, and Shen (see bibliography). Unfortunately, the problem is so complicated that, even after the standard linearization, there exists no practicable, exact solution. Moreover, the independent lines of approach initiated by various authors have led to answers that cannot be shown to be entirely equivalent. Experimental results available at present, due to their scatter, cannot discriminate definitely which one of these theories approximates best the physical reality, although generally the methods of Reissner and that of Biot and Boehnlein are favored.

For a compressible fluid, the linearized theory in a supersonic flow is quite advanced. A number of exact solutions have been obtained for some special wing planforms and modes of motion. A few solutions are known also in the high subsonic and transonic speed range, but, generally speaking, the three-dimensional oscillating-airfoil theory is still a subject for future research.

15.3   EXPERIMENTAL DETERMINATION OF UNSTEADY LIFT*

Many methods have been used in measuring the unsteady aerodynamic forces (lift, moment, or pressure distribution) acting on an airfoil which is moving or is situated in a nonuniform flow. One may speak of "direct" and "indirect" measurements. If the forces are measured directly by dynamometers, manometers or strain gages, the measurement is said to be direct. If they are determined from their effect on the motion of the airfoil, air density, or other quantities, it is said to be indirect.

The requirements imposed on direct-measuring instruments are not easy to meet. The influence of the measuring instruments on the phenomena under examination must be kept as small as possible. The installation must not affect the flow to any appreciable degree. The variation of forces with time must be recorded with sufficient accuracy. In particular,

---

* Most of the experimental investigations are restricted to the determination of the forces normal to the plane of the airfoil, i.e., lift, moment, and pressure distribution. Little is known about the unsteady drag force, which plays only a minor role in aeroelastic problems. Only the normal forces will be considered here.

the time lag in the recording instrument due to the mass inertia of the sensing elements must be kept small.

In some of the indirect methods of measurement, the mass-inertia time-lag problem is completely eliminated. These methods are listed as follows:

1. *Indirect Methods.* (*a*) Determination of circulation from photographs of the flow. In Walker[15.119] and Farren's experiments,[15.112] a wing model is drawn through a glass vessel filled with water. The flow is made visible by small drops of olive oil and ethylene dibromide, and can be photographed. The velocity of the fluid particles can be determined from the time of exposure and from the streak lines of the oil drops. The circulation about the wing is then obtained from the velocity field by a numerical integration.

(*b*) Determination of pressure from the variation of air density. Since the index of refraction of light varies with the density of a fluid, which in turn is related to the pressure field, the pressure can be determined optically. The variable index of refraction is measured by light interference. Zender-Mach interferometer can be used for this purpose. This method is effective for high-speed flow, in the transonic and supersonic ranges.

(*c*) Determination of unsteady forces from the motion influenced by these forces. The lift acting on an airfoil in entering a gust can be calculated from the motion of the airfoil. The gust response may be verified by measuring the trajectory of an airfoil in passing through a jet stream (e.g., by dropping an airfoil through the test section of an open-jet wind tunnel). This method has been applied with certain degree of success by Küssner.[15.116]

A more commonly used method is based on forced oscillations. The wing model is excited by harmonic external forces, and the aerodynamic forces are determined from the kinematic quantities involved. Generally, the aerodynamic forces are not large compared to the inertia and elastic forces. Hence, the model must be built as light as possible (yet sufficiently rigid to prevent appreciable distortions) or a water tunnel must be used. See the works of Cicala,[15.125] Dresher,[15.126] Greidanus,[15.128] etc.

Flutter experiments may be considered as another approach. But, owing to the large number of parameters involved, it is unsuitable for an exact determination of the aerodynamic forces.

2. *Direct Methods.* (*a*) Spring balances. The transient force to be measured is opposed by a spring whose deflection is converted, for instance, into a rotation of a mirror, which is recorded by a pencil of reflected light. Various versions of spring balances are used by Farren,[15.111] Silverstein,[15.134] Scheubel,[15.117] Reid and Vincenti,[15.130] etc.

(b) Electric measuring elements.   Forces can be measured by a number of electromagnetic devices, such as: (i) Piezoelectric gages (Kramer[15.115]), (ii) Wire-resistance strain gages, (iii) Inductance transducers (Wieselsberger[15.120]).

A piezoelectric gage measures the electric charges (or their voltages) produced on the end surfaces of certain crystal (e.g., quartz) when it is subject to pressure.   It has been used for stationary models in a flow whose direction is changed at constant velocity.

A wire-resistance strain gage measures the change of resistance of a fine wire (e.g., tungsten) due to elastic strain.   It can be used to measure the elastic deflections of springs.   If the wire is attached to a thin metal membrane which deflects under pressure, it can be used as a pressure gage.   Wire-resistance gages are used extensively both in wind-tunnel and in flight testing.

There are many types of inductance transducers.   Either the air gap or the position of an iron core may be varied, and the corresponding change in inductance is measured.   The displacement of the air gap or the iron core can be made proportional to the deflection of a spring, thus measuring a force.   Very high accuracy can be achieved in certain designs.

Use of electrical means for measuring forces often involves a complicated electronic system.   To improve the accuracy and ease the analysis, various ingenious schemes have been invented.   As an example, one may name the "wattmeter" harmonic analyzer of Bratt, Wight, and Tilly,[15.121] in measuring the aerodynamic damping for pitching oscillations.   The modulated output from the stress indicator is first rectified and then analyzed electrically by means of an electronic wattmeter; the damping coefficient is obtained directly from meter readings.

The electrical measurements have the important advantage of minimizing the mass-inertia effect of the sensing elements.   With proper electronic equipment and circuits, it is probably the most convenient, accurate, and versatile of all methods.

(c) Manometers.   Pressure distribution over the model surface can be measured by manometers.   Drescher[15.110] describes a successful multiple manometer used in measuring unsteady pressure distribution over an airfoil in a water tunnel.

15.4   EXPERIMENTAL RESULTS

Measurements of responses to sudden start of motion, and to gusts, reveal many interesting phenomena.   Due to experimental difficulty, most of the results available so far are qualitative in nature.   In the following, whenever comparison with theoretical results is mentioned, it is meant that the results given by the two-dimensional linearized theory for infinitely thin wings, as exposed in the preceding chapters, are being compared. All the experiments quoted below are performed at low air speeds; hence, only the theory of incompressible flow are checked.

Change of Circulation due to Sudden Start of Motion. Figure 15.1

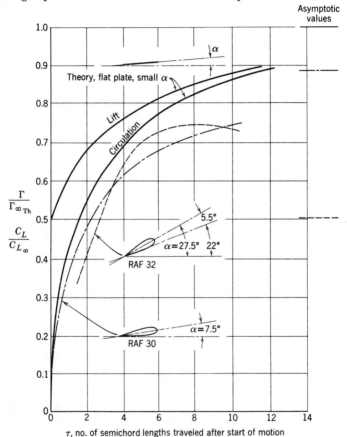

τ, no. of semichord lengths traveled after start of motion

Fig. 15.1. Growth of circulation and lift after a sudden start of motion. The theoretical values $\Gamma_{\infty Th}$ and $C_{L\infty}$ are for two-dimensional flat plate of zero thickness. The steady-state circulation and lift of the experimental airfoils are smaller than $\Gamma_{\infty Th}$ and $C_{L\infty}$, respectively. If the ratio of the experimental instantaneous circulation to the corresponding experimental steady-state value (shown as asymptote) were plotted as a function of time, the experimental curves will appear to be in better agreement with the theory. (From data given by Walker, Ref. 15.119, and Francis, Ref. 15.113. Cambridge Aeronautical Laboratory. Wing chord 4 in. Reynolds number 1.4×10^5 (water). Span 6 in. between plane walls.)

shows the results obtained by Walker[15.119] and Francis[15.113] by photographing the flow patterns in a water tunnel.   The airfoil was suddenly moved with constant velocity in still water.   In one case the angle of attack was so small ($\alpha = 7.5°$) that the flow remained unseparated   In the other case, however, the angle of attack was so large ($\alpha = 27.5°$ measured from the zero-lift line) that the flow began to separate after a flight-path length of 2.8 wing chord.   In Fig. 15.1 the measured values

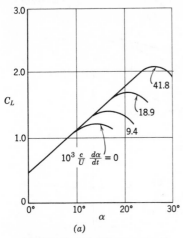

(a) Airfoil: Gö 398, Aspect ratio 5, Reynolds Number 3.6×10^5.
From Wieselberger, FB 266 (1935). Ref. 15.120.   Airfoil   stationary. Direction of flow varied

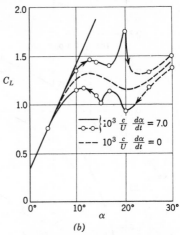

(b) Airfoil: Clark-YH,   Effective aspect ratio ∞, Reynolds number 1.2×10^5.
From Farren, R. & M. 1648 (1935). Ref. 15.111.   Airfoil angle of attack varied.   Direction of flow constant

Fig. 15.2.   Effect of rate of change of angle of attack on the maximum lift coefficient.   (Courtesy of Dr.-Ing. H. Drescher of Max-Planck-Institut für Strömmungsforschung.)

of circulation Γ and lift coefficient C_L are compared with Wagner's theoretical values (§ 15.1), $\Gamma_{\infty Th}$ and $C_{L\infty}$ being the theoretical limiting value of Γ and C_L, respectively, as time $t \to \infty$.   The limiting values of the experimental curves were determined from steady lift measurements.

It is seen that the experimental values of Γ are always lower than the theoretical ones.   In the small-angle-of-attack case the shape of the experimental curve resembles closely that of the theory.   In the large-angle-of-attack case the shape differs considerably.

Increase of the Maximum Lift Coefficient—Kramer's Effect. The maximum lift coefficient $C_{L\,max}$ of a moving airfoil is different from that of a stationary airfoil in a steady flow.   Kramer[15.115] first showed that

$C_{L\,max}$ increases when the angle of attack increases with time. In Fig. 15.2 are shown the results of Wieselsberger[15.120] for a stationary airfoil situated in a flow the angle of which increases with time, and those of Farren[15.111] when the angle of attack is first increased and then decreased. The rate of change of angle of attack $d\alpha/dt$ is made nondimensional by multiplying with the wing chord c and dividing by the speed of flow U.

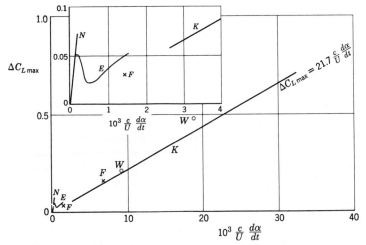

Fig. 15.3. Increase of $C_{L\,max}$, over the stationary values with the rate of change of angle of attack. The curve marked K is given by Kramer, Ref. 15.115, for airfoils Gö 398, Gö 459, aspect ratio 5, Reynolds number 1.2×10^5 to 4.8×10^5. Curve N is given by Silverstein, Katzhoff, and Hootman, Ref. 15.118, for a high-wing monoplane Fairchild 22 (airfoil profile NACA 2 R_1 12), Reynolds number 10×10^5 to 30×10^5. Curve E is Ehrhardt's result given in Ref. 15.117, Reynolds number 10^5 to 2.8×10^5. The points marked W are given by Wieselsberger, Ref. 15.120. Points F are given by Farren, Ref. 15.111. (Courtesy of Dr.-Ing. H. Drescher.)

The increment of the maximum lift coefficient, $\Delta C_{L\,max}$ over the stationary values is plotted against $\dfrac{c}{U}\dfrac{d\alpha}{dt}$ in Fig. 15.3, where the results of other authors are also presented. The values found by Kramer are represented by the straight line

$$\Delta C_{L\,max} = 21.7\,\frac{c}{U}\frac{d\alpha}{dt} \qquad (1)$$

which is seen to be invalid for small values of angular speed parameter. For $\dfrac{c}{U}\dfrac{d\alpha}{dt} < 0.21 \times 10^{-3}$, the NACA data obtained by Silverstein,

Katzhoff, and Hootman[15.118] gives a straight line whose slope is about 17 times higher than that of Kramer's Eq. 1. In the intermediate range, Ehrhardt's[15.117] test results show a curious transition from the NACA curve to Kramer's curve.

Motion of a Wing Encountering a Gust. By dropping a wing model through the jet of an open-section wind tunnel, and photographing the trajectories of two little lamps attached to the wing leading and trailing edges, Küssner[15.116] obtained a good qualitative agreement with his theory of gust response. The theory predicts that the lift created by a gust is due to circulation and that the resultant force acts through the forward aerodynamic center. In Küssner's tests a rectangular wing (chord 0.204 meter, span 0.415 meter), with and without tip plates, fell (at zero angle of attack) into the horizontal free jet of the wind tunnel. When it reached the geometric jet boundary, its velocity was of order 6.13 meters per second. The ratio of the wind velocity w to the falling velocity of the airfoil U could amount to $w/U = 0.359$ without separation of the flow. A ratio $w/U = 0.359$ corresponds to an angle of attack $\alpha = 19.5°$, at which separation will occur in a steady flow. This delay of separation is another revelation of Kramer's effect. In experiments with higher wind speed, the wing turned (pitched) to the wind, which means that the center of pressure moved backward because of the separation of flow.

For tests in which $w/U \leqslant 0.359$ the wing with its center of gravity at 0.236 chord length behind the leading edge did not turn (pitch) when passing the jet boundary. Assuming the aerodynamic center of this rather short wing to be located at 0.236 chord (the theoretical value is 0.25 for a two-dimensional wing without thickness), the experiments may be taken as confirming the theoretical location of the center of pressure.

From measurements of the curvature of the trajectories of the airfoil, the lift force acting on the wing can be calculated. However, the accuracy of such measurements was rather low. Within the experimental error, no significant difference between the theory and experiment has been found with respect to the transient lift force, provided that separation did not occur.

Flap Motion—Adherent Flow. An example of the measurements of transient pressure distribution over an airfoil due to a sudden deflection of a flap is shown in Fig. 15.4, which is given by Drescher[15.110] from experiments in a water tunnel. The pressure distribution was measured by a multiple manometer. At the top of this figure is a curve of the normal force coefficient C_n vs. time, C_n being the force (normal to wing chord) divided by $q \times$ (wing area). Next is a curve of the flap force coefficient C_F vs. time; C_F, like C_n, is referred to the main-wing area (including flap) and is the component of force acting on the flap in the

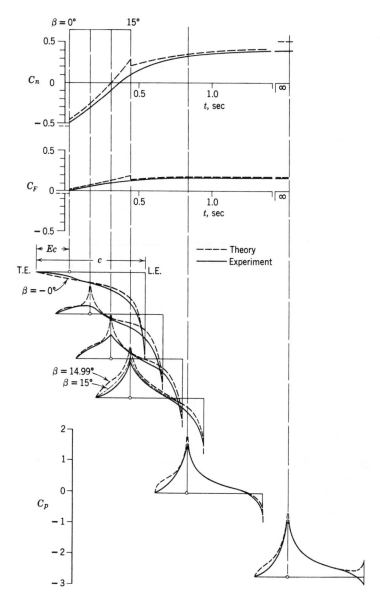

Fig. 15.4. Transient normal force coefficient, flap force coefficient, and pressure distribution over an airfoil following a sudden deflection of flap. (Courtesy of Dr.-Ing. H. Drescher.)

direction normal to the main wing chord. In the lower part of Fig. 15.4 are plots of pressure distribution over the airfoil and the flap. The pressure coefficient C_p is defined as the actual pressure jump across the airfoil divided by the dynamic pressure $q = \frac{1}{2}\rho U^2$. The angular speed of flap Ω was a constant in each experiment, and was expressed in non-dimensional parameter $\Omega c/2U$. The wing model had a symmetric profile Gö-409; it spanned wall to wall in the tunnel, so that in the mid-span section, where the measuring holes were arranged, the flow

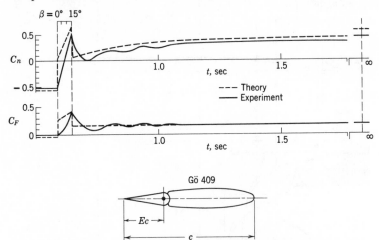

Fig. 15.5. Transient normal force and flap force coefficients for a higher angular speed of flap deflection. Reynolds number 6×10^5. $\alpha = -5°$. $\beta = 0° \rightarrow 15°$. $\Omega c/2U = 0.356$. (Courtesy of Dr.-Ing. H. Drescher.)

approximated well a two-dimensional one. The flap was mounted to the main wing without a slot.

In the experiment of Fig. 15.4, the angle of attack was $\alpha = -5°$, and the flap was deflected from $\beta = 0°$ to $\beta = 15°$ at constant angular speed Ω. In general, $C_n(t)$, $C_F(t)$, and $C_p(x)$ curves all agree fairly well with theoretical values (dotted curves) based on infinitely thin two-dimensional plate. The airfoil profile shape causes a discrepancy in the pressure distribution, particularly at the leading edge, where a sharp edge is assumed in the theory.

In Fig. 15.5 is plotted another result by Drescher[15.110] for a higher angular speed of the flap motion. A periodic oscillation in $C_n(t)$, which gradually dies out, appears after the flap has ceased to move. This is connected with the flow picture of Fig. 15.6. The vortex surface deflected by the flap motion is followed by periodic vortices of decreasing intensity.

Such periodic vortices had been observed after sudden starting or sudden stopping of the flap motion.   One may conclude that the circulation about the wing cannot instantaneously assume the value which is given by the kinematic conditions, but that it oscillates about the prescribed value with appreciable amplitude.

Flap Motion—Separated Flow.   Experimental pressure distribution on an airfoil with flow separation is also reported by Drescher.[15.110]   The beginning of the separation process depends on the Reynolds number. Low Reynolds number will favor the separation.   But the most important factor influencing the separation process is the angular velocity with which the flap moves.   If the main-wing angle of attack is small, and the flap

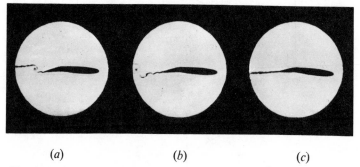

(a)   (b)   (c)

Fig. 15.6.   Flow following a sudden deflection of flap. (*a*) Immediately after stopping the flap motion. (*b*) Next instant. (*c*) Approaching steady state. (Courtesy of Dr. Ing. H. Drescher.)

is deflected to a large angle, the flow separates in the flap region, and in later stages a Kármán vortex street is developed.   For Gö–409 airfoil, $\alpha = 0$, β moves from 0 to 60°, separation occurs at $\beta = 12°$ when $\Omega c/2U$ is 0.014, but it occurs after $\beta = 60°$ when $\Omega c/2U$ is 1.16.   In the latter case very high suction is obtained at the flap nose, and the C_L and C_F values may exceed the corresponding theoretical values at the instant before separation occurs.   Very large oscillations in $C_L(t)$ and $C_F(t)$ curves are often observed after separation.

If the main-wing angle of attack α is small and β decreases from 60° to 0, the separated flow becomes adherent again, but often after an appreciable time delay, which is required to scavenge away the dead water accumulated behind the wing.   More complicated motions of the flap and the main wing induce more complicated responses; but such responses can generally be understood on the basis of the facts mentioned above: oscillation in circulation before it reaches a steady value, and time delay required for the scavenging process.

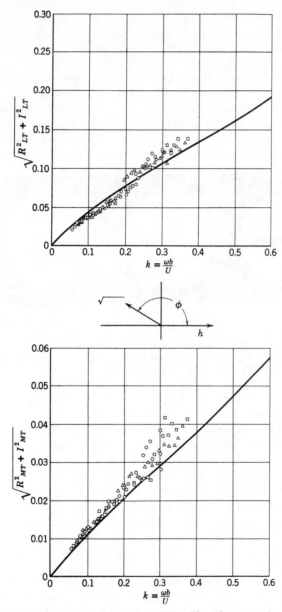

Fig. 15.7. **Response to harmonic motion—lift and moment in pure trans-lation.** Experiments by Halfman, Ref. 15.129. (Courtesy of the NACA.)

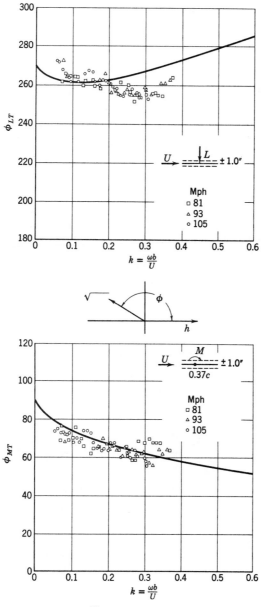

Fig. 15.7—*continued*

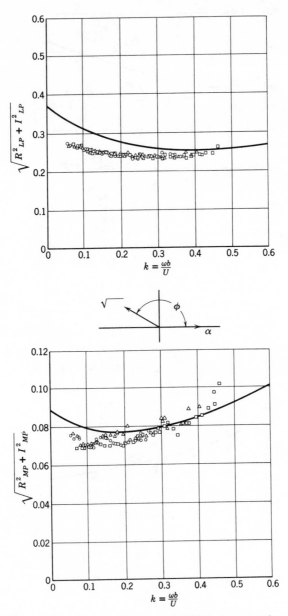

Fig. 15.8. Response to harmonic motion—lift and moment in pure pitch. Experiments by Halfman, Ref. 15.129. (Courtesy of the NACA.)

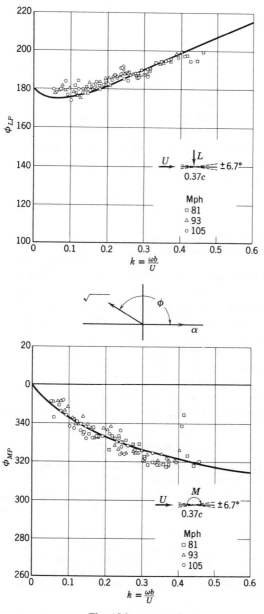

Fig. 15.8—*continued*

Harmonic Oscillations. Results of oscillating-wing experiments have been published by many authors. (See bibliography.) For a wing without a flap, the experimental results in general agree with the theoretical. There are, however, some small quantitative difference between theory and experiment, and among various authors. Typical results obtained by Halfman[15.129] are shown in Figs. 15.7 and 15.8. The airfoil (NACA 0012 section, chord 1 ft, span 2 ft) was tested in a 5×7-ft wind tunnel with plates shielding the wing tips so that two-dimensional flow condition was assured. Oscillations in two degrees of freedom were imparted to the airfoil: h and α. The expressions of aerodynamic force and moment corresponding to the translational motion $h = h_0 e^{i\omega t}$ (h_0 real) are, respectively,

$$\frac{L_T}{4qb} = \sqrt{R_{LT}^2 + I_{LT}^2}\, e^{i(\omega t + \phi_{LT})}, \qquad \phi_{LT} = \tan^{-1}\frac{I_{LT}}{R_{LT}}$$

$$\frac{M_T}{4qb^2} = \sqrt{R_{MT}^2 + I_{MT}^2}\, e^{i(\omega t + \phi_{MT})}, \qquad \phi_{MT} = \tan^{-1}\frac{I_{MT}}{R_{MT}}$$

Similarly, those corresponding to pure pitch $\alpha = \alpha_0 e^{i\omega t}$ (α_0 real) are given by the same formulas except that the subscript T is replaced by P. The force L, as well as the displacement h, is taken positive downward in these figures. Both the magnitude and phase angle are plotted in Figs. 15.7 and 15.8. The solid curves are theoretical values.

More serious differences between theoretical lift and moment coefficients and experimental ones are found for oscillating flaps. But test results are meager. See papers by Drescher[15.126] and Walter.[15.135,15.136]

Experimental results for oscillating wings having a mean angle of attack near or greater than the static stalling angle have been reviewed in § 9.4.

BIBLIOGRAPHY

Extensive bibliography are found in the following general reviews of the unsteady airfoil theory, and in the general reviews 7.1–7.4.

15.1   Cicala, P.: Present State of Development in Nonsteady Motion of a Lifting Surface. *NACA Tech. Memo.* **1277**. Translated from *Aerotecnica* **21** (Sept.–Oct. 1941).

15.2   Küssner, H. G.: A Review of the Two-Dimensional Problem of Unsteady Lifting Surface Theory during the Last Thirty Years. *Univ. Maryland Inst. Fluid Dynam. & Applied Math. Lecture Ser.* **23** (Apr. 1953).

15.3   Lyon, H. M.: A Review of Theoretical Investigations of the Aerodynamical Forces on a Wing in Non-uniform Motion. *Aeronaut. Research Com. R. & M.* **1786** (1937).

15.4   Frankl, F. I. and E. A. Karpovich: *Gas Dynamics of Thin Bodies.* Translated from the Russian by M. D. Friedman. Interscience Publishers, Inc., New York (1953).

15.5 Temple, G.: Unsteady Motion. Chapter IX of *Modern Developments in Fluid Dynamics, High Speed Flow* (edited by L. Howarth). Oxford Univ. Press, London (1953).

General treatment of unsteady flow problems (see also Ref. 14.40):

15.6 Evvard, J. C.: A Linearized Solution for Time-Dependent Velocity Potentials near Three-Dimensional Wings at Supersonic Speeds. *NACA Tech. Note* **1699** (1948). See also *NACA Tech. Notes* **1429, 1484**.

15.7 Gardner, C.: Time Dependent Linearized Supersonic Flow Past Planar Wings. *Communs. Pure and Applied Math.* **3**, 33–38 (1950).

15.8 Gardner, C., and H. F. Ludloff: Influence of Acceleration on Aerodynamic Characteristics of Thin Airfoils in Supersonic and Transonic Flight. *J. Aeronaut. Sci.* **17**, 47–59 (1950).

15.9 Garrick, I. E.: On Moving Sources in Nonsteady Aerodynamics and Kirchhoff's Formula. *Proc. 1st U.S. Natl. Congr. Applied Mech.*, 733–740 (1951).

15.10 Heaslet, M. A.: The Application of Green's Theorem to the Solution of Boundary Value Problems in Linearized Supersonic Wing Theory. *NACA Tech. Note* **1767**.

15.11 Kármán, Th. von, and J. M. Burgers: *Aerodynamic Theory* (edited by Durand), Vol. II, Chapter V. Julius Springer, Berlin (1934).

15.12 Küssner, H. G.: General Airfoil Theory. *NACA Tech. Memo.* **979** (1941). Translated from *Luftfahrt-Forsch.* **17**, 370–378 (1940).

15.13 Lomax H., M. A. Heaslet, and F. B. Fuller: Three-Dimensional Unsteady-Lift, Problems in High-Speed Flight—Basic Concepts. *NACA Tech. Note* **2256** (1950).

15.14 Lomax, H., M. A. Heaslet, and L. Sluder: The Indicial Lift and Pitching Moment for a Sinking or Pitching Two-Dimensional Wing Flying at Subsonic or Supersonic Speeds. *NACA Tech. Note* **2403** (1951).

15.15 Reissner, E.: On the General Theory of Thin Airfoils for Nonuniform Motion. *NACA Tech. Note* **946** (1944).

Reciprocal relations (see also Ref. 14.5):

15.16 Brown, C. E.: The Reversibility Theorem for Thin Airfoils in Subsonic and Supersonic Flow. *NACA Rept.* **986** (1950).

15.17 Garrick, I. E.: On Some Reciprocal Relations in the Theory of Non-stationary Flow. *NACA Rept.* **629** (1938). See also *Proc. 5th Intern. Congr. Applied Mech. Cambridge* (1938).

15.18 Harmon, S. M.: Theoretical Relations between the Stability Derivatives of a Wing in Direct and in Reverse Supersonic Flow. *NACA Tech. Note* **1943** (1949).

15.19 Hayes, W. D.: Reversed Flow Theorems in Supersonic Aerodynamics. *Proc. 7th Intern. Congr. Applied Mech.* **2**, 412–424 (1948).

15.20 Heaslet, M. A., and J. R. Spreiter: Reciprocity Relations in Aerodynamics. *NACA Rept.* **1119** (1953). Supersedes *NACA Tech. Note* **2700**.

15.21 Miles, J. W.: Some Relations between Harmonic and Transient Loading of Airfoils. *J. Aeronaut. Sci.* **17**, Reader's Forum, 671 (1950).

Indicial responses (gust, sudden motion), two-dimensional flow, incompressible fluid:

15.22 Brandt-Møller, P. N.: The Growth of Circulation around Aeroplane Wings. *Proc. 7th Intern. Congr. Applied Mech.* **2**, 140–154 (1948).

15.23   Glauert, H.: The Accelerated Motion of a Cylindrical Body through a Fluid. *Aeronaut. Research Com. R. & M.* **1215** (1929).

15.24   Kármán, Th. von, and W. R. Sears: Airfoil Theory for Nonuniform Motion. *J. Aeronaut. Sci.* **5**, 379–390 (1938). See also *J. Aeronaut. Sci.* **8**, 104–108 (1941).

15.25   Küssner, H. G.: Das zweidimensionale Problem der beliebig bewegten Tragfläche unter Berücksichtigung von Partialbewegungen der Flüssigkeit. *Luftfahrt-Forsch.* **17**, 355–361 (Dec. 1940).

15.26   Luke, Y. L., and M. A. Dengler: Tables of the Theodorsen Circulation Function for Generalized Motion. *J. Aeronaut. Sci.* **18**, 478–484 (1951). See discussions by A. I. Van de Vooren, E. V. Laitone, W. P. Jones, M. Goland, and C. C. Chang in *J. Aeronaut. Sci.* **19**, 209–213, 717 (1952).

15.27   Sears, W. R.: Operational Methods in the Theory of Airfoils in Non-uniform Motion. *J. Franklin Inst.* **230**, 95–111 (1940).

15.28   Söhngen, H.: Bestimmung der Auftriebsverteilung für beliebige instationäre Bewegungen (Ebenes Problem). *Luftfahrt-Forsch.* **17**, 401–420 (Dec. 1940).

15.29   Schwarz, L.: Berechnung der Funktionen $U_1(s)$ and $U_2(s)$ für grössere Werte von s. *Luftfahrt-Forsch.* **17**, 362–369 (1940).

15.30   Wagner, H.: Über die Entstehung des dynamischer Auftriebes von Tragflügeln. *Z. angew. Math. u. Mech.* **5**, 17–35 (1925).

Indicial responses, two-dimensional flow, subsonic speed:

15.31   Mazelsky, B.: Numerical Determination of Indicial Lift of a Two-Dimensional Sinking Airfoil at Subsonic Mach Numbers from Oscillatory Lift Coefficients with Calculations for Mach Number 0.7. *NACA Tech. Note* **2562** (1951). For similar treatment of pitching airfoil at $M = 0.7$, see *NACA Tech. Note* **2613** (1952). $M = 0.5$ and 0.6, see *Tech. Note* **2739** (1952).

Indicial responses, two-dimensional flow, supersonic speed:

15.32   Biot, M. A.: Loads on a Supersonic Wing Striking a Sharp-Edged Gust. *J. Aeronaut. Sci.* **16**, 296–300 (1949).

15.33   Chang, C. C.: The Transient Reaction of an Airfoil Due to Change in Angle of Attack at Supersonic Speed. *J. Aeronaut. Sci.* **15**, 635–655 (1948).

15.34   Chang, C. C.: Transient Aerodynamic Behavior of an Airfoil Due to Different Arbitrary Modes of Nonstationary Motion in a Supersonic Flow. *NACA Tech. Note* **2333** (1951). See discussion by J. W. Miles. *J. Aeronaut. Sci.* **19**, 138 (1952).

15.35   Heaslet, M. A., and H. Lomax: Two-Dimensional Unsteady Lift Problems in Supersonic Flight. *NACA Tech. Note* **1621** (1948); *NACA Rept.* **945** (1949).

15.36   Miles, J. W.: Transient Loading of Airfoils at Supersonic Speeds. *J. Aeronaut. Sci.* **15**, 592–598 (1948).

15.37   Watkins, C. E.: On Transient Two-Dimensional Flow at Supersonic Speeds. *J. Aeronaut. Sci.* **16**, Reader's Forum, 569–570 (1949).

Oscillating wing, finite span, incompressible fluid: (for a number of papers by Cicala, Possio, and von Borbély, see review in Ref. 15.1).

15.38   Biot, M. A., and C. T. Boehnlein: Aerodynamic Theory of the Oscillating Wing of Finite Span. *GALCIT Flutter Rept.* **5** (1942). California Institute of Technology.

15.39 Dengler, M. A., and M. Goland: The Calculations of Spanwise Loadings for Oscillating Airfoils by Lifting Line Techniques. *J. Aeronaut. Sci.* **19**, 751–759 (1952).

15.40 Dingel, M., and H. G. Küssner: Beitrage zur instationären Tragflächen theorie: VIII, Die schwingende Tragfläche grösser Streckung. *ZWB ForschBer.* **1774** (1943). *U.S. Air Force Translation* **F-TS-935-RE**.

15.41 Fettis, H. E.: A Note on the Evaluation of a Definite Integral. Related to the theories of Cicala, Dingle, Küssner, and Reissner. *J. Aeronaut. Sci.* **17**, Reader's Forum, 184 (1950). Comments by Luke and Ufford, **18**, 429 (1951).

15.42 Hildebrand, F. J., and E. Reissner: Studies for an Aerodynamic Theory of Oscillating Swept-Back Wings of Finite Span, parts II, III, and IV. *Chance Vought Aircraft Co. Engr. Repts.* **6733** (1947); **7039** (1948); and **7949**.

15.43 Jones, R. T.: The Unsteady Lift of a Wing of Finite Aspect Ratio. *NACA Rept.* **681** (1940).

15.44 Jones, W. P.: Aerodynamic Forces on Wings in Simple-Harmonic Motion. *Aeronaut. Research Council R. & M.* **2026** (1945). See also Jones, *R. & M.* **2142**; Jones and S. W. Skan, *R. & M.* **2215**; Jones, *R. & M.* **2470** (1946).

15.45 Kinner, W.: Die Kreisförmige Tragfläche auf potentialtheoretischer Grundlage. *Ing. Arch.* **8**, 47–80 (1937). *RTP Translation* **2345**. See also *Z. angew. Math. u. Mech.* **16**, 349–352 (1936).

15.46 Kochin, N. E.: The Steady Vibrations of a Wing of Circular Planform. In Russian. *Prik. Mat. i. Mekh. T.* **VI** (1942).

15.47 Lyon, H. M., W. P. Jones, and S. W. Skan: Aerodynamical Derivatives of Flexural-Torsional Flutter of a Wing of Finite Span. *Aeronaut. Research Council R. & M.* **1900** (1939).

15.48 Reissner, E. : Effect of Finite Span on the Airload Distributions for Oscillating Wings. I, Aerodynamic Theory of Oscillating Wings of Finite Span. *NACA Tech. Note* **1194** (1947). II, Methods of Calculation and Examples of Application, by Reissner and J. E. Stevens. *Tech. Note* **1195** (1947).

15.49 Reissner, E.: A Problem of the Theory of Oscillating Airfoils. *Proc. 1st U.S. Natl. Congr. Applied Mech.*, 923–926 (1951).

15.50 Schade, T.: Theorie der schwingenden Kreisförmigen Tragfläche auf potential-theoretischer Grundlage. I, Analytischer Teil. *Luftfahrt-Forsch.* **17**, 387–400 (1940). II, Numerischer Teil, by T. Schade and K. Krienes. *Luftfahrt-Forsch.* **19**, 282–291 (1942). Translation, *NACA Tech. Memo.* **1098** (1947).

15.51 Sears, W. R.: A Contribution to Airfoil Theory for Non-uniform Motion. *Proc. 5th Intern. Congr. Applied Mech. Cambridge*, 483–487 (1938).

15.52 Shen, S. F.: A New Lifting Line Theory for the Unsteady Lift of a Swept or Unswept Wing in an Incompressible Fluid. *U.S. Air Force Tech. Rept.* **6358**, Part X (1953). Wright Air Development Center.

15.53 Turner, M. J.: Aerodynamic Theory of Oscillating Sweptback Wings. *J. Math. & Phys.* **28**, 280–293 (1950).

15.54 Wasserman, L. S.: Aspect Ratio Corrections in Flutter Calculations. *U.S. Air Force Memo. Rept.* **MCREXA-5-4595-8-5**. AMC (1948).

15.55 Zartarian, G., P. T. Hsu, and H. Ashley: Analysis of Tip Effects on the Aerodynamic Forces on an Oscillating Wing. *MIT Aeroelastic Lab. Rept.* on Contract NOa(s) 8790, Vol. 8 (1950).

Oscillating wing, finite span, subsonic flow:

15.56   Küssner, H. G.: A General Method for Solving Problems of the Unsteady Lifting Surface Theory in the Subsonic Range. *J. Aeronaut. Sci.* **21**, 17–26 (1954). Comments by Miles, p. 427.

15.57   Miles, J. W.: On the Compressibility Correction for Subsonic Unsteady Flow. *J. Aeronaut. Sci.* **17** Reader's Forum, 181 (1950).

15.58   Reissner, E.: On the Theory of Oscillating Airfoils of Finite Span in Subsonic Compressible Flow. *NACA Tech. Rept.* **1002** (1950). Supersedes *NACA Tech. Note* **1953** (1949). Extension of the theory, *Tech. Note* **2274** (1951).

15.59   Watkins, C. E., H. Runyan, and D. Woolston: On the Kernel Function of the Integral Equation Relating the Lift and Downwash Distributions of Oscillating Finite Wings in Subsonic Flow. *NACA Tech. Note* **3131** (1954).

Oscillating wing, finite span, supersonic flow:

15.60   Chang, C. C.: The Aerodynamic Behavior of a Harmonically Oscillating Finite Swept-Back Wing in Supersonic Flow. *NACA Tech. Note* **2467** (1951).

15.61   Froehlich, J. E.: Nonstationary Motion of Purely Supersonic Wings. *J. Aeronaut. Sci.* **18**, 298–310 (1951).

15.62   Garrick, I. E., and S. I. Rubinow: Theoretical Study of Air Forces on an Oscillating or Steady Thin Wing in a Supersonic Main Stream. *NACA Tech. Note* **1383** (1947).

15.63   Goodman, Th.: The Quarter-Infinite Wing Oscillating at Supersonic Speeds. *Quart. Applied Math.* **10**, 189–192 (1952).

15.64   Haskind, M. D., and S. V. Falkovich: Vibration of a Wing of Finite Span in a Supersonic Flow. *NACA Tech. Memo.* **1257** (1950). Translated from *Prik. Math. i. Mekh.* **11** (1947).

15.65   Hipsh, H.: Harmonic Oscillations of Narrow Delta Wing in Supersonic Flow. Ph.D. thesis, California Institute of Technology (1951).

15.66   Jones, W. P.: Supersonic Theory for Oscillating Wings of Any Plan Form. *Aeronaut. Research Council R. & M.* **2655** (1953).

15.67   Li, T. Y.: Purely Rolling Oscillations of a Rectangular Wing in Supersonic Flow. *J. Aeronaut. Sci.* **18**, 191–198 (1951).

15.68   Miles, J. W.: On Harmonic Motion at Supersonic Speeds. *J. Aeronaut. Sci.* **16**, Reader's Forum, 378 (1949). See also remarks on oscillating aileron, p. 511, and on damping in pitch for delta wings, p. 574.

15.69   Miles, J. W.: On the Reduction of Unsteady Supersonic Flow Problems to Steady Flow Problems. *J. Aeronaut. Sci.* **17**, Reader's Forum, 64 (1950).

15.70   Miles, J. W.: On the Oscillating Rectangular Airfoil at Supersonic Speeds. *J. Aeronaut. Sci.* **16**, Reader's Forum, 381 (1949). Errata, p. 702. *Quart. Applied Math.* **9**, 47–65 (1951).

15.71   Nelson, H. C.: Lift and Moment on Oscillating Triangular and Related Wings with Supersonic Edges. *NACA Tech. Note* **2494** (1951).

15.72   Nelson, H. C., R. Rainey, and C. E. Watkins: Lift and Moment Coefficients Expanded to the Seventh Power of Frequency for Oscillating Rectangular Wings in Supersonic Flow and Applied to a Specific Flutter Problem. *NACA Tech. Note* **3076** (1954).

15.73   Robinson, A.: Rotary Derivatives of a Flat Delta Wing at Supersonic Speeds. *J. Roy. Aeronaut. Soc.* **52**, 735–752 (1948).

15.74 Robinson, A.: On Some Problems of Unsteady Supersonic Aerofoil Theory (Delta Wing). *Proc. 7th Intern. Congr. Applied Mech.* **2**, 500–514 (1948). *Coll. Aeronaut. Cranfield Rept.* **16** (1948).

15.75 Stewart, H. J., and T. Y. Li: Periodic Motions of a Rectangular Wing Moving at Supersonic Speed. *J. Aeronaut. Sci.* **17**, 529–539 (1950).

15.76 Stewart, H. J., and T. Y. Li: Source-Superposition Method of Solution of a Periodically Oscillating Wing at Supersonic Speed. *Quart. Applied Math.* **9**, 31–45 (1951). Results differ from Miles, Ref. 15.70; Stewartson, Ref. 14.40; and Goodman, Ref. 15.63.

15.77 Wasserman, L. S.: Aspect Ratio Corrections for Flutter Calculations in Compressible Flow. *U.S. Office of Air Research Tech. Rept.* **4** (1950).

15.78 Watkins, C. E.: Effect of Aspect Ratio on the Air Forces and Moments of Harmonically Oscillating Thin Rectangular Wings in Supersonic Potential Flow. *NACA Rept.* **1028**. Supersedes *Tech. Note* **2064** (1950).

15.79 Watkins, C. E., and J. H. Berman: Air Forces and Moments on Triangular and Related Wings with Subsonic Leading Edges Oscillating in Supersonic Potential Flow. *NACA Rept.* **1099** (1952). Supersedes *NACA Tech. Note* **2457**. See also *Tech. Note* **3009** (1953).

Indicial responses, wings of finite span, incompressible fluid:

15.80 Jones, W. P.: Aerodynamic Forces on Wings in Non-uniform Motion. *Aeronaut. Research Council R. & M* **2117** (1945).

15.81 Scalan, R. H.: Correction for Aspect Ratio of the Indicial Lift Function of Wagner and Küssner. *J. Aeronaut. Sci.* **19**, Reader's Forum, 357–358 (1952).

Indicial responses, wings of finite span, compressible fluid:

15.82 Goodman, Th. R.: Aerodynamics of a Supersonic Rectangular Wing Striking a Sharp-Edged Gust. *J. Aeronaut. Sci.* **18**, 519–526 (1951).

15.83 Lomax, H., M. A. Heaslet, and F. B. Fuller: Three-Dimensional Unsteady Lift Problems in High-Speed Flight—The Triangular Wing. *NACA Tech. Note* **2387** (1951).

15.84 Lomax, H.: Lift Developed on Unrestricted Rectangular Wings Entering Gusts at Subsonic and Supersonic Speeds. *NACA Tech. Note* **2925** (1953).

15.85 Miles, J. W.: Transient Loading of Supersonic Rectangular Airfoils. *J. Aeronaut. Sci.* **17**, 647–652 (1950). See also **19**, Reader's Forum, 418 (1952).

15.86 Miles, J. W.: Transient Loading of Wide Delta Airfoils at Supersonic Speeds. *J. Aeronaut. Sci.* **18**, 543–554 (1951). See also, **16**, Reader's Forum, 568 (1949).

15.87 Miles, J. W.: A Note on Subsonic Edges in Unsteady Supersonic Flow. *Quart. Applied Math.* **11**, 363–367 (1953).

15.88 Rott, N.: On the Unsteady Motion of a Thin Rectangular Wing in Supersonic Flow. *J. Aeronaut. Sci.* **18**, Reader's Forum, 775 (1951).

15.89 Strang, W. J.: Transient Lift of Three-Dimensional Purely Supersonic Wings. *Proc. Roy. Soc. London A.* **202**, 54–80 (1950).

Wings of low-aspect ratio:

15.90 Halfman, R. L., and H. Ashley: Aeroelastic Properties of Slender Wings. *Midwestern Conf. Applied Mech. Chicago* (1951).

15.91 Jones, R. T.: Properties of Low-Aspect-Ratio Pointed Wings at Speeds below and above the Speed of Sound. *NACA Rept.* **835** (1946).

15.92   Lawrence, H. R., and E. H. Gerber: The Aerodynamic Forces on Low Aspect Ratio Wings Oscillating in an Incompressible Flow. *J. Aeronaut. Sci.* 19, 769–781 (1952).

15.93   Merbt, H., and M. Landahl: Aerodynamic Forces on Oscillating Low Aspect Ratio Wings in Compressible Flow. *Roy. Inst. Technol. Stockholm Rept. KTH Aero.* TN 30. Results and Tables of Auxiliary Functions, TN 31 (1954).

15.94   Miles, J. W.: On the Low Aspect Ratio Oscillating Rectangular Wing in Supersonic Flow. *Aeronaut. Quart.* 4 231–244 (Aug. 1953).

Slender bodies (see also Refs. 15.5 and 14.40):

15.95   Dorrance, W. H.: Nonsteady Supersonic Flow about Pointed Bodies of Revolution. *J. Aeronaut. Sci.* 18, 505–511 (1951).

15.96   Laitone, E. V.: The Linearized Subsonic and Supersonic Flow about Inclined Slender Bodies of Revolution. *J. Aeronaut. Sci.* 14, 631–642 (Nov. 1947).

15.97   Miles, J. W.: On Non-steady Motion of Slender Bodies. *Aeronaut. Quart.* 2, 183–194 (1950). See also further analyses in *J. Aeronaut. Sci.* 19, Reader's Forum, 280 (1952). *Quart. J. Mech. Applied Math.* 6, 286–289 (1953).

Wing-body interference:

15.98   Henderson, A., Jr.: Pitching-Moment Derivatives C_{mq} and $C_{m\dot\alpha}$ at Supersonic Speeds for a Slender-Delta-Wing and a Slender-Body Combination and Approximate Solutions for Broad-Delta-Wing and Slender-Body combination. *NACA Tech. Note* 2553 (1951).

15.99   Spreiter, J. R.: Aerodynamic Properties of Slender Wing-Body Combinations at Subsonic, Transonic, and Supersonic Speeds. *NACA Tech. Note* 1662 (1948).

15.100 Vandrey, F.: Zur Theoretischen Behandlung des gegenseitigen Einflusses von Tragflügel und Rumpf. *Luftfahrt-Forsch.* 14, 347–355 (July 1937).

Wind-tunnel wall interference:

15.101 Goodman, T. R.: The Upwash Correction for an Oscillating Wing in a Wind Tunnel. *Cornell Aeronaut. Lab. Rept.* AD-744-W1 (1951).

15.102 Jones, W. P.: Wind Tunnel Interference Effect on the Values of Experimentally Determined Derivative Coefficients for Oscillating Aerofoils. *Aeronaut. Research Council R. & M.* 1912 (1943).

15.103 Reissner, E.: Wind Tunnel Corrections for the Two-Dimensional Theory of Oscillating Airfoils. *Cornell Aeronaut. Lab. Rept.* SB-318-S-3 (1947).

15.104 Runyan, H. L., and C. E. Watkins: Considerations on the Effect of Wind Tunnel Walls on Oscillating Air Forces for Two-Dimensional Subsonic Compressible Flow. *NACA Tech. Note* 2552 (1951).

15.105 Timman, R.: The Aerodynamic Forces on an Oscillating Aerofoil between Two Parallel Walls. *Applied Sci. Research* A3, no. 1, 31–57 (1951). The Hague, Netherlands.

15.106 Woolston, D. S., and H. L. Runyan: Some Considerations on the Air Forces on a Wing Oscillating between Two Walls for Subsonic Compressible Flow. Paper presented at the 22d annual meeting, Institute of Aeronautical Sciences (1954). *J. Aeronaut. Sci.* 22, 41–50 (1955).

For propulsion and drag, see review by Cicala, Ref. 15.1.

Oscillating wedge with shock wave:

15.107   Borg, S. F.: On Unsteady Nonlinearized Conical Flow. *J. Aeronaut. Sci.* **19**, 85–92 (1952).

15.108   Carrier, G. F.: The Oscillating Wedge in a Supersonic Stream. *J. Aeronaut. Sci.* **16**, 150–152 (1949).

15.109   Carrier, C. F.: On the Stability of the Supersonic Flows Past a Wedge. *Quart. Applied Math.* **6**, 367–378 (Jan. 1949).

Experiments. For a general review of German works, see Küssner, Ref. 7.3, where an extensive bibliography of less accessible German papers can be found. British works are reviewed by Fage, Ref. 4.43.

Experiments, indicial motion:

15.110   Drescher, H.: Untersuchungen an einem symmetrischen Tragflügel mit spaltlos angeschlossenem Ruder bei raschen Aenderungen des Ruderausschlags (ebene Strömung). *Mitt. Max-Planck-Inst. Strömungs-Forsch.* Nr. 6 (1952). Göttingen.

15.111   Farren, W. S.: Reaction on a Wing Whose Angle of Incidence Is Changing Rapidly. *Aeronaut. Research Com. R. & M.* **1648** (1935).

15.112   Farren, W. S.: An Apparatus for the Measurement of Two-Dimensional Flow at High Reynolds Numbers with an Application of the Growth of Circulation Round a Wing Started Impulsively from Rest. *Proc. 3d Intern. Congr. Applied Mech. Stockholm* (1930).

15.113   Francis, R. H., and J. Cohen: The Flow Near a Wing Which Starts Suddenly from Rest and Then Stalls. *Aeronaut. Research Com. R. & M.* **1561** (1933).

15.114   Harper, P. W., and R. E. Flanigan: The Effect of Rate of Change of Angle of Attack on the Maximum Lift of a Small Model. *NACA Tech. Note* **2061** (1950).

15.115   Kramer, M.: Die Zunahme des Maximalauftriebes von Tragflügeln bei plötzlicher Anstellwinkelvergrösserung (Böeneffekt). *Z. Flugtech. u. Motorluftschif.* **23**, 185–189 (1932).

15.116   Küssner, H. G.: Untersuchung der Bewegung einer Platte beim Eintritt in eine Strahlgrenze. *Luftfahrt-Forsch.* **13**, 425–429 (1936).

15.117   Scheubel, N.: Some Tests on the Increase of the Maximum Lift of Aerofoils Whose Angle of Incidence Changes at Constant Angular Velocity. *Mitt. deut. Akad. Luftfahrt-Forsch.* **1** (1942).

15.118   Silverstein, A., S. Katzhoff, and I. A. Hootman: Comparative Flight and Full-Scale Wind Tunnel Measurements of the Maximum Lift of an Aeroplane. *NACA Rept.* **618** (1938).

15.119   Walker, P. B.: Experiments on the Growth of Circulation about a Wing with a Description of an Apparatus for Measuring Fluid Motion. *Aeronaut. Research Com. R. & M.* **1402** (1931).

15.120   Wieselsberger, C.: Electric Measurement of Forces by Varying an Inductivity. *ForschBer.* **266** (1935).

Experiments, oscillatory motion: See Ref. 7.3 for additional references.

15.121   Bratt, J. B., and C. Scruton: Measurements of Pitching Moment Derivatives for an Airfoil Oscillating about the Half-Chord Axis. *Aeronaut. Research Com. R. & M.* **1921** (1938). See also Bratt, K. C. Wight, and V. J. Tilly, *R. & M.* **2063**; Bratt and Wight, *R. & M.* **2064**; Bratt, Wight, and A. Chinneck, *R. & M.* **2214**; Bratt and Chinneck, *R. & M.* **2680**.

15.122 Bratt, J. B., and G. J. Davis: The Influence of Aspect Ratio and Taper on the Fundamental Damping Derivative Coefficient for Flexural Motion. *Aeronaut. Research Council R. & M.* **2032** (1945).

15.123 Bratt, J. B.: Flow Patterns in the Wake of an Oscillating Aerofoil. *Aeronaut. Research Council R. & M.* **2773** (1953).

15.124 Buchan, A. L., K. D. Harris, and P. M. Somervail: Measurements of the Derivative Z_w for an oscillating aerofoil. *Coll. Aeronaut. Cranfield Rept.* **40** (June 1950). *Aeronaut. Research Council Current Papers* **52** (1951).

15.125 Cicala, P.: Ricerche sperimentali sulle azione aerodynamiche sopra l'ala oscillante. *Aerotechnica* **17**, 405–414, 1043 (1937); 46 (1951).

15.126 Drescher, H.: Eine experimentelle Bestimmung der aerodynamischen Reaktionen auf einen Flügel mit schwingenden Ruder. *Österreichisches Ing.-Arch.* **4**, 270–290 (1950).

15.127 Gracey, W.: The Additional-Mass Effect of Plates as Determined by Experiments. *NACA Rept.* **707** (1941).

15.128 Greidanus, H., A. I. van de Vooren, and H. Bergh: Experimental Determination of the Aerodynamic Coefficients of an Oscillating Wing in Incompressible, Two-Dimensional Flow. *Natl. Luchtvaartlab. Amsterdam Repts.* **F101, F102, F103, F104** (1952).

15.129 Halfman, R. L.: Experimental Aerodynamic Derivatives of a Sinusoidally Oscillating Airfoil in Two-Dimensional Flow. *NACA Rept.* **1108** (1952). Supersedes *NACA Tech. Note* **2465**. See also Ref. 9.22.

15.130 Reid, E. G., and W. Vincenti: An Experimental Determination of the Lift of an Oscillating Airfoil. *J. Aeronaut. Sci.* **8**, 1–6 (1940).

15.131 Scruton, C.: Some Experimental Determination of the Apparent Additional Mass Effect for an Aerofoil and for Flat Plates. *Aeronaut. Research Com. R. & M.* **1931** (1941).

15.132 Scruton, C., and W. G. Raymer: Measurements of the Direct Elevator and Fuselage Vertical Bending Derivatives for Decaying Oscillations. *Aeronaut. Research Council R. & M.* **2323** (Monograph), App. IV (1943).

15.133 Scruton, C., W. G. Raymer, and D. V. Dunsdon: Experimental Determination of the Aerodynamic Derivatives for Flexure-Aileron Flutter of B.A.C. Wing Type 167. *Aeronaut. Research Council R. & M.* **2373** (1945).

15.134 Silverstein, A., and U. T. Joyner: Experimental Verification of the Theory of Oscillating Airfoils. *NACA Rept.* **673** (1939).

15.135 Voigt, H., and F. Walter: Trial of a Measuring Method and an Experimental Installation for the Determination of Unsteady Aerodynamic Forces (experiments of an oscillating control surface). *Deut. Luftfahrt-Forsch. ForschBer.* **1575** (1942).

15.136 Walter, F., and W. Heger: Results of Wind Tunnel Tests for the Determination of the Moments of the Aerodynamic Forces on an Oscillating Control Surface. *Deut. Luftfahrt-Forsch. UM* **1207** (1944).

Appendix 1

ON DEFINITIONS OF SHEAR CENTER

There exists several interpretations for the term "torsion-free bending" on which the definition of shear center (§ 1.2) is based.

Consider a uniform cantilever beam (Fig. A 1.1) of length l, built-in at the right end ($z = l$), and loaded at the left end by a force S which acts in the negative y-axis direction. The origin is taken at the centroid of

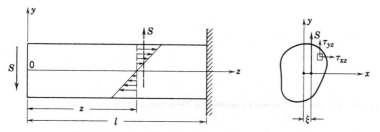

Fig. A 1.1.   A uniform cantilever beam.   Notations.

the cross section at the free end.   The horizontal axis x and the vertical axis y are the principal axes of the cross section.

First Analytical Definition.   The stresses and deflection of this beam can be found by Saint-Venant's theory of bending.   Let u, v represent the components of displacement in the x, y axes' directions, respectively. It is well known that the "rotation" ω of an element in the cross section of the beam is expressed by the equation

$$\omega = \frac{1}{2}\left(\frac{\partial v}{\partial x} - \frac{\partial u}{\partial y}\right) \tag{1}$$

The rate of change of ω in the axial direction is

$$\frac{\partial \omega}{\partial z} = \frac{1}{2}\frac{\partial}{\partial z}\left(\frac{\partial v}{\partial x} - \frac{\partial u}{\partial y}\right) = \frac{1}{2}\frac{\partial}{\partial x}\left(\frac{\partial v}{\partial z} + \frac{\partial w}{\partial y}\right) - \frac{1}{2}\frac{\partial}{\partial y}\left(\frac{\partial u}{\partial z} + \frac{\partial w}{\partial x}\right)$$

$$= \frac{\partial e_{yz}}{\partial x} - \frac{\partial e_{xz}}{\partial y}$$

471

where e_{xz} and e_{yz} are the shearing-strain components. By Hooke's law, one obtains

$$\frac{\partial \omega}{\partial z} = \frac{1}{2G}\left(\frac{\partial \tau_{yz}}{\partial x} - \frac{\partial \tau_{xz}}{\partial y}\right) \tag{2}$$

where τ_{xz} and τ_{yz} are shearing-stress components.

It turns out that the right-hand side of Eq. 2 can be expressed very simply in terms of the stress function used in Saint-Venant's theory. The result indicates that the "local twist" $\partial \omega/\partial z$ at different points in the cross section has different values. It is impossible to have this zero for all elements of the cross section.

It seems natural to define the torsion-free bending by the condition that the average value of the local twist over the whole section vanishes.* Hence, the first analytic definition of a torsion-free bending is

$$\iint \frac{\partial \omega}{\partial z}\, dx\, dy = 0 \tag{3}$$

By using Eq. 3, the problem of "torsion-free" bending can be solved in a classical way. The shear center is then determined from the fact that the resultant of the shearing stresses in the section and the load S must be equal and opposite and have the same moment arm about the z axis. The distance ξ of the resultant from the y axis is therefore the abscissa of the shear center. Equating the moment about the z axis of the shearing stresses τ_{xy} and τ_{yz} with that of S, one obtains

$$\xi = \frac{1}{S}\iint (x\tau_{yz} - y\tau_{xz})\, dx\, dy \tag{4}$$

In this formulation, the position of the shear center depends on the Poisson's ratio μ.

Second Analytical Definition. Trefftz proposes† another definition of torsion-free bending on the basis of energy considerations. If a beam is twisted by a couple M at the free end, so that that end rotates through an angle α, the elastic strain energy stored in the beam is equal to the work done by the couple during deformation

$$A_1 = \tfrac{1}{2}M\alpha \tag{5}$$

---

* This is the view taken by J. N. Goodier, *J. Aeronaut. Sci.* **11**, 272–280 (1944). R. D. Specht, in a note to *J. Applied Mech.* **10**, A-235–236 (1943), attributed this definition to A. C. Stevenson (*Phil. Trans. Roy. Soc. London, A.* **237**, 161–229 (1938–1939).) This definition agrees with the works of Timoshenko.

† E. Trefftz, *Z. angew. Math. u. Mech.* **15**, 220–225 (1935). For a different point of view, which leads to results agreeing with Trefftz's definition, see P. Cicala, *Atti. R. Acc. Sci. Torino* **70**, 356–371 (1935), and A. Weinstein, *Quart. Applied Math,* **5**, 97–99 (1947).

On the other hand, if a single force S is applied at the free end where the deflection is δ, then the elastic strain energy is

$$A_2 = \tfrac{1}{2}S\delta \tag{6}$$

In combined action of the couple M and the force S, the elastic strain energy stored in the beam is, in general, different from the sum $A_1 + A_2$. Let us apply the torsional moment first, so that it does the work A_1. Then apply the bending load S, keeping M fixed. The load S does the work A_2, while the moment M must do additional work corresponding to the angle of rotation of the section induced by the action of S. Trefftz defines the torsion-free bending by the condition that this energy of "interaction" be zero. Alternately, if a shear acts through the "shear center," so that the beam is in "pure" bending, and then a torque is added, the elastic energy in the beam is simply the sum of elastic energies due to the torsion and the "pure" bending alone. The order of application of S and M is evidently immaterial to this definition.

Trefftz derives an expression for the coordinates of the shear center with the aid of the "warping" function in St.-Venant's torsion theory.

The shear center so determined is independent of the Poisson's ratio μ; and the average value of the local twist over the cross section, given by the integral in Eq. 3, does not always vanish.

When the section is symmetrical about the y axis, the two definitions yield the same location of the shear center. It can also be shown that, for a single-cell closed thin-walled section, the two definitions agree, whereas, for a multicell thin-walled section, the two definitions in general disagree.

Example. For a semicircular cylinder the abscissa of the shear center is

(a) By the first definition,*

$$\xi = \frac{8}{15\pi}\frac{3+4\mu}{1+\mu}a$$

(b) By the second definition,†

$$\xi = \frac{8a}{5\pi}$$

For a Poisson's ratio $\mu = 0.3$; the locations of the shear center P are shown in Fig. A 1.2.

* Timoshenko, *Theory of Elasticity*, p. 301, McGraw-Hill (1934).
† Trefftz, *op. cit.*

Thin-Walled Sections—Third Definition. For a single-cell closed thin-walled tube, it is shown in § 1.2 that the shear flow q at a point s is

$$q = q_0 + \frac{S}{I} \int_{s_0}^{s} yt \, ds \tag{7}$$

where q_0 is the value of q at s_0, and s is the distance measured along the wall. In order to find the shear flow, it is necessary to determine q_0. In aeronautical literature it is customary to define a "pure" bending of the tube as one in which the value of q_0 is so chosen that the strain energy

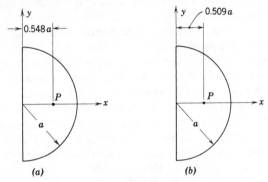

Fig. A 1.2.   Locations of shear center according to different definitions.

stored in the tube is a minimum. Now, the elastic strain energy per unit length of the tube, due to the shearing strain, is

$$U = \oint \frac{\tau^2}{2G} t \, ds = \oint \frac{q^2}{2Gt} \, ds \tag{8}$$

where the integral is taken over the entire section. By definition of the pure bending,

$$\frac{\partial U}{\partial q_0} = \oint \frac{2q}{2Gt} \frac{\partial q}{\partial q_0} \, ds = 0 \tag{9}$$

But $\partial q/\partial q_0 = 1$ according to Eq. 7; hence, the condition for pure bending is

$$\oint \frac{q}{t} \, ds = \oint \tau \, ds = 0 \tag{10}$$

The last result can be generalized to multicelled tubes. For an *n*-celled tube it is necessary to determine n integration constants of the nature of

q_0 above. It can be shown that an equation like 10 holds for every possible closed circuit drawn along the walls. Since there are n independent circuits for an n-celled tube, the n integration constants can be uniquely determined.

Having determined the shear flow, the abscissa of the shear center is obtained from Eq. 18, § 1.2.

Appendix 2

ROUTH-HURWITZ METHOD

The problem of the stability of a dynamic system is often reduced to determining whether the real parts of all the roots of a polynomial are negative.

For a polynomial of fourth degree with real-valued coefficients,

$$a_4x^4 + a_3x^3 + a_2x^2 + a_1x + a_0 = 0 \qquad (a_0 > 0) \qquad (1)$$

Routh* shows that the real parts of all the roots are negative if and only if all the coefficients $a_0, a_1, \cdots, a_4$, and the discriminant

$$R = a_1a_2a_3 - a_0a_3{}^2 - a_4a_1{}^2 \qquad (2)$$

are of the same sign. For a polynomial of nth degree, Hurwitz established a criterion in terms of a series of determinants. Hurwitz's determinants for the fourth-degree equation 1, with $a_0 > 0$, are

$$|a_1|, \quad \begin{vmatrix} a_1 & a_0 \\ a_3 & a_2 \end{vmatrix}, \quad \begin{vmatrix} a_1 & a_0 & 0 \\ a_3 & a_2 & a_1 \\ 0 & a_4 & a_3 \end{vmatrix}, \quad \begin{vmatrix} a_1 & a_0 & 0 & 0 \\ a_3 & a_2 & a_1 & a_0 \\ 0 & a_4 & a_3 & a_2 \\ 0 & 0 & 0 & a_4 \end{vmatrix} \qquad (3)$$

These must all be positive if the real parts of all the roots are negative, and vice versa. For a fourth-degree polynomial, determinants of order 1 to 4 are involved. The diagonal from the upper left to lower right contains coefficients with increasing indices beginning with 1. The indices of the coefficients decrease from left to right in each row. Negative indices and indices greater than the degree of the polynomial involved are replaced by zero. These rules are sufficient to establish the determinants for a polynomial of any degree n.

It is easy to show that Routh and Hurwitz conditions are equivalent. The proof for Routh's conditions for polynomials of third and fourth degrees can be found in Kármán and Biot's book;† that for Hurwitz can be found in Uspensky.‡ The literature related to this algebraic problem is very extensive. In Bateman's review§ on this subject, more than 100 papers are quoted.

---

* Routh, *Advanced Rigid Dynamics*, Vol. II, Macmillan Co., London (1930).

† *Mathematical Principles in Engineering*, McGraw-Hill, New York (1940).

‡ Uspensky, *Theory of Equations*, McGraw-Hill (1948).

§ H. Bateman, *The Control of Elastic Fluids*, Bull. Am. Math. Soc. **51**, 601–646 (1945).

Appendix 3

ON DAMPING COEFFICIENT

Consider a single-degree-of-freedom system such as a mass-spring-dashpot model. An idealized equation of motion is (§ 1.8)

$$m\ddot{x} + \beta\dot{x} + Kx = X \tag{1}$$

where the constants m, β, K are such that the kinetic energy, potential energy, and the dissipation function of the system are given, respectively, by

$$\tfrac{1}{2}m\dot{x}^2, \qquad \tfrac{1}{2}Kx^2, \qquad \tfrac{1}{2}\beta\dot{x}^2$$

The dissipation function has the dimensions of energy per unit time. β is interpreted as viscous damping factor. X is a forcing function.

In § 2.4, p. 69, it is shown that Eq. 1 can be conveniently written as

$$\ddot{x} + 2\gamma\omega_0\dot{x} + \omega_0{}^2x = \frac{1}{m}X \tag{2}$$

where $\omega_0{}^2 = K/m$, and γ is the ratio of actual damping β to the critical damping factor $\beta_{cr} = 2m\omega_0$.

If the forcing function is a harmonic function, the solution of Eqs. 1 and 2 has already been given in § 1.8.

It is instructive to compare the form of damping assumed in Eq. 1 with that commonly used in flutter analysis. Following § 6.9, we write

$$\ddot{x} + (1 + ig)\omega_0{}^2x = \frac{1}{m}X \tag{3}$$

which is applied to harmonic oscillations only. But if a motion is harmonic so that $x = x_0e^{i\omega t}$, Eq. 2 may be written as

$$\ddot{x} + \left(1 + 2i\gamma\,\frac{\omega}{\omega_0}\right)\omega_0{}^2x = \frac{1}{m}X \tag{4}$$

This is identical in form with Eq. 3 if, and only if,

$$\omega = \omega_0, \qquad g = 2\gamma \tag{5}$$

It is therefore clear that the mechanisms of damping suggested by Eqs. 1 and 3 are entirely different if $\omega \neq \omega_0$, and if the motion is not a harmonic one.

As is discussed in § 11.4, the detailed mechanism of damping in structures concerned with in aeroelasticity is yet unknown.   Hence, a choice of the particular form of damping is open to question.   For the same reason, the value of g is rarely accurately determined.   It is known, however, that g is small for metal airplanes.

When g is small, a free oscillation of a system described in Eq. 2 following an initial disturbance will be almost sinusoidal with a frequency close to ω_0.   Assuming that Eq. 3 is applicable also to this case, then we have approximately

$$g = 2\gamma$$

as in Eq. 5.   It is then easily derived from the solution given in § 1.8 that, when $g \ll 1$, $\gamma \ll 1$, the logarithmic decrement δ in a free oscillation following an initial disturbance is

$$\delta \doteq \pi g \doteq 2\pi\gamma$$

More generally, g can be deduced, approximately, from the rate of decay of free oscillations,

$$g = \frac{\log_e N}{\pi \Delta n}$$

where Δn is the number of cycles required for the oscillation to reach an amplitude equal to $1/N$ times the base amplitude.

It is often more convenient to measure g by a resonance vibration test. Let the exciting force be harmonic so that $\dfrac{1}{m} X = Ae^{i\omega t}$, then, according to the method of § 1.8, the steady-state solution of Eq. 3 is

$$x = \frac{Ae^{i\omega t}}{-\omega^2 + (1 + ig)\omega_0^2}$$

Thus the amplitude of response if

$$|x| = \frac{|A|}{\omega_0^2 \sqrt{\left(1 - \dfrac{\omega^2}{\omega_0^2}\right)^2 + g^2}}$$

The resonant peak is reached at $\omega = \omega_0$, at which

$$|x|_{\max} = \frac{|A|}{\omega_0^2 g}$$

When ω differs slightly from ω_0, so that

$$\omega = \omega_0 \pm \frac{\Delta\omega}{2}, \qquad (\Delta\omega \ll \omega_0)$$

' we have

$$|x| = \frac{|A|}{\omega_0^2 \sqrt{\left[1 - \frac{(\omega_0 \pm \Delta\omega/2)^2}{\omega_0^2}\right]^2 + g^2}} \doteq \frac{|A|}{\omega_0^2 \sqrt{\left(\frac{\Delta\omega}{\omega_0}\right)^2 + g^2}}$$

or

$$\frac{|x|^2}{|x|^2_{\text{max}}} \doteq \frac{g^2}{(\Delta\omega/\omega_0)^2 + g^2}$$

Solving for g, we obtain

$$g \doteq \frac{\Delta\omega/\omega_0}{\sqrt{\frac{|x|^2_{\text{max}}}{|x|^2} - 1}}$$

If $|x|/|x|_{\text{max}} = 1/\sqrt{2}$, then

$$g \doteq \frac{\Delta\omega}{\omega_0}$$

This result offers a practical way of determining the value of g by resonance oscillation test. Assume that a vibrator, whose exciting force is independent of the frequency, is used. If one measures the width (difference in frequencies) of the resonance curve at which the resonance amplitude equals $1/\sqrt{2} \doteq 0.707$ of the peak amplitude, then g is equal to that width divided by the resonance frequency (Fig. A 3.1).

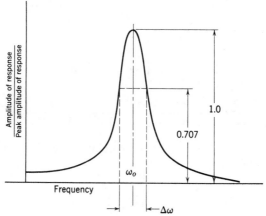

Fig. A 3.1. An amplitude response curve.

Appendix 4

EVALUATION OF AN INTEGRAL OCCURRING IN § 14.3

$$I = \int_0^\infty e^{-i\omega\kappa x} H_0^{(2)} \left(\frac{\omega}{\beta^2 a} x \right) dx$$

Let

$$\omega\kappa x = \frac{\omega}{U\beta^2} x = u$$

Then

$$I = \frac{U\beta^2}{\omega} \int_0^\infty e^{-iu} H_0^{(2)}(Mu) \, du$$

The integral representation of $H_0^{(2)}(x)$ is

$$\int_1^\infty e^{-ix\xi} \frac{d\xi}{\sqrt{\xi^2 - 1}} = -\frac{i\pi}{2} H_0^{(2)}(x) \qquad (x > 0) \quad \text{(p. 170, Watson,}$$
$$\text{\textit{Bessel Function})}$$

Hence,

$$I = \frac{U\beta^2}{\omega} \frac{2}{\pi} i \int_0^\infty e^{-iu} \, du \int_1^\infty e^{-iMu\xi} \frac{d\xi}{\sqrt{\xi^2 - 1}}$$

Change the order of integration

$$I = \frac{2U\beta^2}{\pi\omega} i \int_1^\infty \frac{d\xi}{\sqrt{\xi^2 - 1}} \int_0^\infty e^{-i(1 + M\xi)u} \, du$$

$$= \frac{2U\beta^2}{\pi\omega} i \int_1^\infty \frac{d\xi}{\sqrt{\xi^2 - 1}} \left[\frac{ie^{-i(1+M\xi)u}}{1 + M\xi} - \frac{i}{1 + M\xi} \right]_{u \to \infty}$$

When $u \to \infty$, the first term vanishes according to the Riemann-Lebesque lemma on Fourier integral. Hence,

$$I = \frac{2U\beta^2}{\pi\omega} \int_1^\infty \frac{d\xi}{\sqrt{\xi^2 - 1}(1 + M\xi)}$$

Let $\xi = \cosh \eta$; the integral becomes

$$I = \frac{2U\beta^2}{\pi\omega} \int_0^\infty \frac{d\eta}{1 + M \cosh \eta}$$

From Pierce, *A Short Table of Integrals*, formula 472,

$$\int \frac{dx}{1 + \cos \alpha \cosh x} = 2 \csc \alpha \tanh^{-1} \left(\tanh \frac{x}{2} \tan \frac{\alpha}{2} \right)$$

Hence,

$$I = \frac{2U\beta^2}{\pi\omega} \frac{2}{\sqrt{1 - M^2}} \tanh^{-1} \left(\sqrt{\frac{1 - M}{1 + M}} \tanh \frac{x}{2} \right) \Bigg|_0^\infty = \frac{4U\beta}{\pi\omega} \tanh^{-1} \sqrt{\frac{1 - M}{1 + M}}$$

$$= \frac{2U\beta}{\pi\omega} \log \frac{1 + \sqrt{1 - M^2}}{M}$$

AUTHOR INDEX

SUBJECT INDEX

Acceleration potential, 388, 396
  general form in airfoil theory, 398
  linearized equations of, 389–391
  of a doublet, 400, 422
Admittance, 272, 278, 280, 300
Aeolian harp, 62
Aerodynamic center, 33, 130
Aerodynamic derivatives, classical derivative coefficients, 228
  L_h, L_α, etc., 229–232
  quasi-steady, 192, 193
  sign conventions for, 229
Aerodynamic hysteresis, 322, 323, 453
Aerodynamic operator, 136–137, 375
Aerodynamics, fundamental equations,
    equation of continuity, 382
  Eulerian equation of motion, 382
  unsteady ·motion, 462, 463, 469, 470
    arbitrary motion, 208, 381, 395,
      444–446
    impulsive motion, 206, 446, 451–452
  see also Airfoil theory, unsteady motion, oscillating
Aileron efficiency, 114, 117, 118
  optimum aileron chord ratio for, 119
Aileron reversal, 140, 156
  critical speed of, 113
  general case, 119
  general equations for, 121
  matrix method, 140
  method of successive approximations, 124
  relation with divergence, 117, 125
  semirigid theory, 122
  two-dimensional case, 116
Airfoil efficiency factor, 32
Airfoil theory, boundary conditions,
    392–394, 397, 401, 406
  conformal transformation of, 399
  finite span wing, 125–130, 448
  of planar wing, 393

Airfoil theory, quasi-steady theory, 191
  thin airfoil in steady flow, 187
  see also oscillating airfoils
AMC method, 241
Anelasticity of solids, 373
Angle of attack, 30, 32
Apparent mass, 70, 210, 406, 447

Beam theory, 5–10, 471
Bessel functions, 404, 410
Biot and Arnold's low-speed flutter criterion, 171, 184
Birnbaum's oscillating-wing theory, 179, 182, 415
Block diagram, 366
Bode diagram, 353
Boundary layer, 63–64
Bratt's oscillating-wing experiments, 322, 323, 450, 469
Buffeting, bibliography, 330
  boundaries of, 313, 316
  cause of, 313
  control by fillet, 312
  Duncan's intensity contour of, 311
  historical notes on, 310–312
  phenomenon of, 310
  theories of, 317
Buffeting flutter of a wing, 328–330

Castigliano's theorem, 22, 253
Cauchy-Riemann differential equations, 387
Cauchy's theorem, 356
Center of twist, 18
Circulation, 384, 385
  due to sudden start of motion, 451
Complementary-energy method, 264
Complex eigenvalue problem, 238, 239, 378
Complex potential, 386–388, 392, 397
Complex representation of harmonic motion, 45

A CATALOG OF SELECTED
DOVER BOOKS
IN SCIENCE AND MATHEMATICS

A CATALOG OF SELECTED

DOVER BOOKS

IN SCIENCE AND MATHEMATICS

QUALITATIVE THEORY OF DIFFERENTIAL EQUATIONS, V.V. Nemytskii and V.V. Stepanov. Classic graduate-level text by two prominent Soviet mathematicians covers classical differential equations as well as topological dynamics and ergodic theory. Bibliographies. 523pp. 5⅜ × 8½. 65954-2 Pa. $10.95

MATRICES AND LINEAR ALGEBRA, Hans Schneider and George Phillip Barker. Basic textbook covers theory of matrices and its applications to systems of linear equations and related topics such as determinants, eigenvalues and differential equations. Numerous exercises. 432pp. 5⅜ × 8½. 66014-1 Pa. $9.95

QUANTUM THEORY, David Bohm. This advanced undergraduate-level text presents the quantum theory in terms of qualitative and imaginative concepts, followed by specific applications worked out in mathematical detail. Preface. Index. 655pp. 5⅜ × 8½. 65969-0 Pa. $13.95

ATOMIC PHYSICS (8th edition), Max Born. Nobel laureate's lucid treatment of kinetic theory of gases, elementary particles, nuclear atom, wave-corpuscles, atomic structure and spectral lines, much more. Over 40 appendices, bibliography. 495pp. 5⅜ × 8½. 65984-4 Pa. $12.95

ELECTRONIC STRUCTURE AND THE PROPERTIES OF SOLIDS: The Physics of the Chemical Bond, Walter A. Harrison. Innovative text offers basic understanding of the electronic structure of covalent and ionic solids, simple metals, transition metals and their compounds. Problems. 1980 edition. 582pp. 6⅛ × 9¼. 66021-4 Pa. $15.95

BOUNDARY VALUE PROBLEMS OF HEAT CONDUCTION, M. Necati Özisik. Systematic, comprehensive treatment of modern mathematical methods of solving problems in heat conduction and diffusion. Numerous examples and problems. Selected references. Appendices. 505pp. 5⅜ × 8½. 65990-9 Pa. $11.95

A SHORT HISTORY OF CHEMISTRY (3rd edition), J.R. Partington. Classic exposition explores origins of chemistry, alchemy, early medical chemistry, nature of atmosphere, theory of valency, laws and structure of atomic theory, much more. 428pp. 5⅜ × 8½. (Available in U.S. only) 65977-1 Pa. $10.95

A HISTORY OF ASTRONOMY, A. Pannekoek. Well-balanced, carefully reasoned study covers such topics as Ptolemaic theory, work of Copernicus, Kepler, Newton, Eddington's work on stars, much more. Illustrated. References. 521pp. 5⅜ × 8½. 65994-1 Pa. $12.95

PRINCIPLES OF METEOROLOGICAL ANALYSIS, Walter J. Saucier. Highly respected, abundantly illustrated classic reviews atmospheric variables, hydrostatics, static stability, various analyses (scalar, cross-section, isobaric, isentropic, more). For intermediate meteorology students. 454pp. 6⅛ × 9¼. 65979-8 Pa. $14.95

CATALOG OF DOVER BOOKS

RELATIVITY, THERMODYNAMICS AND COSMOLOGY, Richard C. Tolman. Landmark study extends thermodynamics to special, general relativity; also applications of relativistic mechanics, thermodynamics to cosmological models. 501pp. 5⅜ × 8½. 65383-8 Pa. $12.95

APPLIED ANALYSIS, Cornelius Lanczos. Classic work on analysis and design of finite processes for approximating solution of analytical problems. Algebraic equations, matrices, harmonic analysis, quadrature methods, much more. 559pp. 5⅜ × 8½. 65656-X Pa. $12.95

SPECIAL RELATIVITY FOR PHYSICISTS, G. Stephenson and C.W. Kilmister. Concise elegant account for nonspecialists. Lorentz transformation, optical and dynamical applications, more. Bibliography. 108pp. 5⅜ × 8½. 65519-9 Pa. $4.95

INTRODUCTION TO ANALYSIS, Maxwell Rosenlicht. Unusually clear, accessible coverage of set theory, real number system, metric spaces, continuous functions, Riemann integration, multiple integrals, more. Wide range of problems. Undergraduate level. Bibliography. 254pp. 5⅜ × 8½. 65038-3 Pa. $7.95

INTRODUCTION TO QUANTUM MECHANICS With Applications to Chemistry, Linus Pauling & E. Bright Wilson, Jr. Classic undergraduate text by Nobel Prize winner applies quantum mechanics to chemical and physical problems. Numerous tables and figures enhance the text. Chapter bibliographies. Appendices. Index. 468pp. 5⅜ × 8½. 64871-0 Pa. $11.95

ASYMPTOTIC EXPANSIONS OF INTEGRALS, Norman Bleistein & Richard A. Handelsman. Best introduction to important field with applications in a variety of scientific disciplines. New preface. Problems. Diagrams. Tables. Bibliography. Index. 448pp. 5⅜ × 8½. 65082-0 Pa. $12.95

MATHEMATICS APPLIED TO CONTINUUM MECHANICS, Lee A. Segel. Analyzes models of fluid flow and solid deformation. For upper-level math, science and engineering students. 608pp. 5⅜ × 8½. 65369-2 Pa. $13.95

ELEMENTS OF REAL ANALYSIS, David A. Sprecher. Classic text covers fundamental concepts, real number system, point sets, functions of a real variable, Fourier series, much more. Over 500 exercises. 352pp. 5⅜ × 8½. 65385-4 Pa. $10.95

PHYSICAL PRINCIPLES OF THE QUANTUM THEORY, Werner Heisenberg. Nobel Laureate discusses quantum theory, uncertainty, wave mechanics, work of Dirac, Schroedinger, Compton, Wilson, Einstein, etc. 184pp. 5⅜ × 8½. 60113-7 Pa. $5.95

INTRODUCTORY REAL ANALYSIS, A.N. Kolmogorov, S.V. Fomin. Translated by Richard A. Silverman. Self-contained, evenly paced introduction to real and functional analysis. Some 350 problems. 403pp. 5⅜ × 8½. 61226-0 Pa. $9.95

PROBLEMS AND SOLUTIONS IN QUANTUM CHEMISTRY AND PHYSICS, Charles S. Johnson, Jr. and Lee G. Pedersen. Unusually varied problems, detailed solutions in coverage of quantum mechanics, wave mechanics, angular momentum, molecular spectroscopy, scattering theory, more. 280 problems plus 139 supplementary exercises. 430pp. 6½ × 9¼. 65236-X Pa. $12.95

CATALOG OF DOVER BOOKS

ASYMPTOTIC METHODS IN ANALYSIS, N.G. de Bruijn. An inexpensive, comprehensive guide to asymptotic methods—the pioneering work that teaches by explaining worked examples in detail. Index. 224pp. 5⅜ × 8½. 64221-6 Pa. $6.95

OPTICAL RESONANCE AND TWO-LEVEL ATOMS, L. Allen and J.H. Eberly. Clear, comprehensive introduction to basic principles behind all quantum optical resonance phenomena. 53 illustrations. Preface. Index. 256pp. 5⅜ × 8½.
65533-4 Pa. $7.95

COMPLEX VARIABLES, Francis J. Flanigan. Unusual approach, delaying complex algebra till harmonic functions have been analyzed from real variable viewpoint. Includes problems with answers. 364pp. 5⅜ × 8½. 61388-7 Pa. $8.95

ATOMIC SPECTRA AND ATOMIC STRUCTURE, Gerhard Herzberg. One of best introductions; especially for specialist in other fields. Treatment is physical rather than mathematical. 80 illustrations. 257pp. 5⅜ × 8½. 60115-3 Pa. $5.95

APPLIED COMPLEX VARIABLES, John W. Dettman. Step-by-step coverage of fundamentals of analytic function theory—plus lucid exposition of five important applications: Potential Theory; Ordinary Differential Equations; Fourier Transforms; Laplace Transforms; Asymptotic Expansions. 66 figures. Exercises at chapter ends. 512pp. 5⅜ × 8½. 64670-X Pa. $11.95

ULTRASONIC ABSORPTION: An Introduction to the Theory of Sound Absorption and Dispersion in Gases, Liquids and Solids, A.B. Bhatia. Standard reference in the field provides a clear, systematically organized introductory review of fundamental concepts for advanced graduate students, research workers. Numerous diagrams. Bibliography. 440pp. 5⅜ × 8½. 64917-2 Pa. $11.95

UNBOUNDED LINEAR OPERATORS: Theory and Applications, Seymour Goldberg. Classic presents systematic treatment of the theory of unbounded linear operators in normed linear spaces with applications to differential equations. Bibliography. 199pp. 5⅜ × 8½. 64830-3 Pa. $7.95

LIGHT SCATTERING BY SMALL PARTICLES, H.C. van de Hulst. Comprehensive treatment including full range of useful approximation methods for researchers in chemistry, meteorology and astronomy. 44 illustrations. 470pp. 5⅜ × 8½. 64228-3 Pa. $10.95

CONFORMAL MAPPING ON RIEMANN SURFACES, Harvey Cohn. Lucid, insightful book presents ideal coverage of subject. 334 exercises make book perfect for self-study. 55 figures. 352pp. 5⅜ × 8¼. 64025-6 Pa. $9.95

OPTICKS, Sir Isaac Newton. Newton's own experiments with spectroscopy, colors, lenses, reflection, refraction, etc., in language the layman can follow. Foreword by Albert Einstein. 532pp. 5⅜ × 8½. 60205-2 Pa. $9.95

GENERALIZED INTEGRAL TRANSFORMATIONS, A.H. Zemanian. Graduate-level study of recent generalizations of the Laplace, Mellin, Hankel, K. Weierstrass, convolution and other simple transformations. Bibliography. 320pp. 5⅜ × 8½. 65375-7 Pa. $8.95

THE ELECTROMAGNETIC FIELD, Albert Shadowitz. Comprehensive undergraduate text covers basics of electric and magnetic fields, builds up to electromagnetic theory. Also related topics, including relativity. Over 900 problems. 768pp. 5⅜ × 8¼. 65660-8 Pa. $18.95

FOURIER SERIES, Georgi P. Tolstov. Translated by Richard A. Silverman. A valuable addition to the literature on the subject, moving clearly from subject to subject and theorem to theorem. 107 problems, answers. 336pp. 5⅜ × 8½. 63317-9 Pa. $8.95

THEORY OF ELECTROMAGNETIC WAVE PROPAGATION, Charles Herach Papas. Graduate-level study discusses the Maxwell field equations, radiation from wire antennas, the Doppler effect and more. xiii + 244pp. 5⅜ × 8½. 65678-0 Pa. $6.95

DISTRIBUTION THEORY AND TRANSFORM ANALYSIS: An Introduction to Generalized Functions, with Applications, A.H. Zemanian. Provides basics of distribution theory, describes generalized Fourier and Laplace transformations. Numerous problems. 384pp. 5⅜ × 8½. 65479-6 Pa. $9.95

THE PHYSICS OF WAVES, William C. Elmore and Mark A. Heald. Unique overview of classical wave theory. Acoustics, optics, electromagnetic radiation, more. Ideal as classroom text or for self-study. Problems. 477pp. 5⅜ × 8½. 64926-1 Pa. $12.95

CALCULUS OF VARIATIONS WITH APPLICATIONS, George M. Ewing. Applications-oriented introduction to variational theory develops insight and promotes understanding of specialized books, research papers. Suitable for advanced undergraduate/graduate students as primary, supplementary text. 352pp. 5⅜ × 8½. 64856-7 Pa. $8.95

A TREATISE ON ELECTRICITY AND MAGNETISM, James Clerk Maxwell. Important foundation work of modern physics. Brings to final form Maxwell's theory of electromagnetism and rigorously derives his general equations of field theory. 1,084pp. 5⅜ × 8½. 60636-8, 60637-6 Pa., Two-vol. set $19.90

AN INTRODUCTION TO THE CALCULUS OF VARIATIONS, Charles Fox. Graduate-level text covers variations of an integral, isoperimetrical problems, least action, special relativity, approximations, more. References. 279pp. 5⅜ × 8½. 65499-0 Pa. $7.95

HYDRODYNAMIC AND HYDROMAGNETIC STABILITY, S. Chandrasekhar. Lucid examination of the Rayleigh-Benard problem; clear coverage of the theory of instabilities causing convection. 704pp. 5⅜ × 8¼. 64071-X Pa. $14.95

CALCULUS OF VARIATIONS, Robert Weinstock. Basic introduction covering isoperimetric problems, theory of elasticity, quantum mechanics, electrostatics, etc. Exercises throughout. 326pp. 5⅜ × 8½. 63069-2 Pa. $7.95

DYNAMICS OF FLUIDS IN POROUS MEDIA, Jacob Bear. For advanced students of ground water hydrology, soil mechanics and physics, drainage and irrigation engineering and more. 335 illustrations. Exercises, with answers. 784pp. 6⅛ × 9¼. 65675-6 Pa. $19.95

NUMERICAL METHODS FOR SCIENTISTS AND ENGINEERS, Richard Hamming. Classic text stresses frequency approach in coverage of algorithms, polynomial approximation, Fourier approximation, exponential approximation, other topics. Revised and enlarged 2nd edition. 721pp. 5⅜ × 8½.
65241-6 Pa. $14.95

THEORETICAL SOLID STATE PHYSICS, Vol. I: Perfect Lattices in Equilibrium; Vol. II: Non-Equilibrium and Disorder, William Jones and Norman H. March. Monumental reference work covers fundamental theory of equilibrium properties of perfect crystalline solids, non-equilibrium properties, defects and disordered systems. Appendices. Problems. Preface. Diagrams. Index. Bibliography. Total of 1,301pp. 5⅜ × 8½. Two volumes. Vol. I 65015-4 Pa. $14.95
Vol. II 65016-2 Pa. $14.95

OPTIMIZATION THEORY WITH APPLICATIONS, Donald A. Pierre. Broad-spectrum approach to important topic. Classical theory of minima and maxima, calculus of variations, simplex technique and linear programming, more. Many problems, examples. 640pp. 5⅜ × 8½. 65205-X Pa. $14.95

THE MODERN THEORY OF SOLIDS, Frederick Seitz. First inexpensive edition of classic work on theory of ionic crystals, free-electron theory of metals and semiconductors, molecular binding, much more. 736pp. 5⅜ × 8½.
65482-6 Pa. $15.95

ESSAYS ON THE THEORY OF NUMBERS, Richard Dedekind. Two classic essays by great German mathematician: on the theory of irrational numbers; and on transfinite numbers and properties of natural numbers. 115pp. 5⅜ × 8½.
21010-3 Pa. $4.95

THE FUNCTIONS OF MATHEMATICAL PHYSICS, Harry Hochstadt. Comprehensive treatment of orthogonal polynomials, hypergeometric functions, Hill's equation, much more. Bibliography. Index. 322pp. 5⅜ × 8½. 65214-9 Pa. $9.95

NUMBER THEORY AND ITS HISTORY, Oystein Ore. Unusually clear, accessible introduction covers counting, properties of numbers, prime numbers, much more. Bibliography. 380pp. 5⅜ × 8½. 65620-9 Pa. $9.95

THE VARIATIONAL PRINCIPLES OF MECHANICS, Cornelius Lanczos. Graduate level coverage of calculus of variations, equations of motion, relativistic mechanics, more. First inexpensive paperbound edition of classic treatise. Index. Bibliography. 418pp. 5⅜ × 8½. 65067-7 Pa. $11.95

MATHEMATICAL TABLES AND FORMULAS, Robert D. Carmichael and Edwin R. Smith. Logarithms, sines, tangents, trig functions, powers, roots, reciprocals, exponential and hyperbolic functions, formulas and theorems. 269pp. 5⅜ × 8½. 60111-0 Pa. $6.95

THEORETICAL PHYSICS, Georg Joos, with Ira M. Freeman. Classic overview covers essential math, mechanics, electromagnetic theory, thermodynamics, quantum mechanics, nuclear physics, other topics. First paperback edition. xxiii + 885pp. 5⅜ × 8½. 65227-0 Pa. $19.95

CATALOG OF DOVER BOOKS

HANDBOOK OF MATHEMATICAL FUNCTIONS WITH FORMULAS, GRAPHS, AND MATHEMATICAL TABLES, edited by Milton Abramowitz and Irene A. Stegun. Vast compendium: 29 sets of tables, some to as high as 20 places. 1,046pp. 8 × 10½. 61272-4 Pa. $24.95

MATHEMATICAL METHODS IN PHYSICS AND ENGINEERING, John W. Dettman. Algebraically based approach to vectors, mapping, diffraction, other topics in applied math. Also generalized functions, analytic function theory, more. Exercises. 448pp. 5⅜ × 8¼. 65649-7 Pa. $9.95

A SURVEY OF NUMERICAL MATHEMATICS, David M. Young and Robert Todd Gregory. Broad self-contained coverage of computer-oriented numerical algorithms for solving various types of mathematical problems in linear algebra, ordinary and partial, differential equations, much more. Exercises. Total of 1,248pp. 5⅜ × 8½. Two volumes. Vol. I 65691-8 Pa. $14.95
Vol. II 65692-6 Pa. $14.95

TENSOR ANALYSIS FOR PHYSICISTS, J.A. Schouten. Concise exposition of the mathematical basis of tensor analysis, integrated with well-chosen physical examples of the theory. Exercises. Index. Bibliography. 289pp. 5⅜ × 8½. 65582-2 Pa. $8.95

INTRODUCTION TO NUMERICAL ANALYSIS (2nd Edition), F.B. Hildebrand. Classic, fundamental treatment covers computation, approximation, interpolation, numerical differentiation and integration, other topics. 150 new problems. 669pp. 5⅜ × 8½. 65363-3 Pa. $14.95

INVESTIGATIONS ON THE THEORY OF THE BROWNIAN MOVEMENT, Albert Einstein. Five papers (1905–8) investigating dynamics of Brownian motion and evolving elementary theory. Notes by R. Fürth. 122pp. 5⅜ × 8½. 60304-0 Pa. $4.95

CATASTROPHE THEORY FOR SCIENTISTS AND ENGINEERS, Robert Gilmore. Advanced-level treatment describes mathematics of theory grounded in the work of Poincaré, R. Thom, other mathematicians. Also important applications to problems in mathematics, physics, chemistry and engineering. 1981 edition. References. 28 tables. 397 black-and-white illustrations. xvii + 666pp. 6⅛ × 9¼. 67539-4 Pa. $16.95

AN INTRODUCTION TO STATISTICAL THERMODYNAMICS, Terrell L. Hill. Excellent basic text offers wide-ranging coverage of quantum statistical mechanics, systems of interacting molecules, quantum statistics, more. 523pp. 5⅜ × 8½. 65242-4 Pa. $12.95

ELEMENTARY DIFFERENTIAL EQUATIONS, William Ted Martin and Eric Reissner. Exceptionally clear, comprehensive introduction at undergraduate level. Nature and origin of differential equations, differential equations of first, second and higher orders. Picard's Theorem, much more. Problems with solutions. 331pp. 5⅜ × 8½. 65024-3 Pa. $8.95

STATISTICAL PHYSICS, Gregory H. Wannier. Classic text combines thermodynamics, statistical mechanics and kinetic theory in one unified presentation of thermal physics. Problems with solutions. Bibliography. 532pp. 5⅜ × 8½. 65401-X Pa. $11.95

CATALOG OF DOVER BOOKS

ORDINARY DIFFERENTIAL EQUATIONS, Morris Tenenbaum and Harry Pollard. Exhaustive survey of ordinary differential equations for undergraduates in mathematics, engineering, science. Thorough analysis of theorems. Diagrams. Bibliography. Index. 818pp. 5⅜ × 8½. 64940-7 Pa. $16.95

STATISTICAL MECHANICS: Principles and Applications, Terrell L. Hill. Standard text covers fundamentals of statistical mechanics, applications to fluctuation theory, imperfect gases, distribution functions, more. 448pp. 5⅜ × 8½. 65390-0 Pa. $9.95

ORDINARY DIFFERENTIAL EQUATIONS AND STABILITY THEORY: An Introduction, David A. Sánchez. Brief, modern treatment. Linear equation, stability theory for autonomous and nonautonomous systems, etc. 164pp. 5⅜ × 8¼. 63828-6 Pa. $5.95

THIRTY YEARS THAT SHOOK PHYSICS: The Story of Quantum Theory, George Gamow. Lucid, accessible introduction to influential theory of energy and matter. Careful explanations of Dirac's anti-particles, Bohr's model of the atom, much more. 12 plates. Numerous drawings. 240pp. 5⅜ × 8½. 24895-X Pa. $6.95

THEORY OF MATRICES, Sam Perlis. Outstanding text covering rank, nonsingularity and inverses in connection with the development of canonical matrices under the relation of equivalence, and without the intervention of determinants. Includes exercises. 237pp. 5⅜ × 8½. 66810-X Pa. $7.95

GREAT EXPERIMENTS IN PHYSICS: Firsthand Accounts from Galileo to Einstein, edited by Morris H. Shamos. 25 crucial discoveries: Newton's laws of motion, Chadwick's study of the neutron, Hertz on electromagnetic waves, more. Original accounts clearly annotated. 370pp. 5⅜ × 8½. 25346-5 Pa. $10.95

INTRODUCTION TO PARTIAL DIFFERENTIAL EQUATIONS WITH APPLICATIONS, E.C. Zachmanoglou and Dale W. Thoe. Essentials of partial differential equations applied to common problems in engineering and the physical sciences. Problems and answers. 416pp. 5⅜ × 8½. 65251-3 Pa. $10.95

BURNHAM'S CELESTIAL HANDBOOK, Robert Burnham, Jr. Thorough guide to the stars beyond our solar system. Exhaustive treatment. Alphabetical by constellation: Andromeda to Cetus in Vol. 1; Chamaeleon to Orion in Vol. 2; and Pavo to Vulpecula in Vol. 3. Hundreds of illustrations. Index in Vol. 3. 2,000pp. 6⅛ × 9¼. 23567-X, 23568-8, 23673-0 Pa., Three-vol. set $41.85

CHEMICAL MAGIC, Leonard A. Ford. Second Edition, Revised by E. Winston Grundmeier. Over 100 unusual stunts demonstrating cold fire, dust explosions, much more. Text explains scientific principles and stresses safety precautions. 128pp. 5⅜ × 8½. 67628-5 Pa. $5.95

AMATEUR ASTRONOMER'S HANDBOOK, J.B. Sidgwick. Timeless, comprehensive coverage of telescopes, mirrors, lenses, mountings, telescope drives, micrometers, spectroscopes, more. 189 illustrations. 576pp. 5⅜ × 8¼. (Available in U.S. only) 24034-7 Pa. $9.95

CATALOG OF DOVER BOOKS

SPECIAL FUNCTIONS, N.N. Lebedev. Translated by Richard Silverman. Famous Russian work treating more important special functions, with applications to specific problems of physics and engineering. 38 figures. 308pp. 5⅜ × 8½.
60624-4 Pa. $8.95

OBSERVATIONAL ASTRONOMY FOR AMATEURS, J.B. Sidgwick. Mine of useful data for observation of sun, moon, planets, asteroids, aurorae, meteors, comets, variables, binaries, etc. 39 illustrations. 384pp. 5⅜ × 8¼. (Available in U.S. only)
24033-9 Pa. $8.95

INTEGRAL EQUATIONS, F.G. Tricomi. Authoritative, well-written treatment of extremely useful mathematical tool with wide applications. Volterra Equations, Fredholm Equations, much more. Advanced undergraduate to graduate level. Exercises. Bibliography. 238pp. 5⅜ × 8½.
64828-1 Pa. $7.95

POPULAR LECTURES ON MATHEMATICAL LOGIC, Hao Wang. Noted logician's lucid treatment of historical developments, set theory, model theory, recursion theory and constructivism, proof theory, more. 3 appendixes. Bibliography. 1981 edition. ix + 283pp. 5⅜ × 8½.
67632-3 Pa. $8.95

MODERN NONLINEAR EQUATIONS, Thomas L. Saaty. Emphasizes practical solution of problems; covers seven types of equations. ". . . a welcome contribution to the existing literature. . . ."—*Math Reviews.* 490pp. 5⅜ × 8½. 64232-1 Pa. $11.95

FUNDAMENTALS OF ASTRODYNAMICS, Roger Bate et al. Modern approach developed by U.S. Air Force Academy. Designed as a first course. Problems, exercises. Numerous illustrations. 455pp. 5⅜ × 8½.
60061-0 Pa. $9.95

INTRODUCTION TO LINEAR ALGEBRA AND DIFFERENTIAL EQUATIONS, John W. Dettman. Excellent text covers complex numbers, determinants, orthonormal bases, Laplace transforms, much more. Exercises with solutions. Undergraduate level. 416pp. 5⅜ × 8½.
65191-6 Pa. $9.95

INCOMPRESSIBLE AERODYNAMICS, edited by Bryan Thwaites. Covers theoretical and experimental treatment of the uniform flow of air and viscous fluids past two-dimensional aerofoils and three-dimensional wings; many other topics. 654pp. 5⅜ × 8½.
65465-6 Pa. $16.95

INTRODUCTION TO DIFFERENCE EQUATIONS, Samuel Goldberg. Exceptionally clear exposition of important discipline with applications to sociology, psychology, economics. Many illustrative examples; over 250 problems. 260pp. 5⅜ × 8½.
65084-7 Pa. $7.95

LAMINAR BOUNDARY LAYERS, edited by L. Rosenhead. Engineering classic covers steady boundary layers in two- and three-dimensional flow, unsteady boundary layers, stability, observational techniques, much more. 708pp. 5⅜ × 8½.
65646-2 Pa. $18.95

LECTURES ON CLASSICAL DIFFERENTIAL GEOMETRY, Second Edition, Dirk J. Struik. Excellent brief introduction covers curves, theory of surfaces, fundamental equations, geometry on a surface, conformal mapping, other topics. Problems. 240pp. 5⅜ × 8½.
65609-8 Pa. $7.95

CATALOG OF DOVER BOOKS

ROTARY-WING AERODYNAMICS, W.Z. Stepniewski. Clear, concise text covers aerodynamic phenomena of the rotor and offers guidelines for helicopter performance evaluation. Originally prepared for NASA. 537 figures. 640pp. 6⅛ × 9¼.
64647-5 Pa. $15.95

DIFFERENTIAL GEOMETRY, Heinrich W. Guggenheimer. Local differential geometry as an application of advanced calculus and linear algebra. Curvature, transformation groups, surfaces, more. Exercises. 62 figures. 378pp. 5⅜ × 8½.
63433-7 Pa. $8.95

INTRODUCTION TO SPACE DYNAMICS, William Tyrrell Thomson. Comprehensive, classic introduction to space-flight engineering for advanced undergraduate and graduate students. Includes vector algebra, kinematics, transformation of coordinates. Bibliography. Index. 352pp. 5⅜ × 8½.   65113-4 Pa. $8.95

A SURVEY OF MINIMAL SURFACES, Robert Osserman. Up-to-date, in-depth discussion of the field for advanced students. Corrected and enlarged edition covers new developments. Includes numerous problems. 192pp. 5⅜ × 8½.
64998-9 Pa. $8.95

ANALYTICAL MECHANICS OF GEARS, Earle Buckingham. Indispensable reference for modern gear manufacture covers conjugate gear-tooth action, gear-tooth profiles of various gears, many other topics. 263 figures. 102 tables. 546pp. 5⅜ × 8½.   65712-4 Pa. $14.95

SET THEORY AND LOGIC, Robert R. Stoll. Lucid introduction to unified theory of mathematical concepts. Set theory and logic seen as tools for conceptual understanding of real number system. 496pp. 5⅜ × 8¼.   63829-4 Pa. $10.95

A HISTORY OF MECHANICS, René Dugas. Monumental study of mechanical principles from antiquity to quantum mechanics. Contributions of ancient Greeks, Galileo, Leonardo, Kepler, Lagrange, many others. 671pp. 5⅜ × 8½.
65632-2 Pa. $14.95

FAMOUS PROBLEMS OF GEOMETRY AND HOW TO SOLVE THEM, Benjamin Bold. Squaring the circle, trisecting the angle, duplicating the cube: learn their history, why they are impossible to solve, then solve them yourself. 128pp. 5⅜ × 8½.   24297-8 Pa. $4.95

MECHANICAL VIBRATIONS, J.P. Den Hartog. Classic textbook offers lucid explanations and illustrative models, applying theories of vibrations to a variety of practical industrial engineering problems. Numerous figures. 233 problems, solutions. Appendix. Index. Preface. 436pp. 5⅜ × 8½.   64785-4 Pa. $10.95

CURVATURE AND HOMOLOGY, Samuel I. Goldberg. Thorough treatment of specialized branch of differential geometry. Covers Riemannian manifolds, topology of differentiable manifolds, compact Lie groups, other topics. Exercises. 315pp. 5⅜ × 8½.   64314-X Pa. $8.95

HISTORY OF STRENGTH OF MATERIALS, Stephen P. Timoshenko. Excellent historical survey of the strength of materials with many references to the theories of elasticity and structure. 245 figures. 452pp. 5⅜ × 8½. 61187-6 Pa. $11.95

GEOMETRY OF COMPLEX NUMBERS, Hans Schwerdtfeger. Illuminating, widely praised book on analytic geometry of circles, the Moebius transformation, and two-dimensional non-Euclidean geometries. 200pp. 5⅜ × 8¼.
63830-8 Pa. $8.95

MECHANICS, J.P. Den Hartog. A classic introductory text or refresher. Hundreds of applications and design problems illuminate fundamentals of trusses, loaded beams and cables, etc. 334 answered problems. 462pp. 5⅜ × 8½. 60754-2 Pa. $9.95

TOPOLOGY, John G. Hocking and Gail S. Young. Superb one-year course in classical topology. Topological spaces and functions, point-set topology, much more. Examples and problems. Bibliography. Index. 384pp. 5⅜ × 8¼.
65676-4 Pa. $9.95

STRENGTH OF MATERIALS, J.P. Den Hartog. Full, clear treatment of basic material (tension, torsion, bending, etc.) plus advanced material on engineering methods, applications. 350 answered problems. 323pp. 5⅜ × 8½. 60755-0 Pa. $8.95

ELEMENTARY CONCEPTS OF TOPOLOGY, Paul Alexandroff. Elegant, intuitive approach to topology from set-theoretic topology to Betti groups; how concepts of topology are useful in math and physics. 25 figures. 57pp. 5⅜ × 8½.
60747-X Pa. $3.50

ADVANCED STRENGTH OF MATERIALS, J.P. Den Hartog. Superbly written advanced text covers torsion, rotating disks, membrane stresses in shells, much more. Many problems and answers. 388pp. 5⅜ × 8½. 65407-9 Pa. $9.95

COMPUTABILITY AND UNSOLVABILITY, Martin Davis. Classic graduate-level introduction to theory of computability, usually referred to as theory of recurrent functions. New preface and appendix. 288pp. 5⅜ × 8½. 61471-9 Pa. $7.95

GENERAL CHEMISTRY, Linus Pauling. Revised 3rd edition of classic first-year text by Nobel laureate. Atomic and molecular structure, quantum mechanics, statistical mechanics, thermodynamics correlated with descriptive chemistry. Problems. 992pp. 5⅜ × 8½. 65622-5 Pa. $19.95

AN INTRODUCTION TO MATRICES, SETS AND GROUPS FOR SCIENCE STUDENTS, G. Stephenson. Concise, readable text introduces sets, groups, and most importantly, matrices to undergraduate students of physics, chemistry, and engineering. Problems. 164pp. 5⅜ × 8½. 65077-4 Pa. $6.95

THE HISTORICAL BACKGROUND OF CHEMISTRY, Henry M. Leicester. Evolution of ideas, not individual biography. Concentrates on formulation of a coherent set of chemical laws. 260pp. 5⅜ × 8½. 61053-5 Pa. $6.95

THE PHILOSOPHY OF MATHEMATICS: An Introductory Essay, Stephan Körner. Surveys the views of Plato, Aristotle, Leibniz & Kant concerning propositions and theories of applied and pure mathematics. Introduction. Two appendices. Index. 198pp. 5⅜ × 8½. 25048-2 Pa. $7.95

THE DEVELOPMENT OF MODERN CHEMISTRY, Aaron J. Ihde. Authoritative history of chemistry from ancient Greek theory to 20th-century innovation. Covers major chemists and their discoveries. 209 illustrations. 14 tables. Bibliographies. Indices. Appendices. 851pp. 5⅜ × 8½. 64235-6 Pa. $18.95

DE RE METALLICA, Georgius Agricola. The famous Hoover translation of greatest treatise on technological chemistry, engineering, geology, mining of early modern times (1556). All 289 original woodcuts. 638pp. 6¾ × 11.

60006-8 Pa. $18.95

SOME THEORY OF SAMPLING, William Edwards Deming. Analysis of the problems, theory and design of sampling techniques for social scientists, industrial managers and others who find statistics increasingly important in their work. 61 tables. 90 figures. xvii + 602pp. 5⅜ × 8½.

64684-X Pa. $15.95

THE VARIOUS AND INGENIOUS MACHINES OF AGOSTINO RAMELLI: A Classic Sixteenth-Century Illustrated Treatise on Technology, Agostino Ramelli. One of the most widely known and copied works on machinery in the 16th century. 194 detailed plates of water pumps, grain mills, cranes, more. 608pp. 9 × 12.

25497-6 Clothbd. $34.95

LINEAR PROGRAMMING AND ECONOMIC ANALYSIS, Robert Dorfman, Paul A. Samuelson and Robert M. Solow. First comprehensive treatment of linear programming in standard economic analysis. Game theory, modern welfare economics, Leontief input-output, more. 525pp. 5⅜ × 8½.

65491-5 Pa. $14.95

ELEMENTARY DECISION THEORY, Herman Chernoff and Lincoln E. Moses. Clear introduction to statistics and statistical theory covers data processing, probability and random variables, testing hypotheses, much more. Exercises. 364pp. 5⅜ × 8½.

65218-1 Pa. $9.95

THE COMPLEAT STRATEGYST: Being a Primer on the Theory of Games of Strategy, J.D. Williams. Highly entertaining classic describes, with many illustrated examples, how to select best strategies in conflict situations. Prefaces. Appendices. 268pp. 5⅜ × 8½.

25101-2 Pa. $7.95

MATHEMATICAL METHODS OF OPERATIONS RESEARCH, Thomas L. Saaty. Classic graduate-level text covers historical background, classical methods of forming models, optimization, game theory, probability, queueing theory, much more. Exercises. Bibliography. 448pp. 5⅜ × 8¼.

65703-5 Pa. $12.95

CONSTRUCTIONS AND COMBINATORIAL PROBLEMS IN DESIGN OF EXPERIMENTS, Damaraju Raghavarao. In-depth reference work examines orthogonal Latin squares, incomplete block designs, tactical configuration, partial geometry, much more. Abundant explanations, examples. 416pp. 5⅜ × 8¼.

65685-3 Pa. $10.95

THE ABSOLUTE DIFFERENTIAL CALCULUS (CALCULUS OF TENSORS), Tullio Levi-Civita. Great 20th-century mathematician's classic work on material necessary for mathematical grasp of theory of relativity. 452pp. 5⅜ × 8½.

63401-9 Pa. $9.95

VECTOR AND TENSOR ANALYSIS WITH APPLICATIONS, A.I. Borisenko and I.E. Tarapov. Concise introduction. Worked-out problems, solutions, exercises. 257pp. 5⅜ × 8¼.

63833-2 Pa. $7.95

CATALOG OF DOVER BOOKS

THE FOUR-COLOR PROBLEM: Assaults and Conquest, Thomas L. Saaty and Paul G. Kainen. Engrossing, comprehensive account of the century-old combinatorial topological problem, its history and solution. Bibliographies. Index. 110 figures. 228pp. 5⅜ × 8½. 65092-8 Pa. $6.95

CATALYSIS IN CHEMISTRY AND ENZYMOLOGY, William P. Jencks. Exceptionally clear coverage of mechanisms for catalysis, forces in aqueous solution, carbonyl- and acyl-group reactions, practical kinetics, more. 864pp. 5⅜ × 8½. 65460-5 Pa. $19.95

PROBABILITY: An Introduction, Samuel Goldberg. Excellent basic text covers set theory, probability theory for finite sample spaces, binomial theorem, much more. 360 problems. Bibliographies. 322pp. 5⅜ × 8½. 65252-1 Pa. $8.95

LIGHTNING, Martin A. Uman. Revised, updated edition of classic work on the physics of lightning. Phenomena, terminology, measurement, photography, spectroscopy, thunder, more. Reviews recent research. Bibliography. Indices. 320pp. 5⅜ × 8¼. 64575-4 Pa. $8.95

PROBABILITY THEORY: A Concise Course, Y.A. Rozanov. Highly readable, self-contained introduction covers combination of events, dependent events, Bernoulli trials, etc. Translation by Richard Silverman. 148pp. 5⅜ × 8¼.
63544-9 Pa. $5.95

AN INTRODUCTION TO HAMILTONIAN OPTICS, H. A. Buchdahl. Detailed account of the Hamiltonian treatment of aberration theory in geometrical optics. Many classes of optical systems defined in terms of the symmetries they possess. Problems with detailed solutions. 1970 edition. xv + 360pp. 5⅜ × 8½.
67597-1 Pa. $10.95

STATISTICS MANUAL, Edwin L. Crow, et al. Comprehensive, practical collection of classical and modern methods prepared by U.S. Naval Ordnance Test Station. Stress on use. Basics of statistics assumed. 288pp. 5⅜ × 8½.
60599-X Pa. $6.95

DICTIONARY/OUTLINE OF BASIC STATISTICS, John E. Freund and Frank J. Williams. A clear concise dictionary of over 1,000 statistical terms and an outline of statistical formulas covering probability, nonparametric tests, much more. 208pp. 5⅜ × 8½. 66796-0 Pa. $6.95

STATISTICAL METHOD FROM THE VIEWPOINT OF QUALITY CONTROL, Walter A. Shewhart. Important text explains regulation of variables, uses of statistical control to achieve quality control in industry, agriculture, other areas. 192pp. 5⅜ × 8½. 65232-7 Pa. $7.95

THE INTERPRETATION OF GEOLOGICAL PHASE DIAGRAMS, Ernest G. Ehlers. Clear, concise text emphasizes diagrams of systems under fluid or containing pressure; also coverage of complex binary systems, hydrothermal melting, more. 288pp. 6½ × 9¼. 65389-7 Pa. $10.95

STATISTICAL ADJUSTMENT OF DATA, W. Edwards Deming. Introduction to basic concepts of statistics, curve fitting, least squares solution, conditions without parameter, conditions containing parameters. 26 exercises worked out. 271pp. 5⅜ × 8½. 64685-8 Pa. $8.95

CATALOG OF DOVER BOOKS

TENSOR CALCULUS, J.L. Synge and A. Schild. Widely used introductory text covers spaces and tensors, basic operations in Riemannian space, non-Riemannian spaces, etc. 324pp. 5⅜ × 8¼. 63612-7 Pa. $8.95

A CONCISE HISTORY OF MATHEMATICS, Dirk J. Struik. The best brief history of mathematics. Stresses origins and covers every major figure from ancient Near East to 19th century. 41 illustrations. 195pp. 5⅜ × 8½. 60255-9 Pa. $7.95

A SHORT ACCOUNT OF THE HISTORY OF MATHEMATICS, W.W. Rouse Ball. One of clearest, most authoritative surveys from the Egyptians and Phoenicians through 19th-century figures such as Grassman, Galois, Riemann. Fourth edition. 522pp. 5⅜ × 8½. 20630-0 Pa. $10.95

HISTORY OF MATHEMATICS, David E. Smith. Nontechnical survey from ancient Greece and Orient to late 19th century; evolution of arithmetic, geometry, trigonometry, calculating devices, algebra, the calculus. 362 illustrations. 1,355pp. 5⅜ × 8½. 20429-4, 20430-8 Pa., Two-vol. set $23.90

THE GEOMETRY OF RENÉ DESCARTES, René Descartes. The great work founded analytical geometry. Original French text, Descartes' own diagrams, together with definitive Smith-Latham translation. 244pp. 5⅜ × 8½. 60068-8 Pa. $6.95

THE ORIGINS OF THE INFINITESIMAL CALCULUS, Margaret E. Baron. Only fully detailed and documented account of crucial discipline: origins; development by Galileo, Kepler, Cavalieri; contributions of Newton, Leibniz, more. 304pp. 5⅜ × 8½. (Available in U.S. and Canada only) 65371-4 Pa. $9.95

THE HISTORY OF THE CALCULUS AND ITS CONCEPTUAL DEVELOPMENT, Carl B. Boyer. Origins in antiquity, medieval contributions, work of Newton, Leibniz, rigorous formulation. Treatment is verbal. 346pp. 5⅜ × 8½. 60509-4 Pa. $8.95

THE THIRTEEN BOOKS OF EUCLID'S ELEMENTS, translated with introduction and commentary by Sir Thomas L. Heath. Definitive edition. Textual and linguistic notes, mathematical analysis. 2,500 years of critical commentary. Not abridged. 1,414pp. 5⅜ × 8½. 60088-2, 60089-0, 60090-4 Pa., Three-vol. set $29.85

GAMES AND DECISIONS: Introduction and Critical Survey, R. Duncan Luce and Howard Raiffa. Superb nontechnical introduction to game theory, primarily applied to social sciences. Utility theory, zero-sum games, n-person games, decision-making, much more. Bibliography. 509pp. 5⅜ × 8½. 65943-7 Pa. $12.95

THE HISTORICAL ROOTS OF ELEMENTARY MATHEMATICS, Lucas N.H. Bunt, Phillip S. Jones, and Jack D. Bedient. Fundamental underpinnings of modern arithmetic, algebra, geometry and number systems derived from ancient civilizations. 320pp. 5⅜ × 8½. 25563-8 Pa. $8.95

CALCULUS REFRESHER FOR TECHNICAL PEOPLE, A. Albert Klaf. Covers important aspects of integral and differential calculus via 756 questions. 566 problems, most answered. 431pp. 5⅜ × 8½. 20370-0 Pa. $8.95

CHALLENGING MATHEMATICAL PROBLEMS WITH ELEMENTARY SOLUTIONS, A.M. Yaglom and I.M. Yaglom. Over 170 challenging problems on probability theory, combinatorial analysis, points and lines, topology, convex polygons, many other topics. Solutions. Total of 445pp. 5⅜ × 8½. Two-vol. set.

Vol. I 65536-9 Pa. $7.95
Vol. II 65537-7 Pa. $6.95

FIFTY CHALLENGING PROBLEMS IN PROBABILITY WITH SOLUTIONS, Frederick Mosteller. Remarkable puzzlers, graded in difficulty, illustrate elementary and advanced aspects of probability. Detailed solutions. 88pp. 5⅜ × 8½.
65355-2 Pa. $4.95

EXPERIMENTS IN TOPOLOGY, Stephen Barr. Classic, lively explanation of one of the byways of mathematics. Klein bottles, Moebius strips, projective planes, map coloring, problem of the Koenigsberg bridges, much more, described with clarity and wit. 43 figures. 210pp. 5⅜ × 8½. 25933-1 Pa. $5.95

RELATIVITY IN ILLUSTRATIONS, Jacob T. Schwartz. Clear nontechnical treatment makes relativity more accessible than ever before. Over 60 drawings illustrate concepts more clearly than text alone. Only high school geometry needed. Bibliography. 128pp. 6⅛ × 9¼. 25965-X Pa. $6.95

AN INTRODUCTION TO ORDINARY DIFFERENTIAL EQUATIONS, Earl A. Coddington. A thorough and systematic first course in elementary differential equations for undergraduates in mathematics and science, with many exercises and problems (with answers). Index. 304pp. 5⅜ × 8½. 65942-9 Pa. $8.95

FOURIER SERIES AND ORTHOGONAL FUNCTIONS, Harry F. Davis. An incisive text combining theory and practical example to introduce Fourier series, orthogonal functions and applications of the Fourier method to boundary-value problems. 570 exercises. Answers and notes. 416pp. 5⅜ × 8½. 65973-9 Pa. $9.95

THE THEORY OF BRANCHING PROCESSES, Theodore E. Harris. First systematic, comprehensive treatment of branching (i.e. multiplicative) processes and their applications. Galton-Watson model, Markov branching processes, electron-photon cascade, many other topics. Rigorous proofs. Bibliography. 240pp. 5⅜ × 8½. 65952-6 Pa. $6.95

AN INTRODUCTION TO ALGEBRAIC STRUCTURES, Joseph Landin. Superb self-contained text covers "abstract algebra": sets and numbers, theory of groups, theory of rings, much more. Numerous well-chosen examples, exercises. 247pp. 5⅜ × 8½. 65940-2 Pa. $7.95

Prices subject to change without notice.
Available at your book dealer or write for free Mathematics and Science Catalog to Dept. GI, Dover Publications, Inc., 31 East 2nd St., Mineola, N.Y. 11501. Dover publishes more than 175 books each year on science, elementary and advanced mathematics, biology, music, art, literature, history, social sciences and other areas.